BOLD VENTURES

Volume 2

Case Studies of U.S. Innovations in Science Education

W0245841

Other Volumes in the Series: Bold Ventures

Volume 1: Patterns of U.S. Innovations in Science and Mathematics Education

Study Background
Senta A. Raizen

The General Context for Reform
Senta A. Raizen

Changing Conceptions of Science, Mathematic, Teaching, and Learning
J. Myron Atkin, Jeremy Kilpatrick, Julie A. Bianchini, Jenifer V. Helms, Nicole I. Holthuis

Changing Roles of Teachers
Norman L. Webb

Changing Conceptions of Reform
Senta A. Raizen, Douglas B. McLeod, Mary Budd Rowe

Assessing the Implementation of Reforms and Innovations
Michael Huberman

Underplayed Elements in These Reform Efforts
Robert E. Stake

Appendix: Case Studies in Other OECD Countries

Volume 3: Case Studies of U.S. Innovations in Mathematics Education

Setting the Standards: NCTM's Role in the Reform of Mathematics Education
Douglas B. McLeod, Robert E. Stake, Bonnie P. Schappelle, Melissa Mellissinos, Mark J. Gierl

Teaching and Learning Cross-Country Mathematics: A Story of Innovation in Precalculus
Jeremy Kilpatrick, Lynn Hancock, Denise S. Mewborn, Lynn Stallings

The Urban Mathematics Collaborative Project: A Study of Teacher, Community, and Reform
Norman L. Webb, Daniel J. Heck, William F. Tate

BOLD VENTURES

Volume 2

Case Studies of U.S.
Innovations in Science
Education

edited by

Senta A. Raizen
Edward D. Britton

from

**The National Center for Improving
Science Education**

a division of
The NETWORK, Inc.

KLUWER ACADEMIC PUBLISHERS
DORDRECHT / BOSTON / LONDON

A C.I.P. Catalogue record for this book is available from the Library of Congress

ISBN-13:978-0-7923-4236-6   e-ISBN-13:978-94-009-0341-8

DOI: 10.1007/978-94-009-0341-8

Published by Kluwer Academic Publishers,
P.O. Box 17, 3300 AA Dordrecht, The Netherlands.

Kluwer Academic Publishers incorporates
the publishing programmes of
D. Reidel, Martinus Nijhoff, Dr W. Junk and MTP Press.

Sold and distributed in the U.S.A. and Canada
by Kluwer Academic Publishers,
101 Philip Drive, Norwell, MA 02061, U.S.A.

In all other countries, sold and distributed
by Kluwer Academic Publishers Group,
P.O. Box 322, 3300 AH Dordrecht, The Netherlands.

Printed on acid-free paper

DEDICATION

We dedicate this volume to Mary Budd Rowe, our dear colleague and friend. Mary's study of Chemistry in the Community in chapter 5 is her last research publication. In this project and all her endeavors, Mary inspired and counselled many of the project's graduate students, not just those in her own research team. With her typical quiet intensity, Mary offered her peers some sage insights into patterns among the cases. We value each personal and professional moment we had with Mary. She was a warm and exceptional colleague and friend to the authors and editors, and generously mentored several of us throughout our education and careers. It pains us and all who knew her that Mary Budd will no longer be here to help researchers, teachers and students explore science.

The National Center for Improving Science Education

The National Center for Improving Science Education (NCISE) is a division of The NETWORK, Inc., a nonprofit organization dedicated to educational reform. The Center's mission is to promote change in state and local policies and practices in science curriculum, teaching, and assessment. To further this mission, we carry out research, evaluation, and technical assistance. Based on this work, we provide a range of products and services to educational policymakers and practitioners to help them strengthen science teaching and learning across the country.

We are dedicated to helping all stakeholders in science education reform, preschool to postsecondary, to promote better science education for all students.

Advisory Board, U.S. Case Studies

Contents

Contributors

The Case Study Teams

Brief biographical information about team members is found in appendix A of their respective chapters. Below, team leaders are listed first.

Building on Strength: Changing Science Teaching in California Public Schools
J. Myron Atkin, Jenifer V. Helms, Gerald L. Rosiek, Suzanne A. Siner
School of Education, Stanford University

The Different Worlds of Project 2061
J. Myron Atkin, Julie A. Bianchini, Nicole I. Holthuis
School of Education, Stanford University

The Challenges of Bringing the Kids Network to the Classroom
James W. Karlan, Michael Huberman, Sally H. Middlebrooks
National Center for Improving Science Education, Harvard University

Science, Technology, and Story: Implementing the Voyage of the Mimi
Sally H. Middlebrooks, Michael Huberman, James W. Karlan
National Center for Improving Science Education, Harvard University

ChemCom's Evolution: Development, Spread, and Adaptation
Mary Budd Rowe, Julie E. Montgomery, Michael J. Midling, Thomas M. Keating
Stanford University

The Editors

Senta A. Raizen, Director of The National Center for Improving Science Education, is principal investigator and editor of the U.S. case studies discussed in this and two companion volumes. Raizen is the primary author of a number of books, reports, and articles on science education in elementary, middle, and high school; indicators in science education; preservice education of elementary school teachers; and technology education. Her work also includes educational assessment and program evaluation, education policy, reforming education for work, and linking education research and policy with practice. She is principal investigator for NCISE research for the Third International Mathematics and Science Study (TIMSS) and serves on the TIMSS International Steering Committee. Raizen directs NCISE evaluations of several federal programs that support science education. She serves in an advisory capacity to—among others—the National Assessment of Educational Progress, the National Goals Panel, the National Institute for Science Education, and the National Research Council.

Edward D. Britton, Associate Director of NCISE, serves as project director for several international studies, including the work presented in this volume. He was lead editor of *Examining the Examinations: An International Comparison of Science and Mathematics Examinations for College-Bound Students*. Britton also works on several aspects of TIMSS, including the U.S. and international curriculum analyses and the international teacher and student questionnaires. In addition, he has managed development of CD-ROM disks and videotapes designed to help elementary teachers enhance their science knowledge and pedagogy. Britton has written on indicators for science education, dissemination of innovations, and evaluation.

Preface

This book presents comprehensive results from case studies of five innovations in science education that have much to offer toward understanding current reforms in this field. Each chapter tells the story of a case in rich detail, with extensive documentation, and in the voices of many of the participants—the innovators, the teachers, the students. Similarly, Volume 3 of *Bold Ventures* presents the results from case studies of five innovations in mathematics education. Volume 1 provides a cross-case analysis of all eight innovations.

Many U.S. readers certainly will be very familiar with the name of at least one if not all of the science innovations discussed in this volume—for example, *Project 2061*—and probably with their general substance. Much of the education community's familiarity with these arises from the projects' own dissemination efforts. The research reported in this volume, however, is one of the few detailed studies of these innovations undertaken by researchers outside the projects themselves.

Each of the five studies was a large-scale effort involving teams of researchers over three years. These teams analyzed many documents, attended numerous critical project meetings, visited multiple sites, conducted dozens of individual interviews. The team leaders (Atkin, Huberman, Rowe), having spent much time with science education over long careers, looked at these innovations through many lenses. It was a daunting task for each team to sift through the mountains of detail in order to bring the most compelling themes to the surface. But through some exciting—if exhausting—meetings two or three times each year as well as ongoing exchanges, the key storylines did become clear. Deliberately, so as to let the stories emerge for the reader as they did for the researchers, we have not enforced a uniform format on the case study reports.

The Introduction gives a more detailed overview of the eight U.S. innovations we studied and the resulting case studies. It is important to note here, however, that this substantial case study effort was part of a larger international endeavor. Both the science and mathematics cases constituted the participation of the United States in a study of educational innovations in science, mathematics and technology education that was undertaken by 13 member countries of the Organisation for Economic Co-operation and Development (OECD). The genesis of the case study project was member countries' shared dissatisfaction with the state of education in these school subjects; this led them to seek a contextualized understanding of what innovations were under way to address this concern. The result was the largest qualitative research project ever undertaken across countries.

Just as we in the United States are learning from reform efforts in other countries, so too will the stories and findings of the U.S. projects be beneficial for the educational systems of the other OECD member nations. Innovation in a large, decentralized, and diverse country such as the United States is of particular interest for countries in which centralized government bureaucracies experiment with decentralized curriculum reform on the local or regional level, as several of the industrialized countries are now doing. Educators in other countries will be able to compare policy priorities in the United States to issues considered important in their own countries. Also, the documentation of these reform efforts available through the three volumes in this series will complement and help interpret both for this country and for an international audience the quantitative results generated on student achievement, current curriculum, and teacher and student background by the Third International Mathematics and Science Study (TIMSS).

Audiences

Bold Ventures is intended primarily for people working to improve science and mathematics education. We include in this audience:

- policymakers in a position to influence schools and educational systems;

- teacher educators and staff developers working with prospective or already practicing teachers;

- science, mathematics, and engineering professionals;

- school administrators; and

- teachers of science and mathematics.

The three volumes in the series unravel the origins, development, and implementation stories of eight of this country's major reform initiatives. They address questions surrounding the "what" as well as the "how" of these innovations. Curriculum specialists and teachers of science in particular will find the detailed case studies in this volume of practical utility in the teaching and learning of science. Having the case stories before them should help educators reflect on their own practices and programs.

Organization of this Book

We anticipate that some readers will concentrate on the case studies of the greatest interest to them. They will find the two comprehensive reforms discussed in the first two chapters; the state-level California Science Education Reform in the first chapter, the national-level Project 2061 in the second chapter. The next three chapters are devoted to innovations focused on producing curricular materials. They are ordered by grade level; chapter 3 (on the Kids Network) and chapter 4 (on The Voyage of the Mimi) deal with curricular innovations for elementary and middle schools; chapter 5 discusses an innovative chemistry course (Chemistry in the Community, ChemCom) for high school. We hope many in our audience will find the case studies sufficiently engaging and relevant to their work to read several or all of them.

Acknowledgments

We gratefully acknowledge the many individuals who contributed to the research effort reported in this and the two companion volumes. Just counting the researchers, advisors, and support staff, over 40 U.S. professionals invested substantial amounts of their time in this effort over the last four or more years. Most of these individuals are cited on other pages of this volume: advisors are listed just before the contents page, and short biographies of research team members are included in each of the individual studies. We thank the indispensable administrative assistants to this project, especially those who spent large amounts of time organizing meetings, preparing briefing books, and working with manuscripts: Susan Callan and LaDonna Dickerson at NCISE, Sally Lesher at the University of Wisconsin, and Sunny Toy at Stanford University. Nita Congress and her colleagues toiled to help us line edit this rather large body of work; Shelley Wetzel of Marketing Options handled the manuscript layout.

The Department of Education and the National Science Foundation have supported this work on an equal basis, as administered by NSF under grant number RED-9255247. We are grateful for their funding and supportive monitoring. The content of this report does not necessarily reflect the views of the Department or NSF. Their funding extended beyond the research to support dissemination efforts including a 1993 workshop to acquaint potential audiences with the goals of the U.S. case studies; and a 1996 conference in Washington, "Getting the Word Out," to release the U.S. and international results. Eve Bither, Director of the National Educational Research Policy and Priorities Board, spearheaded the Department of Education's support of this project and has steadfastly advanced our work. Several current and former officers at NSF have generously assisted us over the years: Daryl Chubin, Director of Research, Evaluation, and Communication; David Jenness, now an independent consultant; Conrad Katzenmeyer, senior program officer; Iris Rotberg, program officer; and Larry Suter, deputy division director.

In closing, we wish to applaud and thank those who have made our work possible—the reform-minded professionals who worked in and out of the classroom to make their visions of better science and mathematics education come to life. Many people were gracious enough to let us witness their endeavors again and again and to speak with them at length. When busy educators and innovators welcome researchers into their world, it behooves these researchers to take the greatest care in depicting their efforts. We and our colleagues hope the many individuals who generously gave of themselves and their time will find *Bold Ventures* to be respectful, accurate, and helpful in advancing their important work.

Senta A. Raizen   *Washington, D.C.*
Edward D. Britton   *July 1996*

Introduction
Study Background

Senta A. Raizen

The last dozen years have been ones of both challenge and excitement for science education—challenge because, on the whole, students' science achievement continues to be disappointing, excitement because of the significant reform efforts that have been undertaken. This volume of *Bold Ventures* presents case studies of five such reform efforts: the California Science Education Reform initiative, the Chemistry in the Community (ChemCom) course, the Kids Network units, Project 2061, and the Voyage of the Mimi materials.[1] Because of their promise for improving student learning in science, while representing different approaches to reform, these five innovations were selected for intensive study. Along with the three mathematics innovations discussed in volume 3, they comprise the participation of the United States in the largest cross-national case study project ever undertaken, the international context of which is described later in this chapter. This volume provides the full, rich detail of the stories these science innovations have to tell and what lessons can be learned from them. Volume 1 of *Bold Ventures* presents our cross-case analysis of all eight studies.

The U.S. Studies

In selecting the project's mathematics and science innovations for intensive study, we aimed for a varied and balanced set with respect to subject areas, grade levels, nature of the interventions demanded by the innovation, and the origins and main driving force behind the innovation. Table 1 provides an overview of the range of the selected innovations.

As noted, three of the innovations are in mathematics education: the Contemporary Precalculus course, the National Council of Teachers of Mathematics (NCTM) *Standards*, and the Urban Mathematics Collaboratives (UMC). Two of the science innovations (Kids Network and Voyage of the Mimi) use communication technology and multimedia approaches for instruction. Four of the choices are focused on innovative curriculum materials; four are more comprehensive in nature. Two of the materials development projects are aimed

---

[1]The Voyage of the Mimi was designed to be used as a tool in both mathematics and science instruction, but in practice teachers typically have emphasized science much more than mathematics.

Table 1.   Characteristics of U.S. Innovations Studied

	Curriculum Development		Comprehensive		
	K-8	9-12	District	State	National
Mathematics		Contemporary* Precalculus	Urban Math Collaborative		NCTM Standards
Science	Kids Network* Mimi*	ChemCom		California Science Ed. Reform	Project 2061

*While the primary focus of these three innovations was new approaches to science or mathematics, they also developed new uses of technology in instruction.

at high school (ChemCom, Contemporary Precalculus); one is intended for middle school and upper elementary (Voyage of the Mimi) and one for upper elementary (Kids Network).

The origins of the innovations also are varied: two are large national initiatives (NCTM *Standards*, Project 2061); one was developed as a result of a National Science Foundation (NSF) grants competition (Kids Network); two came about through dynamic leadership within a professional school or organization and subsequently received federal and foundation funding (Voyage of the Mimi and ChemCom); one represents a state initiative (California Science Education Reform); one was conceived by a private foundation and focuses on large urban school systems (UMC); and one came into existence through the initiative of teachers at a magnet science and mathematics school (Contemporary Precalculus). The selection of the U.S. cases was made by the case study researchers based on advice from a widely representative group of experts, i.e., the U.S. project's advisory board convened by The National Center for Improving Science Education (NCISE). (Members are listed in the front of this volume.)

In what ways does each of the selected innovations represent a change? How notable are these changes for U.S. education? Why is the innovation worth studying? What can be learned from each case? These and many other questions find responses in the richly told and documented stories of the innovations in this volume.

Recent data confirm that the four nationally applicable cases discussed in this volume are well known across the country and, therefore, very relevant selections for a national study of major innovations in science education. In their 1993 survey, Weiss, Matti, and Smith (1994) found that 68% of elementary and middle school teachers had heard of the Kids Network; 49% of teachers for grades 5-8 knew of The Second Voyage of the Mimi; and 58% of all high school science teachers (not just chemistry teachers) knew Chemistry in the Community. Ten percent of teachers at every type of school (elementary, middle, and high) reported making "considerable use" of Project 2061's Science for All Americans, a strong influence for something that is not a textbook nor a specific pedagogical technique.

To provide the necessary background for the rest of this book, we briefly summarize below some of the most innovative features of the science cases.

Building on Strength—Changing Science Teaching in California Schools. California has led the way in systemically influencing science instruction by concerted, congruent actions in four policy arenas: reconceptualizing the science curriculum in an innovative state framework, developing new statewide assessments that support appropriate science instruction, creating means for helping teachers change their science instruction, and requiring state-approved instructional materials to be consistent with the framework. The case study analyzes the changes in approach to science content and teaching required by the reform and, especially, describes in detail California's reliance on teacher networks to effect changes.

The Different Worlds of Project 2061. This initiative, led by the American Association for the Advancement of Science, endeavors to fundamentally restructure science, mathematics, and technology education across all of precollegiate schooling. *Science for All Americans*, the project's first publication, set out a vision for what science was important for all students to know. The second publication, *Benchmarks*, suggests in detail what students should know and be able to do at various grade levels. The case study examines the development of the publications and their national impact, and emphasizes their relationship to curriculum designs formulated in six partnering school districts.

The Challenges of Bringing the Kids Network to the Classroom. Through telecommunications, Kids Network enables upper elementary school students across the country— and in other countries—to exchange scientific data they collect during investigations of real-world environmental problems. The Kids Network was developed by Technical Education Research Centers (TERC) and is distributed by the National Geographic Society. The case study describes how the project provides a medium of change for teachers and classrooms, how teachers adopt and implement the curriculum, the reaction of students, and incentives and barriers related to the use of the technology.

Science, Technology and Story: Implementing the Voyage of the Mimi. The Voyage of the Mimi is a curriculum program for 9- to 14-year olds that combines videos or videodiscs, computer software, and print materials to present an integrated set of concepts in mathematics, science, social studies, and language arts. The project uses a story-line of expeditions involving a research ship called Mimi. The case study relates how the course units were introduced into schools, how teachers collaborated to teach them, and how they adapted these materials. It describes teachers' reactions to the materials and analyzes factors that affected implementation.

ChemCom's Evolution: Development, Spread, and Adaptation. This course, popularly known as ChemCom, is an alternative to traditional high school chemistry. The curriculum focuses on chemistry as needed to understand sociotechnological problems, and attempts to attract to the study of chemistry nontechnical college-bound students as well as students not planning to attend college. The case study details the development of the course; its implementation; its varied uses in the high school curriculum; and the continuing commitment to ChemCom of the American Chemical Society, the sponsoring society.

To give a little more comprehensive view of the U.S. research, following are short descriptions of the three case studies of innovations in mathematics education.

Setting the Standards: The National Council of Teachers of Mathematics and the Reform of Mathematics Education. The case study tells the story of how a professional organization of mathematics teachers assumed national leadership in the field of mathematics education and influenced national and state policy in the movement to develop high educational standards. The standards documents emphasize a new conception of mathematics as problem solving, reasoning, communication, and connection. An extensive dissemination effort has brought the mathematics standards to national attention, one result being that they have served as a model for the development of standards documents in science and other school subjects. NCTM developed standards on its own initiative, over a long period of time, without government resources.

Teaching and Learning Cross-Country Mathematics—A Story of Innovation in Precalculus. In this "grassroots" effort, a group of teachers at a science and mathematics magnet high school in North Carolina have developed a precalculus course based on the modeling of real phenomena. It emphasizes applications, data analysis, and matrices; the use of graphing calculators is integral to the course. Through word of mouth, teachers in other locales have learned of the course and begun to use it. This in turn has enabled the North Carolina teachers to obtain NSF funding for polishing the materials and also to secure a commercial publisher.

The Urban Mathematics Collaborative Project: A Study of Teacher, Community, and Reform. Funded by the Ford Foundation, the UMC project is aimed at improving mathematics education in 16 urban school districts, while at the same time identifying new models for meeting the ongoing professional needs of teachers. The case study reports on the interaction among business, industry, higher education, school systems, and the teachers to create change in the teaching of mathematics. It chronicles the personal growth of teachers as they assumed leadership roles as well as the new conceptions and practices of classroom mathematics that emerged.

The International Project

The origin of the international Science, Mathematics and Technology Education case study project dates back to 1989, when a meeting of interested researchers resulted in an issues paper, further refined during a 1990 meeting of experts from 18 member countries (including the United States) of the Organisation for Economic Co-operation and Development (OECD). The idea for conducting international case studies developed out of a growing concern shared by OECD member nations for more effective programs in science, mathematics and technology education to serve their populations. Moreover, the participating countries recognized the need to have a better understanding of the policies, programs, and practices likely to lead to more successful outcomes in these educational fields. What is the nature of the changes being formulated? How are such programs, policies, and practices developed? How are they implemented in settings where they are successful? What processes support implementation? What roles are played by whom? What outcomes are attained?

The desire to have answers to these questions led 13 countries to carry out one or more case studies of innovations in science, mathematics, and technology education.[2] The participating countries included, in addition to the United States, Australia, Austria, Canada (two studies), France, Germany, Ireland, Japan (two studies), The Netherlands, Norway (two studies), Scotland, Spain, and Switzerland.

Countries agreed on main research themes that each case study should seek to address. These themes are:

- context (historic, social, political, educational) within which the innovation was formulated;

- processes by which change was implemented, both as envisioned in planning and as experienced in reality;

- goals and content of the innovation;

- perspectives of the students participating in the innovation;

- methods, materials, equipment, and settings for learning;

- teachers and teacher education; and

- assessment, evaluation, and accountability.

---

[2]Three of these countries studied the introduction into precollegiate education of technology as a school subject. As the U.S. work included no such case—though three of the cases involve the use of educational technology in teaching science or mathematics—we generally refer to the U.S. cases as innovations in science and mathematics education.

In terms of the three subject areas, researchers were to address:

- how the scope and structure of science content was being redefined;

- the role of problem-solving approaches and applications in building mathematical knowledge and skills;

- the extension of science and mathematics education to more diverse students;

- ways in which technology education is being implemented in the general curriculum of elementary and secondary schools; and

- the interrelationships among science, mathematics, and technology in the school curriculum.

The outcome of the international project is *Changing the Subject: Innovations in Science, Mathematics and Technology Education* (Black and Atkin 1996). This book distills the major findings from all 23 cases, concentrating on the themes of greatest concern to the participating countries. The insights provided complement those of the Third International Mathematics and Science Study (TIMSS) which has collected data on a number of contextual factors as well as data on student achievement in science and mathematics in over 40 countries. TIMSS was not able, however, to examine in detail how these factors operate at specific sites.

Below are very brief descriptions of those countries' studies that focused on science education only, or in combination with mathematics or technology education. More detailed descriptions are available in the appendix of Volume 1 of *Bold Ventures*.

The state of Tasmania in **Australia** initiated the introduction of technology education, and concomitant changes in science and mathematics instruction for pupils from ages 4 to 16. The case study report, Science, Mathematics and Technology in Education (SMTE) Project, notes that the enthusiasm and leadership of a key teacher were imperative for successful introduction; other factors that were important included using approaches that emphasized student-centered learning and encouraged collegiality among teachers. The study was led by John Williamson of the School of Education at the University of Tasmania.

From **Canada,** A Case Study of the Implementation of the Ontario Common Curriculum in Grade Nine Science and Mathematics looks at a system undergoing extensive change. Two statewide requirements were introduced simultaneously: the integration of courses in science and mathematics, and the de-tracking of ninth grade classes. Detailed learning outcomes were specified. The case study team was led by John Olson of Queen's University.

Another study from **Canada** was built around a single classroom on a single aspect of physics teaching, described in Gender-Equity in Science Instruction and Assessment a Case Study of Grade 10 Electricity in British Columbia. The innovation was intended to promote better outcomes for females, with students themselves involved in shaping the course unit on electricity to incorporate social issues. Girls indicated they were more inclined to enroll in physics after they had participated in the unit. Jim Gaskell of the University of British Columbia carried out this research.

Practicing Integration in Science Education (PING), an Innovation Project for Science Education in **Germany**, was initiated because the establishment of comprehensive secondary schools led teachers to want a student-centered, integrated science curriculum. Supported by the German National Institute for Science Education (IPN), teachers in Schleswig-Holstein have developed materials for grades 5 to 10. The curriculum has spread voluntarily and widely to other German states. K.H. Hansen of the Institute for Science Education in Kiel led this study.

Researchers in **Ireland** studied In-career Development in Equity and Science (IDEAS), a Ministry initiative concerned with establishing physics or chemistry teaching in 33 high schools with no previous tradition of teaching the subject, notably girls' or mixed schools. Science teachers in these schools were trained in their own classrooms by experienced physics or chemistry teachers from another school. The case study is available from study leader Dearbhál Ní Charthaigh of the University of Limerick.

Japan is introducing a revised science curriculum for ages 9 to 15 that focuses on human responsibility, individuality and resourcefulness; a particular feature in its introduction is the development and refinement of lesson plans. Case Studies of the Implementation of a New School Science Course in Japan examines the effectiveness of the implementation; the findings will be used in the next round of national curriculum revision. Toshiyuki Fukuoka and colleagues at Yokohama National University conducted this study.

In **Norway**, A Case Study of Science Teaching in Grade 8 followed three classes. As part of the study, an electricity course was developed and taught during one term, to be followed by oral examination based on laboratory and field experience untried before this experiment. The teaching materials developed for the study were based on observation and experiment, so that science knowledge could not be isolated from science process. The investigation was led by Doris Jorde of the University of Oslo.

Spain has adopted an integrated science curriculum in the context of raising the school-leaving age to 16. Students' Diversity and the Changes in the Curriculum an Evaluation of the Spanish Reform of Lower Secondary Education investigates how the curriculum has been developed and what teachers want to achieve. One outcome of collaborative work among teachers is the closer relationship forged between the school as an educational community and actual instruction in individual classrooms. Maria J. Sáez of the University of Valladolid headed this study.

The study from **Switzerland** tells of the use of computers as the medium whereby 13- to 16-year-old students gain understanding of real-life phenomena.

The innovations of a small group of teachers and researchers are reported in The Representation, the Understanding and the Mastering of Experience—Modelling and Programming in a Transdisciplinary Context. Bruno Vitale of the Centre de Recherches Psycho-Pédagogiques in Geneva conducted this research.

Two additional studies are in the related field of technology education.

Research from the **Netherlands**, An In-depth Study of Technology as a School Subject in Junior Secondary Schools, addresses the introduction of technology as a separate school subject in response to government policy. All Dutch schools are providing instruction in technology to students aged 12-14 years. The study found that the more academically oriented schools emphasized knowledge, whereas the pre-vocational schools emphasized practical activities. A.M. Houtveen of Utrecht University led this case study.

In **Scotland**, the National Guidelines for schools have introduced technology as a new school subject, as discussed in A Report on Technology in Case Study Primary Schools in Scotland. The primary schools (for ages 5 to 12) observed in the study all adopted different approaches. Key individuals proved crucial to managing change in each of the schools. Open communication between the school and the community was another essential. The research was headed by Alistair Marquis of Her Majesty's Inspectors of Schools in Edinburgh.

Methodology

We chose the case study approach as our main methodology for studying the innovations. This is the method of choice for a close-up, in-depth understanding of their origins, progress, and implementation. Case studies also allow researchers to deal with latent or underlying issues that would escape the survey analyst or the field observer. For example, an understanding of site dynamics can evolve only gradually, from the time when a project is just under way to the time when it has spread widely enough to allow the formulation and testing of hypotheses about what facilitates or inhibits successful implementation. All the same, case study research admits that there can be several, varying interpretations, both among actors at the research sites and between these actors and the field researchers. We have tried to reflect these differences in interpretation in our individual case reports.

Each case report represents a *snapshot in time*. The case study researchers summarized, to the extent possible based on documents and interviews, critical events that occurred before they picked up the case; they were in the field for a year (and in some cases longer) following project activities and developments as they took place that year at selected sites and events associated with the project. Nevertheless, each of the case reports documents a "work in progress"—the situation and context as they were observed and recorded during a specific period in the life of each project.

The development of the U.S. case studies occurred in two phases, both funded by the U.S. Department of Education and the National Science Foundation. In the first phase, NCISE and the National Center for Research in Mathematical Sciences Education selected eight innovations for case study development and prepared 20-page descriptions of each. These descriptions were published by OECD (1993) under the title *Science and Mathematics Education in the United States: Eight Innovations.* The second phase consisted of the in-depth study of each of the innovations as reported in Volumes 2 and 3 of *Bold Ventures* and the cross-case analyses summarized in Volume 1.[3]

The Research Questions

Each of the five case studies in this volume addressed some general research questions; a second group of questions was particularly pertinent to the more comprehensive efforts represented by Project 2061 and the California Science Education Reform; a third group consisted of some more detailed questions applicable to implementation issues—for example, regarding the Voyage of the Mimi. The three groups of questions are given below. Specific questions addressed by each case study are elaborated in the individual chapters.

General Questions .

- What has motivated the genesis of the innovation? Who were and are the key movers? What were the contexts of development—social, historic, institutional, political?

- What is the problem or dilemma addressed by the innovation?

- What are the content, design and underlying assumptions of the innovation? How much variation in point of view is there—or has there been—in different settings?

- For each of the innovations, how is progress or success defined, assessed, and known?

- Which criteria for success are used by different key actors or institutions?

- What were—or are—the principal facilitators and constraints in the design and execution of the innovation?

---

[3]Seven of the innovations studied in depth are the same as those summarized in the OECD 1993 report; for the eighth innovation, the precalculus project was substituted for the California Restructuring of Mathematics Education (Webb 1993). The reason for the substitution was that the precalculus project represents a unique grassroots effort by teachers, whereas a state initiative already was represented in the California Science Education Reform case study.

- What is the involvement of teachers in each innovation? How does teacher involvement affect the introduction and acceptance of innovations?

Additional Questions for Comprehensive Efforts. This next set of questions is particularly appropriate for such comprehensive efforts as represented by Project 2061 and the California Science Education Reform.

- What policies are being advocated? What is the rationale for these policies? Which of the policies have been instituted?

- Has the effort generated a favorable climate for science education reform— nationally, in states, in districts?

- What roles are envisioned in the effort for business and industry, institutions of higher education, and other nonschool organizations?

- Did collaboration occur among professionals working in different contexts (scientists in higher education or industry, teachers, school administrators)?

- Did local users modify the intended innovations and, if so, for what reasons?

Additional Implementation Questions. These questions are relevant to Chemistry in the Community, the Kids Network, and the Voyage of the Mimi.

- What are the key marker events of each phase of implementation, and how have these events affected the outcomes of local implementation?

- What resources and working conditions have schools or districts provided to facilitate use of project materials?

- What do the innovations "look like" on a daily basis? Do teachers' and students' perceptions of what goes on in the classroom match or vary?

- What changes do teachers make in their instructional practices during successive uses of materials? What training and other support are available?

- Are schools and classrooms equipped technologically and with needed supplies to carry out the new curriculum or the innovation's teaching strategies?

The Case Study Research Cycle

The overall work of studying the eight U.S. innovations took three years, as summarized in table 2.

The first year was devoted to three tasks: The first task was to identify, within the OECD framework, the common research questions covering all the cases as well as project specific ones. The second task was to identify potential sites for in-depth study; quite different approaches were needed for the individual cases, ranging from developing a sampling plan for innovations already in wide use (as was the case for several of the curriculum materials projects) to selecting

Table 2.   **Research Cycle for Eight Individual Cases**

1989	OECD meeting: design of international study, including research themes.
1991-1992	U.S. pilot study: Researchers identify and describe potential U.S. cases; Advisory board selects eight innovations for extensive case studies; brief descriptions of U.S. innovations published by OECD (1993).
1992-1993	Refinement of research questions, analysis of documents, development of instruments, sampling of sites, start of field work.
1993-1994	Main data collection and analysis: observation of classrooms and key events, interviews of key individuals—initiators, developers, and teachers. ED and NSF sponsor an NCISE "dissemination" workshop to alert potential audiences to the research and elicit suggestions for directions of special interest and importance.
1994-1995	Completion data analysis, draft reports.
1995-1996	Editing and publication of *Bold Ventures* in three volumes. Release of synopsis of results in *Education Week's* "Forum" series, April 10, 1996; ED, NSF, OECD and NCISE conduct "Getting the Word Out," a conference in April 1996 to highlight international and U.S. results.

from a few extant sites (for example, for the Project 2061). The third task consisted of developing specific methodology: identifying key project informants to be interviewed, formulating appropriate surveys and questionnaires, and creating interview and observation protocols. Each of the case study reports provides further methodological details germane to the individual case.

The second year was devoted to intensive field work, development of analytical methods including coding of observations and interviews, analysis of critical documents, preliminary analyses of field data to generate further field work, and interaction with potential audiences to generate dissemination plans for the study results. The Introduction to Volume 1 illustrates the scale and nature of activity that went into data collection and analysis. Intermediate analyses, reviews, and feedback from the projects and sites as well as from outside consultants, final analyses, and writing of each case consumed the third year.

The entire effort was characterized by close collaboration among the research teams. Teams met collectively several times each year, first to decide on the common research questions and general approaches to the field work and site selection; and later to exchange early findings, develop common themes for the cross-case analysis, and decide on the length and style of the case reports. Still later, project meetings discussed key chapters of the cross-case volume and addressed dissemination issues. Several of the meetings also involved reviews and critiques by the project advisors (listed in the front of this volume). This unusually dense collaboration among the eight research teams proved highly motivating and intellectually stimulating to all involved, especially to the graduate students serving as research associates. This mode of working may also have enriched the quality of both the cross-case volume and the individual case studies reported in the following chapters.

References

Black, P. and J.M. Atkin, eds. 1996. *Changing the Subject: Innovations in science, mathematics and technology education.* London and New York: Routledge.

Organisation for Economic Co-operation and Development (OECD). 1993. *Science and mathematics education in the United States: Eight innovations.* Paris.

Webb, N. 1993. *State of California: Restructuring of Mathematics Education.* In Science and Mathematics Education in the United States: Eight Innovations. Paris: Organisation for Economic Co-operation and Development.

Weiss, I., M. Matti, and P. Smith. 1994. *Report of the 1993 national survey of science and mathematics education.* Chapel Hill, NC: Horizon Research.

Chapter 1

Building on Strength
Changing Science Teaching
in California Public Schools

J. Myron Atkin
Jenifer V. Helms
Gerald L. Rosiek
Suzanne A. Siner

School of Education
Stanford University

Contents

Building on Strength

Changing Science Teaching in California Public Schools

Setting the Stage

This is the story of California's attempt to improve science education from kindergarten through grade 12 between 1989 and 1995. The period was eventful. A new and different *Science Framework for California Public Schools* (California Department of Education 1990) was prepared and released; it took an innovative approach both to science content and to the instructional process. Efforts were undertaken to encourage commercial publishers to produce materials compatible with the *Framework*. Teacher networks —at both elementary and secondary school levels—were expanded and strengthened to enable teachers to come to grips directly with the spirit as well as the shape of the changes that were sought. The state's universities established new links with teachers and schools and strengthened existing ones.

These manifestations of reform were only the most visible part of the story. Beneath them was a particular philosophy of change—comprehensive rather than piecemeal, facilitative rather than prescriptive, built on identified strength in the system rather than focused on weakness. Based on patterns that had begun to emerge well before the period encompassed by this study, science education reform in the 1990s acknowledged and capitalized on existing teacher initiative. More precisely, procedures were devised that dispersed opportunities for leadership (and the associated responsibilities) broadly among key figures in the state Department of Education, a few knowledgeable people in the public universities, supportive school administrators, and dozens of teachers who already had demonstrated their inclination and ability to change things.

This approach was not, and is not, risk free. While everyone has a chance to help, and personal commitment is thereby enhanced, broad participation by people who operate an enterprise is not usually a formula for significant change. Indeed, the California science reform between 1989 and 1995 is not a story of trouble-free progress or unalloyed success. But to conclude that is to leap too far ahead of events. It is necessary to go back nearer the beginning to find out what happened and why, to identify some of the obstacles and accomplishments, and—finally—to return to some of the lessons that may have been learned.

In 1990, the California State Department of Education published its third *Science Framework for California Public Schools*, its statement of curricular priorities and textbook adoption standards that is rewritten or amended every seven years. This document, significantly different from its predecessors in both content and design, was to prove pivotal in what followed: the story related here never strays far from that publication and its effects, planned and unplanned.

The most significant difference between past *Frameworks* and the 1990 version lies as much in its use as in its content. It was intended by Tom Sachse, then administrator of the Mathematics, Science, and Environmental Education Unit of the California Department of Education and a key figure in the reform effort, to be "provocative," to stimulate serious consideration and extended discussion. As such, the 1990 *Framework* turned out to be one element—albeit the central one—in a multifaceted statewide education improvement strategy that has increasingly come to be thought of as *systemic*. The relatively open-ended text required interpretation to be useful. It also required involvement not only by teachers, textbook publishers, and school administrators, but also by teacher educators, parents, and school board members. In turn, these participants in the statewide effort derived direction and legitimation from the 1990 *Framework*. This relationship between a key document and its many audiences is sometimes subtle and easy to miss.

It may be particularly difficult to grasp this connection outside the state. California's 1990 *Framework* enjoys a national reputation as an uncommonly effective education policy document. Elizabeth Stage, the chair of California's Science Framework Committee and in the early 1990s a member of the National Research Council's (NRC's) Science Education Standards staff, reports on what is often overlooked about the *Framework*.

> I went to a meeting in Delaware with their science framework commission, where I walked in with a copy of the National Research Council draft. I was supposed to be talking about it. [But] in front of every person . . . was a California *Framework*.

> Many other states are devising frameworks for the first time. One of the impacts of California *Science Framework* that is pretty far reaching and sometimes scary is how many states are using it as a model . . . [However,] most people don't understand the whole context: that it is a piece of a systemic reform effort. It's hard to tell people that . . . You can't break the hearts of those people by telling them that they've only just begun [when they have developed the framework document].

Systemic Reform, California Style

California is the nation's most populous state and one of its most ethnically diverse. Nearly 5 million students attend California public schools. Of these, more than half are nonwhite, and one-fifth speak 1 of 150 different languages as their mother tongue. The state is stressed financially. In the late 1970s, a voter initiative, Proposition 13, drastically limited local governments' ability to raise taxes. The effect on public schools has been severe. School buildings are in disrepair. Supplies are limited. Class size is increasing. Most people believe, with reason, that the quality of public education is declining. These conditions comprise some of the background for the efforts described here.

What else frames the *Framework*? What are some of the national and local circumstances against which the California science reform was, and is, played out, and that therefore influenced its shape and content? The American Association for the Advancement of Science's (AAAS's) *Project 2061: Science for All Americans* had been launched in 1985. Tom Sachse was close to Project 2061 leadership, and some of that project's underlying philosophy ("less is more") and curriculum detail (the use of "themes") found resonance in California. The National Science Teachers Association's Scope, Sequence, and Coordination (SS&C) Project, a high school curriculum reform movement seeking to see "every science" taught to "every student, every year," had also become prominent; and it, too, was to have its mark in California.

Additionally, systemic approaches to educational reform were being discussed and defined around the country (and were to eventuate into a five-year $10 million state systemic grant from the National Science Foundation [NSF]) to the state of California in 1991), and California developed its distinctive version. The basic premise of the earliest systemic reform theories was to recognize that attempts to improve educational practice solely by improving one element of the educational system missed that system's complexity. Teachers were central. But textbooks, the assessment system, and the higher education system were powerful influences, too. Furthermore, the efforts to improve all elements of the education system had to be consistent conceptually. Tests for students had to be compatible with instructional goals. Instructional goals had to be compatible with teacher education. And instructional materials had to be compatible with everything else. You couldn't fix anything unless you fixed everything was the key insight of the systemic view of change that began to take shape in the late 1980s.

Size, though, makes a difference. Whatever else characterizes California, it is big—with more than 30 million people, including 5,200,000 of school age. As a direct result, the state has certain opportunities and certain constraints. For example, California has been in an unusually potent position for affecting certain aspects of reform. Most notably, the size of the California market for textbooks—upwards of 10 percent of the nation's total—allows state education leaders to exert disproportionate influence on publishers. Its size has also won it influence in the assessment community, such as with the Educational Testing Service's development of the National Teacher Exam (NTE) for science teachers. Stage recounts:

So the California Commission on Teacher Credentialing said it wouldn't buy the NTE unless the NTE changed to reflect California's changed notion of the subject area . . . You want to talk about leverage! You want to talk about far-reaching effects. You sit there at Princeton, and you see these people who look scared because if your customer who accounts for 15 percent of your sales is bailing all of a sudden . . . [If] you've also heard from other quarters that they may actually be anticipating where the rest of the country is going, then you think, "Oh, well . . . maybe we should listen."

A crucial correlate of size, however—at least in the minds of key education leaders in the state—is that when trying to make changes in schools, it is difficult to implement sweeping mandates from Sacramento if there is to be much hope that they will be followed faithfully. There are too many forces at work, too many school districts, too many schools, and too many teachers to expect fidelity. Nationally, the process of systemic reform in its early formulations was often conceptualized, even if implicitly, as top down. State leaders around the country were encouraged to target relevant features of the system and make sure these efforts were "aligned." Systemic approaches to reform were to be given both their form and force by state mandates. Presuming states would be able to effect compatibility in a top down manner, federal granting agencies began making evidence of moving toward "alignment" as a criterion for receiving federal funds.

Interviews with California education leaders such as Bill Honig, Tom Sachse, Elizabeth Stage, Kathy DiRanna, and Kathy Comfort reveal that, while systemic alignment of educational improvement efforts was embraced enthusiastically, a centralized approach to systemic reform was never seriously considered. One reason is that a top-down approach to systemic reform in a place the size of California would involve an impossible and intolerable level of micromanagement.

The alternative vision of systemic implementation that emerged was less a totally grassroots approach than a blending of centralized and decentralized initiatives. It can be characterized as consisting of three elements: (1) building a consensus about a general set of priorities for improving California science education among key constituencies, (2) designing an implementation structure flexible enough to be responsive to the evolving needs of the reform, and (3) employing that structure to put the resources and incentives in place to aid educators at all levels in constructing their own path to the shared goals.

The 1990 *Framework*'s Place in Systemic Reform

California had had earlier science frameworks. The 1978 version had been amended in 1984, but as the decade neared its close, leaders in the science education community had become dissatisfied with its laundry-list approach to the science curriculum. It specified important facts and concepts, but seemed not to have a unifying direction. Commenting on the origins of the 1990 *Framework*,

Stage observes: "The science education community didn't know what you could do with a good framework, because they'd just done an addendum with 597 discrete facts to be learned over the course of 13 years. They didn't know what you'd expect or want to have in a good framework."

Bill Honig, California's state superintendent of public instruction who commissioned the 1990 *Framework* and is widely credited with the vision of what a good framework could accomplish, describes its role in the following terms:

> The *Framework* has two jobs. First, it is an announcement of a total philosophy of a discipline. Less is more. A total system . . . an overarching statement of the type of learning we are seeking. Second, it lays out . . . all the things you have to do to see this actually happen. It is a planning document for teaching, publishers, assessment.

One can think of the 1990 *Framework* as the keystone for systemic efforts to improve science education in California: it is a central and supporting element of all the other components of the reform. Also like a keystone, the *Framework* does not stand alone; it is itself supported by many organizations and people committed to making the vision embedded in the *Framework* a reality.

These interdependent relationships between the 1990 *Framework* and other elements of the reform take form as the various groups involved in education improvement try to fathom the meaning of the *Framework* document and figure out its consequences. In the process, they shape it. The deliberative process actually was set in motion at the outset as the *Framework* was planned and written. By 1988, as Honig began to assemble a framework writing committee, a number of organizations and state leaders with experience and interest in science education had already begun efforts to make significant changes in California science. These people were known to one another. Their interest was high and their goals broad. Says Honig, "The aim of *Framework* was to give focus [to this group] without giving explicit direction."

To help forge consensus, the actual writing of the 1990 *Framework* document had to be an activity that brought together the many able and influential actors, loosely connected though they were. As noted years later by a group charged in 1993-94 with evaluating activities undertaken under an NSF grant to strengthen California's State Systemic Initiative (SSI):

> Authority and responsibility for science and mathematics education policy [in California] is distributed among several dozen key individuals and agencies who are known to each other and who have sometimes worked together and in opposition to one another. The result of this distributed system of state policy making for science and mathematics education influences schools throughout California via a handful of social, institutional, and material avenues that together forms a powerful network (Far West Laboratory 1994, p. 102).

Stage describes her role in building consensus:

By being an outsider, I didn't have to play the either/or games. I could play both/and games. So the compromises didn't have to be "Do you love Larry Lowery and the process skills *or* do you think the content is important?" It was, "Well, let's talk about this idea of themes." It's the connections among the big ideas. That resonated with everybody. So that was what I brought to the table.

All of us were taking on the opportunities of the *Framework* to develop some piece of work we wanted and needed to develop professionally and to which we could contribute. So for a whole bunch of individuals, it could hang together in that fashion. But having done it, this idea that we all bring something to the table establishes a way of interacting that still goes on today. It's a bunch of people who all do favors for each other and respect each other because we can see the advantages of working hard for no pay, no credit, no nothing—and then the good things happen.

There were at least two consequences of bringing a diverse group together to reach consensus wherein no one had particular authority over another. One was the need to establish unusual levels of confidence and trust. Stage illustrates the issue of trust with the following story:

> I remember the day that the *Science Framework* passed the State Board of Education, and Tom Sachse, and Kathy Comfort, and Kathy DiRanna and I went out to lunch at Frank Fats in Sacramento. We said, "Well, we've just started. Now what are we going to do?" There was an agreement at that time, quite explicit, that each of us had to go from where we were, [with] the resources that we had, and the institutional position that we had, and leverage that we had, and the ideas we had, and goals we had; we had to do what we could do . . . [Later], when I was away, it was hard to transfer some of that trust . . . to a stranger.

Different Chapters for Different Audiences

A second consequence of constituting a writing team with few formal links among them was that the final form of the 1990 *Framework* reads more like a collage of loosely assembled educational improvement visions, with a few themes in common, than like a tightly woven statement of priorities. Each chapter had its own primary author and its own intended audiences. Reinforcing the multiple-audience aspect of the document, the *Framework* was also a response to political pressures that were strong at the time, one in particular.

A year before the *Framework*, the State Board of Education adopted a Policy Statement on the Teaching of Natural Sciences. This policy is the first piece of text in the 1990 *Framework*, and it includes such statements as the following:

> Nothing in science or in any other field shall be taught dogmatically . . . Compelling belief is inconsistent with the goal of education.

> Philosophical and religious beliefs are based, at least in part, on faith and are not subject to scientific test and refutation. Such beliefs should be discussed in the social science and language arts curricula.

> Neither the California nor the United States Constitution requires, in order to accommodate the religious views of those who object to certain material . . . in science classes, that time be given in the curriculum to those particular religious views (*California Department of Education* 1990, pp. xi -xii).

The *Framework* goes on to press for special status of "fact" and "theory." The first chapter, which follows the policy statement, an executive summary, and an introduction, contains such statements as:

> Theories are replaced in their entirety infrequently and then only when a new theory is proposed that subsumes the old theory. The new theory does so by explaining everything that the old theory explained as well as other evidence that might not have fit very well in the old theory . . . A **theory** is not a half-baked idea nor an uncertain fact but a large body of continually refined observation, inference, and testable hypotheses. Terms such as **fact** and **theory** are used differently in scientific literature than they are in supermarket tabloids or even in the normal conversation of well-educated people (*California Department of Education* 1990, p. 17, emphasis original).

The first sentence of the 1990 *Framework* states,

> This first chapter is about science itself: what it is, what it is not, what its philosophies and methods are, and how these differ from those of other intellectual activities . . . This chapter also guides teachers on how to deal with socially relevant and sensitive scientific issues and how to ensure that the beliefs of all students are treated with respect (*California Department of Education* 1990, p. 12).

To understand the juxtaposition of a discussion of the nature of science with practical advice about handling sensitive issues, as well as the close attention to facts and theories, one must realize the importance of creationist pressures in the state at the time the *Framework* was written. Creationists were avid in their views about evolution in the curriculum, and powerful. As the new science framework was being planned, creationists demanded that the science curriculum treat evolution as a theory, not a fact, and that creationist views be taught alongside Darwin and other mainstream scientific perspectives—either that, or eliminate the teaching of evolution altogether.

In approaching this issue through a general and expansive discussion of the nature of science, the 1990 *Framework* takes a view that appears to be minimally informed about developments in the history and philosophy of science over the last three decades. Most scholars in the field, certainly since Thomas Kuhn (1962) addressed the subject with his *The Structure of Scientific Revolutions*, accept the view that progress in science is not as orderly as scientific method or as clear cut as the *Framework* suggests. How one theory comes to replace another is now seen as complex and highly contextualized, involving politics, culture, human nature, and available technology, as well as experimental results.

It is unlikely that the *Framework* writers were unaware of these contemporary views about the nature of science, especially in light of the backgrounds and qualifications of the members of the group. Almost certainly, the adopted stance was a conscious one in response to the creationist pressures. This interpretation seems supported by the fact that the *Framework*'s first chapter's opening discussion of the nature of science flows quickly to illustrations of certain scientific theories that are emphatically "always testable, objective, and consistent." Examples, however, come primarily from one field—evolution. The last half of the chapter mentions Darwin more than any other scientist. The longest of the scientific examples concerns the Piltdown man hoax and why this "discovery" does not discredit evolutionary biology. In the concluding section, "Socially Sensitive Issues Have a Place in the Classroom," 3 paragraphs are devoted to environmental conservation, 3 paragraphs are devoted to animal experimentation, and 14 paragraphs are devoted to defending the teaching of evolution in public school science classes.

Stage confirms the relation between the practical realities of California school politics and the philosophy of science included in the *Framework*:

> That's what the argument was about [in the first chapter]. Most science teachers are pre-Kuhnian, and many scientists are pre-Kuhnian . . . When you've got a social credibility problem, it's not the time to have the post-modernist argument and set the boundaries of science; in fact, one of the ways that the people on the [National Council of Teachers of Mathematics] standards [project] have shot themselves in the foot was to go overboard on the socially constructed knowledge bit and cause the scientists to come down off the rafters.

> You can only have so many fights on so many fronts at the same time. So let's take the mainstream view of science and show that it's not incompatible with the mainstream view of religion. If you get to messing around with the socially constructed knowledge . . . , you get into big problems with religious stuff. I'm not equipped to argue that. I can *follow* that argument, but I'm not equipped to *have* it.

Affirming the fact that the *Framework* covers a range of audiences, by design, the document states:

> Different audiences will be interested in different sections of this *Framework*. Part I answers the question, What is science? and should be read by those interested in science and science curriculum. Part II outlines the required content of physical, earth, and life science programs. Part III has three chapters: Chapter 6 is intended for teachers and supervisors who are interested in the processes of teaching and the processes of science; Chapter 7 is intended for supervisors, department chairs, principals, and other administrators who have responsibility for implementing the science curriculum; and Chapter 8 is intended for publishers and other developers and reviewers of instructional material who collectively form the other primary audience for the *Framework* (*California Department of Education* 1990, p. 8).

What did this melange of visions and purposes result in? The following sections highlight the most salient material in the major chapters of the *Framework* document, organized by main ideas rather than chapter.

Science Themes

In a 1995 interview, Bill Honig asserted that the original intent of the *Framework* was largely corrective: "The 1990 *Science Framework* starts off as a response to what was wrong . . . [For example,] the books at the time were compendiums of facts. And the teaching: . . . even where hands-on labs were required, they were considered add-ons to the curriculum, as opposed to its core."

The remedy prescribed by the *Framework* can be found in its second chapter, where it is recommended that science teaching be thematically organized. The introduction to the second chapter reads:

> This science framework differs from previous frameworks and addenda in its emphasis on themes of science . . . themes should be a major emphasis of the science curricula in order to reinforce the importance of understanding ideas rather than the memorization of seemingly isolated facts (*California Department of Education* 1990, p. 26).

The *Framework* maintains that this thematic organization will require a rethinking of science teaching priorities. One of the principles of incorporating themes listed in this chapter's concluding section states: "The emphasis on themes in science requires a reconsideration of how much detailed material should be included in the science curricula . . . To some extent the omission of various traditional terms and concepts that are not well integrated in current instructional material will be required" (*California Department of Education* 1990, p. 35).

This favoring depth of coverage over breadth of coverage, which is implicit in the advocacy of thematic teaching, pervades the *Framework*. It is to this fundamental preference that the popular teaching slogan "less is more" refers.

The six themes the *Framework* offers as a means of organizing the science curriculum are:

- energy,
- evolution,
- patterns of change,
- scale and structure,
- stability, and
- systems and interactions.

The *Framework* avoids giving definitions of these themes, providing instead examples of how the themes emerge in various disciplines. For example, of the theme "energy" it says:

Defined in physical terms, energy is the capacity to do work or the ability to make things move; in chemical terms, it provides the basis for reactions between compounds; and in biological terms it provides living systems with the ability to maintain their systems, to grow, and to reproduce (*California Department of Education* 1990, p. 28).

It goes on to expand on how the theme of energy is employed in the physical, earth, and biological sciences. It closes with a comment on the importance of the theme of energy to the ethical use of science and technology (e.g., energy conservation).

The *Framework* distinguishes themes from theories. It states more than once that theories "unify and make sense of facts and hypotheses related to a particular natural phenomenon." Themes are also "pedagogical tools that cut across disciplines." In another place, themes are referred to as "ideas that integrate the concepts of science across disciplines in ways that are useful to the presentation and teaching of scientific content." According to the *Framework*, themes are essential to good science teaching.

The second chapter emphasizes repeatedly that the six themes outlined are only examples, and other sets of themes are possible. It notes the danger in treating themes superficially and turning them into scientific vocabulary definitions of the sort the *Framework*'s authors wanted to move beyond. Other possible sets of themes, including those in AAAS's *Science for All Americans*, are listed and endorsed to further emphasize this point.

The Subject Matter Chapters

The 1990 *Framework* did not seek curricular integration across the traditional boundaries of science subject matter with themes. This can be seen in the organization of its content chapters. Entitled "What is the Content of Science?" the content section of the *Framework* comprises three chapters, respectively, "Physical Sciences," "Earth Sciences," and "Life Sciences." Most space is given to the physical sciences.

These chapters are relatively detailed, but considerable effort has been put into making them something other than a list of topics to be covered. Each chapter is broken down into a set of content areas that are related to scientific theory. For example, the chapter on physical sciences is organized into the following content areas:

- matter,

- reactions and interactions,

- force and motion,

- energy: sources and transformations,

- energy: heat,

- energy: electricity and magnetism,

- energy: light, and

- energy: sound.

The *Framework* stresses that these are not the only possible ways to subdivide a subject matter into theoretical content areas.

Each of these content areas is in turn further subdivided into a set of questions that are central to the content area. For example, the content area of "matter" is accompanied by the following questions:

- What is matter, and what are its properties?

- What are the basic units of matter, and where did matter come from?

- What principles govern the interactions of matter? How does chemical structure determine the physical properties of matter?

The descriptions of the content areas appropriate to various grade levels are then developed in the form of narrative responses to these questions. Each question is answered at four grade ranges: K-3, 3-6, 6-9, and 9-12. These descriptions were authored with the intent to emphasize the conceptual and not the factual content of science. For example, an excerpt of the answer to the first question above considered appropriate for grades 3-6 reads:

> The properties of matter depend very much on the scale at which we look. For example, sand flows through your hands almost like a liquid. But an ant carries a single grain of sand very much as you might carry a rock. Very often, however, the properties of matter at the larger scales depend on its properties at smaller scales. Solid objects and liquids are made of particles that stick together; gases are made of particles that do not stick together (*California Department of Education* 1990, p. 42).

At the end of each grade-level-appropriate response is a bracketed list of the 1990 *Framework*'s themes that apply to the concepts described.

Science Processes

Chapter 6 is titled "Science Processes and the Teaching of Science." This chapter strongly advocates the use of developmentally appropriate scientific processes as part of science teaching. While not using the term "inquiry" favored in the recent national standards, the concept is similar (National Research Council 1996).

Like the themes, an emphasis on teaching science as a scientific process is presented as being consonant with the "less is more" philosophy of teaching. In this case, hands-on, discovery-based lessons are presented as requiring more time but resulting in deeper and longer lasting learning.

In this chapter, scientific thinking processes are subdivided into levels of increasing sophistication. These levels of thinking processes and their developmentally appropriate grade ranges are:

- observing (K-12),

- communicating (K-12),

- comparing (K-12),

- ordering (K-12),

- categorizing (K-12),

- relating (3-12),

- inferring (6-12), and

- applying (9-12).

Each of these is accorded a few paragraphs of explanation and then illustrated with brief examples. Scientific processes are not covered in nearly the depth as are themes.

Achieving the Desired Curriculum

In addition to an explication of science processes, over half of chapter 6 is devoted to presenting specific goals for achieving desirable science programs in kindergarten through grade 12. For example, recommendations for elementary school include expanded explanations of: "Provide a balanced curriculum in the physical, earth and life sciences;" "Reinforce conceptual understanding rather than rote learning;" and "Arrange the classroom setting and student grouping to optimize positive attitudes for learning science."

For high school, the document advises teachers to: "Help students understand the nature of science—in particular, its experimental, non-dogmatic nature and the methods by which progress is made;" "Develop in students a strong sense of the interrelationships between science and technology and an understanding of the responsibility of scientists and scientifically literate individuals to both present and future societies;" and "Provide an expanded view of science-related careers." It is in this section that the *Framework* alludes to the concept of coordinating the curriculum. The message, however, is not strong and is buried in one of the seven recommendations.

The latter few pages of this chapter outline recommendations for achieving equity, or ways to ensure that all students are taught a quality science program. This section underscores strategies for teaching science to historically underrepresented groups and to students with limited English proficiency.

Implementation

Chapter 7 of the *Framework* emphasizes the importance of staff development, and the active participation of teachers and administrators in the implementation process over a sustained period of time. The *Framework* explains:

> It is not enough to focus on science for the one adoption year out of every seven. The adoption of materials should be preceded by extensive planning and staff development and followed by more staff development and assessment . . . Teachers and administrators need to experience the changes themselves and have the time to plan, to experiment, to revise their plans, and to implement the changes at a pace that allows for adequate reflection and internalization (*California Department of Education* 1990, p. 173).

A significant portion of the chapter describes and provides examples of a relatively mature version of a school-based implementation model developed by Kathy DiRanna, executive director of the California Science Implementation Network (CSIN); Herb Strongin; and others at the Science Curriculum Implementation Centers (CIC's) in the years preceding the 1990 *Framework*. The following section of this report, "Elementary School Science," provides the details of this process.

Instructional Materials Criteria

The final chapter of the *Framework* lays out the standards by which the state will judge instructional materials in its statewide adoption process for grades K-8. Twelve criteria are outlined. The most notable of these is, perhaps, number 10, the explicit recommendation that 40 percent of the instructional materials involve hands-on experiences.

The criteria, and the relative weights they were to be given in the decision-making process expressed as percentages, are as follows:

Content (50 percent)
1. Material discussed in content sections is present (5 percent).

2. Material is treated accurately and correctly (15 percent).

3. Material is treated thematically (15 percent).

4. Depth of treatment is adequate (10 percent).

5. Emphasis is on how scientific knowledge is gained (5 percent).

Presentation (25 percent)
6. Language is made accessible to students (5 percent).

7. Prose is considerate and engaging; scientific language is respected (10 percent).

8. Science is open to inquiry and controversy and is presented nondogmatically (5 percent).

9. Science is shown as an enterprise connected to society (5 percent).

Pedagogy (25 percent)

10. Hands-on experience is emphasized; hands-on experiences should represent at least 40 percent of class time, and this priority should be reflected in instructional materials (15 percent).

11. Instructional materials recognize cultural diversity (5 percent).

12. Assessment reflects experience, integration, and creativity (5 percent).

Next Steps

With respect to design, the 1990 *Framework* is considerably more open-ended than its preceding editions. Most notably, it does not mandate specific factual content for the curriculum. In fact, it is written in a style intended to discourage teachers from using the examples it offers as prescriptions. The *Framework* was not to be the final statement of the reform effort, but the beginning. Bill Honig assumes as much when he points out, "after the *Framework*, the next stage was the staff development networks, CSIN and SS&C."

◆

Elementary School Science

California employs two teacher networks as the main vehicles for science education reform: the California Science Implementation Network for elementary school teachers and Scope, Sequence, and Coordination for secondary school teachers. However, there are many other regionally based professional development initiatives that have generated increased interest in light of the *Framework's* curricular goals for K-12 science education. A number of these efforts, such as the California Science Project (CSP), are supported by institutions of higher education around the state. While some of these initiatives predate CSIN, others are newer to the reform scene. This chapter focuses on CSIN because of it's history of involvement with statewide science education reform, and its efforts to address both the content and the structural context of elementary school science programs. There is much to be learned not only from the scale on which CSIN has worked, but from the ways in which it has attempted to reach teachers.

CSIN's Origins: From Little Steps to Big Ideas

CSIN has its roots in organizations that predate by several years its formal establishment by the California Department of Education in 1988. Kathy DiRanna, executive director of CSIN from its inception, had played a major role working with elementary school teachers in Southern California in a state-funded, but regionally bounded initiative, called Processes and Concepts in Elementary School Science (PACES). Says DiRanna of this staff development effort: "Get good people and let them create. Get out of their way, remove the roadblocks, and you'll get fantastic kinds of things. I think it is a credit that CSIN has actually tried to continue with [the approaches we took in PACES]."

Establishing a teacher network was one ingredient in CSIN. Another had been identified and demonstrated by Herb Strongin and Gary Nakagiri at the Science Curriculum Implementation Center in San Francisco. Similar to PACES in many ways and operating at about the same time, CIC took a "little steps" approach: attempt modest initiatives with teachers in trying to make improvements in the classroom, rather than make lots of changes at once. But toward what end? Strongin believed it essential to identify a clear direction for improvement of curriculum and teaching that would focus on major science concepts. DiRanna says, "He was the first one to bat around the idea that there should be 'big ideas' in science, that we should get away from the little factoid approach to our teaching."

While PACES and CIC worked on approaches to educational change and a new vision of the subject matter in elementary school science, educators at the University of California at Irvine were developing a model for university professors to collaborate with K-12 classroom teachers on professional development in science. This model proved to be influential in the later structure of programs such as CSIN and the CSP summer institutes.

In 1986, Tom Sachse at the California Department of Education capitalized on the strength and experience evident in PACES, CIC, and at the University of California at Irvine by bringing these groups together to seek further common ground and help establish a base from which to make deeper and broader changes in the state curriculum. Sachse wanted to move past California's major curricular document of the time, the *Science Addendum* of 1984, which detailed hundreds of facts and concepts that children were expected to learn, but with little sense of a central purpose. Strongin's "big ideas" seemed attractive and educationally important.

Additionally, Sachse and many others were beginning to view long-term backing of professional development for teachers as essential to reform: unless teachers understood and believed in the new program, any changes would likely be merely cosmetic. DiRanna had helped create a structure for teacher support that also recognized the importance of schoolwide change. When the 16-member Curriculum Framework Committee was created, DiRanna and Nakagiri were asked to serve. DiRanna also became a key player on the nine-person group that did much of the actual writing of the 1990 *Framework*.

This instance is but one of several whereby Sachse built on ongoing efforts in the state that seemed sound and effective. He thus solidified and integrated separate initiatives that were conceptually consistent. An added benefit was that the initial CSIN staff knew one another through previous professional associations and so brought an easy collegiality, a common vision, and basic levels of trust to the new enterprise. This approach to educational change reflected a central feature of the California science education reform—look for strength in the system, and build on it.

CSIN and the *Science Framework*

The *Framework* writing committee included CSIN teachers and others who had worked closely with the network. In particular, DiRanna and CSIN teachers played a significant role in designing chapter 7, "Implementing a Strong Science Program." The most obvious CSIN contribution to the *Framework* document was inclusion of CSIN's own program elements matrix (PEM) and content matrix, grassroots procedures for an entire school community to examine its science program goals and curriculum. PEM and the content matrix had emerged within CSIN from discussions about how the big ideas translate into specific grade-level science concepts. A description of the technique, with examples, takes more than 14 pages of the 220-page *Framework*. DiRanna notes on the development of the Content Matrix: "CSIN has never been a top-down organization. The content matrix didn't come from us first. It came from a brainstorm session with teachers, and we went back and put the formality to it, and the structure to it . . . That's been a really important part of most of what we've done. We're learning with the schools."

DiRanna's comment highlights another feature of the reform, a sort of consistency in the approach to educational change: just as the state's efforts enhanced the momentum already generated in PACES and CIC, CSIN took the same approach with teachers and school administrators.

With the incorporation of CSIN's actual procedures, PEM and the content matrix, the 1990 *Framework* became immediately recognizable to many elementary school teachers and administrators. In turn, direct reliance on CSIN in the *Framework* gave the network added prestige and direction. DiRanna explains:

> It became an enormous selling point for people to join the network because [CSIN] was in the *Framework*. Here is a major push. If we weren't in the *Framework* we would never have gotten the push . . . CSIN's primary job was to translate what the philosophy and goals of the *Framework* look like at your school.

The Implementation Process

Schoolwide Change. Before CSIN, the almost universal model of professional development in elementary school science had been to create opportunities for teachers to hear from well-recognized speakers who had provocative ideas, usually by attending a lecture or a one- or two-day workshop. Teachers frequently had a passive role in this approach to staff development. They were then usually on their own to act (or not act) on what they had heard. If changes were attempted by individual teachers, it was not uncommon for them to receive little support in their schools or to face active resistance.

In contrast to the previous model, participation in the network requires a schoolwide commitment to implementation of an articulated K-6 or K-8 science program. In addition, the network approach to staff development is built around teachers teaching teachers. CSIN currently employs approximately 150 staff developers, themselves full-time teachers, who work with other teachers on a release-time or after-hours basis. Teachers are eligible to be staff developers if they have previously participated in CSIN or have science and staff development backgrounds. Many staff developers are already science mentors or science resource specialists. Staff developers provide district leadership for between three and five CSIN schools. Each individual CSIN school also has a lead teacher who coordinates on-site staff development. Site-level activity centers on developing the PEM and the content matrix, selecting and implementing appropriate instructional materials, and developing teachers' and students' conceptual understanding of science.

When CSIN was founded in 1988, it was funded by state funds from the U.S. Department of Education's Eisenhower program. However, since 1991, CSIN has had additional assistance of State Systemic Initiative funds from the National Science Foundation. A portion of the SSI funds have been used to employ 12 teaching consultants, distributed across state regions, who support the staff developers and coordinate the entire effort on a full-time basis. All teaching consultants are former teachers. They each work with approximately 11 staff developers and an average of about 55 schools in their designated regions.

Schools are involved in the network for an average of about three years. Involvement occurs at different levels depending on a school's stage in its development of a science program. CSIN Level 1 is considered the planning phase and prepares the staff developer, lead teacher, and principal to work as a collaborative team for schoolwide change. PEM, part of the Level 1 planning phase, is an opportunity for a school to begin the process of collaboration, to define group goals, and to build on the knowledge of the teachers at the school site. Notes one staff developer: "If you use the CSIN model, you have teacher involvement all along the way. You have better buy-in. It's a wonderful model, wonderfully applicable. There is a great community support element. We have general parent meetings where we discuss with them wishes and needs. It's very positive." DiRanna underscores the point:

> You cannot say, "Here is the change, go do it," because in the past the school plan would [already] have been written, so you were either on the train or off the train . . . What we really recognized is that we needed to have stopping points all along the way. We have to have places where people could get on [and] maybe only travel this far, but for them that is tremendous growth.

Once schools have developed their PEM, and have articulated and agreed upon goals for their science program, the next step is to build the content matrix. This process entails teachers creating a program that connects specific grade-level science concepts to big ideas and unifying concepts. (See appendix B for examples of content matrices.) Kathy DiRanna explains the function of the content matrix: "It's a scaffolding on which to select, adapt, or adopt instructional materials." The goal is for the science program to be driven by a teacher-developed content matrix rather than solely by the instructional materials. In cases where the content matrix is developed at the district level, teachers often feel less ownership of the implementation process.

After developing an implementation plan, schools are ready for the more intensive CSIN 2 training, which provides staff development in science content and pedagogy associated with the content matrix and the instructional materials selected by the school or district. CSIN 2 also includes a program for school administrators to develop, organize, and provide ongoing support for program implementation. CSIN 2 provides a school with a process for schoolwide change that includes:

- 36 hours of professional development for the entire school faculty, including grade-level-specific assistance with instructional materials, science content, and pedagogy;

- on-site assistance from a team consisting of a teaching consultant, a science staff developer, a scientist, and a high school science teacher; and

- an on-site expert—the lead teacher—who is trained in the 20-day institute that emphasizes the big ideas in science, instructional materials, and change strategies.

Schools use a variety of funding sources to participate in CSIN, such as Eisenhower, Mentor, School Improvement Program, Title VII, Chapter 1, Migrant Education, SB 1274 Restructuring, and local school support. Network participation requires a school fee of $2,500 which covers the staff developers stipend. Compensation for the lead teacher, which is $750 to $1,000 a year, is an additional cost to the school beyond the school fee. On average, participation in CSIN costs a school $8,000 to $10,000 a year including compensation for staff developers, release time for all teachers, and materials for staff development. CSIN's fee scale is negotiable and often is determined on a case-by-case basis.

CSIN-organized professional activities occur at several different levels, from statewide professional meetings to in-service sessions at the school site for the entire faculty. Each year, CSIN staff developers and lead teachers attend 20-day institutes that are coordinated by the teaching consultants. The first half of the institute (days 1-10) occurs during the summer and is an opportunity for teachers to work intensively together on content, pedagogy, instructional materials, diversity issues, assessment, and strategies for implementation of the *Framework*. There were eight regional institutes in the summer of 1994; the one at Mills College in Oakland was attended by more than 150 staff developers and lead teachers from the San Francisco Bay Area. Staff developers and lead teachers who have participated in CSIN since 1992 have had the opportunity to attend 10-day summer institutes in physical science in 1993, earth science in 1994, and life science in 1995. The 1996 summer institute will include an environmental education strand.

Content sessions at the institutes are led by teaching teams comprised of teaching consultants, staff developers, high school science teachers, and scientists. This model for teaching content is also used at the school site for staff development. Staff development days that occur at the district and school site are led by staff developers and lead teachers. Lead teachers continue with the 20-day institute during the school year (days 11-20) to help them prepare for the 36-hour on-site program they conduct with their school's staff.

Not only do staff developers and lead teachers focus on science content and pedagogy in CSIN staff development, they also learn about educational change strategies. Teachers have reported that understanding these strategies has helped them work more effectively with administrators and teachers who are resistant to participating in the science program. They recognize and attempt to work with both the structural and emotional barriers to change. The structural elements are addressed by involving administrators and other school and district staff. Staff developers report, "We create a balance between administration and classroom teacher. We help the classroom teacher feel free to plan and take action without the administration feeling threatened." Throughout the process of program implementation and staff development, the network leadership tries to help teachers and administrators understand that long-lasting change is difficult, that it "hurts," and that it takes time.

Structure Versus Flexibility. CSIN has been criticized for being highly centralized, monolithic, and excessively formal. A rigid model is implied by PEM and the content matrix. Every school has a staff developer or lead teacher, and most schools go through two phases—CSIN Level 1 and CSIN 2. All the staff developers and teaching consultants appear to operate in about the same way. The teaching consultants' prominent leadership positions, and their distribution throughout the state, seem to convey further desire for uniformity. In CSIN's defense, Sachse says:

> Kathy DiRanna started working and structured things with the content matrix and with the PEM because she needed to have a roadmap for herself as well as the crew. And then we started seeing 75 schools, and then 150, and then 300 and then 1,500. On it goes. Kathy will tell you, and she is the first one to say it unless I say it first, that CSIN really does have a lot of flexibility, but when people see something like the content matrix or the PEM they get so frightened [of the] boxes into which their creativity is cycled, they run from it, even though they don't have to use it. So while it is an aid to Kathy and some of the people who know it well, it is a straightjacket for others. But Kathy has benefited and so has CSIN from at least one model out there. [You can] then tear it apart and make your own.

As evidence of flexibility, a teaching consultant pointed to an area of rural Monterey County where small, multigrade classrooms have special difficulties in devising instructional plans that fit the *Framework*. The schools were already part of a small co-op, and so pooled their resources to form a consortium. The consortium then approached CSIN for staff development and support, which it received. CSIN tailored its program and adapted the content matrix to meet the needs of these teachers of multigrade classrooms. DiRanna addresses the matter of local autonomy and flexibility, as follows:

> I think people who criticize us for being centralized misunderstand what we are about. The whole point is: here's a process, a way to start thinking. If this does not fit for you, then modify it, change it, throw it out. Do whatever you want to do, but you've got to start somewhere, so here's a place to start. So we tried really hard to say this is a model, not a mold. It doesn't mean you all have to march to the tune of the same drummer. The whole idea of content matrices and other things is you design it at your school for what makes sense, and we'll help you implement it.

CSIN at Howe Avenue

Howe Avenue, an elementary school in Sacramento, is ethnically and linguistically diverse. A high percentage of its students come from low-income families, and the school has a large population of students. Fourteen different languages are spoken by the students with limited English proficiency.

Teresa Ramirez was a staff developer for four years at Howe Avenue. In discussing what she sees as a largely successful program, she identifies one particular person as critical:

> I was supported along the way [by the school administrator] . . . That is such a key factor. Part of the team must include the administrator. It's a must, because that person will either make it or break it. They allowed us to try new things without being critical and allowed us to make mistakes as well. I think that was the foundation that empowered many of us teachers.

Howe Avenue joined CSIN in June 1991 and in the first year developed a site leadership team which included the principal, one intermediate teacher, one primary teacher, and a preservice teacher. The school used part of a state grant for restructuring and part of a separate School Improvement Program allocation from the state to cover the $2,500 annual CSIN program fee. The two lead teachers at the school site were each paid $750 a year.

There are five designated CSIN "release days" during each school year. The staff development days at the school site include the Overview Day at the beginning of the year, when the school reexamines its content matrix. There is also a Seminar Day at the school site, when a science teacher from a local high school conducts an in-service program at the school site. The remaining three days are for regional meetings attended by staff developers and lead teachers.

The content matrix at Howe Avenue has given classroom teachers an opportunity to play a central role in the development of an articulated, developmentally appropriate K-6 science program.

> We revisit [the content matrix] every year, and it's been really nice because we go back, and at the end of the year Michelle [another Howe Avenue lead teacher] and I meet with all the grade-level teachers and we say, "How did it go this year? What would you do differently on your matrix? What would you

change?" We redo the matrix, and then we come back in September and start again and say, "Now this is what we decided last June and these are the changes that were made" . . . We work with it for another year, so the part that I think our staff is really excited about is that the Matrix is ever-changing. It's not static. It moves. It can meet the needs of your school.

Not only do teachers revisit and continually adapt the content matrix each year, students revisit and build on scientific concepts as they progress through the grade levels. For this reason, lead teachers at Howe Avenue describe their curriculum as "spiraling":

The idea about the CSIN model is that you get introduced to one concept and then it spirals. Three years later you hear it again, because you might not have caught it the first time around. So the notion is that we've been able to . . . establish a program that starts in one grade level and spirals back two grade levels later, and then spirals back again, so that you're covering matter in second grade, and then again in fourth grade, and then again in sixth grade and each time being able to build on it.

It is important to note that a spiral curriculum does not just circle over the same material. As students progress through the grade levels, the hope is that they will not only be reintroduced to concepts, but that they will explore these concepts again using more advanced skills and a more complex approach. Table 1 is an excerpt from the earth science strand of the Howe Avenue content matrix which reflects this notion of a spiral curriculum.

While the school has made significant accomplishments in developing an articulated K-6 science program, staff developers and lead teachers at Howe Avenue note that the change process has occurred slowly. A lead teacher says, "It's taken five years for them to really do science and hands-on activities." Another explains:

It took a while, but some of these traditional teachers came from not teaching any science at all to now . . . trying, which I think is a great success. What we saw take place at Howe were different degrees of success . . . There was a teacher who thought, "I will not teach science. These kids cannot have science because they don't know how to read!" [while others teachers were more enthusiastic about the program]. The neat thing about being Howe Avenue was that if it could be done at a school like [ours] where we had so many challenges, it can be done anywhere . . . That's what my district was so proud of.

CSIN at the Cotati/Rohnert Park District

Upon the suggestion of a kindergarten teacher with special interest and experience in science, the Cotati/Rohnert Park District in Northern California decided on affiliation with CSIN. The district used part of a $90,000 Hewlett-Packard three-year grant to join CSIN and develop a K-6 hands-on science program. One provision of the Hewlett-Packard grant was support for a leadership team from the district to spend a week at the National Science Resource Center in Washington, D.C.

Table 1.   Earth Science Unifying Concept: Howe Avenue School
The earth within its universe is constantly changing.

Grade	Grade-Level Concept	Subconcept
K	I can see patterns of change in nature.	• Growth changes living things on earth. • Weather changes on my earth.
1	Natural forces change and shape the earth.	• Seasons change on my earth. • Seasons cause changes on my earth.
2	The surface of the earth changes over time.	• Wind and water cause erosion. • Erosion wears down the land.
3	The earth has a changing atmosphere.	• The atmosphere of the earth is unique. • The weather changes our earth. • Weather is the state of the atmosphere.
4	The planet earth changes over time.	• Rocks and minerals have identifiable characteristics. • Erosion and tectonics are the forces that most affect the earth's surface.
5	Oceans affect or are affected by the changing earth.	• Water has observable properties. • There are forces which affect the ocean. • There is life within the ocean. • The ocean is affected by humans.
6	The earth's surface is changed by universal forces.	• The earth's surface is composed of elements. • Impacts and meteorite showers cause permanent surface features.

Prior to joining CSIN 2, the district had revised its K-12 science program along the lines of the *Framework* and developed a content matrix. Teachers asked themselves questions like, "What are the big ideas, and what is important for all students to be able to know and be able to do?" Greta Viguie, the assistant superintendent of elementary education for the district, describes the process: "We developed some bigger, broader outcomes that involved process as well as content [and] skills, and asked ourselves, 'How can we bring this curriculum to life in our district?'"

Nine lead teachers and one staff developer provided the leadership core for the district of 162 K-6 elementary school teachers. This team attended CSIN professional development activities, including summer institutes and regional meetings, and reviewed the state-adopted instructional materials. Where there were no available materials to help their students learn specific concepts, members of the team developed their own. The district undertook extensive adaptation and development of curriculum units, particularly in the physical and earth sciences. The result was a compilation of both commercial publishers' units, such as Lawrence Hall of Science's Full-Option Science System units, and adaptations of Project Storyline units (a K-6 curriculum development project; see p. 30). One teacher reflects on the new units, "At one time I might have said this is matter. Now the students have to decide for themselves what matter is. Now instead of a few days spent on matter it takes us four weeks."

The district has attributed its success in creating a strong K-6 hands-on science program to a consistent group of committed teacher-leaders who have become invested in the process of developing the district's program. However, there have also been some drawbacks to a teacher-based curriculum development project. The district's staff developer explained that, while the lead teachers had an interest in science, in many cases they did not have adequate content background to revise and develop the physical science instructional units. As a result, the staff developer had to rewrite many of the units during the first year of curriculum development. This proved to be time consuming. Despite these initial difficulties, the development of the earth science strand was a much smoother process because the lead teachers had gained experience and knowledge from the first year of instructional materials development.

One potential barrier to continuing CSIN program implementation in Cotati is that there is only one districtwide staff development day per year in science. The result is that lead teachers are on their own to conduct staff development at the school site. They are often reluctant to teach content to other teachers because they feel they do not have sufficient science background.

In the area of administrative support, there has been a significant difference between the approach at Howe Avenue and the one in Cotati. Viguie explains: "The one thing that I would have done more of, that I didn't do, was involve the principals more. I think there would have been more understanding, commitment, and interest in CSIN and the professional development program had [that happened]."

In addition, teachers characterized the districts' approach to reform as the "jumping-on-every-bandwagon-that-comes-along mentality." As one teacher explained with frustration, "Our district seems to not give one reform sufficient time for systemic change before going on to the next reform."

Conceptions of Science Content and Pedagogy

Staff Development: The Priority of Process. While the network focused initially on the change process, the leadership realized that part of teachers' aversion to change was lack of content background. The *Framework* recommends "a balanced program in the physical, earth, and life sciences." Traditionally, elementary school programs have focused primarily on the life sciences, and teachers are often inadequately prepared to expand their teaching into the physical sciences.

Teaching consultants and staff developers strive to create a supportive climate for teachers that encourages them to take the risks necessary to learn the scientific content. This is facilitated by the network's implementation process and staff development structure, described earlier, which allows teachers to model for other teachers effective approaches to learning scientific content and pedagogy. Teachers who are more experienced in specific areas of science content and pedagogy assist teachers who are less familiar with the material. The goal is for less experienced teachers to gradually assume more leadership roles, even in the early stages. DiRanna explains:

> So we've had this philosophy all along that no matter who you are, you're part of the training group. You're up there in front of people sinking and swimming . . . so from the beginning, instant responsibility, instant accountability . . . I refuse to let go of this philosophy . . . If you want change, people have to be able to see themselves as capable of doing this. If all your trainers are top professionals, then the low person on the totem pole, or the average person, says, "That's them," and "I could never do that."

This approach to teachers seems to mirror the hands-on pedagogical philosophy for children advocated by the *Framework*. In a staff development session entitled, "What Is Energy?" a teaching consultant initiated the lesson by asking the teachers to imagine that they were in third grade and to reconstruct how they might have conceived of energy at that age. Teachers were then encouraged to explore energy sources by examining a variety of wind-up and battery-powered toys. At the end of the lesson, the teaching consultant asked the teachers how they would describe her pedagogical style. They responded: "Listening, facilitating, not giving us any answers." She emphasized the difference between her approach and a straightforward lecture. This point led to a discussion about the importance of developing a nonthreatening learning environment for students. The teaching consultant then drew a parallel between her approach to staff development and one teachers might use with children in the classroom: "If we do nothing else as primary teachers, we must provide time for children to explore and discover. It is through play that children connect prior understandings to new discoveries and construct ideas which they will be examining and reconstructing for, hopefully, the rest of their lives."

An important organizing strategy used by CSIN staff developers when teaching science content to other elementary school teachers is the notion of "story." Stories are something with which elementary teachers are comfortable, and they provide a framework around which teachers can imagine organizing their science teaching. The following excerpt from a CSIN outline of a possible content workshop captures the emphasis on conceptual understanding, assisted by the notion of story. It also illustrates CSIN's sensitivity toward teachers' comfort with scientific content.

Objective: To help teachers understand the "story" of their section, i.e. the story of matter, energy or forces *through hands-on lessons and activities from their instructional materials or other sources which support the physical science story.* This is also an opportunity for teachers to use their journals as a resource for questions to ask the scientist.

Procedure: Select hands-on activities to exemplify the matter, energy, and forces story spanning primary grades, upper grades, and adult learning situations. Lead teachers facilitate the hands-on activities while the scientist weaves the physical science story connecting the big ideas of physical science.

Possible matter story:
>Matter has properties.
>Properties can change; be changed.
>Change in matter requires energy.
>Properties are related to structure.

Concept Construction: What did I hear? What did I think about? What did I learn? What do I want to learn? Is the science correct? Do I still have misconceptions? (Self reflection exercise to be conducted in journals).

As the participating teachers complete the exercise, they are encouraged to share their reflections with the group. At this point, the scientist is often asked to leave the room so that teachers do not feel constrained. As noted on the CSIN Seminar Day agenda, "Explain to scientists as well as to the participants that this may lower the anxiety level of participants, allowing more freedom to state what they feel they learned." When the discussion is completed, the scientist returns to respond to teachers' questions and listen to their concerns. The goal for the day is for teachers to gain enough content information to be able to make their own connections between scientific ideas. Teaching consultant Karen Cerwin observes: "Teachers will say to me, 'If I just had the book, it would provide the links.' But it isn't that the book provides the links, it's the teacher who provides the links. And some of them don't know that yet because they don't have enough information."

Bill Honig discusses this shift from the focus on the instructional materials to the teacher's own role in understanding conceptual connections:

>It is an active approach. Learning is shifted from the materials to the process of wrestling with the subject matter. Here are the concepts, and here is how you learn them. I think those two go together. CSIN is an interesting case . . . It is a shift from the elementary teacher assigning a chapter about light, assigning a quiz on it, throwing in some cute activities, [with] none of it conceptually hooked to a concept-driven curriculum.

Seminar Days also focus on issues such as performance-based assessment. At one such day in Santa Clara County, an elementary school science specialist and CSIN staff developer conducted a session titled, "Science and the Role of Authentic Assessment (K-8)." They began by asking the assembled group of teachers, "What does assessment mean to you?" The teachers' responses revealed that they had varied backgrounds and prior experiences with regard to assessment.

The staff developer then explained how "assessment should bridge the gap between instruction and testing." She described some of the ways that different pedagogical approaches and learning styles can be addressed in assessment. For example, the idea of "multiple intelligences" can be incorporated into performance-based assessments. This assessment strategy addresses each student's unique strengths: if a student is spatial/visual dominant, she could design a physical model of an ecosystem, for example.

One teacher lamented, "It is so difficult to break the mold, because students are so used to previous forms of assessment. It has taken me five years to break the mold." Teachers stressed how students are often resistant to new forms of assessment. "Some students come to school expecting to have information poured into their heads. Kids are so worried about getting the 'correct' answer." A high school science teacher said, "In the beginning of my biology lab class, the students wouldn't give much when I tried to get them to think for themselves. They were so worried about getting the . . . "—a chorus of teachers interrupted, "the wrong answer" One second grade teacher said, "So, what if they go to another class or another subject and it's back to the pouring-in mode?"

During this assessment workshop, a number of teachers also expressed concerns about their weak scientific backgrounds and frustration with the lack of time available to learn from what other teachers were doing in their classrooms. "Teachers are not afforded sharing time to talk about lessons and how we do them."

This emphasis on pedagogy and process has implications for not only the elementary school teachers but the scientists and high school science teachers who participate in CSIN. Professors have been forced to examine their own teaching styles when faced with the challenge of teaching elementary school teachers, who typically have a different conception of effective instructional strategies. One teaching consultant remarked on how the scientists are adopting the language of the classroom teachers and using expressions such as "teaching for understanding." In another discussion, a teaching consultant recounted a conversation she had with a scientist who was initially quite apprehensive about teaching science to elementary school teachers. After he had some experience teaching, he reflected, "This experience is really going to change me."

Constructivism in CSIN. Discussions of effective pedagogy focus on constructivism and its role in teaching and learning. In staff development presentations on constructivism, teachers are led through a four-step process. The first stage is "invitation" or "awareness"; teachers invite student interest by posing a problem or question. In the next stage, "exploration," teachers ask open-ended questions, facilitating and encouraging students to develop "personal meaning." During the "inquiry" stage, teachers help students refine their thinking by asking them more focused questions and helping them make connections.

The final stage is "application" or "utilization," where students are encouraged to apply their knowledge to new situations. The model is one in which students are encouraged to think for themselves and take ownership of their ideas. It views the teacher's role in the classroom as a facilitator. The response of teachers supporting this model to students is described as "respectful, empowering, eliciting, nonjudgmental."

Despite the rather formulaic presentation of constructivism in staff development, interpretations of this pedagogical philosophy vary widely. At a Seminar Day in the Cotati/Rohnert Park School District, a CSIN staff developer began a session on pedagogy by recounting a discussion she had had with another staff developer about the meaning of constructivism. "One thing that totally surprised me was that our ideas about constructivism did not match. She was talking about skills. We were supposed to be talking the same language, but we really were not talking the same language at all."

While interpretations may vary, constructivism is used in CSIN as a general label that suggests an approach to learning in which students are provided with direct experience and encouraged to think for themselves. When CSIN teachers are asked what constructivism means to them, they frequently respond, "Not giving the answers." CSIN is indeed clear and consistent in discouraging teaching-by-telling: lectures are anathema. But, says an SSI evaluator, "experiential or hands-on lessons are often mistaken for constructivist lessons." While experiential lessons are highly encouraged, they are not constructivism until they acknowledge that the uniqueness of each child's meaning helps determine the group meaning.

An "anti-telling" sentiment unmistakably dominates CSIN. The continual invocation of constructivist rhetoric, particularly an emphasis on the importance of direct experience in science for the students and the opportunity for children to talk about their science ideas, provides a touchstone for teachers to critique their own classroom styles. A large number of CSIN teachers attribute changes in their pedagogy to "constructivism." They say that they now allow instruction to be more student-led, that students' questions guide the learning process more than in the past, and that they believe constructivism is the reason for these changes.

In a third grade classroom in Northern California, a CSIN staff developer was exploring the behavior and characteristics of meal worms with her students. When questioned about her pedagogical approach to this activity, she responded that it was much different from previous years because she had incorporated more constructivist strategies. Rather than present all her students with the same experiment, she encouraged each one to develop his or her own individual questions and determine how they would conduct an appropriate investigation. The teacher explained to her students, "There is no one way to go about this process. There are as many ways as you can think of to conduct this investigation."

The processes of teaching and learning seem to have priority, at least in the CSIN program. They are often separated into content sessions and pedagogy sessions. However, a constructivist approach to learning, even CSIN's, does not always characterize the in-service training on content. CSIN's approach is sometimes aimed at content coverage; teachers are not accorded the opportunity to construct their own meaning or examine their prior knowledge. Also, teaching consultants have expressed concern that, "Constructivism is treated like an event, rather than a journey" and that teachers still engage students in hands-on activities without helping them make conceptual connections.

The Inverness Research Associates' evaluation of Eisenhower science education projects in the state confirmed that the quality of the science content is sometimes questionable—in part due to the heavy emphasis on process.

> Eisenhower projects focus heavily on the transformation of teaching and learning. Consequently they emphasize "process" and "constructivism." Few projects attempt to teach teachers science and mathematics content in traditional ways, such as one might find in college courses. Rather, most projects see the learning of science and mathematics as the outcome of inquiry-oriented learning experiences that the teachers themselves engage in. Most of these experiences focus on learning the content in activities that more or less take place at the level of knowledge appropriate to their students . . . The approach taken by most Eisenhower projects is clearly to focus on the pedagogical shifts and then let the science and math content be learned by teachers in that context (Inverness Research Associates 1994, p. 15-16).

While constructivism is not directly mentioned in the 1990 *Framework*, the emphasis on process in staff development is supported by the importance of students learning by doing highlighted in the *Framework*. This philosophy also supports the rationale for teachers' assuming significant leadership roles in teacher education.

Topics as Themes. As described, one of the goals of staff development is to help teachers develop a conceptual understanding of scientific ideas. One approach is outlined in chapter 2 of the *Framework*; it stresses the importance of themes in the teaching of science and cites six as illustrative: energy, evolution, patterns of change, scale and structure, stability, and systems and interactions. "Educators and developers of instructional materials are encouraged to weave these *or alternative* thematic strands into science curricula. The main criterion of a good theme is its ability to integrate facts and concepts into overarching constructs" (*California Department of Education* 1990, p. 27, emphasis added).

Many elementary school teachers, however, try to integrate elements of the curriculum through topics rather than through "big ideas, overarching concepts, and unifying constructs." Notes DiRanna:

Many of them organize the curriculum around subjects like "pumpkins," "whales," and "spring." They study the anatomy of whales and learn science. They study their economic role and learn social studies. They read stories about whales, and talk about them, thus improving language skills. They sing songs about whales and learn music...

We need to be clear that topical approaches, while motivational, most often shortchange science. This is due to many factors: time, comfort level of the teachers, lack of science equipment. Even if science is addressed in the topic, scientific concepts are often not explored in depth . . . If we want elementary students to see and make connections, they must be presented with a solid, conceptually based science program, built on connections (DiRanna 1992, p. 43).

Other educational authorities have questioned the utility of themes at the elementary school level. A 1991 issue of *2061 Today*, the newsletter of Project 2061, encouraged delaying emphasis on themes until the secondary grades (Project 2061 1991). Kovalik and Olsen (1992), in "Kids Eye View of Science," caution that, while science themes are effective outcome goals, they should not be curricular organizers. "The stuff of themes must come from students' experience and be solidly in place before you attempt to build on it."

In a similar vein, DiRanna writes:

There seem to be several difficulties in implementing the thematic goals of the *Framework*. One difficulty is simply that the ideas expressed in the *Framework* require a major shift in how science is taught. It requires the facilitators (teachers) of the content to have a depth of understanding of big ideas that they may not currently possess. Thus, it is important for us to realize that full implementation of the *Framework* may be many years away (Di Ranna 1992, p. 41).

She elaborates on this point in a later interview:

CSIN as a group said our job is to take teachers and move them to the next step . . . That is an important part of what we are all about. The goal may be out there but we need to give them [ways] to get there. [We cannot] jump from no science at all to teaching integrated science . . . from no science content background to connecting ideas . . . from not understanding what a concept is to themes.

Because of confusion between "topics" and the *Framework*'s "themes" among many elementary school teachers, CSIN has tended to drop both terms. Instead, it encourages teachers to conduct staff development and to develop curriculum around "unifying concepts" and "big ideas." Unifying concepts and big ideas, which are often used interchangeably, are the major ideas that unite and connect the scientific disciplines. These concepts and big ideas are based on the *Framework*'s conceptual organization and include: "living things are diverse, life evolves, the earth is constantly changing, energy is never lost or created but does change form." Appendix C contains examples of the big ideas. As DiRanna explains:

Unifying concepts can be thought of as the "umbrella" concepts that we want students to understand when they leave elementary, middle, and high school. Helping to support unifying concepts are grade-level concepts and subconcepts. These concepts must be developmentally appropriate—the curriculum needs to build up, not water down, from the university to the kindergarten level (CSTA 1992, p. 41).

Project Storyline. Project Storyline, a K-6 curriculum development project funded for three years by the California Post-Secondary Education Commission, focused directly on science content. It was an attempt to make the content matrix a working reality for teachers and provide instructional materials to support the *Framework*. The commission's grant funded the development of approximately 20 units in the earth, life, and physical sciences. The project was directed by Eloy Rodriguez, professor of biological sciences at the University of California at Irvine, and Kathy DiRanna. Not surprisingly in view of the leadership, Project Storyline is closely aligned with CSIN, and Storyline units have been field tested through the network. The curriculum development teams included 35 mostly CSIN-trained teachers and a number of scientists who reviewed the units for scientific accuracy. Says DiRanna of the significance of Project Storyline:

> Project Storyline is important because a lot of the thinking of that group impacted more and more what these [content] matrices look like, and, in all honesty, I think the people that understand the matrix most . . . are the Storyline writers. They have had to make it real, and so they have really learned that you are never done with a matrix because there is always a better way to write it, a better concept that kids might be capable of understanding. We still don't know enough about how kids learn to really know what is the best content for them in elementary school.

The writing of the Storyline units was a collaborative process with consultation among teachers at different grade levels and between teachers and scientists. Storyline attempted to identify the big ideas in units titled "Energy Through Time, Forces, Matter, Geology" and "Weather"; and to identify concepts and subconcepts for the various grade levels. The *Framework*'s suggested themes are addressed in every unit.

Storyline is an attempt to link concepts and subconcepts into a cohesive unit. In the Storyline unit "Energy Through Time" at the fourth grade level, the big idea is "Energy may be observed in many forms, transformed, transferred, stored, and used." The three subconcepts of the unit from the 1993 Project Storyline and their suggested *Framework* themes are:

1. Energy exists in different forms from various sources; the ultimate source of most energy used by people is the sun (scale and structure; systems and interactions).

2. Energy can change from one form to another (transform) and move from one place to another (transfer) (systems and interactions).

3. Energy is not lost in any change or move, but in almost every instance part of the energy is transferred as heat.

Each subconcept is accompanied by several hands-on activities, and student assessments are embedded throughout the unit.

The energy unit begins with a preassessment that allows students to explore their preconceptions about energy sources and transformations. One strategy for presenting the unit is through a story that incorporates the use of energy by people through time. Energy sources (food, fire, mechanical, wind, water, steam, electricity, storage devices, and solar) are presented in the unit in the sequence that they were discovered and used by humans from ancient times to the present. One goal of this unit is to allow students to construct devices and test and retest the application of the concepts as they improve the device they are building. "Thus it is important to allow the students time to experiment and rebuild rather than rushing to complete all projects perfectly" (California Science Implementation Network 1993). The following is an excerpt from the energy unit.

> *Fire*: Harnessing heat was an early achievement of man. The discovery of fire in Africa about a million years ago provided an excellent resource. Fire needs fuel (which has stored chemical energy), heat for ignition, and oxygen. By controlling fire, mankind was able to use energy that was stored in living things to produce heat for cooking and warmth. Heat can also be produced by rubbing two objects together. In "Turn 'Til You Burn," students transform their chemical energy (food) to mechanical energy (to turn the drill) into heat energy as a product of friction. That heat in turn is transferred to ignite a fire. Chemical energy in food is one way to store energy; fire is a way to release that energy (California Science Implementation Network 1993).

Units like this one were intended to be models that schools would use in developing their own curricular materials. For the most part, however, the schools use the units with little modification and as a supplement to district-adopted instructional materials.

Project Storyline began when the publishers' materials were still in development and *Framework*-compatible instructional materials did not yet exist. Project Storyline invited publishers to observe the development of a content matrix. As a direct result, some publishers incorporated the idea of a storyline to link scientific ideas in their instructional materials. In turn, a number of Storyline writers participated in the activities of commercial publishers, including the development of instructional materials and assessment techniques.

The professional development networks have been a vehicle for involving more teachers in curriculum development activities and decisions about the adoption and implementation of instructional materials. As one Storyline writer explains:

> Because of the professional development networks, teachers have developed a critical eye. Teachers have developed a set of . . . program standards. Teachers have been on the curriculum adoption teams. They are well-informed teachers

who can discriminate quickly at a glance, and that hasn't happened before. We have a more informed professional group. It empowers us. The publishers did not expect it to affect the final [publishing] decisions to the degree that it did. Before, it would have been a more administrative decision.

One Storyline writer, CSIN staff developer, and writer for *Scholastic* reflects on the importance of her experience as a curriculum developer. She highlights the shift from viewing the instructional materials as the curricular authority to the view of the teacher as an expert:

One of the most significant aspects has been my own professional growth. The views of ourselves as teachers have just blossomed. We have examined some of our misconceptions about science and teaching and the methodology that we use. It's like looking through another side of the prism. Having been involved in curriculum development, assessment, publishing, and instruction, it's so nice to put all the pieces together and see how you can make a difference teaching teachers and empowering them to feel that excitement that you feel inside. It's going to take a lot of time for teachers to change from "the book drives the curriculum" to "the teacher drives the curriculum." Ideally, we see the big ideas and concepts, and then the materials are just used to complement those ideas.

While the quality of the content of some Soryline units has been criticized—and, in some instances, compared unfavorably to the commercial unit—the above comments from teachers point to the value of curriculum development as an opportunity for teacher collaboration and in-depth questioning about scientific content. The storyline mode of curriculum development has given teachers an occasion to critique, debate, and adapt the 1990 *Framework*'s recommendations. And as demonstrated, teacher involvement in Project Storyline, even on a small scale, has affected the process of commercial instructional materials' development and selection.

Effecting Educational Change

Focus on Change Versus Focus on Content: A Necessary Tradeoff? The preceding sections illustrate CSIN's evolving conception of professional development. As described, the network has been primarily concerned with addressing the experience of the change process for teachers. Rather than focusing on the individual teacher, as was the approach with so many previous staff development models, CSIN viewed the teacher in the context of the entire school. The result has been to enlist the commitment of the entire school faculty, including administrators and, in some instances, even complete districts. This has been accomplished by developing an implementation structure that helps schools designate the time and develop both the short-term and long-term goals necessary for establishing a science program. The structure has facilitated teachers' central roles in the implementation process and provided the on-site support for teachers and schools attempting curricular changes.

The structure was essential, but it was not enough. How could CSIN actively engage teachers directly with the scientific content? CSIN believed that this was integral to establishing lasting changes in elementary school science. One strategy was to create a supportive environment that encouraged teachers to assume significant leadership roles in staff development. Another was to involve teachers in curriculum and assessment development projects. Teacher involvement in both Project Storyline and the California Learning Assessment System's (CLAS) development of performance-based assessments demonstrated the value of teacher participation in these processes. In these situations teachers, had to engage collectively in critical discussion about the subject matter. The process was viewed as being equally valuable and perhaps integral to both the utility and the quality of the product.

Some of the implications of these strategies are already apparent. As noted earlier, teachers involved in the curriculum development process have seen the professional benefits, including a deeper subject matter understanding and the ability to make more informed decisions about the adoption and implementation of instructional materials. Many of the teachers involved in the development and subsequent scoring of the CLAS performance-based assessments are beginning to examine their students' work in new ways.

However, these strategies have also been criticized because of concern that elementary school teachers' lack of adequate content preparation significantly reduces the quality of both staff development and the curricular materials produced. Network leaders would concur. Yet CSIN also saw involving teachers directly in these processes as an essential component of educational change.

While CSIN believes these strategies will ultimately enhance the quality of both professional development and classroom science teaching and learning, how are these decisions evaluated? How does one weigh the short-term versus the long-term benefits of these strategies? CSIN's primary criterion at this stage is the teachers' educational and professional experience in what they view as a long-term process of change in elementary school science. They believed this was an important place to begin.

One teaching consultant reflected on the critical role of time in assessing educational change:

> Some people want to immediately measure student achievement, but you've got to work with the teachers for a long time before you can see that. Teachers won't change after even one year. Some will, but how many? How many teachers that are on the staff that are not lead teachers try these new ideas, even though there is a lead teacher there to help them and talk to them? If they're resistant to change, what's going to change them?

Leadership: Neither Top-Down nor Bottom-Up. The network structure reflects a continuing tension between hierarchical, centralized leadership and teacher-centered initiatives. CSIN is based on the premise of shared leadership—neither top down nor bottom up, but a combination of the two features with initiatives from both outside and inside the classroom. Thus, while professional development activities are based on teachers' needs and interests, it is also understood that educational change can only be achieved by building leadership capacity at the school site, at the district level, and in Sacramento.

Kathy DiRanna and the 12 teaching consultants who each are responsible for a specific district or region in the state form a crucial leadership core for the network. The teaching consultants provide not only critical ongoing support to classroom teachers but also the links to many of the other agencies and people that influence reform. They have been involved in instructional materials evaluation, development of performance-based assessments and assessment manuals for teachers, preservice reform in the state's universities, articulation and collaboration between CSIN and SS&C, and review of national curricular documents such as the draft NRC standards.

Teaching consultants also play an important role in facilitating collaboration among classroom teachers. Here is a description of how one teaching consultant views this role:

> And these discussions [with teachers] go on for a very long time, but they give us enough time to really refine our very best thinking on how to help teachers interact to make their classrooms better . . . I don't think we have answers, I don't think anyone has answers. I think we have to develop processes that can respond quickly and respond to the needs of what teachers are saying. So to me that's what's the most professional of what we do. It's not maintaining. It's not management. It's bringing out the best in the best people in the profession.

Teaching consultants often use the term "facilitation" to describe their work. It is a term, also, that classroom teachers use to describe their role with students. The same term is also used by key figures at the state level when they discuss their responsibilities. Just as facilitation is used to describe staff development roles from the state level to the classroom, the network emphasizes the importance of modeling at all levels beginning with Kathy DiRanna. One teaching consultant explains:

> Kathy would always encourage us to present sections of the training that we had never presented before. She was there to help us, encourage us to grow and stretch, and she continues to do that to this day. That's one thing that I do now as a teaching consultant. I encourage the lead teachers to grow and stretch. We model collaboration.

This ethos of collaboration and collegiality, with no pronounced sense of hierarchy, seems to contribute to strong professional associations among those involved in the California reform. And it also brings personal as well as professional rewards. How otherwise to explain the kind of commitment people make to the reform effort?, Sachse asks:

> Kathy DiRanna does the networking as I imagine it should be done . . . It is a person-to-person thing. It's professional to professional. And you build this family structure. She is the matriarch of the family. Everybody wants to be

closer not to a hierarchy, but closer to personal and professional ties that come from the greater association of being a staff developer or teaching consultant. I think that is a huge magnet that keeps people going into it and spending more of their time and more of their professional effort. There is a real sense of reward about it.

CSIN leadership believes that change at the classroom level depends on the creation of a professional community of committed teachers. The network serves as a forum for the learning and exchange of content-specific and peda-gogical ideas for teachers and by teachers. One staff developer from the Sacramento area describes the network as "an umbrella of support that allows us to take risks."

This established professional community validates teachers' experience and knowledge and empowers them to find solutions to their questions within the community. Says one teaching consultant: "There are opportunities within CSIN to have extremely professional discussions with other teachers to make the pro-fession better. And to believe that teachers can do it themselves and that it does not require an outside person. It's maybe a link to other teachers, but it is not an outside person."

This teacher community enhances a sense of professionalism, that teachers have contributions to make as well as responsibilities to assume. One teaching consultant notes:

> I look at intellectual growth as really key . . . and also contributing to the pro-fession with your own experiences. You have to give back somehow, and maybe that's why teachers keep coming back, maybe because they give back to the profession and to CSIN. That could be part of the professionalism for them. They might never get to give back to the profession in their own school district. They might not be viewed as a professional in their own school district.

An enhanced sense of professionalism is seen by some in the CSIN leader-ship as an antidote to burnout. There is little question that change is arduous, that teachers committed to new and nontraditional approaches to the teaching of science make a heavy investment of their own time. They also become deeply committed emotionally. Opposition from parents, resistance from administrators, and other challenges can hurt. Burnout is a constant concern. Cerwin, a teaching consultant, puts it this way:

> The energy that it takes the staff developers and the lead teachers is beyond belief. I think a couple of things help them. One is the professionalism. They like being connected with something that makes them feel like they are doing something important. And they are. And that's the real reward for them. And that's why their districts need to keep sending them to the summer institutes, and for the camaraderie within the group. I think you have time and energy and materials to work with, and the energy is much more difficult to replace than the materials. This is the kind of atmosphere we try to create in CSIN, which I

think is rather unique . . . It's not accidental that people have a need to come together and enjoy each other's company and work hard and play hard. So that's something that we build in.

DiRanna reflects on the contribution of substantial teacher commitment to the self-sustaining quality of the network in somewhat different terms:

[CSIN] was a valuable thing to be a part of. People who came in walked away feeling honored. They learned great things. They learned from one another. [CSIN] took on a life of its own . . . There is a lot of honoring involved and a lot of trust, partially because this thing grew faster than anybody could have ever kept a lid on, so you had to give away control early on.

Sachse puts it this way: "The structural part of the networks is easy to get across to people. But the spirit of personal relations, the trust and the way it is built between people—that is the challenging part to capture."

Obstacles to Change

While teachers and the CSIN leadership attest to the substantial professional rewards that result from commitment to the network, their efforts to effect educational change have faced a number of obstacles. Some of these are due to the network's approaches to staff development; others are structural issues related to school, district, and statewide educational policies. Many obstacles are the intersection of these two structures and sometimes their conflicting priorities and goals.

Elementary schools struggle with competing priorities, trying to juggle the ever-changing curricular demands. In California, frameworks in each subject area are adopted approximately every seven years. For elementary school teachers, this means they are contending with a new framework in a different subject area every one to two years. The implications of this are that schools can only devote limited time to examining one curricular area in depth, then they must move on. This problem is especially acute in science, which is not even considered core in many elementary school curricula, and is still viewed as an add-on or elective. After a year or two, schools often feel they are done with science.

Administrators, time, and cost are cited as some of the biggest barriers to science program implementation. The role of administrative support in the process has proven to be crucial. When teachers do not receive adequate support from their administrators, their rate of burnout is much greater.

Participation in CSIN requires a substantial time and financial commitment. Funds are required to maintain materials for quality hands-on science programs. Additional staff development days in science are required. Release time for staff developers and lead teachers must be paid for beyond their annual compensation, all financed by the school or district. CSIN requires the commitment of the entire faculty at a participating school. Some staff developers feel this is an unrealistic expectation. In addition, in many cases, professional development in science is still not seen as a district priority.

Many of the teachers that become staff developers and lead teachers have also assumed leadership roles in other subject areas. Thus, many have significant responsibilities beyond CSIN. Many of them find it a struggle to balance the challenges of their role as teacher—leaders and the demands of their own teaching schedules. In a district in Northern California, most of the CSIN lead teachers were also mentor teachers for the district. After one year in CSIN 2, all five lead teachers found the process exhausting, and the entire district pulled out. The district also cited cost and the need to move funds in staff development from science to mathematics as reasons for leaving CSIN.

Finally, elementary school teachers' exposure to scientific content in undergraduate and preservice preparation remains weak. While CSIN attempts to provide teachers with both the breadth and the depth of the scientific knowledge required to teach the physical, earth, and life sciences effectively, profound understanding of these disciplines only comes with time and experience.

Systemic Change: Expanding CSIN's Impact

Despite the numerous barriers to implementing long-term professional development and curricular changes in elementary school science, CSIN has grown both within itself and systemically in the years following the publication of the *Framework*. For example, CSIN teachers assisted in the CLAS development of new performance-based science assessments (CLAS funding was withdrawn in the fall of 1994); participated in the instructional materials evaluation panel; and assisted in drafting the Program Quality Review (PQR) document. CSIN teaching consultants and classroom teachers now aid in the PQR process in science at elementary schools. This expansion of the network's role provides teachers with opportunities to influence many areas of statewide science education reform. DiRanna notes:

> CSIN was used as the CLAS training arm and went out and did the implementation of the performance-based assessment. The actual workforce was CSIN people. We labeled leaves, we stuffed baggies, we field tested; and that very first year, it was all CSIN people that trained teachers to use this. Again, CSIN was seen as something that was right in with assessment, and if you were doing the CSIN program you got first knowledge of what was happening. And that was another real shot in the arm to making this whole thing move along.

Elizabeth Stage reflects on the broader impact of teacher networks' involvement with statewide efforts to develop performance-based assessment:

> The year that the three networks [CSIN, SS&C, CSP] cosponsored the performance [assessment] piloting for the state, there were 3,500 teachers who participated. Half identified themselves as being part of a network. Half of them did not. There's the ticket. I'm going, want to come. So this thing about being part of a community, and then you realize the power of that.

The award of the five-year $10 million SSI grant from NSF in 1991 proved to be a critical point for CSIN. These funds were used to develop an umbrella organization, the California Alliance for Math and Science, which includes CSIN and Math Renaissance. In addition, the funds provided CSIN with the opportunity to hire 12 teaching consultants. Prior to the award of the grant, all CSIN staff development had been conducted by full-time classroom teachers. The teaching consultants were now able to provide essential support and guidance to the staff developers, participate in statewide reform activities, and help the network expand and grow. DiRanna explains the significance of the SSI funds:

> Statewide, there was no way to be funded, because there was no incredible amount of money coming into this. It was back to schools having to put in their fair share—on top of which elementary science itself is a difficult struggle. It's not considered core. The teachers don't have the background, and they don't have the materials. You're fighting a tremendous uphill battle to say you need staff development dollars in science. One measure of the success of the Systemic Initiative would simply be to go back to districts that for the first time in their history had a staff development day in science.

Of the nearly 6,000 elementary and middle schools in the state of California, CSIN has reached approximately 2,500 of them since 1987, distributed from one end of the state to the other. During the 1994-95 school year, approximately 670 K-8 schools and 1,000 staff developers and lead teachers participated in CSIN. While CSIN schools are exclusively elementary and middle level, many high school teachers and scientists are closely involved with network activities by teaching content at staff development institutes and providing advice to Project Storyline. In addition, CSIN has recently added a mathematics liaison position to the staff development program to assist teachers with the mathematics textbook adoption under way in the state. The goal of adding this position was to build on the strength of schoolwide change already in motion in science.

However, as educational priorities and goals change, will CSIN be able to sustain the significant teacher and administrative commitment to improve elementary school science developed in recent years? While this question can only be answered over time, the network has demonstrated that ongoing professional development and support are an essential component of educational change. With this understanding of the change process and growing involvement in systemic reform efforts, elementary school teachers are now making significant contributions to the science education community. DiRanna explains the shift in recent years: "The science education community pre-1989 was pretty much a network of high school people . . . The involvement and focus of elementary school brought a new dimension."

Secondary School Science

While the elementary reform effort has focused primarily on the process of change, reform in high schools centers on rethinking the science content. In particular, the major move in California has been toward integrating or coordinating the science disciplines. Some schools are also attempting to consolidate science with other subjects, as in one Northern California school's offering of an integrated physics and geometry course for the ninth grade, and a "science, technology, and society" block for the 11th and 12th grades that combines science, mathematics and social science.

Other schools have increased their commitment to project-oriented work as a way to relate science with other subjects and enhance student interest. In one Northern California high school, a ninth grade teacher collaborated with the local water board and city council to develop a project to investigate the restoration of a salt marsh in San Francisco Bay. The project involved the compilation of data on a freshwater pond built by the city in an attempt to restore the area after 70 years of industrial and residential development. Students were asked to choose a site in the marsh area and complete a year-long "documentary" of their chosen territory. Among other activities, students met with the environmental compliance engineer for the city and toured the area, learned how to monitor the returning bird and salt marsh harvest mouse populations, and became skilled at taking water samples to perform chemical analyses. This partnership allowed students the opportunity to design an original study and collect and analyze their data for the purpose of providing their community with needed scientific information. The data were collected and reported directly to the local water quality control board—data that the city would not have had the resources to collect were it not for the students' efforts.

While these kinds of projects have been increasing in number all over the state, most schools have been involved primarily in developing integrated science courses for the 9th and 10th grades. Some schools have integrated, or are in the process of integrating, their entire science program. Why the move to integration? Several explanations have been offered by people in the field: integrated science is more like "real" science, integrated science will reach more students, and integrated science is simply the "wave of the future." While there does not seem to be consensus as to why integrated science is better than the traditional programs, or the best way to do it, teachers and others agree that anything is better than the current "layer cake" curriculum: first general or earth science, then biology, then perhaps chemistry, and finally physics.

How is integrated science different from the scope and sequence of traditional science? In one Northern California school, all three years of science are integrated into courses titled Science and Technology 1/2, Science and Technology 3/4, and Science and Technology 5/6. These courses are an amalgamation of concepts from earth science, physics, chemistry, and biology. For example, the large units of material for the first year fall into the following categories: planet earth, motion, forces and energy, air and water, energy and atoms, chemical reactions, chemical reactions in living organisms, and food and digestion. Within each of these units, the students experience a phenomenon drawing on examples from all of the sciences. The section on forces and energy, for instance, includes lessons on heat transfer and flow, kinetic and potential energy, friction, and equilibrium. Students might analyze the evolution of warm-blooded animals to study heat flow, use a pendulum to investigate potential and kinetic energy, and compare the relative grip of the rubber on sneakers soles to learn about friction. In a series of lessons on equilibrium, students study models and case studies of how living organisms manage to maintain a relatively steady internal state.

The move toward integration or coordination in California has no single character; the sciences are consolidated in almost as many different ways as there are science departments willing to do it. While most programs are based at a general level on the California *Science Framework,* Project 2061's *Science for All Americans,* and other national and local reform guides, each school's program has its distinctive character. Teachers and other leaders in the reform view a reconceptualization of the content as an answer to many of the problems in science education today. They also admit to the myriad challenges they have faced in their attempts to develop and implement integrated programs.

Autonomy and pluralism are championed as the best ways to enlist the commitment necessary to make substantive and lasting change. However, greater independence at the local level can also lead to such potential problems as overburdening the teachers, lack of control of the quality of the science taught, and a general confusion about what is important. An added difficulty when subject boundaries are blurred—especially for high school teachers—is the threat to one's hard-won sense of expertise in the special subject in which one has been trained.

How are these challenges confronted? One way that California has attempted to alleviate some of the difficulty associated with change at the high school level has been to structure a network for teachers to communicate with each other and receive the kinds of professional development assistance they need to develop, implement, and sustain a new science program. The remainder of this section on reform at the secondary level examines the most organized and visible effort to restructure the high school curriculum: California's Scope, Sequence, and Coordination Project.

Origins of Scope, Sequence, and Coordination

The California SS&C project, one of six national projects attempting to reform science education along the lines of the National Science Teachers Association's recommendations, has a story of improvisational and opportunistic beginnings. In 1989, the California Department of Education received a grant from the U.S. Department of Education to support schools in their efforts to implement new science programs. Taking a lead from what was understood as general practice in Europe and assumed to be a better way of organizing curriculum, the basic focus of SS&C was to emphasize relationships among the science subjects by teaching each of them every year in some sort of coordinated or integrated fashion rather than as four separate subjects: earth science, biology, chemistry, and physics.

The underlying concept of the original SS&C grant to California was to fund the creation of model schools where every student would be taught every science, every year. The California Department of Education sent out a call for proposals where 100 schools would be awarded from $8,000 to $10,000. From those 100 schools, 10 were to be picked as model programs to be showcased and to receive further funding and support.

The call for proposals stimulated over 300 responses. According to those involved (both in submitting and evaluating the proposals), the quality of a great many of these proposals was extremely high, and the eventual programs that resulted were beyond expectation. Tom Sachse, at the time in charge of science education in the state, suggested to Zack Taylor, then-president of the California Science Teachers Association, that the development of model schools was not a very effective implementation strategy and that perhaps another approach would better serve the schools involved. Model schools fail to stimulate reform, Sachse believed, because they are created by pouring vast amounts of resources into individual schools. Other schools are invited to emulate the models without commensurate resources or support.

Sachse's alternative was to further support the 100 schools that received initial funds. In 1990, the NSF awarded $1.5 million to the California Department of Education to provide support for the schools as they made the transition from planning to implementing. This grant was comprised of five components with funding for: (1) groups of high schools and participating middle schools engaged in the reform, (2) the brokering of university faculty to provide in-service training for the participating schools, (3) a preservice component (see the next section of this paper, "The Reform and Higher Education"), (4) an action research project for selected participants, and (5) documentation and evaluation.

The first component, a network eventually called The 100 Schools Project, was to be organized into 10 geographically distributed regions, or hubs. Each hub would have a teacher/coordinator who would organize monthly meetings of site representatives from nearby schools. There would also be annual regional and statewide meetings. Their mission would be to collaborate on the development and implementation of a coordinated or integrated science curriculum for grades 6-12.

The advantage of this approach, it was thought, was that it would provide the initial infrastructure and models for an implementation strategy that was affordable on a larger, statewide scale. Another anticipated advantage was the prospect of increased teacher commitment as a consequence of the greater autonomy and decreased isolation this implementation strategy promised. With support from Taylor, the California Science Teachers Association president, the SS&C hub structure for California was born.

Since its creation, the hub network has expanded to include nearly 300 (out of a total of about 800) schools. What began as a vision to help 100 California high schools overturn their layer cake approach to teaching science has become a much larger, more complex project than Sachse anticipated. In fact, he now feels that there is nothing to stand in the way of SS&C taking over as the dominant mode of curriculum organization: "I think it will become the dominant vehicle for teaching high school science. It will take over everything else; it's just going to take a few years. It could take 5 to 10 years."

A New View of the Subject Matter: Integrated and Coordinated Science

Science, as seen in the real world, doesn't happen in little compartments, does it? You don't see a physics thing happen here and a biology thing happen there. It is integrated in real life, so I think by integrating science we're teaching more like it really is, like it really happens. — J. Clift, Natural Science I & II teacher

Every student, every science, every year — California SS&C Project

Teaching coordinated or integrated science has become the defining feature of the SS&C project in California. While the science concepts and processes taught are drawn from the 1990 *Framework*—the basis for the "Scope" of SS&C—the sequence and coordination of the science curriculum are based on the vision of Bill Aldridge, the executive director of the National Science Teachers Association and leader of the national SS&C project. The SS&C calls for "essential changes" in science education, which he interprets as a true spiraling and articulation of science content and processes. The content should then be coordinated, such that the four traditional disciplines—biology, chemistry, physics, and earth/space science—are connected within and among units of instruction. The result for California is the development and implementation of science programs that coordinate, where each of the four disciplines are treated individually and equitably each year of high school. The coordination is facilitated by the use of overarching ideas which are explicated in terms of the four disciplines. Another option, one that is becoming more prevalent, is the integrated course. Integrated science deliberately blurs the boundaries between the science disciplines: "Integrated curricula speak to how earth, life, and physical sciences are so intertwined that these fields (and the disciplines which comprise them) cannot be separated" (California SS&C 1993).

Connections to the *Framework*. Aldridge's vision of the spiraling and articulation of science concepts and processes stands out as a major influence of secondary science reform in California. However, the role of the 1990 *Framework* should not be underestimated. In the early stages of the effort and through to the present, teachers have relied heavily on the *Framework* for the design of their science programs. Many schools have developed a content matrix, a grid that distributes science concepts by discipline and then arranges them by grade level, based on *Framework* recommendations. Moreover, the *Framework*'s challenge to provide students with hands-on experiences for 40 percent of instructional time and spend less time on "factoids" has been taken quite literally by most teachers involved in the reform. Indeed, many teachers believe that these may be the *Framework*'s most important messages. One SS&C teacher responds this way when asked about the role of the *Framework* in the program at her school:

> Well, that's kind of the basis for our whole reform, just looking at less facts, going into the processes of science more. If you look at any of the textbooks from recent years, if you look at their tests, it's like "What is a chloroplast?" "What is mitosis?" It's completely factoid-based. We've pushed trying to get our students thinking in science. It's a lot harder to do than it is to memorize what a chloroplast is or what mitosis is, and the *Framework* is the document that has given me the freedom to push this type of approach, because this is what the state is mandating; it's not what I'm asking you to do.

While the connection to the 1990 *Framework* seems clear in the case of the emphasis on hands-on science and science processes, the origin of the idea to integrate the sciences is slightly more mysterious. One could point to the *Framework* and its focus on cross-disciplinary themes as a likely candidate. However, people in the field suggest additional possibilities. Tom Sachse saw integrating the sciences as "baby steps" toward the integration of all of the disciplines, a vision of education he believes is a long way down the road. Some teachers believe that integrating science is a reaction to the less-than-adequate science teaching that has been going on for a long time. Rick Rule, a science teacher and hub coordinator suggests this, relative to teaching the body systems:

> [M]ost of us acknowledge that none of the subjects, and especially not within science, exist in some kind of vacuum . . . [as if we were] going to teach biology today and we're not going to teach about pressure as it applies to anything else, even though it has everything to do with the respiratory or circulatory systems . . . but we won't really get into the physics of that at all. It was sort of ludicrous, and maybe this just gave confirmation to the ludicrousness of that and allowed us a way to get out of that kind of thing.

When asked pointedly where the idea to integrate the sciences came from, one SS&C teacher responds this way:

> I think integrated science is a response to something that hasn't worked for the past 30 years. We've been teaching these traditional, secular sciences. Here's sectioned-off sciences and it hasn't worked. We get no return of students in the upper level science class, test scores are dropping, and things like that. So I think the integration is a response [to that].

Developing an integrated curriculum is no small feat. Though they had an idea of where they wanted to go, many teachers found themselves without a roadmap. The case of Washington High, a semi-urban, diverse Northern California high school, offers an example of how teachers within a department came together to develop a new program. The whole process began when Ron Ulrich, the department chair with 25 years at Washington High, received a call for proposals in the mail. In 1991, Washington became one of the original 100 schools of the California SS&C project to receive a grant to develop a program that complemented the 1990 *Framework* and the other reform policy documents available at the time. The process was long and arduous, but the group soon discovered that the most efficient way to finish the task was to "divide and conquer":

> We finally decided, and went about writing that first unit of six six-week blocks. We soon found that all of us writing one unit would not work. We asked for contributions after about the fourth week from everybody, put them together, and then two people would work jointly on producing a particular six-week block. It would include almost daily activities that would support the concept that we're trying to present. That seemed to work pretty well. We had people generally strong in the physical sciences and strong in the biological sciences on a team, so we would have a cross. Our framework was really designed around earth, physical, life, and the chemical or physics portion; and we try to encompass all those things within a topic. So if the theme would be energy, we try to encompass all things in energy as we go, so we're comparative across, whether it be plate tectonics or photosynthesis or motion, things like that. So we integrated all the subjects into one area.

When asked what went into their decisions to include the kinds of things they have in their curriculum, Ulrich responds, "That [process] went on forever, at least two Sunday afternoons. We finally just took the state *Framework*, because we're pulling out the major ideas and trying to synthesize them to our program currently."

What Drives Them? Many teachers have spent extra time on weekends, after school, and during summers to work on their programs. Schools involved in the 100 Schools Project received a grant for this work. Today, schools that are interested in joining the project must do so on their own time and money. Small grants do become available, but they are rarely commensurate to the task. What drives the teachers?

To rally his department, Ulrich stressed the possibility that integrated science might mitigate the pitfalls of tracking and at the same time provide coherence. Ulrich explains:

> The incentive was we felt tired of the old program. It was tracked, so you would have a class with failures or D students, [and] you weren't going any place with them.

> [We] wanted to change the program [because] it had been there forever, nobody had any feel for it, any ownership of it. So, we felt that if we all got into it and worked . . . we would all have ownership. And we'd all know what the other person had taught. As [the students] go through as a freshman and as a sopho-

more, we know what they've had, and that was a benefit, and continues to be a benefit over time to be able to understand and know what [the students] have had previous[ly].

One issue that seems to play a large role as a stimulus to change is the desire to reach more students. One teacher says that, as a traditional science teacher, he is at best meeting the needs of his most motivated students; at worst, turning off those that don't typically gravitate toward science. He feels that being a part of SS&C has provided the necessary support for him to reflect on what he was teaching, how he was teaching it, and why.

> I taught traditional biology for nine years and always thought I was doing a really good job. I put a lot of energy into what I did and I thought I was exciting my students, but no matter what I was doing I still wasn't reaching all of them. The biggest thing that this project has done is caused me to reevaluate how I approach teaching, especially teaching the masses. How do I reach the majority of my students? Well, I'll probably never reach all of them, but how can I reach a majority of them? Through that reflection and reevaluation of what I'm doing, I've undergone a lot of different strategy changes in the way I present things to classes, and I think I'm a much better teacher because of that. I think I'm doing a much better job with my students.

Changing Pedagogy. In addition to the change in organization of the subject matter, teachers' pedagogy has changed. This has been due primarily to the *Framework*'s urging that teachers design more hands-on learning experiences for their students. One teacher, recalling his previous job teaching "traditional" science, remembers feeling "jealous" of the time that labs took away from transmission of the content. He said he used to "tuck in" labs because his courses were billed as lab-based, but they were never a very important part of his curriculum.

Helen Kota, project manager of the California SS&C project, points to the benefits of integrated science to teachers and their new approach to instruction:

> Teachers need to be ongoing learners. I think that oftentimes, you become a chemistry teacher and you get stuck . . . You're in a situation where you keep on teaching the same thing over and over and over again. You're not a lifelong learner; you don't expand yourself. I think that [integrated science] is causing teachers to look at all the different arenas, get updated, [and] change their pedagogy and change the way they relate to students. If you're up in front of a class just lecturing, that's a whole different ballgame than if you're facilitating lots of hands-on activities. [You are] facilitating the students' input, rather than just giving your output.

Another teacher, when asked if teaching integrated science has changed his pedagogy to any degree, responds: "Oh, absolutely . . . It's completely transformed me . . . With this lab approach—with this hands-on approach—it's a variety of stimulus every day. . . so it's totally transforming, as far as my teaching."

When asked which was more difficult for teachers, changing the curriculum or changing their pedagogy, Kota answers:

> Teaching differently. We're so used to [the way] we were taught, by a lecture mode. We were encouraged to keep the kids quiet and teach that way . . . To change the kinds of ways you address and bring kids into a much more personal aspect of your teaching, that's a real change, and I think that that's been the hardest. You can start looking at your curriculum and say, "Oh, yeah, I can do a little bit of earth science here, a little bit of chemistry here;" but when you start talking about how you're going to present that, if you're going to use a constructivist way of dealing with that, then that becomes a real hard issue for teachers.

This "new view" of science has been separated from a different approach to pedagogy in describing the California reform, but the two are not so disconnected in the minds of the teachers or the project manager. As the above quotes exemplify, when asked why integrated science was more sensible, teachers reply in teaching terms: the new approach to the science curriculum went hand in hand with changes in the way they presented science to their students. This, they believe, allows them to attain their goal of reaching more students. For these teachers, the purposes and practices that accompany the integrated perspective are related in intimate and important ways.

Integrated Science: Up Close and Personal

You don't cover stuff, you actually teach things. — E. Young, science teacher

What happened to the themes as outlined in the 1990 *Framework*? Many of the integrated and coordinated programs do use themes to organize their content, but they do not seem to be the kind advanced in the document. In fact, the *Framework* does not recommend collapsing the traditional disciplines, and teaching the themes explicitly was not the intent of the writers.

In a series of articles—led by one from Doug Martin at California State University (CSU) Sonoma—various leaders within the reform published their views in the *California Science Teachers Association Journal* in an attempt to clarify the role of themes in science instruction (CSTA 1992). The main message was that themes are simply a way to approach teaching that emphasizes big ideas in science and the connections among them. They are not, according to Martin, the content of science or the intended focus of instruction, nor are they appropriate for the early grades. Still, there are many SS&C schools that gravitate toward the idea of overarching topics and are using them to organize their curriculum. This section highlights the salient features of integrated science courses and the different ways in which schools in California are using themes in science instruction.

Integrated Science. Since each school has developed a program uniquely suited to its student body and teachers' expertise, no generalizations can be made confidently about what integrated science looks like in practice.

However, one common element that seems to have emerged in several school programs is the emphasis on selected organizing concepts that hold together the various topics within a course.

For example, in the first year of a two-year sequence of integrated science in a Southern California high school, students begin with some basics—science skills (including using metric measurements, graphing and data tables, and the scientific method) and inorganic chemistry (including such topics as atomic theory, the periodic table, and compounds and bonding). Building from this point, the remainder of the year is organized around the more inclusive topics "water," "air," "earth," and "our physical environment." Again, while not the kind described in the *Framework*, teachers refer to these topics as themes.

In the section on water, students might study the chemistry of water (including atomic structure, solutions and mixtures, ionic bonding, and water as a solvent); the physical properties of water (including ideas such as density, the effects of temperature and solutes on density, buoyancy, changes in state, and surface tension); water quality (including concepts like dissolved solutes, turbidity, hardness, pH, microorganisms, and pollution); the effects of water on the earth (such as groundwater and erosion); and life in the water (primarily adaptations to aquatic life). In the context of water, then, the students are exposed to a variety of scientific ideas from the earth, physical, and life sciences. Within the section, students perform laboratory exercises that demonstrate such phenomena as the hydrolysis of water, how to measure water's specific gravity, how water can be boiled in a paper cup, the capillary action of water, microorganisms that live in water, and the effects and causes of water pollution.

On the whole, the lessons that comprise this school's integrated program include something of the four science major fields. However, lesson by lesson, they often remain separated. For example, a lesson on currents and circuits was part of a section of material on electricity and magnetism that involved the students building circuits with wires, bulbs, and batteries. One of the goals of this particular lesson was to understand the idea of resistance. During the first several minutes of the lesson, the students were told what they were about to do, what resistance was, how to recognize its effects, and what their hypothesis was. The students then moved to the lab tables and began assembling the elements of the circuit to produce the effects their teacher told them about. There was no attempt in this particular case to connect the concepts of current and resistance to other areas of science.

Another school in Southern California offers a coordinated science sequence, called Biophysical Science. In this program, life and physical science units are taught on alternate weeks by a teacher with the appropriate background. In this scheme, teachers plan together and team teach the material, drawing on each others' expertise. The students use three separate texts throughout the year.

Coordinated science can take various forms. One SS&C school in Southern California follows an agenda similar to the one just described, where physical, life, and earth sciences are explicitly treated each semester. The outline describing the course states the following with regard to the scope and sequence of the program:

> Coordinated Science is a four-semester course designed to meet the graduation requirement of the state of California. Students are instructed with a thematic approach which integrates life, earth, and the physical sciences in a program called "Science in a S. A. C. K. (Spiral Approach to Content Knowledge)." Two of the key features in this program are (1) exposure of students to multiple disciplines of science in a given unit of study and (2) the fact that no one topic is studied to a maximum depth in any one semester. The topics are spiraled throughout the four semesters . . . introduced one semester, then expanded upon sequentially in succeeding semesters of study.

In the first semester of this program, students study the overarching concept of water quality. Through the use of discussions, labs, other hands-on activities and lecture, students study the physics associated with water-like molecular structure, static electricity, and super freezing. They also study how water affects the earth through the formation of rivers, aquifers, and aquitards. The chemical concepts addressed in this unit are surface tension, chemical nomenclature, water purification, symbols and formulas, and the periodic table. Biogeochemical cycles, biomes, viruses, and respiration in invertebrates and fish are examples of the biological topics.

Any of these lessons in integrated or coordinated science courses could stand alone as they do in a traditional physical science, basic chemistry, or sophomore biology class. It is not apparent that the individual lessons are connected to ideas from other areas of science. So, one might ask, what is different about integrated/coordinated science? Is integrated/coordinated science simply the same cake without the layers? Or a more finely layered cake? To what extent must the science be integrated? Is it even possible (or necessary) to have a truly integrated curriculum, where every lesson includes ideas from all areas of science?

These questions are answered differently by different science teachers involved in this reform. Sachse believes true integration is very difficult, something of a "holy grail." In the end, he asks, does it matter what it's called? What these schools are doing now is different from what they did before. And what teachers are doing now gives them greater satisfaction and makes them feel that they are better teachers, he says. SS&C is characterized by the variety of interpretations of what scope, sequence, and coordination look like. The numerous opportunities within the reform make room for different contexts of meaning for the goal of "Every student, every science, every year." At the same time, the reform provides enough of a boundary, however loose, to permit the schools to develop a program that makes sense to them and still enables them to talk about it with others in the reform and be understood.

Still, the labels "integrated" and "coordinated" convey to many people a degree of connectedness that may not exist in many places. Also, whether integrating the sciences as an end in itself is worthwhile is yet to be determined. What do the students really gain from an integrated approach? What may be lost through dissolving the disciplinary boundaries? Should teachers integrate the sciences purely for the sake of making connections among the different disciplines, or should an integrated approach result from a more authentic model of scientific inquiry or from a more project-oriented approach? If students can't decipher the point of integration, or if they don't make the connections themselves, what is gained? These are difficult questions, and there is by no means agreement within the field as to the answers.

In California, there seems to be no clear way to demonstrate connections in science teaching. As noted, themes proved to be not very useful to teachers, and those schools that are teaching integrated science may not be successful in helping students see the connections among the traditional disciplines. While there apparently is some coalescence around the use of themes, operationally there is no consensus about what "theme" means. Perhaps it is less important that the curriculum is integrated, coordinated, thematic, or interdisciplinary; and more important that students engage in longer term, more purposeful activities that relate science to their own lives and communities. Rather than integrating the subject matter for its own sake, it may be more important to provide students with authentic, community-based experiences that help them build their own meanings and make their own connections.

Assessing Students in Integrated Science. Another important piece to the integrated or coordinated curriculum is student assessment. Teachers realized that if they were going to emphasize hands-on inquiry and draw on concepts from the different science disciplines in their daily lessons, they would have to change the way they tested their students. Also, many of the hub coordinators and SS&C teachers were involved in such state projects to reform assessment as CLAS and the Golden State Exam (GSE). Thus, many of the hub meetings focused on how teachers could design assessments that measured something more than memorization of science facts. Teachers began designing assessment tasks for their students that reflected the kinds of assignments they were performing daily.

For example, on a semester final examination for one SS&C school, students were asked to design an experiment that would test the effects of variable temperatures on the respiration rate of goldfish. The students set up the experiment and made a prediction as to what would happen. They were also asked to explain their predictions. They then organized their data into a table, constructed a line graph of their data, listed the possible variables that they might have encountered in the experiment, and wrote a conclusion.

These tasks required the students to go beyond recalling facts about the phenomena at hand. They were asked to construct an original experiment (original in that they had not done this experiment or studied this problem before), make predictions, collect and organize data, and suggest conclusions. The students were assessed on the soundness of their experimental design, their ability to defend their predictions, their skill at taking and organizing data, and their understanding of experimental variables and how they affect the outcome. Finally, they were assessed on their ability to write a conclusion that followed logically from the experiment they designed and carried out.

In an SS&C school, teachers are encouraged to view testing as an opportunity not just to find out what the student knows, but what the student can do with what he or she knows; it is an occasion for learning. This orientation to assessment—as innovative in conception as the *Framework* itself—is what has been driving California's standardized assessment projects in the last several years; and, many believe, is what will ultimately drive the reform. As more and more schools adopt the GSE[1], it is hoped that it will be increasingly difficult for teachers to escape the changes necessary for preparing their students for the examination.

Recently, a subgroup of SS&C teachers developed model assessment tasks for coordinated science programs. The resulting publication, *Models for Student Assessment in Coordinated Science*, provides teachers with sample performance-based tasks as well as a variety of open-ended essay-type questions. *Models* also helps teachers with developing their own assessment items as well as formulating and using scoring rubrics (California SS&C forthcoming).

As this section highlights, developing and implementing integrated science programs is a huge task involving educators at many levels. In addition to monetary support, teachers interested in changing their curriculum and instruction practices need collegial interaction, planning time, and additional content expertise. Moreover, the leaders in the reform recognized that change can be difficult, and if they wanted more schools to get involved in the project, they needed to establish some form of professional development that could ease the pain of change. The California SS&C project is attempting to meet these needs through a network of secondary science teachers, called hubs.

SS&C in California

I think one of the biggest strengths in the SS&C approach is the empowerment of teachers to have some ownership in the development of programs . . . I think that teacher buy-in is created through ownership of your program; and nobody knows their students, their resources, and their capabilities better than the teachers themselves. — T. Ritter, science teacher and hub coordinator

Again, the unique feature of the California SS&C effort is that there is no model curriculum; each school is free to develop a program that specifically suits its students and staff. Tom Sachse, the project director, preferred this tactic from the beginning: "I never wanted to have a model for SS&C; I never pushed one, partly because I didn't think any of us knew what the right model was, [and] I

---

[1]GSE is a voluntary science performance assessment that focuses on coordinated science concepts. It is described in more detail in the section "Assessment."

thought that the professionalism of the teachers would be compromised if we were to say that there is a model." Kota echoes this sentiment and reports that thinking of it as a process, rather than a single product, is the key: "While teachers may develop curriculum that they believe everyone should teach, they feel free to implement it in very flexible ways, depending on the strengths and loves of the particular teacher."

For example, at one high school in Northern California, the science department committed itself early to work on a weekly basis to transform the entire curriculum to a fully integrated (grades 10-12) program. As a result, the teachers have managed to change the scope, sequence, and coordination of their curriculum completely in three years. Their approach to staff development is focused on obtaining the additional content background each teacher needs in order to teach their new courses effectively.

Another school in Southern California has integrated only the two years of required science and has no plans to go farther. It would offer traditional science courses as electives. The department head at this school, who also happens to be the hub coordinator for the region, defends this plan:

> We're very content with the two-year program, and the reason is . . . the first two years of high school [should be] more core . . . years where students really establish their foundation, and the third and fourth years I want to be more exploratory, to allow students to begin to explore those things that they find more relevant for them in their lives.

Hub meetings provide a forum for teachers to connect and share what they are learning. The establishment of these formal structures for professional development has provided the space that Sachse anticipated for teachers to develop ideas, critique the ideas of others, and—through that process—get closer to their vision of good science teaching.

The Hub Structure. The state is divided into 10 geographic regions or hubs, each lead by a teacher/coordinator. Some regions include both urban and rural districts, such as Hub 2 in Northern California; others are strictly urban, such as Hub 7, which serves the Los Angeles and Pasadena Unified School Districts. Meetings of hub member schools are held on a monthly basis. The primary purpose of the hub meetings has been to provide teachers with the opportunity to network with other teachers who have been experiencing similar problems or successes and to collaborate on curriculum ideas or implementation strategies. Sometimes hub coordinators arrange for speakers or discuss some aspect of the change process such as obtaining University of California approval for the new courses. For some schools, hub meetings are the only source of professional development available during the school year. In general, hub meetings are attended by a few representatives (site coordinators) from each high school or middle school within the hub. Depending on the hub, anywhere from 10 to 50 people attend meetings regularly.

The hubs provide an opportunity for science teachers to work closely with teachers from other schools and within their own departments on reevaluating their science programs and making substantive changes. One teacher's hub experience has been a powerful force in helping her department through the change process:

> [Being a part of SS&C] has been a real process of pulling teachers together, and I think that's probably an understatement of what SS&C is all about. It really helped pull our department together . . .

> I see [hub meetings] as a valuable tool in just touching base with other teachers that are in the same status, or close to it, [and] just having time to check in and see where they're hitting snags and what they're doing about it. How we hit that snag and what we did about it and sharing that information helps make it a little bit easier when you're going through as much change as we are.

Other departments have had a similar experience. Although not as active in the hub in recent months, one teacher from a Northern California high school reports that the process of changing the curriculum and being part of a larger reform effort has helped to bring the teachers in the department to a level of professional collegiality that had not existed previously:

> The teamwork here has been the most help to all of us. If I wasn't very good in physics, I knew where to go to ask the question, and they were right there, and nobody was ever put down. In fact, this has probably made this department the strongest in the school because of that relationship. Because we had to get in and fight and dig and [say things like] "I want this, well I want this." We can come to understandings, rather than saying, "Screw you, I'm going to do something else."

One teacher and hub coordinator involved in SS&C since the beginning thinks that the hub structure was crucial to the success of the reform:

> Basically, my philosophy since I've been involved in this was to be really evangelical. I felt like it was really important to go out and spread the word about what was going on and bring more people into the flock, so to speak. I didn't want it to be an elitist group or an elitist project. I've always felt . . . that we should be going out there and bringing other people along and sharing our resources, not isolating them.

Another hub coordinator, who briefly held a leadership position at the project headquarters, puts it this way: "The people [in the hub] you associate with, who are agents of change at all of those places, get together fairly regularly and so that's sort of a positive boost so you didn't always feel like you were isolated on the campus with only one or two other people."

Not all schools have experienced such unqualified success, however. One SS&C teacher reports that, in her view, a major shortcoming of the SS&C hub structure and staff development is that they do not provide teachers with the kinds of leadership skills necessary for working with those back at their school site who may be resistant to changing the curriculum. In her experience, the process of change has been a deeply personal one. Some members of her department and parents of her students have opposed the changes to the point of undermining her efforts and those of the other teachers trying to change. This has caused her great personal anguish and has affected the way she interacts with other people who may be interested in the program at her school.

The Hub Coordinator. The functioning of the SS&C hub network is largely the result of the work of the hub coordinator. Hub coordinators are paid a stipend of $4,200 per year and are chosen by the participating schools within each hub. They are all full-time teachers. Their role in the reform has been defined as the project developed; there was no clear job description in the beginning for the chosen leaders to follow (Kota 1992). As the hub idea was a new approach to change in California, nobody knew what to expect, including the directors of the project. As time went on and people began to identify the important issues, coordinators responded in ways they deemed most appropriate and efficient for the teachers they served.

Today, the general responsibilities of the hub coordinator include:

- administrative duties, such as planning hub meetings, coordinating districtwide SS&C implementation, visiting schools within the hub, and attending state hub coordinator meetings;

- staff development projects, such as workshops for SS&C teachers within the hub, assistance with SS&C summer institute planning, and finding outside people to come to the hub meetings;

- leading and participating in local, state, and national conferences and institutes; and

- communication with project head office staff, teachers, administrators, parents, students, industry, and interested schools and their staff.

As a result of their position within the California reform, the hub coordinators have been invited to participate in a variety of additional professional opportunities. One Northern California hub coordinator, in addition to planning hub meetings and conducting visits to hub member schools, spends a great deal of time on public relations. He talks with faculties of interested schools and their school boards; writes articles for the SS&C newsletter, *Restructuring Science*; speaks at local, state, and national science teachers' conferences; and has appeared on a local cable television show to "smooth some ruffled feathers where a nonpreparatory integrated course is being upgraded to a [college] preparatory course without majority faculty buy-in." This coordinator also has helped write GSEs.

The coordinator of Hub 7, as a result of her involvement with SS&C, has secured a position at the University of California at Los Angeles in the Project Issue Program, a project offering professional development opportunities for science teachers in the process of change. She also has become deeply involved with the Urban Systemic Initiative efforts under way in Los Angeles. She has been asked to codirect other NSF-funded projects within her area districts. The coordinator of Hub 8 spends several days each month talking to interested schools in his geographical area. He is also heavily involved with the local California state university in constructing a more meaningful experience for the science student teachers who earn their teaching credentials through both course work at the university and student teaching experiences in his department. He has worked very closely with the lead professor to restructure the preservice experience such that the teachers can spend the entire year in the same school with the same students.

While SS&C has made some attempt to prepare hub coordinators to be leaders within the reform, most coordinators report that they essentially learned their role "on the job." This has not been much of a concern to the project leaders, who feel that the hub coordinators already have leadership qualities and most likely would have initiated change on their own accord—at least within their own schools—regardless of SS&C. Tom Sachse suggests that SS&C merely provides the sense of professionalism they deserve for their efforts:

> A lot of folks realize that if they really want to work hard, SS&C can give them a sense of professionalism and professional abilities and opportunities that [they] wouldn't have otherwise, so they are many times willing to put in the effort [and] become recognized as a leader in the state. What does that bring them? It's professionalism in one sense, it's also opportunity to write textbooks, to write tests, to do staff development, to become a consultant. We will continue to see more opportunities put into the SS&C system.

As Kota describes it: "[The hub coordinators] might have found the project to a certain extent because they are the ones that are . . . out there in front and empowered. Although many were in classrooms and not very empowered, [the project] has . . . made them incredibly enlightened as well as . . . very dedicated."

In sum, the hub coordinators drive the SS&C project. And each has responsibility within a classroom or district as well as SS&C leadership duties. As in CSIN, the reform leaders saw the advantage of building on the existing strengths of teachers within the state. As teachers, hub coordinators have been sensitive to the personal and professional difficulties associated with the transition to an integrated or coordinated science curriculum. They themselves have faced the difficulties of relinquishing their favorite sophomore biology or honors physics course to try something novel, untested, and possibly risky; they are not solely trying to get other people to change. They also serve as mentors who provide classroom-based knowledge about the reform within their own departments and districts. As leaders, they must have the kind of goal-oriented drive that keeps them persevering in the face of obstacles—from resistance of the community to a new science sequence, to assessment dilemmas, to coping with University of California admissions requirements. In fact, the distinction between teacher and leader is not sharply delineated: the hub coordinators are gradually exemplifying a new and integrated professional role in the state—that of teacher-leaders.

Staff Development

Within Hubs. Staff development programs within the hubs have rarely been simply about content and teaching methods: they have also entailed having conversations about what is important as well as opportunities for teachers to air their concerns about what they were trying to do and why. Consider the following example. At a Northern California hub meeting, a Fulbright exchange teacher suggested that perhaps a bit more centralization was necessary for the success of a curricular reform. The following discussion ensued:[2]

> "I disagree. The networks allow teachers to collaborate and so aren't totally isolated. In fact, if you look at their curriculum, you will see it is not totally diverging. Part of this is due to the statewide assessment system, such as Golden State. I think there is a great deal of richness in the diversity that is there."
>
> "Why then can't you get together on the curriculum?" asked the visiting teacher.
>
> "Buy-in," replied another voice.
>
> Another teacher offered, "It's the thinking involved that's important. It forces you to examine what you are teaching and why."
>
> "But not everyone is doing that, are they?" the visitor asked.
>
> "Admittedly, no," replied the teacher. "There are some laggards that may be later forced to accept these now voluntarily explored changes. Maybe a critical mass will be hit and some bandwagon effect will draw them in or maybe not, and we'll be forced back."
>
> "I still think that you should be looking to the universities to develop curriculum," said the visitor.
>
> "Many teachers in my district are objecting to the time demands of writing new curriculum," one teacher said.
>
> "It seems to me," another teacher interjected, "that the goal of this reform is itself a moving target. As a minority teacher and Republican it might seem odd for me to quote Mao Tse Tung but . . . he once said, 'Revolution must never stop. You can't stop it or you will just recreate the bureaucracy you just replaced.' SS&C is in a constant state of growth."
>
> "This isn't growth, this is an earthquake!" another teacher responded.

Most hub activity has centered on opening up this kind of dialog among science teachers. As the change to integrated or coordinated science can be long and difficult, many hub coordinators recognize that allowing teachers to vent their frustrations or share their successes is equal in importance to learning new science content or pedagogical strategies. Therefore, the focus of many hub meetings is to provide diverse kinds of support for teachers as they try teaching

---

[2]The text is not a word-for-word transcription of the actual conversation.

integrated or coordinated science. Each hub coordinator assesses the needs of the various schools in his or her hub region and organizes meetings around them. Not unlike the overall approach of the California SS&C project, the concept of staff development in California is flexible and needs-specific.

For example, one hub coordinator in Northern California arranges for guest speakers for 80 percent of her hub meeting time; other coordinators spend little or no time with outside speakers. Some hubs spend concentrated meeting time on articulation issues with middle and elementary schools; others do not feel that they are at the stage where this is important. The flexibility of this model allows for hubs that serve primarily urban—or primarily rural—schools to address needs specific to that setting. Some hubs serve urban, suburban, and rural communities. This situation gives rise to specific challenges and opportunities for the participating teachers.

The main goal of each hub has been to help teachers face the challenges of changing their science programs. As illustrated, this can happen in a variety of ways. Certainly, not all teachers have been equally served by their hub; some schools reported little or no involvement with the hub after an initial couple of meetings. Their reasons for this varied depending on the school. One school stopped attending meetings because the teachers felt that their school was working toward different aims with a different student population and a different kind of curriculum than the majority of schools in their hub. They were not confident that their hub could be of much assistance. Another school felt that, as its program was up and running smoothly, it did not need assistance from the hub any longer.

Between Hubs. In addition to staff development within hubs, SS&C (with a professional development grant from NSF) has sponsored institutes for SS&C teachers during the summer. For example, in the summer of 1994, SS&C ran two 10-day institutes for teachers, one in Northern California and one in Southern California. At these institutes, teachers received in-service training from university and Sandia Laboratory scientists on the general topic of global warming. They designed and shared curriculum based on the information in the content sessions and worked on implementation strategies.

Further, they had the opportunity to be involved in the scoring of GSEs for students in their second year of a coordinated science program. This process was included in the program because of the prevailing view within the state reform projects that assessment is crucial for influencing change in school programs. A consultant for the science GSE believes that "assessing student work is important and an eye-opening experience for teachers . . . [They] have to examine what works and doesn't work about open-ended [questions] and lab activities." She feels that GSE can be a helpful tool for teachers who are trying to figure out what they value in a coordinated science program.

One teacher in a Bay Area SS&C school listed awareness and time (for which funds must be provided) as the two most valuable benefits of being involved in SS&C staff development institutes. For her, these networking opportunities had greatly influenced the way she thought about the future of the program at her school. She was particularly excited by the institute and the teachers she met there, as many of them had progressed in the reform of their curriculum to the same place as she and her colleagues at school. Hearing about their concerns, their puzzlements, and their successes inspired her as well as validated her impressions about what she was doing with her students. SS&C has increased her awareness of what is going on at other sites, which she claims helps her in her own planning.

In addition to organized staff development opportunities such as the summer institutes, SS&C teachers and hub coordinators are offered a variety of professional growth opportunities. Teachers also have become involved in activities outside of project activities. Many have made presentations on the programs they have developed at their sites at science teaching conferences. Others have been involved in in-service training for other schools interested in changing. Being an SS&C site brings attention from visitors from other schools within California as well as officials from other states looking for ways to change. Teachers in these schools have been involved in informing visitors about the work they have been doing.

Between Networks. By far, the most pressing issue addressed at SS&C hub meetings is implementation of new science programs. More recently, many of the hubs have been involved in making more substantial connections with feeder elementary and middle schools that are linked to other statewide reform projects to learn how other networks are approaching the change process. Teachers experienced in the two main elementary projects—CSIN and CSIP— have begun to be invited guests—and, in some cases, integral to the program of many hub meetings. SS&C is also making an organized and substantive attempt to articulate with the feeder middle schools. Kota reports,

> Last year we formed a collaboration with CSIN to meet the needs of the middle school around the state. We now have SPAN, Science Partnerships for Articulation and Networking. It's following the CSIN model of implementation and the SS&C emphasis of coordination and spiraling of science content . . . That's what happening with our collaboration.

The funds the original 100 schools received were supposed to be used to address the needs of grades 7-12, therefore including the middle grades. As the high schools geared up to make changes, many of the middle schools—according to Kota—"got lost in the shuffle." For the directors of CSIN and SS&C, SPAN has become the arena for addressing the special needs of the middle schools. Middle school teachers are encouraged to participate in CSIN regional meetings and attend hub meetings. There have been three SPAN institutes—in the summers of 1993, 1994, and 1995. At the request of the participating teachers, the 1995 institute included content along with issues of leadership and change. High school and middle school teacher-leader teams attended this conference, a practice Kota intends to continue. The 1995 SPAN institute also involved other projects, including the Fresno and Los Angeles Urban Systemic Initiatives. At this time, the schools pay for their involvement in SPAN. Kota and DiRanna have submitted a proposal to NSF for additional financial support.

In addition to the benefit of working on a program solely for the middle school, the joining of forces between SS&C and CSIN is seen by the state leaders as an opportunity to learn from each other. The leaders of SS&C recognize their lack of attention to the emotional aspect of change, one of CSIN's strengths. CSIN has capitalized on the content expertise of the high school teachers at their content workshops. This kind of sharing has become increasingly important to the functioning of both networks.

Program Evaluation. SS&C has its own program review system for secondary schools. As the project developed, teachers began to become concerned about how they could evaluate their programs. In response to the teachers' demands and to a request by the NSF team that visited the California project in 1992, Kota and others within the SS&C leadership assembled a manual in which they outlined several options for program assessment.

For example, departments could opt to administer the Golden State Exam in coordinated science.[3] The results would presumably give them some indication of the quality of their program. Or, schools could choose to conduct a self-study followed by an extensive peer review. In this case, the school would compare their students' work with the vision set forth in the *Framework*. The department would discuss and analyze student work, performances, and outcomes. Teachers would attempt to identify any problems with their science program, their causes, and possible solutions. The results of these discussions would then serve to orient the peer review team. The peer review team, consisting of three science teachers from other SS&C schools, spend one day at the school and provide an outsider's prospective.

The foreword to the manual describes the purpose of the assessment options this way:

> This effort to provide program assessment options was part of a dual-edged strategy. On one edge was the continued respect for the creativity and originality of participating SS&C teachers. It was not considered appropriate for all SS&C programs in California to be evaluated in light of a single technique or tool. Instead, the creation of these nine assessment options was meant to give SS&C sites some measure of control over their own program evaluation destiny. On the other hand (or edge) it was seen as necessary to provide program assessment support for schools and teachers that wanted to properly evaluate the nature of their SS&C program (California Scope, Sequence, and Coordination 1992).

This degree and kind of support from state leadership is in keeping with SS&C's site-based theory of change. The SS&C project in California "continues

---

[3]See Chapter on "Assessment" for more information about the Golden State Exams in science.

to favor local control in both the design and evaluation of SS&C courses" (California Scope, Sequence, and Coordination 1992).

At this writing, SS&C leadership is assembling a small group of people within the state who have experience and an interest in rethinking their program evaluation strategy. The plan is to highlight those components of the project that are working well, including—but not limited to—an analysis of GSE scores from SS&C schools from 1993, a thorough case study of the functioning of the hubs, and a quantitative look at students' grades and the numbers of students who go on in science after the 10th grade.

Department in Transition: A Case Story

The switch to integrated or coordinated science can be a difficult, slow, and deeply personal process. This section characterizes one department that is currently undergoing such change. Teachers at Middlefield High School, a large, diverse, semi-urban school in Northern California, highlight the issues the consider to be the most salient. This school faces challenges familiar to many within the state; Middlefield represents a typical pattern of development and implementation of the California reform.

The Middlefield High School Program. The central focus for Middlefield is to teach science in a way that the students will find relevant to their everyday lives. This has resulted in curricular themes such as "Criminology"; "Across California"; "Jurassic Park"; "Sex, Drugs, and Rock and Roll"; and "Housology," relating science to things around the house and garden. These themes are "catchy," the teachers say, "At least we get their attention." They want the students to make connections between what they see and do every day and the world of science:

> When we did Jurassic Park, we showed the first 20 minutes of the movie, and then we had them do a writing assignment. [We asked] "what's the science in Jurassic Park?" All of a sudden they're thinking a little bit. They come in every day and ask, "Are we going to see the end of the movie?" But at least they're thinking. When we did this biotech unit, I said, "Can you think back?" And they said, "If we look back at the movie, I'll bet we'll see micropipets," and so they're making connections now. I think that's more important than anything, because these children are not [all] going to Harvard.

> Another teacher notes: "[The students] will say, 'I just saw something on the news that had something to do with DNA,' and I'll understand what they're talking about. They've had it in context, not just, 'Okay, this is a DNA molecule.'" And a third teacher points out "But in the book, you would do chapters 19 or 20, and it follows 18, and they answer some questions, and they don't get it. At least they're getting it a little better, but it takes much more time . . . They're getting that it relates to them, that it has some connection to their life.

When these teachers were asked if they were concerned that these kinds of themes diluted the quality of the content in any appreciable way, they responded with the following statements:

I think the science is the same, it's just connected now.

What do you consider content?

Instead of going to the beach and studying cliffs when you're studying earth science and then next year going to the beach and studying tide pools, now you can go to the beach and do the geology of the cliffs and do the tide pools and do the pH of the ocean and talk about the moon and the tides. You connect it all. Of course, we don't have any money to go to the beach, but if we could, that's the idea. It's the same science [as the traditional courses].

I don't think the quality is any less at this level.

I think you have to make the choice whether you want process or whether you want content. If you believe in process, then you let some content go. If you want content, you can't do this process.

Difficult Beginnings. Middlefield High School was an original recipient of the $10,000 grants for the SS&C 100 Schools Project. It received the funding in the 1989-90 school year. A team of teachers worked that first year on a visionary project that included an articulated integrated science program for grades 8-12. Over the next three years, however, all but one of the teachers from the original planning team left the school. This teacher then pulled together a small subset of the department to try to continue the work that was started in 1989. For three years, this subset worked on developing a 9th and 10th grade program, trying to bring in new teachers as they arrived. This proved to be problematic, as new teachers have unique concerns. One teacher explains: "We tried to get the new people involved. We tried to do some meshing of the old philosophy, maybe where the new people were, and what I saw at the end of it was it was too overwhelming for these new people to jump in."

For a project that required a team effort, the high turnover at Middlefield that first year acted as a major roadblock to progress. SS&C leaders soon made it clear that the school needed to show something for its initial funding. At the behest of the project head office, the planning team worked in the summer of 1993 to develop an integrated science course to replace ninth grade general science. The new course, like the old course, was intended for those students who were not expected to go on to college. Their main impetus for developing the course was to put something in place that was interesting to teach as well as to learn. One teacher involved in the process describes the old course this way: "The old courses were terrible . . . The students were not passing, they were not interested. Teachers didn't want to teach it because it was such a low-level course." A second techer adds "It was irrelevant. Nobody wanted to teach it or learn it."

The first year of implementation (1993-94) was difficult. An original member of the initial planning group explains:

> So we went ahead and we taught [the ninth grade integrated course]. And we had struggles. We had people that needed more guidance; they needed counseling to do this. Each teacher needed something different. I was the leader per se, but I had no extra time to be the leader. There were so many different needs: somebody would say, "I need to have a meeting to run through the lab;" and somebody else would say, "I want to figure out what we did last week and how it went for when we do this again;" and the other person says, "What do we do tomorrow?" So that was a concern. And what happened was, people that were uncomfortable with this type of, I guess I want to say philosophy, got really hung up with what *wasn't* working. And the people that were pro this philosophy tried to grab onto the good side of it, even though there were problems.

Why was it so difficult for some of the teachers? Could it simply be an issue of clashing philosophies of science teaching? Perhaps, but the teachers of Middlefield suggest other, more pragmatic, reasons.

> Or, they were new teachers . . . and they had so much to deal with— they were running from classroom to classroom [and] they didn't know where the stuff was. The kids are tough anyway, so you've got one more thing. *Then* you're dealing with the content that you're not familiar with, because we really tried to integrate instead of coordinate, and now all of a sudden you have somebody having to teach something that they have no clue about, so they can't wing it. And so they felt that they had to learn things all over again.

> [The experienced teachers] go back to what they feel comfortable doing. If they don't have the extra time to get the support, [they] resort back to the easier way out, and it's not the easier way out because it's all hard. It's what you know and what you can do. The other thing is there is a lot of discipline problems with our kids, and some people don't like to give these kids, when they are discipline problems, an opportunity to do a lab, because it's more trouble . . . [P]art of the advantage is that their behavior is better [with the integrated program], but to get to that point, it's painful.

The move to develop integrated courses, then, became a very complex problem, one that involved issues of personal and professional identity. Some of the teachers were attentive to their own comfort levels in terms of the content as well as instructional style, and they did not recognize themselves in the reflection of the new content and pedagogy. Moreover, there were incredible logistical problems associated with finding common planning time and sharing materials. What was the result?

> What happened was, towards the end of the year, due to everybody being I guess stressed out, it ended in not a positive way. Was the cup half full or was the cup half empty? Those of us who believe in the program, even though there were problems with it, we said there are things that are fixable. It wasn't a total disaster, but you can't run something for the first time and expect it to be wonderful. And then there are people that want to have things: tell me what to do on

Wednesday, what are the objectives, what page am I going to give them for homework, all those things. You can't do that the first time you run something. So, by the end of the year, everybody was upset.

Next Steps. The ninth grade course was finished in the summer of 1993 and implemented that school year. At the end of the summer of 1994, the team that had pushed the ninth grade course through worked to develop a 10th grade course. This course was implemented in the 1994-1995 school year. As a result of the group leaving the 9th grade to teach the 10th, the 9th grade course has changed from the original vision.

> The whole course has changed around. They decided they wanted to buy a book, and they bought a junior high school book . . . And it's no longer integrated, it's not even coordinated. It's six months physical science and six months of life science. There's nobody [teaching the ninth grade] that really cares about [the integrated curriculum]. Our original plan was to get all the materials together to implement [Science 1] last year, and then spend some time redoing [Science 1] last summer. [We were hoping] we could rewrite the whole curriculum and turn the cookbook labs into real science labs, but that didn't happen.

The teachers involved in and supportive of the reform at Middlefield were frustrated for the primary reason that they didn't think the rest of the department seriously considered who the students were when they made the decision to go back to the traditional ninth grade class with a traditional and low-level textbook. One teacher says,

> The problem is that they're not putting the needs of the students first. These kids are not going to go to college, maybe some of them will go to junior college for a while. If you're going to teach traditional science classes and the kids are going to go on to Berkeley or Stanford or Harvard or Yale, that's one thing. But to take kids that are going to graduate from high school and go out into the world, they need a different kind of science class. They don't need to know the Krebs cycle, they don't need that stuff. They really need to learn how science works in the everyday world; that's what they need to understand.

The only way to make a program like this work, say the teachers, is to have full and open cooperation and collaboration. "This course requires a lot of working with each other. If you're going to coordinate what you're doing, and we're doing the same kinds of things and the same labs, and if you don't get along and have the same philosophy, [it's not going to work]. We're up and down stairs, and at different ends of the wing."

What is the future of integrated science at Middlefield? Without further funding, their goal to expand into the 11th grade will not be realized. Without the support of the rest of the faculty, the 9th grade program—and even the 10th grade program—may be in danger of becoming another set of dusty binders in the back room.

The news is not all bad, however. The teachers have received positive feedback on their new courses from their district. They were recognized for their efforts in the form of additional funding, approximately $7 per pupil in the integrated courses to help pay for consumable materials and photocopying. In addition, they have applied for and received $2,500 in planning money from an outside source to revise and augment the 10th grade program. They have also, with leftover SS&C funding, brought a roving, hands-on biotechnology program, called The Gene Connection, to their science students.

The teachers who have remained faithful to the reforms look forward to a bright future. Even though they have suffered setbacks from various sources, they continue to believe that they are doing the right thing.

> Even though there are problems, I think we need to grab onto the good in this, rather than the bad. How to fix the [personnel] problem? I'm not sure that you can change peoples' philosophy on this unless you give them lots of time to work with each other, and that you give them lots of training to help them and support them. It never would have split as badly as it did if we had a common prep time last year, if we were given extra money to work on this, if we had less kids in the classrooms. There are all sorts of things that would have lessened the strength of the problem, but I don't think it would have eliminated the problem. I think there are teachers that like to lecture and like to give tests, and it's just a different philosophy in how you teach science.

How did their membership in SS&C help Middlefield? The teachers cite start-up money and recognition with district administrators as the most important factors. They have not been very active in their hub, partly due to the geographical distance between their school and the location of the majority of hub meetings, and partly because of their perception that their situation is so unique that other schools in the hub could not relate to what they are and have been going through. Finding like-minded people who teach a similar curriculum and face similar issues is very important to these teachers; they are eager to belong to a community of professionals that care about science education reform.

> The first time that I really found out who was doing what was last summer at the [SS&C summer] Institute. When I found out that there were people in Bakersfield doing the same themes that I was, I was really excited. And there are several of us, because I met a gal from Nebraska doing the same things in her restructuring . . . So the money should be put in the sharing . . . And then we could work on where we are and share some of those units. But if you don't even know what everybody is doing. . .

This case story speaks to the staying power in the SS&C project. Kota reports that most of the original schools still have programs in place. In her view, schools that have survived have done so because they have continued to inform their constituencies of the need for change. Schools that fail to continue the public relations often find themselves without support, and the reforms eventually fade out. Kota also cites the problem of neglecting to introduce new teachers properly to the SS&C philosophy as a threat to schools staying with the new programs.

Kota relates that she is continually impressed with the level of dedication she has seen over the years and the increasing quality of the science programs that are being developed in SS&C schools. While the project has suffered temporary setbacks recently, such as the resignation of Tom Sachse as project director, Kota hopes for a long and bright future for the California SS&C project.

◆

The Reform and Higher Education

While the direction of influence has historically moved from higher education to K-12 education, the California science reform effort provides evidence of influence in the other direction. This section highlights three as examples: the Science Teaching Development Project (STDP), University of California course approval, and the California Science Project.

Preservice Teacher Education

A large majority of teachers in California receive their teaching credentials from one of the 20 California state university teacher education programs. The state universities also provide most of the professional development and supplementary credential opportunities for California teachers. As part of the SS&C project, the state university teacher education programs have accepted the challenge of preparing teachers for classes in integrated and coordinated science. Much of the leadership for this initiative—the Science Teaching Development Project— STDP has been assumed by Herbert and Bonnie Brunkhorst at CSU-San Bernardino.

According to Bonnie Brunkhorst, the project requires major adjustments in the responsibilities of the universities for science teacher preparation, and faculty have been charged specifically with providing collaborative leadership in the effort to reform secondary science education. Each university is free to make changes in ways that fit its individual traditions and mission, provided there is significant collaboration among the science faculty, education faculty, and high school science teachers. All 20 campuses have participated in the project, each with different funding structures, student populations, and faculty. Each campus helps realize the SS&C vision of reform in science education either through its in-service or preservice offerings.

Bonnie Brunkhorst says that one of the goals of this new teacher education program is to "lay a framework for ongoing professional development." She believes that the key to the success of reforming teacher education is the collaboration of the various experts on each campus. The focus of the collaboration is to provide the necessary facilitation to allow the teachers to bring about the changes. "In California," she explains "leadership is placed in the hands of the teachers . . . They are the leaders. The rest of us have to clear the way." Brunkhorst understands that the change process can be slow and painful, and that it should be carried out by the people who "own the systems." However, she also asserts that without organized facilitation "from the top," the programs could not move forward.

As a part of this facilitation, CSU science and education faculty meet regularly with SS&C hub coordinators to discuss how they might help each other with the various SS&C projects—specifically, how the universities might be of assistance to the hub coordinators and their member schools. At these meetings, high school science teachers and university science professors have the opportunity to talk about what each expects from the other, and how they can work together to provide the best education possible for students and teachers. The focus is on preparing teachers to teach integrated or coordinated science.

The obstacles are significant. One science education professor summed them up as lack of good curriculum, other than what is developed on site, and lack of an undergraduate integrated science model. This professor feels that teaching prospective teachers about integrating the sciences in two weeks of a course on methods of teaching science is wholly inadequate.

The Triad Concept. One of STDP's major components is the triad concept. This approach brings together a science educator, a scientist, and a local science teacher to work on some aspect of improving either the preparation of preservice teachers or the continued education of in-service teachers. In planning STDP, the Brunkhorsts set two conditions:

1. The triad is to involve people with the right mixture of expertise and mutual respect for their expertise.

2. Each campus is to find its own funding for the projects.

All 20 CSU campuses are involved to some degree, depending on their needs and available experts. Fifteen of the 20 campuses have set up partnerships with SS&C high schools or are in the process of doing so. Each campus has a different funding source, population, and faculty and, therefore, a different focus. To learn more about the process of this kind of collaboration, each triad engages in action research on some aspect of its project.

For example, the CSU-Northridge campus emphasizes assessment, leading the entire project in getting student teachers involved in scoring GSEs. CSU-Dominguez Hills wrote and implemented an elementary science curriculum, which has attracted the attention of a major publisher. The triad at CSU-Pomona has focused on publishing a newsletter to facilitate discussion and the sharing of ideas. CSU-Fresno and CSU-Stanislaus implemented an elementary science methods course where before there was none. Moreover, CSU-Fresno has received funding for its Project MOST (Minority Opportunities in Science Teaching), aimed at "recruiting, educating, nurturing and credentialing under-represented science teachers" (California SS&C forthcoming). CSU-San Luis Obispo involved student teachers and minority students from the Valley and Southeast Los Angeles in a one-week workshop on the environment.

The CSU-San Bernardino program is perhaps the most elaborate. It focuses directly on preparing teachers for integrated science programs. Student teachers in science are placed in a local school that is active in SS&C. The purpose is to provide them with an initial professional support system. At this school, the student teaching structure has been redesigned to allow the teachers to teach in two integrated/coordinated classes—at least one in their own discipline—and then observe or help in a subject other than their own. Moreover, the teaching methods courses, both general and content-specific, are held on the high school campus.

What do the Brunkhorsts see as the purpose for arranging their teacher education program in this way? Primarily they view it as a way to offer classroom teachers an opportunity to be involved in the preparation of new teachers more broadly and to ensure new teachers an experience within a school that is in the process of reform. Herb Brunkhorst puts it this way:

> In California . . . you have people with a number of priorities. Money is a big issue, time is an issue, and so we wanted to come up with some type of coherent design that [involved student teachers' taking] their course work (or at least part of their course work) on the high school campus, so we could use faculty from the high school in a broader way . . . We also wanted to get teachers involved the very first day, even workshop week, and that's what they did, and they will be there until June. Now that's different than most student teaching programs [where] you sort of pop in for 12 weeks [and] you're done.

One effect the Brunkhorsts have seen as a result of this kind of collaboration has been that many of the teachers who obtain their credential through CSU-San Bernardino have been able to find jobs easily because of their expertise in integrated science. These teachers have become leaders, in the sense that they participate in meetings and other teacher development activities associated with the reform. Herb Brunkhorst believes that this has been a direct result of immersing new teachers in an environment that is responsive to and involved with the reform.

In Herb Brunkhorst's conversations with teachers leading the reforms, they describe how more teachers are asking: "What district can I go to that's doing something?" and "Where am I going to be happy as a leader?" Brunkhorst suggests, "The message in another year or two is going to be pretty powerful to school districts. If you want good teachers, you better start doing something, because people aren't going to want to work for you." Here the universities are putting themselves in a position not only to support the K-12 reform effort, but also to sustain the reform through the preparation of new teachers.

The Problem of the Undergraduate Major. An STDP subgroup has taken on the problem of the undergraduate major for the prospective teacher of science. The group's most recent proposal to NSF to fund this project was rejected, and major efforts to change the undergraduate science major for teachers are currently on hold. However, there has been some movement in this area on some of the CSU campuses. A few are devising new science majors (e.g., a B.A. in natural science) expressly for prospective science teachers. Rather than creating entirely new courses, however, these new majors consist of existing courses, which raises the issue of whether subject integration will be modeled. Other CSU campuses are pulling together a major in science education. At one meeting of CSU faculty and hub coordinators, the issue of rigor was raised with regard to this new major. Will a degree in science education be a watered-down science degree? Will it be possible for teachers to teach integrated science without the deep understanding of science that a "regular" science major would instill? What can be done to combat the perceived low status of these new majors? One faculty member suggested that if a student wants both a strong science background and a teaching credential, he or she will simply have to attend school longer. "Teaching is not a walk-on," notes one hub coordinator. At a few campuses, all science majors are required to take a one-year senior seminar that will be integrated and interdisciplinary. This course will focus on issues of equity as well as the history and philosophy of science.

While these solutions seem innovative and important, the universities face the problem of finding faculty members who will agree to teach the courses and students who will take them. As more funding becomes available to address the problem of the undergraduate science major, leaders in the reform anticipate large-scale changes to accommodate the kinds of changes happening in K-12 science education. Since changing the scope and sequence of higher education course work is an enormous undertaking, many within the reform decided that one place to begin to stimulate change might be in the standards for the science credential.

New Standards for the Science Credential. In 1992, the California Commission on Teacher Credentialing decided to "streamline the structure of science teaching credentials" by combining the two separate teaching credentials in life or physical science into a single credential with specialized preparation in one of the major science disciplines. This is a major departure from the previous model and was strongly supported by the two advisory panels that developed the standards. The impetus for this move stemmed from two major factors: (1) it was recognized that the physical science credential was too broad, requiring prospective teachers to be experts in physics, chemistry, and earth science; and (2) the panelists were well aware and very supportive of the current reform efforts of SS&C and Project 2061.

In the introduction to the handbook *Science Teacher Preparation in California: Standards of Quality and Effectiveness for Subject Matter Programs*, the advisory panel states the following regarding characteristics of future California science teachers:

> For the state to educate its students to become scientifically and technologically literate citizens, California's science teachers must possess many complex skills that will enable them to communicate and demonstrate concepts, principles, processes, attitudes and applications of science and technology equally well to students of both sexes and of diverse ethnic and cultural backgrounds.

To reflect the cooperative nature of progress in science, and to understand the common bases of the science fields, future teachers must be broadly prepared in all major disciplines of science. Furthermore, to be well-prepared to teach their future students how science operates in detail, participants in subject matter preparation programs must study at least one science discipline in considerable depth (CTC 1992, p. 12).

The standards were compiled by two groups of panelists. Both panels consisted of scientists, teachers, science educators, and administrators; and are distinguished by their association with either life or physical science. The handbook describes in great detail 16 standards for subject matter preparation programs. Examples of standards and this rationale follow (CTC, pp. 18-19).

Standard 4. The program reflects science as an integrated entity and emphasizes interrelationships among the disciplines of science. Concepts that occur in all science disciplines are examined, and variations in the structures, content and methods of inquiry in the disciplines are studied.

Examination of phenomena from the perspectives of different science disciplines leads to a more accurate and complete understanding of the natural world. The ideas that define the scientific endeavor are larger than the facts or concepts that belong to any single field in science. It is therefore essential that the program view science as an integrated entity, emphasize relationships among the sciences, include interdisciplinary studies in science, and examine similarities and differences among the disciplines.

Standard 5. The overall program in the sciences is organized to assure that students meeting the requirements of the program acquire sufficient general understanding of science that, as future teachers, they will have the necessary background to impart a high quality general scientific literacy to their students. General studies provide familiarity with the nature of science and major ideas common to all the natural sciences, including: matter and energy, mechanisms and processes of evolution, scientific models and systems, and their interactions. The program provides a foundation for students to engage in further studies of any natural science. General studies in the program familiarize all students with important societal and environmental concerns and the application of scientific principles to everyday phenomenon.

An effective science teacher needs to be broadly educated in science in order to teach a coordinated science curriculum that emphasizes the major themes and concepts of the biological sciences, chemistry, geosciences, and physics, as reflected in the *Science Framework for California Public Schools, Kindergarten Through Grade Twelve.*

While the majority of the standards address scientific content or ways of thinking, only one deals explicitly with instructional strategies. This standard reads: "The program exposes students to a variety of appropriate methods and strategies for teaching, learning, and assessing science" (CTC 1992, p. 29). The rationale refers to the diversity of "learning styles" and the importance of employing a variety of instructional models.

The new credential, it is hoped, will help move the undergraduate programs toward integration of the sciences. Collapsing the two credentials into one has sent a very important message that the developers hope will permeate the system: if the panel members can agree that they have more in common than not, then many others throughout the science education community will do the same.

George Miller, an advisory panel member who has reviewed program proposals, reported that only a few of the proposals for new credential programs were approved. When asked what the main problems are for universities, he answers that many science faculty members felt that they had to develop and teach new courses, which was not the intention. Rather, the advisory panels hoped that the new standards would provide an occasion for science faculty members to come together to discuss how their existing courses could be modified to meet the new criteria. Further, most students don't decide they want to teach until they near the end of their bachelor's degree, or even when they are finished with their undergraduate education. Miller claims that "to assume that [people] enter the university knowing they want to be teachers is pretty unrealistic."

Credential Supplements. What about teachers currently in the schools? How are they to be brought up to speed and maintain a current and legal credential? Currently, there are two hubs dealing with this issue directly. One hub coordinator is working with the local university campuses (both University of California and California State) to structure some of the hub meetings such that the teachers who attend them earn staff development units that count toward a credential supplement. The hub coordinator assesses the needs of the teachers and offers content sessions in those areas. Another hub coordinator has set up a partnership with the local university that is offering in-service training specifically geared to the needs of the teachers within the hub. These are not hub meetings, but in-service workshops or courses that directly address those content areas that the teachers need to supplement their credentials.

Kota reports that this is an issue about which that teachers are very concerned. She believes it is an important and "positive" problem, as it is moving the role of the hubs from basically a networking function to more formalized staff development connected to and supported by higher education. There are many other hubs that are beginning to reach out to the universities to help experienced teachers get the kind of staff development they need to teach the new courses they are developing.

University of California Approval for SS&C Courses

As schools began moving toward developing integrated or coordinated science programs, the immediate question became: what textbook will we use? This question is by no means trivial. In the United States virtually all science courses are defined by a textbook; if there is no textbook for integrated science, what college or university would accept a student who had taken such a course? As more schools changed their programs, it became clear that to gain approval from the state university systems, they needed to identify a textbook (or some other form of instructional materials) and, because the new books and curriculum were unknown to those in higher education, be very specific about outcomes and expectations.

Peter Wilding, a hub coordinator and writer for Cambridge University Press,[4] was asked to look into what kinds of books could be used for integrated curricula and the procedure for obtaining University of California approval. This process took Wilding to England where, he notes, there is a "reasonable selection of integrated/coordinated science texts." Many of the SS&C schools are now using these English texts as references in their classrooms. However, much of the curriculum is still developed by teachers drawing on a variety of sources.

Having a selection of texts, however, is only half the problem. What about university approval of courses? Referring to a University of California booklet, "Statement on Preparation in Natural Science Expected of Entering Freshmen." Wilding learned that schools must submit a report that describes in detail what the courses are and what the students will do. The key is to be explicit, so that the readers of the application will understand the depth and breadth of the course material.

Wilding's examination of the approval process for integrated science courses came at a time when the University of California was increasing the science requirement for incoming freshmean from one to two years. The major committee overseeing this process, the Board of Admissions and Relations With Schools (BOARS), consists of individual representatives from each University of California campus. BOARS members agreed that the two required years should be in different science disciplines, i.e., it would not be acceptable for a student to enter the university with two years of biology or two years of chemistry.

Wilding obtained BOARS approval for his integrated courses. Since then, many schools have become interested in doing the same. The SS&C head office compiled materials for those schools to help them initiate the process and obtain approval for their new courses. (See appendix D.) Schools can receive approval for any course for grades 9-12 in two ways: as college preparatory electives or as laboratory science courses, each one representing a different requirement for entrance into a California university. The distinction is made in terms of whether they provide core experiences in laboratory science content or are either a more introductory or advanced course. Most schools stop restructuring their curriculum at the 10th grade. Currently, however, schools are being encouraged to take a serious look at developing a third-year course. As Miller explains, if schools don't offer a third-year integrated course, then "all they're really doing is getting students ready for the old courses, and [BOARS] is not very impressed with that [kind of] program, [and] they'll get negative messages [from BOARS]."

---

[4]One of Wilder's Cambridge University Press assignments was to "Americanize" the British textbook *Balance Science*, a popular reference work used by many SS&C programs.

While obtaining University of California approval was an important step for the schools involved in the reform, there was still the question of private, out-of-state universities. Both University of California and private school approval must be obtained on a case-by-case basis. Currently, Wilding has letters of confirmation for the integrated science courses from such schools as Stanford University, Harvard University, Yale University, Cornell University, and the University of Michigan. None of his requests has been rejected.

University approval for SS&C courses has been a major boost to the reform. It also illustrates an important way in which higher education institutions in the state have moved to accommodate new K-12 programs.

The California Science Project

The California Science Project is a K-14 professional development program which builds collaborations among classroom teachers, science educators, and scientists. Its main focus is on producing a scientifically literate populace, developing effective teacher leaders, and addressing equity issues in science education. CSP is a permanent, statewide education initiative administered and funded by the University of California Office of the President in conjunction with California State University and the California Department of Education. The stability and institutional context of CSP's funding source allows the project to play a major role in current system reform efforts.

CSP was established by the legislature in 1987 and was then funded by the University of California in 1989. About $250,000 a year comes from the University of California system, and CSP receives about $1.7 million from state legislated funds for professional development. Since its inception, about 1,500 teachers have participated in CSP summer institutes sponsored by its 12 regional sites. While the majority of teacher participants are elementary and middle school teachers, a number of sites, including local community colleges, have developed effective K-14 articulation in their science education reform efforts. For example, along with other networks such as CSIN and SS&C, CSP site directors and leaders have played an important role in statewide efforts to develop performance-based assessment items aligned with the 1990 *Framework*. Activities such as assessment development have further encouraged collaborations among CSP, CSIN, and SS&C.

CSP was based in part on a model of professional development that had its origins in the Bay Area Writing Project and the California Math Project. Central to this model of professional development is the philosophy of "teachers teaching teachers" and immersing teachers in the discipline. There are now analogous projects in all subject fields.

Elizabeth Stage, the first CSP director and previous director of the Bay Area Math Project, played a central role in shaping CSP in its early years. She reflects on the philosophy of the subject matter projects. "So you have to set up a context in which the insights from research and insights from practice can be of equal value. And that philosophy has been propagated through all the subject matter projects."

CSP was established in the years surrounding the development and adoption of the 1990 *Framework*. Stage describes the relationship between the *Framework* and CSP.

> So the deal is that the *Framework* should be a focus of the work of the teachers in the California Science Project, and it could well be with the purpose of disagreeing with it. That's not a problem. The deal is for people to know what this document says as a statement of the profession, blessed by the state Board of Education, and to contend with it.

Individual teachers are selected to participate in CSP who have already demonstrated commitment and expertise in some aspect of teaching. When they join CSP, there is the expectation that they will use these skills and their CSP experiences to help them assume greater leadership roles at their school, district, and state. As the directors of the CSP site in Sacramento explain, "CSP is not just about being better for Monday morning." However, these greater leadership expectations can sometimes be overwhelming for teachers when they return to their schools.

From Common Philosophy to Regional Approach. While the 12 CSP sites all share a common mission of developing "scientific literacy" and are committed to hands-on, inquiry-based instruction, these goals translate into a unique focus at each site. This decentralized approach has advantages and drawbacks. While the flexibility has allowed sites to develop project foci adapted to the needs of the teachers and region they serve, the lack of cohesion and inconsistencies within the project have allowed some sites to flounder. In addition, rather than requiring a schoolwide commitment as does CSIN, CSP's professional development programs are geared toward individual teachers from schools or districts. This approach allows teachers to tailor programs to fit their individual needs—but then they are often on their own to implement changes back at the school site.

The codirector of the Sacramento site reflects on the variability: "I think there are a lot of things that different CSP sites talk about. We use the same language, but that means different things. I think there are different things going on in every site, and I think that is part of the charm of the subject matter projects."

For example, the LIFE Project (LIFE is CSP's environmental education site) in Humboldt County in Northern California has focused primarily on environmental education. The CSP site in Orange County, housed at the University of California at Irvine, has developed a countywide, district-based approach to professional development.

Currently, CSP of Orange County—in collaboration with CSIN and the Orange County community colleges and universities—is working to establish K-6 professional development activities for each of the 28 school districts within the county. This is quite a challenge in a county of 10,000 elementary school teachers where the students speak 40 different languages. Surveys of curriculum coordinators for Orange County K-6 schools indicate that existing statewide initiatives such as CSP and CSIN have reached only about 15 percent of the county's K-6 teachers. In conjunction with local teacher networks, colleges, and universities, CSP of Orange County hopes to target a "second tier" of teachers who have not yet had access to current science education reform programs. CSP's goal is to strengthen teachers' understanding and ability to use district-adopted science materials, increase the number of elementary school teachers who can act as science mentors, and develop science courses and resource centers at the participating community colleges specifically designed for K-6 teachers.

CSP's decentralized approach, and its initial commitment to serving populations traditionally underrepresented in science, has provided a number of CSP sites with the opportunity to develop approaches to addressing the needs of teachers and students in ethnically and linguistically diverse classrooms. While these issues are by no means addressed consistently throughout CSP, the Sacramento site and the Center for Language and Status Issues in Science have progressed beyond merely examining demographics to make significant efforts to engage teachers in discussions about issues of language and status in the science classroom. The equity issues embedded in the approaches to professional development at these two sites are examined later in this chapter; here we describe CSP of Sacramento in some detail to provide one example of the distinctive nature of the CSP sites.

The CSP of Sacramento. In 1991, the University of California at Davis, and CSU-Sacramento established a CSP site to serve schools of the Sacramento area. The site selects teachers who have demonstrated expertise in working with diverse student populations. Thus, professional development activities are focused on the needs of science teachers with multilingual and limited English proficiency students, and explores in depth the application of constructivist teaching strategies. The site is codirected by Wendell Potter, professor of physics at Davis; and former classroom teacher and educational researcher Pam Castori.

The central component of professional development occurs at the four-week intensive summer institute, which is a combination of science content and teaching strategies. During the school year, teachers lead and organize many activities, such as Saturday workshops and teacher research groups. In 1994, CSP of Sacramento received support from the California Department of Education to develop a leadership team of 15 teachers in science assessment. These teachers created a resource manual of effective performance-based assessment tasks and portfolios that they had developed in their own classrooms. The team is now available to work with a school or district to conduct staff development. CSP of Sacramento encourages teachers to determine what their needs are in their science classrooms and provides a forum for examining their questions in depth.

Castori describes the approach to scientific content used in the summer institutes:

> Teachers get immersed in an ongoing scientific experience that's focused on one or more big ideas in science. The goal is to involve teachers in an experience in a classroom setting in science that will be as close to what we would hope would happen in their classroom . . . We believe that if teachers haven't experienced a science instructional setting or science experience that is hands-on, minds-on constructivist, then it's very difficult for them to teach that way in their classroom.

Not only do scientists play an important role in providing content expertise, but also their personal commitment to K-12 education can send an important message to teachers. Teachers have the opportunity to interact and collaborate with a scientist and examine their views of what it means to "do science," or "be a scientist," or "conduct a scientific investigation." Castor notes:

> Having a scientist who is struggling—and is very upfront with the fact that he is struggling—with how to make science accessible in real, conceptual ways is the best possible model you could have. I think for a lot of [teachers], scientists are these removed folks. Science for them is removed, and so here you've got one of those removed people who's decided that science is important, but so is the enterprise of teaching.

A first grade teacher describes how the CSP Sacramento summer institute was based entirely on modeling constructivist teaching strategies. She relates how the translation of constructivism from the summer institute to the classroom was a challenging and, at times, frustrating experience:

> When you come back and you think about what constructivist science looks like at the primary grade, that is a whole different picture. I went through a phase of feeling angry, because I had spent a whole summer doing this. I came back and I was hit with the realities: that number one, I have a first grade class that barely knows how to talk to each other and, number two, 50 percent of my children were bilingual. Over half of them did not even speak English. So it was how do you share ideas, which is what constructivist science is all about, sharing ideas and building on your own ideas and other people's. So I spent a whole entire year working on, "What does constructivist science really look like in a bilingual first grade classroom?"

The Influence of CSP on Higher Education. As a result of the collaborations between university faculty and K-12 school teachers, a distinctive dynamic has emerged. While classroom teachers have access to valuable content expertise, university faculty have reported that, as a result of their interactions with elementary school teachers and their involvement in the current reform effort, their views of teaching and learning have been altered significantly .

Elizabeth Stage comments on some of these effects of the K-12 reforms in science education:

> One thing that I think in the long run is very important is the reform of undergraduate education itself—there's been a big push to get university faculty to change admissions requirements. Well, guess what's happened? The faculty

say, "What's happening here?" Some of them actually learn from what's going on and get involved. So it's very long term. We're talking a decade. There's a chemistry professor involved in the California Science Project, and the first couple of years he is saying, "Why are you talking to me about equity and stuff like that? I want to have better prepared freshmen." Well now he says, "I'm really glad I got involved in the California Science Project, because originally I thought I was going to help those poor teachers of chemistry, but I learned so much *from them* about being a more effective teacher."

Wendell Potter, codirector of CSP-Sacramento, reflects on how university professors view their role in professional development and how they have been affected by their involvement in CSP:

> I think most faculty have the sense that the problem at the lower grades centers around content . . . Give them the right content so they don't screw up the kids. So they're totally off in terms of what the problem is, but they sense there is a problem, and they are interested in getting involved. A lot of them then start changing pretty quickly once they do start getting involved.

To encourage these sorts of changes among university scientists, the CSP-Sacramento site has begun sponsoring seminars on undergraduate science reform for a group of professors at the University of California at Davis. Sacramento site teachers have also developed an Adopt-a-Scientist program which establishes individual collaborations between teachers and scientists. Pam Castori describes the nature of this relationship: "This program has turned around several scientists. It's really changed their conception of what the problem is and of what could be done, the way they start thinking about their teaching, and the changes they start making here in their own courses."

◆

Assessment

A truism in education is, "What you test is what you get."

The design of an assessment program requires as much care and consideration as the design of the instructional program itself. — California Department of Eduction 1990, p. 176

One of the central principles of systemic reform is that improvement of assessment methods must both accompany and be compatible with any curricular reform if change is to last. It is not enough solely to change teacher conceptions of science teaching; the other elements that influence classroom teaching and learning must also be changed.

Efforts to reform assessment in education are usually directed toward gauging what students have learned. If standardized assessment instruments measure only the amount of scientific information students retain, as opposed to the level of critical thinking they have developed or the depth of their understanding of scientific concepts, then teachers and administrators will have difficulty convincing their community that changes aimed at developing these skills are in the interest of their children. In turn, reform leaders will have a harder time convincing teachers and administrators to participate in the reform.

A notable feature of the California science education initiative is that its vision of assessment has been applied to many levels of the reform. Essentially, the underlying conviction is simple: assessment should be an occasion for learning. The various ways in which this philosophy has been applied in the classroom, at the school site, and in the formal evaluations of the reform programs are explored in this section.

Student Assessment: California Learning Assessment System

California law requires that statewide assessment of student learning be aligned with the curriculum frameworks. The California Learning Assessment System (formerly the California Assessment Program) was intended to meet this mandate. Beginning in 1989, CLAS developed, piloted, and field tested assessment items including performance tasks, enhanced multiple-choice items, open-ended and justified multiple-choice questions, essays, and portfolios for four subject areas: English/language arts, mathematics, history/social science, and science. Designed for grades 5, 8, and 10, the science CLAS tests were to focus on big ideas instead of facts, the integration of science content and process, and conceptual understanding. Among the design principles behind the CLAS science tests were:

1. The assessment should model good instruction.

2. Assessment should be coordinated with the curriculum and viewed as an instructional task rather than as an isolated opportunity for evaluation.

3. Assessments will have no single prescribed answer, but will allow for a variety of appropriate responses.

4. Teachers will be heavily involved in the development, piloting, field testing, administration, and scoring of the CLAS tests.

CLAS's performance-based assessment tasks were organized around 40 big ideas based on scientific concepts outlined in the 1990 *Framework*. For example, one big idea in earth science is, "Geological processes explain the evolution of the earth." A K-5 concept that supports this big idea is: "Two processes account for the change in the earth's geologic features over time: one process is the building of surface features by energy released from inside the earth . . . The other process is wearing down, or erosion" (CLAS 1994).

Fifth grade students might be asked to perform the following task to assess their comprehension of the scientific concepts: "Design a model river system and observe the movement of earth materials and the change of the surface.

Compare and explain the differentiation of particle size with a sample set of sedimentary rock, from fine to 'storm deposited'" (CLAS 1994).

In the spring of 1994, a performance-based assessment was administered to fifth graders throughout the state. Like the example here, each performance task on the 1994 Grade Five Science Assessment included one or more of the following skills: generating and organizing data, using and discussing data, using scientific concepts to explain data and conclusions, and applying science beyond the immediate task (CLAS 1994).

The 1994 assessment tasks were scored by classroom teachers, and their observations were compiled and used as an opportunity to educate other teachers. More than 300 teachers were involved in the development and initial trials of the science assessments; many additional teachers volunteered to participate in the scoring and analysis of the 1994 assessments. (It is worth noting that the scoring of the science CLAS assessments was never funded; in contrast, in other content areas, teachers were paid for their efforts. The extent of participation illustrates the degree of teacher commitment to sustaining the science CLAS program.) This broad level of participation was intended to inform the tests with teachers' practical insight. It was also intended as a valuable professional development experience for the teachers. Most importantly, it was intended to cultivate a base of support for the new approach to assessment among science teachers.

The CLAS initiative floundered, however, but not because teacher support for the new assessment instruments was lacking. Rather, the CLAS science tests were never fully funded. In the fall of 1994, funding was withdrawn from all CLAS tests indefinitely by order of the governor following a controversy about the content of the language arts CLAS test. While this was not a direct reflection on the science assessments themselves, several state leaders anticipated public resistance to the science tests. In retrospect, many leaders faulted themselves for not putting more effort into public relations for the science CLAS tests and CLAS in general. Sachse reflects:

> I think we've done a rather poor job of communicating what we're doing in science to a much broader community. I think Kathy Comfort did a super job getting the word out to the science community as much as she did. Thousands of people volunteering, paying their own money, getting involved. It's a huge effort. But as a consequence of that, I don't think we worked with other discipline areas as much as we might have, and we didn't work as much with the professional, political, policymaking groups.

In the absence of the CLAS tests, the elementary science reform effort was left without an assessment instrument aligned with its curricular philosophy. This has been considered a significant loss by elementary science reform leaders, especially within CSIN; CLAS provided an important motivating element for districts to examine their elementary school science programs. Unlike secondary schools, where science is an established part of the core curriculum, elementary schools typically treat science as a peripheral subject.

The suspension of CLAS has been treated as a temporary setback, however. The general sentiment among state-level leaders seems to be that some form of statewide performance-based science assessment is inevitable in California, whether it is called CLAS or not. Under the auspices of a new program initiated in fall 1994, leaders from many of California's science education networks (CSIN, SS&C, SPAN, and CSP) formed a new assessment task force called Science Teachers and Educators Leading Assessment Reform (STELAR). The goal of this group is to follow the "well-placed footsteps of CLAS" and develop K-12 performance-based tasks, teacher manuals, and portfolio criteria.

In addition to STELAR, a group that includes many of the same players but also the larger State Systemic Initiative and Urban Systemic Initiative projects has formed to devise an assessment implementation plan for grades 5, 8, and 10. This project is not yet funded and will have to rely on local district and project money (as opposed to state money). DiRanna reports that this new approach is a "true change in the system," because "people who don't typically make policy are beginning to shape it."

Teacher involvement in assessment development has been facilitated by the close relationship among assessment and network leaders across the state. Teachers who have been involved in the development and scoring of performance-based assessments report that this process has encouraged them to look at their own students' work in new ways. Stage explains:

> CSP, CSIN, [and] SS&C introduced many, many teachers to the idea of performance assessment in a very superficial, very much awareness level, but the responses from the teachers were very positive because it was something that they were asked, "Would you like to try this? . . . Try it, make sense of it. What do you think of your kids' responses?" And it just opened their eyes all over the place and they saw that different kids are doing well. "Am I leaving some kids out? Am I thinking some kids understand when they don't really, and other kids I'm thinking don't understand really do? Maybe I ought to take a second look.

Student Assessment: The Golden State Exam

The Golden State Exam is a voluntary subject-specific performance assessment for secondary students established by the Educational Reform Act of 1983 and reauthorized in 1991. Beginning with mathematics, there are now GSEs available in history, economics, biology, chemistry, and coordinated science. The biology and chemistry examinations began in 1989, and in 1992 the coordinated science examination was piloted. Today, there are over 120,000 students taking the examination. (Appendix E depicts the GSE science program focus.)

While the official purpose of GSE is to "identify and recognize students who demonstrate outstanding performance in the subject matter" (CLAS 1994), Meg Martin, director of the GSE science component, reports that the examination is equally important as a professional development tool. She contends that, as was true with CLAS, teachers' involvement in all levels of the assessment has contributed greatly to their interest and competence in making changes in their own teaching.

> It's important to show best practice for teachers, to give them a sense of process, and how they can go about implementing the reforms. Teachers are involved in all levels of the GSE, from development to scoring. The new component, the portfolio, allows the teachers to get a sense of the growth of students.

Martin was concerned that people would rely on assessments, such as GSE, to drive the reform. She prefers to view these examinations differently: "I see [the new assessments] more like a carrot rather than a stick. I can see how it has been used and can be used as a stick, but I like to think of it as an example of best practice and as a model of the process."

GSE is actually part of the now-defunct CLAS, but still has funding because it is voluntary. In fact, the governor has written GSE in as a special line item in the state's budget. Martin believes that the voluntary nature of the program is important in that it "fits with the grassroots nature of the reform" that is, that teachers both develop the examination and decide whether they want to implement it.

Significance of GSE. How can a voluntary examination carry any weight? The answer to this question changes depending on whether it is asked in the pre- or post-CLAS era. When CLAS was in place, GSE played a minor role in the overall state assessment scheme. However, since the demise of CLAS, state science assessment leaders believe that GSE "looms large for the reform: it's the only game in town." Tom Hinojosa, a former teacher, hub coordinator, and now assessment specialist in the SS&C project, reports that GSE is especially important today, particularly from the teachers' and administrators' perspectives.

> From the teachers' point of view, [GSE] lends credibility to their style of teaching. It gives them guidance as to the rigor and depth of the curriculum. As teachers involved with innovative things, they need some validation of their program. Along with University of California approval, [GSE] validates a lot of these programs. For administrators and their relationship to the community, they can point to [GSE] so they know that the new programs are not a wacko local thing they're doing.

While not high risk, Hinojosa described the GSE as having a great deal of "emotional" impact on the schools and teachers involved in the reform. This, he believes, is crucial to the ongoing support of alternative assessment and to the sustenance of the integrated and coordinated programs at the high school level.

School Site Assessment

At the elementary level, California's primary site-based assessment system is the Program Quality Review (PQR). Throughout the 1980s, PQRs focused on instructional practices and compared them to exemplars describing teacher behavior and curricular organization. State leaders eventually found this system lacking because it did not directly effect changes in how and what students learn. In the 1990s, the PQR process was redesigned to be compatible with the thinking- and meaning-centered curriculum of all the subject matter frameworks.

The new PQRs follow many of the same design principles as the CLAS tests. They focus directly on student work and on one curricular area per year, allowing the school to conduct a more in-depth review and analysis of its program. This process allows the site assessment to be an occasion for learning. The review is conducted by a site-based leadership team comprising a principal, teachers, students, and parents in collaboration with external reviewers. A team member notes that "It is a very teacher-centered process with much more talking about what strategies we are going to use to get where we want to be. Change now comes from within."

The emphasis in PQR is on review from within rather than on external evaluation. This makes the process less threatening and often more productive for teachers. It increases teachers' investment in the process of program improvement and is a more empowering experience for the schools. Because the PQR review manual mentions CSIN as a means of bringing a school into alignment with the 1990 *Framework* goals, the PQR process has been instrumental in motivating school districts to participate in the network.

A similar site evaluation system is in place at the secondary level, the Western Association of Schools and Colleges Accreditation (WASC) review. Again, the design of WASC reviews seeks to maximize the opportunity for learning by school faculty and to lower the stakes for those being assessed by involving them centrally in the evaluation process.

The WASC review process has three distinct phases: the self-study, the on-site visit, and the follow-up. About a year before the date of the on-site visit, schools send a representative for a training session. The school receives an information package from WASC; and this representative helps to form various committees within the school, such as the subject matter departments, a parent committee, and a student committee, to begin the review process. Each committee is assigned specific questions regarding the functioning of the school. Science departments are asked to explore the extent to which their programs are aligned with the *Framework*, and if not, what their plans are for doing so. This process is not meant to admonish those schools that have not yet implemented new programs, but rather to foster conversations around meaning and message.

Following this, the school is visited by a committee made up of administrators, teachers, a university representative, and—sometimes—a student. The school people on the committee always come from a different district. This committee examines the self-study materials and spends approximately two days in the school, visiting every teacher at least once. At the end of the observation period, subsets of the visiting committee meet with each school committee (or department) and assist them in identifying areas that are particularly strong or that need improvement. As one committee member describes it:

The emphasis is validating the self-study. Have they been fair to themselves? It isn't the main focus, but we are encouraged to find things that need improvement at the school. We're looking a lot at process as much as the reality in the sense that they may not be doing something very well, but why don't they have a process that's catching that and improving that?

This accreditation process is an example of the state's commitment to sanctioning practitioners and their knowledge about what they do and what they need. According to the SS&C program assessment guide, "This process enables schools to reflect upon and respond to essential questions: What do we want our students to know and be able to do? What are the learning experiences needed to produce these outcomes?" (California Department of Education 1991).

Evaluation of the Reform Programs

There is no independent external evaluation of the overall reform. However, individual programs are formally evaluated. The design of most of these individual programmatic evaluations was not entirely in the hands of the systemic reform leaders. This, however, has not prevented state leaders from finding the "traditional" present system lacking and applying the new vision of assessment to evaluation of the reform.

For example, student performance on CLAS and GSE tests has been examined as a means of assessing the effects of the reform. In fact, Tom Sachse has commented that it might be useful if CLAS test results were used to assess the reform effort. However, NSF (which provides, among other things, the SSI funds that support CSIN) is more interested in outcomes like, "How many algebra and science courses are students taking, and is this number increasing?" Sachse typically responds by saying "I hope they are taking less algebra courses in four years." He reports that his interlocutors often respond, with surprise, "Less? We were expecting more!" to which he responds, "I would hope by that time they are taking more integrated courses like Math Renaissance." Sachse says he sees such conversations as opportunities to educate external evaluators.

Numbers and types of courses as indicators of reform progress are not the only aspects of programmatic assessment that have been critiqued by the state reform leaders. The assessment of teacher professional development programs has also been discussed. Kathy DiRanna provides an example:

The first piece of evaluation that we had to turn back was a checkpoint list. It asked "How many hours of in-service [training] are you doing?" and there was a place for 3 hours, 5 hours, 6 hours, 10, and over 15. Those were the categories. That immediately says to me they don't know what they're talking about where systemic change is concerned, because three hours shouldn't have been a choice. The minimum should have been 15 hours and increasing. As it was, we were in the oddball category when we had to say we had more hours involved. So, when we talk about "How does this play out?" we are faced with the challenge of how far on the continuum of the change line you are, where are you in relationship to all the other players, and what pressure can you bring to bear on the system because you've got people behind you moving in that same direction.

As emphasized by DiRanna, many state leaders envision a lengthy chain of events that leads to improved science learning. This process begins with changing teachers' conception of science teaching, cultivating leadership among teachers, and then encouraging them to seek instructional materials aligned with their changing conceptions of teaching science. Teachers will need time to critique the new materials and ensure that student assessments are aligned with the changes they are making. Changing teacher practice will require time in the classroom and time to collaborate with other teachers. It will take time to get a majority of teachers in the state involved in such a process. Once this practice has begun on a wide scale in the state, it will then take time for students to become used to changes in teachers' pedagogy and science curricula. Some teachers have suggested it will take a generation before the positive results of this reform effort will begin to show up in standardized student assessments.

Therefore, it is the *time scale* on which reforms are evaluated that seems to be a determining element. DiRanna explains:

> I think we are changing what assessment is all about. We're changing how you deal with teachers. We're changing all of those things. It just doesn't even sound like good science to sit down and say, "Because you did this, you're going to have better student conceptual understanding in a five-year period of time." I think that's an invalid premise to even start with. And, the fact that we're having to face that says to me somebody didn't sit back and say, "If you change a system, what are the appropriate indicators that say the system is changing?" . . . I guess what is disheartening to me is the NSF leadership has tried very hard to promote systemic reform, but it's catching itself in its own evaluation trap, and I think that's why you end up with a lot of the states doing some of the superficial stuff that they're doing right now, because it is difficult to show short-term results [NSF] recognizes.

One of the most frequently offered options for an interim programmatic assessment is a focus on changes in teacher behavior. Again, Kathy DiRanna elaborates:

> I don't think student data, in the first go-round, is the appropriate measure. For CSIN, that's the end result that we need to see. But in order to get that end result, we've got to worry in the meantime about the evaluation of where are we with the teachers in the movement who will eventually cause that change to happen? I just don't know how you can expect, simply because somebody has been in CSIN for a year, that their students will now conceptually understand science.

Just as student assessment instruments are thought to drive teaching practice—sometimes in unconstructive ways—evaluation of the reform programs is perceived to drive many reform practices in ways that detract from the process of change. For example, one state leader with expertise in the area of assessment describes the way in which external evaluations can distract from the goals of the program itself: "It's like you create a cardboard cut-out of yourself for them, but try not to spend too much time doing that so that you can do the real work that you know needs to be done."

External evaluations of reform programs can also have critical consequences on the availability of program funding. Tom Sachse and others have observed that what has been achieved thus far in California would not have been possible without federal support through programs like SSI, Urban Systemic Instiative (USI), Eisenhower funds, and the SS&C grants. Sachse has expressed concern that if short-term federal priorities (documentation of student learning available in a short period of time, minutes per day in science instruction, and other "bean counting" exercises) continue to drive the programmatic assessment of California reform efforts, these assessments will be used to justify the withdrawal of funding from the California systemic reform, thereby "unraveling all the real progress that has been made over the last decade," Sachse notes.

Recently, NSF has started to acknowledge that educational change takes time. It has begun to devise interim benchmarks and to examine a broader selection of evaluation criteria; for example, changes in courses provided and parents' impressions of programs. Nonetheless, NSF is caught in a tension between this more contextual approach to evaluation and the reality of needing to produce concrete results for Congress.

Classroom teachers share concerns similar to the ones expressed by state-level reform leaders. One SS&C high school teacher describes how evaluation and funding cycles can actually interfere or be out of step with the process of change.

> I think it's really hard to work under the funding running out after two years unless you produce this dramatic change, because then you're never able to really work on the change mechanism itself and the process of preparing people and getting them involved in really thinking about the process. A lot of our schools started off with a bang, kind of died, and now they're becoming really motivated again . . . it's a barrier that you hit and then you have to regroup again and go again.

A CSIN teaching consultant involved with Urban Systemic Intiative (USI) expresses many of the same concerns:

> They want you to make these great quantum leaps, but the evaluation is not designed for appropriate benchmarks. You can make quantum leaps, but that takes time. If they would design an evaluation to be realistic with the sequential steps that have to be addressed to achieve this final goal, then I think it can be a very positive thing, but so frequently I'm concerned about that . . . They want

us at point A at the end of year one. Is that an appropriate benchmark? . . . Are they knowledgeable enough about our situation to say that's an appropriate benchmark for that situation? That's a concern because the evaluation determines whether or not you're funded. So it's very critical.

There are styles of evaluation that can encourage innovators to be more reflective, and there are styles that can inhibit the progress of the reform. These approaches have to be examined somehow. Perhaps an ideal method has yet to be invented. A method that strikes a balance between accountability to national concerns and funding agencies, on the one hand, and sensitivity to what the innovators are trying to accomplish, on the other, would better serve the California science education reform programs.

---------------------------------- ◆ ----------------------------------

Equity

The demographic trend of the California school population is on a collision course with the scientific illiteracy rate. — California Deparment of Education 1990, p. 167

Equity issues have been a focus of the rhetoric of the California science education reform since 1990. Slogans such as "Every student, every science, every year"; "Science for all Americans"; "Scientific literacy for all students" saturate the reform's mission statements. By the description of many of the reform leaders, however, this rhetoric has not translated into broad successful action.

Efforts to promote educational equity in California have faced a number of challenges, one of the most prominent being the changing demographics of the state noted earlier in this report. Furthermore, addressing equity in science education raises the additional problem of treating science teaching like most treat scientific knowledge itself—as something that transcends culture, gender, and class. This section examines the ways in which equity has been addressed at various levels of the California science education reform—in the classroom, in the 1990 *Framework*, and in approaches to professional development.

One approach to addressing issues of educational equity is to centralize efforts. Centralization concentrates resources, heightens visibility, and increases accountability to external evaluators. However, this approach also risks separating, and thus isolating, efforts to promote educational equity from the overall development of the reform. This risk of isolation is of special concern in science education reforms, since the California strategy has had so many decentralized features.

Promoting Science Education Equity in California

Many California science educators conceived of equity and science education as two separate issues. Some were of the opinion that redressing issues of inequity and social injustice were beyond the scope of a subject matter reform movement. Generically promoting hands-on, minds-on constructivist teaching methods on a wide scale was thought by these people to be difficult enough without the added burden of discussing how such pedagogy might be tailored to the needs of specific groups of students. Kathy DiRanna, one of the authors of the *Framework*, and the executive director of CSIN, confirms the fact that state leaders conceptually separated the two issues, at least early in the reform: "The equity issues were not a priority in the beginning of CSIN. In the beginning we were fighting just to get good science established as a core part of the elementary school. That was really our focus and target."

This opinion was not shared by every person involved in building the reform effort. Elizabeth Stage, chairperson of the *Framework* committee and widely know as a strong advocate for equity issues, points with pride to five pages in the *Framework* she personally lobbied to have included. Entitled "Teaching All Students," this section includes advice about how to better serve women, minority students, and persons with disabilities in science classes. The section urges schools to have all teachers, including science teachers, attend to the special needs of students with limited English proficiency (California Department of Education 1990, pp. 167-71).

While the inclusion of such a section was a significant departure from previous frameworks, which had not addressed equity as explicitly, the section gives the impression that attention to equity was an afterthought to the mission statement of the document. This might indicate a similar attitude toward issues of educational equity held by science educators and reformers within the state. Elizabeth Stage would likely agree. In a more global assessment of efforts to foster collaboration between science education reformers and educational equity advocates, Stage says, "I think that's our weak suit." In a similar assessment Maria Lopez-Freeman, codirector of the California Science Project's Language and Status Leadership Institute, states, "Equity issues are the Achilles' heel of this reform effort."

The widespread conception of science subject matter as something that transcends culture, gender, and class seems to contribute to the reluctance many science educators have demonstrated toward the idea of integrating equity into their science pedagogy reforms. Since scientific knowledge is believed to be invariant across cultures and personal characteristics, attempts at tailoring science education to the culture of students is often perceived as "watering down" the curriculum. Even some of the strongest advocates of educational equity greet the notion of culturally tailoring science pedagogy with a great deal of suspicion. Such efforts can be seen as "an excuse for lowering the expectations for our students."

Since over 100 different ethnicities and linguistic groups are represented in California classrooms, asking teachers to learn about each cultural group's prior knowledge or learning styles is often thought to be unreasonable. The most common default position among science educators who acknowledge the need to promote equity through pedagogical means is to assume that training in constructivist teaching methods will suffice. The hope is that teachers will be able to fill in the cultural specifics.

Examples of Integrating Science Education and Equity Issues

Despite this pattern of separating science education and equity issues, handfuls of dedicated people have been working in California, some for many years before the current reform, specifically to improve the science education received by women, students of color, and language minorities. Some of these efforts have remained at the district or even the individual classroom level. The broader efforts, however, have operated almost entirely at the level of in-service teacher education. Programs that directly address the integration of equity issues and science pedagogy have been sponsored by organizations such as CSP, the Lawrence Hall of Science EQUALS Project, and the University of California Postgraduate Education Fund (which is directed by educational equity advocate Lynda Barton-White).

For example, CSP addressed equity issues in its charter. Elizabeth Stage, who founded CSP, explains:

> We were able to craft a niche for ourselves when we put out our request for proposals and said this should be based on the *Framework* . . . We also said . . . "It's supposed to be for *all* students. But if you can't do all students and you're going to take a slice of that from the start, work with the least advantaged populations." We explicitly said, "We're going to keep track of your demographics and compare the students served by your teachers and your teachers to the demographics in the counties that you're serving, and you'd better be doing better than average. You better have an overrepresentation of people who are currently underrepresented.

Of the 12 CSP sites, the CSP-Sacramento site and the CSP Language and Status Leadership Institute are most often cited as exemplars in their efforts to integrate attention to science content and to equity issues. The Sacramento site deals primarily with teachers who work with language and cultural minority students. The site's codirector, Wendell Potter, describes the vision: "All of us felt it made sense from both our own experience and theoretical grounds to focus on and combine the science emphasis with language issues. So we started explicitly with that as a core purpose of the CSP-Sacramento program."

Perhaps establishing a climate that facilitates some of the more personal and emotional discussions of equity also encourages risk taking in learning science. In the following dialogue, Pam Castori and Wendell Potter discuss the close relationship between equity issues and science education at their CSP site.

> Castori: Our project has . . . an emphasis [on], a commitment to, learning from the teachers in the project about where they come from, and what they think is important to do, and figuring out ways for them to learn from each other.

Potter: It really carries over into the science we do. I think the connection, doing the two together, is much more powerful, definitely, than doing either one separately. It really works together, the idea that we really do respect, and work with, and legitimate the things they're thinking.

Castori: In terms of the equity issues, what we really try to do is set a climate that says, "We are not about answers and solutions. We are about exposing questions and trying to sift through the possibilities and we are about respecting perspectives and backgrounds."

Potter: The teachers comment about how the same principle applies whether we are talking about science or equity issues.

Castori: And there are always really heated, uncomfortable, and difficult times, and they come about when deep-rooted assumptions and beliefs about kids, about language, about what America is all about . . . [come up].

In addition to the potentially effective results of integrating language, culture, and equity issues into science education, some educators also believe that bilingual teachers are trained in a philosophy of learning that has useful applications in science education. Maria Lopez-Freeman, codirector of CSP's Language and Status Leadership Institute, reflects on this idea:

> I do think that bilingual teachers, by their training, have been grounded in the cognitive views and philosophies of Piaget and Vygotsky. Thus, there is already an established and an explicit cognitive philosophy that is embraced and implemented by bilingual teachers that allows for incorporation of an inquiry-based learning in science as a logical extension and application of their cognitive frameworks.

In support of Lopez-Freeman's observation, Stage describes the enthusiasm she has seen among bilingual teachers discovering this resonance.

> You see those teachers who come in with zero signs of [science] background and zero signs of confidence, and by the time they leave they have a lot of confidence . . . They say, "You're saying the same thing I have heard about language development. You're saying that this is good science teaching. All we have to do is continue our effective inquiry techniques into the realm of science!" . . . Coming from a totally different direction, they have come to the same conclusion.

This is a an important insight. If it has substance, it deserves considerable study. Maria Lopez-Freeman counsels reserve, however. While encountering successes of the type described above with bilingual teachers in the first year of her center, she feels this success was limited in its significance and points out:

> Language discrimination issues are the easiest to deal with. They are the ones we accept guilt for most easily . . . ESL [English as a second language] teachers come with many of the conceptual tools needed for integrating concern for equity into their science teaching. What about all the other teachers for whom this is not the case?

The CSP Language and Status Leadership Institute developed a model that integrates attention both to the culture of English-speaking minority students and to science pedagogy. Lopez-Freeman's approach does not look for culturally specific learning styles, which she feels come dangerously close to negative stereotypes; nor does it encourage teachers to look solely at the students' culturally specific knowledge. Instead, she encourages participants to look at the communication required by constructivist and inquiry-based science teaching. She asks teachers to examine the cultural content both the teacher and the student bring to that communication. She also asks them to reflect on how these cultural elements of the interaction affect a teacher's ability to diagnose and respond to the student's learning experience.

Lopez-Freeman is quick to point out, however, that facilitating such reflection is anything but straightforward. "Context, context, context . . . " is her constant refrain. The context of the teachers' conversation, the students they are concerned with, what district they work in, who they are as people, and how they confront uncomfortable issues change at every institute the center sponsors. The 1995 Language and Status Leadership Institute summer institute was an apt example. Set in Compton, one of the Los Angeles communities most damaged by the 1993 riots, it served 25 teachers—12 from Compton and 13 from other CSP sites—and was intended to address the needs of African-American students in science classes.

As it turned out, Compton teachers were more concerned about the stigma Compton carried with visiting teachers as a result of the riots than with issues having to do with African-American student culture. Emotions ran high with regard to this issue and pre-empted many of the discussions about science pedagogy. It also turned out that certain Compton district policies had left most teachers with very little job security, and center directors felt this contributed to teachers' reticence to take a hard look at the quality of the science instruction they were able to give their students. In addition, the once predominantly African-American Compton district now has a swiftly growing Latino population, which has been an issue of growing concern for teachers. The center directors took this fact into consideration when planning the institute. In this light, Lopez-Freeman's insistence on the importance of context can seem understated.

Overall, the center directors believe they are doing important work and are dealing with real issues; but that their work is, as CSP teacher Norman Brooks put it, "only scratching the surface." Science education equity advocates almost uniformly insist that more needs to be done.

What Has Been Learned?

While a few groups such as CSP have made some promising forays in attempting to deal with equity issues, most science education reform advocates in California have not had positive experiences in their struggles with issues of diversity and equity. Perhaps the most accurate thing that can be said about

California science educators is that they seem to be learning what will not work to systematically promote science education equity.

California science educators appear to have gone through a series of developmental stages in dealing with equity issues. While these stages do not necessarily reflect linear growth, and there would certainly be some dispute about the content of these stages, they can be roughly described in the following way.

First, science educators began with the idea that improving science education and promoting equity were separate enterprises. However, the changing demographics of California's student population in recent years has made addressing issues of equity in the classroom unavoidable. Former state-level SS&C administrator and high school teacher Rick Rule describes how the situation has changed: "Well, it used to be that a science class would have 3 or 4 language or ethnic minority students in a class of 35. It's terrible to say, but then it was easy for some teachers to ignore these students if they were having trouble. Now it's 10, 20, or more. You can't ignore half of your class."

Second, a common approach adopted by science educators concerned with equity issues was to attend solely to the *numbers* of minority students and teachers participating in science classes and professional development programs. Such head counting often made the scope of inequity clearer to reformers, but did not offer constructive solutions. Reformers had difficulty conceptualizing what needed to be done to attract minority students and teachers into their classrooms besides opening the doors.

The ethnic composition of the student population has become increasingly heterogeneous in recent years. However, the state's teachers do not necessarily reflect this diversity, especially in science education. Professional development program directors often attribute their failure to attract minority participants to the lack of minority teachers willing to participate in their programs rather than to their program's culture and features.

Third, once the mere presence of underrepresented students in science classrooms or teacher development programs was recognized as inadequate for improving student learning outcomes, attempts were made to identify better teaching methods. Referred to in retrospect as the "magic bullet" approach to promoting equity in education, this involved identifying learning styles that characterized certain student minority groups and then matching teacher styles and strategies to the characteristics of these groups.

The California experience has shown that framing equity issues as a matter of learning styles makes it easy for equity concerns to get swallowed up in the panacea of "constructivism," thereby rendering equity concerns invisible. It is commonly claimed, if teachers understand and implement constructivism well, they will be equipped to teach science to all of their students equitably. This approach may have resulted in some substantive improvement in the education of linguistic minority students (as was seen in the CSP institutes). However, it seems to have done little for English-speaking cultural minority students. No matter how difficult it may be to acquire, it seems there is no ready substitute for specific understanding about a student's culture and a teacher's critical reflection on his or her own culture.

Finally, as consensus grew that any single teaching method will not be sufficient to ameliorate the inequities experienced by students in California science classes, science educators increasingly found themselves confronting the more personal aspects of the equity issues, particularly the emotional discomfort of white educators. Science educators have begun reluctantly to acknowledge that creating a classroom atmosphere that supports intellectual risk taking by minority students, facilitates the development of relationships with students whose culture is different than their own, and allows constructivist lessons to occur often requires that teachers confront personal and emotional aspects of their relation to teaching and to students.

CSIN leaders were shocked by the level of negative emotion in their first attempt to discuss these issues at their annual staff developer training in June 1992. CSIN, which has demonstrated sensitivity to teacher's comfort levels, acknowledged that the emotions around this issue are like no other change issue it has faced. CSP-Sacramento's leaders Wendell Potter and Pam Castori confirm this experience. Maria Lopez-Freeman reports it as well. The CSP institutes tend to attract teachers already committed to educational equity, yet CSP leaders still find themselves confronting strong emotions around equity issues among participating teachers. CSIN leaders, who deal with a much wider cross-section of teachers, have adopted a vocabulary of talking about "getting at people's belief systems." They further remark that they are not sure if anyone knows a way to facilitate the navigation of the deep emotional responses equity issues elicit.

What can be learned from California's efforts to promote science education equity? Perhaps the answer to this question lies not in the programmatic interventions California educators have devised, but in the growing awareness that equity issues are so context-dependent they may resist "franchise-style" interventions. This term, used by some CSP directors, refers to an approach to professional development that devises a model or program at one school or in one district and then seeks to apply that model to other educational situations with very little adjustment to differing contexts. The alternative would be to see programs develop within communities, refining their identity in response to that community's needs. Moreover, when addressing equity issues, people will not be able to avoid confronting very uncomfortable feelings. As Maria Lopez-Freeman notes, "there will be no magic bureaucratic substitute for commitment on the part of individuals to serving underserved communities."

◆

The *Framework* in the Field

I think what all the frameworks, and things of that sort, try to do is shine the light on best practice and take the shackles off the people who have been wanting to do the right thing. It doesn't move anybody . . . It gives permission to teachers to do the right thing. — Elizabeth Stage, Framework Committee Chair

While the *Framework* writers cite a variety of purposes for the eventual design and message of the 1990 *Framework*, the document is used mostly in two major ways: as a vehicle for science teachers to think through what they want to do once they decide to make changes and to legitimate what many teachers are already doing. There was little evidence that teachers examined the *Framework* and then decided to restructure their curriculum; the *Framework* did not initiate large-scale reform activity in the classroom. Rather, it helped to sustain and magnify reform-oriented work that was already occurring either by acting as a vehicle for conversation and deliberation or as a legitimator of existing practice. It also greatly influenced publishers. George Miller makes this assertion:

> In California, the *Framework* design was utterly immaterial in earlier versions and could have been in this one, if the climate for reform and implementation was not present. I think the design is/was more in line with the perceived climate . . . than that the design created the climate . . . The difference [between the 1990 *Framework* and past *Frameworks*] is in the effort to do something about it, supported by the *Framework*'s models of what could and should be done, rather than driven by the need to "comply" with the *Framework*.

As noted earlier, the *Framework* is a melange of visions, with each chapter written for a different audience. This resulted in an open-ended, elastic policy document, leaving considerable room for interpretation and elaboration. This feature makes it a valuable focus for teachers' deliberations about the means and ends of science education. For example, one of the ways flexibility was built into the *Framework* was in its interpretation of content. The *Framework* advocates a reorganization of the curriculum without mandating specifics. Bill Honig explains: "We wanted to bring widespread collective attention to the quality of teaching, to inform teacher training, assessment, etc. But we didn't think we could direct curriculum."

This open-ended approach to the curriculum made the *Framework* adaptable to local contexts. Moreover, it required teachers to reflect on what matters in science course content. As one hub coordinator puts it: "It's the thinking required that is important. It forces you to examine what you are teaching and why."

The Interpretation of Themes

One important outcome of the lack of clearly mandated curriculum, and one emphasized in this report, has been the interpretation of themes in the reform. The *Framework* states the following with respect to the importance of themes— essentially that they are on a par with scientific theories:

Themes are necessary for the teaching of science because they are necessary for the doing of science . . . [D]iscrete pebbles of particulate knowledge build nothing. They must build an overarching structure. There must be some thematic connection and theoretical integration in order for science to be a philosophical discipline and not merely a collecting and dissecting activity. A thematic basis to a science curriculum reflects what scientists really do and what science really is (California Department of Education 1990, p. 27).

In other places, however, themes are contrasted with theories and their status as pedagogical tools is highlighted, as in these passages:

As opposed to theories, which unify and make sense of facts and hypotheses related to a particular natural phenomenon, themes are pedagogical tools that cut across disciplines . . .

Themes are . . . not the same as theories . . . Theories are organized around content in particular disciplines of science . . . Themes . . . cut across specific content matter. By showing the interrelationships of different facts and ideas, themes serve primarily as pedagogical tools for the presentation of science (California Department of Education 1990, p. 26).

The interpretation of themes provides an occasion to underscore both the role of the *Framework* as a vehicle for conversation as well as an example of how the experiences of educators changed, to some extent, the meaning of the *Framework*. While thematic teaching is arguably the most strongly advocated idea in the document, and its most innovative element, it has very little presence in most classrooms. As noted earlier, most of the original writers of the document have come to reconsider the wisdom of referring to the unifying concepts of science as themes.

Staff developers at the elementary level report the confusion the word "themes" was causing teachers. Rather than attempt to redefine the word for teachers, California leaders instead invoked other words found much less frequently in the *Framework* to describe a conceptually driven science curriculum. As noted earlier, DiRanna reported that in the elementary school science education network the notion of "themes" has been replaced by "big ideas," a term thought less likely to be misinterpreted.

At the high school level, the idea of thematic teaching has been even more of a "nonstarter." Sachse reports that the idea of making connections across the disciplines is alien to high school science teachers:

At the high school level, because we really hadn't come very far along the road of SS&C yet, we ran into problems with, "How can you have a cross-cutting theme when you're only doing biology or you're only doing chemistry?" People hadn't really begun talking among themselves, I don't think, in a really profound way so you could think about energy connecting the biology class and the chemistry class and the physics class. That just wasn't happening. It's happening a little bit more now, but that was tough sledding for a while.

The *Framework* as a Conversation Piece

While actual readership of the *Framework* has proved difficult to estimate, most teachers—whether formally part of the reform or not—have a working knowledge of its message. In this context, the *Framework* has acted as a centerpiece for much deliberation about change. Its presence prompted districts to examine and reflect on their science programs. This function, according to former superintendent Bill Honig, was what the writers of the *Framework* were hoping for:

> We wanted to stimulate discussion. So [the *Framework*] had to be open-ended enough to engender discussion but hard-wired enough to provide a focus for discussion. The state law at the time said districts had to discuss the *Framework*. They did not have to adopt it, but they had to meet and publicly discuss and debate its merits. And if they chose to depart from it, they had to provide reasons. So the state law required discussion.

Elizabeth Stage reflects on the importance of the professional discussions that have followed the publication of the *Framework,* such as the consideration of themes in the *California Science Teachers Association Journal.*

> Doug Martin, a chemistry professor at Sonoma State, took tremendous issue with the themes idea because it takes such a mastery of the subject matter to understand it. Create argument! What happens? People encourage him, put it in the *CSTA Journal,* have a meeting about it, get lots of people talking about it. Those professional conversations, at whatever level, of whatever composition, I think that's more important than anything else.

For another example, at the high schools, the idea of thematic teaching has not been as influential as such notions as integration, coordination, and hands-on inquiry. Again, the *Framework* served not so much as an instigator of change, but an occasion for teachers to talk about what they had already been doing. Deborah Atchley, a high school science teacher, describes a common reaction to the idea of themes at the secondary level:

> I think of all those in-service [training]s. We talk about it. We mull over what this means or what that means: going from six themes to four themes, do you use the Project 2061 four themes? . . . Really it's irrelevant in terms of the curriculum that we are setting up because the way I understand it is that you're not supposed to set up your curriculum so that unit one is dealing with theme one and unit two is dealing with theme two, but that the themes are interwoven through the whole curriculum.

Thus, themes were a central feature of the *Framework* document but diminished in importance in the field. In contrast, constructivism and authentic assessment, while receiving minimal attention in the document, have played a major role in discussions of *Framework* implementation. This provides another example of how teachers projected meaning onto the *Framework* to legitimate their practice.

The *Framework* as Legitimator of Existing Practice

The second major function of the *Framework* in the field has been as a legitimator of previously unrecognized innovative practices. Elizabeth Stage describes this legitimating role:

> So the thing that's important . . . is that there were teachers already around who had been into lots of these things . . . for centuries, without a unifying *Framework*, but with all the substantive ideas. They were considered flakes who went off and did these sort of things in the summer or took time away from their students for professional development. All of a sudden the principal turns around and says, "You're an expert in these things. Please help me!" So this person who has been a loon becomes the expert and now she can take some leadership.

High school teacher Atchley confirms that the *Framework* strengthened the hand of advocates for science education improvement in individual schools.

> [We] . . . refer to the state *Framework* and the philosophy behind the new state *Framework* and say, "Look, we're looking for more process skills." That's not to say that some of our teachers weren't doing that already. Some of them were. But what this has forced then is that all of them come into the mold . . . The *Framework* is the document that has given me the freedom to push this type of approach because this is what the state is mandating. It's not what I'm asking you to do.

Interestingly, this legitimating function has transformed the perception of the *Framework* in an important way: it seems that all good and worthwhile practices are attributed to the *Framework* whether or not they are actually found in the text itself.

The Hazards of an Open-Ended Policy Document

California's open-ended approach to curriculum reform is not without its drawbacks. One hazard is simply the novelty of the approach. A teacher-centered approach to reform is something with which most teachers are unfamiliar and may be misinterpreting. For example, though the *Framework* is not intended to stand alone, this is not clear to many teachers; its lack of specifics is often confusing or frustrating to those not being reached by the networks.

Second, relying on teachers to develop the curricular specifics of the reform effort raises concerns because it puts an increased burden of responsibility on teachers who are already overworked and under-supported. Teacher Janet Brown reports a common reaction to this idea: "Many of the teachers in our district are objecting to the time demands of writing a new curriculum."

It should be noted that the teacher networks in California are not naive about the increased workload this reform asks of teachers and the relation of this to shrinking education budgets. Particularly in CSIN, teachers have their own discourse about burnout, are fatalistic about the fact that the burden of reform will fall on the shoulders of teachers, and spend a great deal of effort trying to provide a supportive professional culture that in some ways compensates for the increased work load. Steve Schneider, CSIN evaluator for the California SSI grant, observes:

> We need to keep in mind that it is unclear whether this grassroots approach is really better than the traditional approach. The California effort is under-funded and most teachers don't have the time to really develop their own materials or to significantly collaborate on the effort. If asked, they might say "give me already developed materials." It still may be the case that the most cost-effective approach to science education reform would be to develop really good curriculum material and give that to teachers. It is an open question.

Third, any effectiveness the *Framework* may be gaining through the stimulation of a broad conversation about science reform may be lost in terms of the depth of the change it precipitates. For example, one of the ways substantive curricular change may be elusive at the high school level is if it is only implemented at the remedial levels. A pattern is emerging in some schools of integrating only the 9th and 10th grade science courses. In several schools, integrated science courses have simply become replacements for non-college-track physical science classes. The identification of integration with remediation would likely be its undoing.

More generally, the looseness of the *Framework*'s message may contribute directly to superficial reform of science teaching. For example, Schneider points out that what the staff development networks in California may have most successfully done is indoctrinate teachers against "telling" and "lecturing" modes of teaching, rather than communicate any clear conception of constructivist or experiential pedagogy. It is unclear how much, if at all, constructivism is understood and/or widely practiced. Honig provides a recent research example of such superficial pedagogical change:

> There is another side of that going on now. People are thinking just organizing a hands-on lesson will make it constructivist, that if the students work on something directly, then they will automatically learn; and that is not true either. They need the discussion of it, with teachers and each other . . . For example, there was a study that examined kids engaged in a lesson on erosion. They were presented with the question of why rocks in a river were smooth, and after a week of working with an artificial stream—placing objects in it and watching what happened—they had no coherent model of what was taking place. Hands-on instruction alone doesn't work.

◆

Lessons From the California Science Education Reform

California science education reform is more than a distant dream. A broad vision of desirable science education has been projected throughout the state in a new and widely accepted *Science Framework for California Public Schools, Kindergarten*

Through Grade Twelve. About a third of the state's 6,000 elementary and middle schools have paid an annual fee of $2,500 to belong to the teacher network for these levels. The network is dedicated to improvements in science teaching that seem to flow from the *Framework*. Ten regional hubs draw in teachers from about 300 of the state's 800 secondary schools for regular meetings, at which teachers discuss detailed changes in their curriculum and teaching styles. Tests for students have been revised to comport with the new programs. Colleges have modified their teacher education programs to make them more compatible with the kind of science education envisioned in the *Framework*. In short—and most of all—an enormous amount of momentum has been generated throughout the state, involving thousands of people, that seems to ensure that serious attention will continue to be devoted to improving science education, regardless of the vicissitudes of external funding and inevitable changes in personnel.

Several features of the California science reform may be especially noteworthy. It is true that programs hardly ever transfer with fidelity from one place to another. People are different, so are external pressures, so are traditions, and available resources. Context counts, more than almost anything else. Nevertheless, it seems to the authors of this report, that certain points about science education reform in California warrant special attention from those facing similar challenges elsewhere.

1. Articulating goals for the science program (in California's case, the 1990 *Framework*) is not enough. The people who subscribe to the broadly stated goals must understand their meaning and figure out how they want to translate them into curriculum. This takes resources, lots of resources. But the prime one is creating opportunities for teachers to talk to one another. In California's case, the vehicle was the strengthened and expanded networks of classroom teachers.

2. It is less important that the guiding vision be clear and consistent than that it be authoritative, attractive, and plausible. What counts is how seriously the newly articulated goals for science education are taken by the people affected by them—particularly the teachers—and how persevering they are in their attempts to translate them into practice.

3. Changes takes time. Assimilating new ideas, figuring out their implications for practice, and having the chance to try things out are steps in educational reform that cannot be compressed.

4. When trying to reshape and redirect a huge system, build on the strengths of people already trying to make some of the changes that seem desirable. Changes are difficult, even small ones: better to begin by starting with people and organizational structures that already exist. With such an approach, one also signals that reform has already begun and is achievable. Such a strategy enhances commitment.

5. The strategy of building on strength, however, seems to accent and exacerbate a serious problem. A major issue for California has been the problematic quality of the science that is being taught. In many classrooms—particularly at the elementary school level but also at the secondary level—the actual science teaching is often episodic, superficial, and inconsistent with the *Framework*. Elementary school teachers generally do not have strong science backgrounds. In choosing to build the reform around the most committed of current teachers, there was a trade-off between the resulting conception of science, even its accuracy, and gaining the emotional commitment of large numbers of teachers. Leaders of the reform in California recognize the seriousness of the problem, but believe involvement and commitment by teachers come before what they see as the even more difficult task of improving their science backgrounds. Other places might choose a different priority.

6. Do everything at once (curriculum, teacher education, assessment), but make sure everyone subscribes to the same goals (in California's case, the 1990 *Framework*). Such an approach increases momentum and the perception of momentum. People sense that they are part of a larger picture in which reform not only is being taken seriously, but the programs are coherent. Furthermore, the reform is not held captive to the successful completion of any single phase or the contributions of any one person; it just keeps moving forward.

7. Be opportunistic but coherent in the use of funds. Try to take money from whatever sources are available (federal, state, private) and direct them toward compatible ends. In California's case, the State Systemic Initiative (federal), assessment design (state), and Eisenhower funds (federal) were used to strengthen different elements of the overall reform.

8. Leadership is critical. California recognized the importance of support from state-level and district-level administrators, without compromising the teacher-based nature of the reform. The balance is a difficult one. It helps also to have the organizational structures in place that enable people to get to know one another and build the necessary levels of trust. It especially requires leaders who remain attuned to evolving classroom practices, preferably by visiting schools and talking with teachers.

There is a sense in which none of these "lessons" is a surprise. They all flow from an attempt to tailor reform to the situation that already exists, to build on what one has. But one more point must be made. Education reform is extraordinarily difficult; change comes slowly. Leaders of the California reform are nudging a particularly massive and complex enterprise: they are dealing with thousands of teachers; they are tuned to the sensibilities of a large, varied, contentious, and active electorate; they operate with fewer funds per student than most other states.

Everyone involved in educational reform needs to recognize that there are no magic paths to a well-marked destination in science education reform. People have dreams of the kind of education they want and need. Sometimes they even agree about them. But no state's education system will ever realize its ideals; it will never arrive at what at any moment seems its desired destination. There are many reasons, but the most important is that the educational vision changes continually as people demand new things from their schools and as teachers invent new and appealing practices.

Just as there is no unvarying destination for reform, there is no one way to make things better. It has been noted that education reform is steady work. It is also demanding—site specific and complex, difficult, and emotionally costly. It would be lamentable if the inevitable inability to fulfill ambitious educational dreams were to detract from what the reformers in California—from classroom teachers to officials in the state Department of Education—already have done in making significant improvements in science education in the state. Science education in California is better than it was 10 years ago, even if it is not ideal. The best news is that able and committed people are still working on it.

References

California Commission on Teacher Credentialing (CTC). 1992. *Science teacher preparation in California: Standards of quality and effectiveness for subject matter programs.* Sacramento: California Department of Education.

California Department of Education. 1990. *Science framework for California public schools, kindergarten through grade twelve.* Sacramento: author.

California Learning Assessment System (CLAS). 1994. *A sampler of science assessment: Elementary.* Sacramento: California Department of Education.

California Science Implementation Network. 1993. *Project Storyline: Science.* Sacramento: California Department of Education.

California Science Teachers Association (Spring, 1992) *California Science Teachers Association Journal,* pp. 23 - 45.

California Scope, Sequence and Coordination (SS&C). 1992. *Program Assessment Options For California SS&C Schools.* Sacramento: California Department of Education.

——. April 1993. *Restructuring Science.*

DiRanna, K. Spring 1992. Responses to themes in science. *CSTA Journal.* pp.40-43.

——. forthcoming. *Models for Student Assessment in Coordinated Science.*

——. forthcoming. *Preservice Reform Programs for Coordinated Science.*

Far West Laboratory. 1994. *California Alliance for Mathematics and Science Evaluation Report* 1993-1994. San Francisco: author.

Inverness Research Associates. 1994. *A study of the California Eisenhower Mathematics and Science Education State Grant Program. A Summary of Issues 1992-1993.* Draft.

Kota, Helen. 1992. Communication and the changing job description of the hub coordinator. *In Action Research: Reports from the field.* Sacramento: California Department of Education.

Kovalik, S. and Olsen, K. Spring 1992. Kids Eye View of Science. *CSTA Journal.* pp. 30-32.

Kuhn, T. 1962. *The Structure of Scientific revolutions.* Chicago: The University of Chicago Press.

National Research Council. 1996. *National Science Education Standards.* Washington, D.C.: National Academy Press.

Project 2061. 1991. Beyond themes. *2061 Today* 1(2):4.

Appendix A: Case Study Team and Research Activities

Case Study Team

J. Myron Atkin is a professor of education at Stanford University and served as Dean of Education from 1979 to 1986. He has chaired the Education Section of the American Association for the Advancement of Science and is a consultant to the Organization for Economic Cooperation and Development in Paris. He co-edited Changing the Subject, OECD's 1996 report on member nations' case studies of innovations in science, mathematics, and technology education. He was a member of the National Committee on Science Education Standards and Assessment and chaired the Committee on Science and Engineering Education of Sigma Xi. His research interests and publications focus on: identification of the science content to be taught in elementary and secondary schools; teacher-initiated inquiry, especially action research; practical reasoning by teachers and children; case methods in educational research; evaluation of educational programs; science education in museums; and development of policies that accord classroom teachers greater influence in determining the educational research agenda.

Jenifer Helms is an assistant professor of science education at the University of Colorado at Boulder beginning fall of 1996. During the case study, she was a doctoral student in curriculum and teacher education at Stanford University. She has taught life sciences at the high school level and master's methods courses at the university level. Her research interests include: teacher collaborative research; the relationship between the subject matter and teacher identity; and social, cultural, and feminist theories of science and science education.

Jerry Rosiek is an assistant professor of science education at Portland State University. At the time of the case study, he was a doctoral student at the Stanford University School of Education. During that time he also worked with Lee Shulman doing case study research on the cultural and affective dimensions of pedagogical content knowledge. His research interests include the nature of teacher knowledge, and the political and philosophical dimensions of research on teacher knowledge.

Suzanne Siner completed her Educational Specialist degree in science education at Stanford University while working as a research assistant on the OECD case study. Her research has focused on elementary school science education and the role of the science specialist. She is interested in innovative approaches to professional development and support for elementary school teachers in science. Suzanne plans to return to the classroom and hopes to bridge her interests in teaching children, research, and professional development.

Research Activities

The data for this case study were collected primarily through interviews and observations over a three-year period from 1992-95. We interviewed state-level participants in the reform; as well as K-12 classroom teachers, university participants, and people who have assumed leadership roles in the central projects of the reform. We attended statewide and regional professional development conferences, leadership meetings, and visited individual school sites. Finally, we developed, distributed, and analyzed surveys of SS&C hub coordinators.

Documents consulted in our analysis include: the California Department of Education's Science Framework for California Public Schools, Kindergarten Through Grade Twelve; CLAS's A Sampler of Science Assessment-Elementary; CSTA's journal and newspaper California Classroom Science; Far West Laboratory's 1993-94 evaluation report of the California Alliance for Mathematics and Science; Inverness Research Associates' 1992-1993 study of the California Eisenhower mathematics and science education projects; CSIN and SS&C newsletters and other publications; action research reports written by various SS&C participants; and curriculum guides from participating CSIN and SS&C schools.

We would like especially to thank the following people for giving their time (around the clock!), sharing their expertise, and telling their stories, without which this report could not have been written: Kathy Comfort, Kathy DiRanna, Helen Kota, George Miller, Tom Sachse, Elizabeth Stage, the CSIN teaching consultants, and the SS&C hub coordinators.

Formal Interviews Conducted

- Debra Atchley, Site Coordinator and Teacher, SS&C (May 1994)
- Wendy Bongber, Staff Developer and Teacher, CSIN (April 1994)
- Norman Brooks, Teacher, CSP (July 1995)
- Bonnie Brunkhorst, Professor, CSU-San Bernardino (March 1994)
- Herb Brunkhorst, Professor, CSU-San Bernardino (October 1994)
- Pam Castori, Co-director, CSP-Sacramento (September 1994, March 1995)
- Karen Cerwin, Teaching Consultant, CSIN (May 1994)
- Jim Clift, Site Coordinator and Teacher, SS&C (November 1994)
- Kathy Comfort, former Director, CLAS-Science (January 1994, September 1994, January 1995)
- Linda De Lucchi, Co-director, Full Option Science System, Lawrence Hall of Science (May 1994)
- Kathy DiRanna, Executive Director, CSIN (January 1994, May 1994, October 1994, March 1995, June 1995, September 1995)
- Esther Garnica, former Teaching Consultant, CSIN (May 1994)
- Judi Gordon, Staff Developer and Teacher, CSIN (August 1994)
- Gary Griffith, Staff Developer and Teacher, CSIN (February 1995)
- Shirley Hall, Lead Teacher, CSIN (February 1995)
- Penny Hamisch, Staff Developer and Teacher, CSIN (March 1994)

- Tom Hinojosa, Assessment Consultant and former Hub Coordinator, SS&C (September 1995, March 1995)
- Linda Holmes, Lead Teacher, CSIN (February 1995)
- Bill Honig, former California Superintendent of Schools (January 1995)
- Sharon Janulaw, Staff Developer and Teacher, CSIN (January 1994, April 1994, September 1995)
- Dolores Jones, Teaching Consultant, CSIN (March 1995)
- Virginia Kammer, Teaching Consultant, CSIN (October 1994)
- Helen Kota, Project Manager, SS&C (October 1993, June 1994, October 1994, March 1995, May 1995, September 1995)
- Erma Larin, Teaching Consultant, CSIN (January 1994)
- Maria Lopez-Freeman, Director, CSP Language and Status Leadership Institute (October 1994, May 1995, July 1995)
- Meg Martin, GSE Science Coordinator (July 1994, September 1995)
- Meg Milani, Site Coordinator and Teacher, SS&C (July 1994)
- George Miller, Co-director, CSP-Orange County; and Professor Emeritus, University of California-Irvine (July 1994, March 1995, June 1995)
- Barbara Novelli, Teaching Consultant, CSIN (November 1993)
- Kathleen OfSullivan, Professor, San Francisco State University (June 1994)
- Michelle Parson, Lead Teacher, CSIN (February 1995)
- Wendell Potter, Co-director, CSP-Sacramento; and Professor, University of California-Davis (September 1994, March 1995)
- Teresa Ramirez, former Staff Developer, CSIN; Teacher; and Assistant Director, CSP (September 1994)
- Tim Ritter, Hub Coordinator and Teacher, SS&C (June 1994, September 1994)
- Sandy Ruehlow, Staff Developer and Teacher, CSIN (June 1994)
- Rick Rule, Hub Coordinator and Teacher, SS&C (February 1994)
- Tom Sachse, former Director, California Alliance of Math and Science, SS&C, and CSP (July 1993, July 1994, March 1995)
- Steve Schneider, Evaluator, Far West Laboratory (December 1993, October 1994, February 1995)
- Ursula Sexton, Staff Developer and Teacher, CSIN (October 1994)
- Karen Shauer, Staff Developer and Teacher, CSIN (April 1994)
- Gary Sokalis, Hub Coordinator and Teacher, SS&C (June 1994)
- Elizabeth Stage, Chair, California Science Framework Committee; and former member, NRC Science Education Standards Staff (October 1994, September 1995)
- Jo Topps, Teaching Consultant, CSIN (May 1994)
- Ginny Trapani, former Principal, Las Niemes Elementary School (May 1994)
- Deborah Tucker, Teaching Consultant, CSIN (March 1994, June 1994, September 1995)
- Ron Ulrich, Teacher, SS&C (February 1995)
- Greta Viguie, Assistant Superintendent for Elementary Education, Cotati/Rohnert Park School District (February 1995)
- Bill Von Felton, Hub Coordinator, SS&C (October 1994)
- Peter Wilding, Hub Coordinator and Teacher, CSIN (January 1993, February 1994, April 1994)
- Ed Young, Teacher, SS&C (November 1994)

Conferences, Meetings, and Workshops Observed

- American Educational Research Association Meeting, San Francisco (April 1995)
- California Science Teachers Association (CSTA) Conference, Palm Springs, California (October 1994)
- CLAS Meeting, Sacramento (August 1994)
- CSIN Lead Teachers Meeting, Cotati/Rohnert Park (February 1995)
- CSIN Northern Regional Meeting, Sacramento (October 1993)
- CSIN Regional Meeting, Hayward (October 1993)
- CSIN/SPAN Regional Meeting, Monterey (April 1994)
- CSIN Seminar Day, Cotati/Rohnert Park (January 1994)
- CSIN Seminar Day, Santa Clara (January 1995)
- CSIN Small Schools Consortium Meeting, Soledo (November 1993)
- CSIN Staff Developers Meeting, Irvine (January 1994)
- CSIN Summer Institute, Oakland (July 1994)
- CSIN Summer Institute, Santa Cruz (August 1993)
- CSIN Teaching Consultants Meeting, Costa Mesa (May 1994)
- CSIN Teaching Consultants Meeting, Palm Springs (October 1994)
- CSIN Teaching Consultants Meeting, Sacramento (January 1994)
- CSP Board of Directors Meeting, Burlingame (September 1994)
- CSP Language and Status Leadership Institute Summer Institute, Compton (July 1995)
- CSTA Conference, Palm Springs (October 1994)
- Get Our Act Together Meeting, Orange County (January 1994)
- National Science Teachers Association Conference, Kansas City, MO (March 1993)
- Project Directors Meeting (CSIN/CSP/SS&C), San Diego (March 1995)
- Project Leadership Day (CSIN/CSP/Project 2061/SS&C), Palm Springs (October 1994)
- Project Storyline Meeting, Burlingame (March 1994)
- Science Education Strategic Planning Meeting, Oakland (January 1995)
- SS&C Hub Coordinators Meeting, Davis (October 1993)
- SS&C Hub Coordinators Meeting, Palm Springs (October 1994)
- SS&C Hub Coordinators/CSU Meeting, Burlingame (January 1994)
- SS&C Hub Meeting, Orange County (May 1994)
- SS&C Hub Meeting, San Rafael (December 1993)
- SS&C Staff Development Summer Institute, Davis (July 1994)
- SS&C/CSU Meeting, Burlingame (April 1993)

Site Visits

- Doyle Park Elementary School, Santa Rosa (April 1994)
- Howe Avenue Elementary School, Sacramento (February 1995)
- La Fiesta Elementary School, Cotati/Rohnert Park School District (April 1994)
- Las Niemes Elementary School, Artesia (May 1994)
- Marguerite Hahn Elementary School Cotati/Rohnert Park School District (April 1994)
- Monte Vista High School, Cupertino (Spring 1993)
- Monte Vista High School, San Diego (February 1994)
- Rancho Cucamonga High School, Rancho Cucamonga (November 1994)

- San Rafael High School, San Rafael (October 1993, February 1994)
- Santa Ana High School, Santa Ana (May 1994)

Individuals Who Provided Feedback on Draft Reports

- Bonnie Brunkhorst, Professor, CSU-San Bernardino
- Herb Brunkhorst, Professor, CSU-San Bernardino
- Kathy DiRanna, Executive Director, CSIN
- Sharon Janulaw, Staff Developer and Teacher, CSIN
- Helen Kota, Project Manager, SS&C
- Maria Lopez-Freeman, Director, CSP Language and Status Leadership Institute
- Meg Milani, Site Coordinator, SS&C
- George Miller, Co-director, CSP-Orange County; and Professor Emeritus, University of California-Irvine
- Tom Sachse, former Director, California Alliance of Math and Science, SS&C, and CSP
- Elizabeth Stage, Chair, California Science Framework Committee; and former member, NRC Science Education Standards Staff
- Jo Topps, Teaching Consultant, CSIN
- Deborah Tucker, Teaching Consultant, CSIN
- Peter Wilding, Hub Coordinator and Teacher, CSIN

Appendix B: *Framework* Content Matrix Example

		Physical science	Earth science	Life science	Local options	
					Ecology, technology, health aviation/space science	
		Unifying concepts				
Grade	Theme(s)	Matter and energy can be changed but not created or destroyed.	Earth systems interact in cyclical patterns.	Life is diverse.	Respect for nature develops from understanding how nature works.	The application of scientific know-ledge changes the world.
K	*Me in My World*	Matter can be observed and classified. "Matter Around Me"				
1	*Systems and Interactions* *Scale and Structure*	Energy comes in different forms. "Kinds of Energy"	Water affects life on earth. "Water in the World"	Similarities and differences in living things. "Diversity of Life"	Human beings affect the environment. "Conservation"	Resources are limited; some can be recycled. "Recycling"
2	*Scale and Structure* *Energy*	Matter has properties and can be changed. "Matter"				
3	*Systems and Interactions* *Patterns of Change*	Forces act on matter and cause motion. "Changes in Motion; Simple Machines"				
4	*Energy* *Systems and Interactions* *Scale and Structure*	Energy can be converted from one form to another. "Energy Transformation"				
5	*Energy* *Scale and Interactions* *Systems and Interactions*	Matter and energy interact at a microscopic level. "Waves, Light, Sound"				
6	*Energy* *Systems and Interactions* *Scale and Structure*	The structure of matter at a microscopic level affects chemical reactions. "Matter"				

Unifying Concept: Matter and Energy Can Be Changed But Not Created or Destroyed		
Grade	Theme(s)	Grade-level concepts and subconcepts
K	*Me in My World*	Matter can be observed and classified. • Everything around me is made of "stuff." The stuff can be described and classified by many characteristics: —Color, texture, shape —Hardness, flexibility —Taste, odor —Sound or light that might be emitted or reflected —State (solid, liquid, or gas) • All things are made of similar structures.
1	*Systems and Interactions* *Scale and Structure*	Energy comes in different forms. • At the macroscopic level, each of the various forms of energy has unique characteristics. Observe and compare the properties of different manifestations of energy (light, sound, static electricity, magnetism, heat, wave motions, and so forth) in order to classify and describe. • Energy can be transmitted, reflected, and absorbed. • Energy can be used to do work and to make changes in matter. Changes in matter sometimes require energy and sometimes release energy.
2	*Energy* *Scale and Structure*	Matter has properties and can be changed. • Matter has definable properties that can be described and reported: —It occupies space. —It has weight and substance. —It sinks and floats. —It exists in three states. • Matter is made of smaller structures. We use tools to measure, perceive, and better understand these structures when they exist at a scale too small, too fast, or too far away for normal perception (telescopes, microscopes, thermometers, scales, clocks, and so forth). The more we understand the structure and function of matter, the better we understand living organisms and how they interact with their environment and the better we understand the changing earth. • Matter undergoes physical and chemical changes.
3	*Patterns of Change* *Systems and Interactions*	Forces act on matter and cause motion. • Forces act on matter. Forces include: —Pushes and pulls by direct contact —Gravity —Friction —Magnetism These forces can affect the motion of matter. • Motion can be measured in terms of distance, time, and weight. —Speed is the distance covered divided by the elapsed time. —The weight of an object is related to the amount of force necessary to change its motion. • Work is done when a force is applied to an object that moves it through a distance. • Simple machines help people do work and change their environment.

Sample Grade-Level Concepts and Subconcepts (continued)		
Grade	Theme(s)	Grade-level concepts and subconcepts
4	*Systems and Interactions* *Scale and Structure*	Energy can be converted from one form to another. • The ultimate source of most of the energy we use is the sun. • Energy can be converted from one form to another. In the process the total energy in the system is conserved but not necessarily in the same form. We use energy from a number of sources (the sun, water, heat within the earth, nuclear reactions, and so forth) to do mechanical work. • Heat energy moves through the environment from warmer to cooler regions by processes called conduction, convection, and radiation. This movement affects meteorologic and geologic processes and ecosystems.
5	*Scale and Structure* *Energy* *Systems and Interactions*	Matter and energy interact at a microscopic level. • Light and sound energy are similar in many ways, but they are not the same thing: —Both travel as waves and can be reflected, refracted, and absorbed. Both have frequences and wavelengths. —Light is an electromagnetic form of energy. —Sound is a mechanical wave arising from vibrating objects. • Infrared radiation is an electromagnetic form of energy. Light carries heat in the form of infrared radiation. When infrared radiation is absorbed by matter, it releases heat energy. • As a mechanical wave, energy can do a force that can do severe damage (earthquakes). • Animals perceive their environment by means of light and sound energy: —Electrochemical impulses arise from light striking the retina and are processed by the central nervous system. —Electromechanical impulses arise from sound in the middle ear and are processed by the central nervous system. • Plants use electrochemical processes in photosynthesis to manufacture food from sunlight.
6	*Systems and Interactions* *Scale and Structure* *Energy*	The structure of matter at a microscopic level affects chemical reactions. • All matter has unique properties that can be observed and measured: —It has weight and mass. —It occupies space and displaces other matter from the same space. —It has density. • The properties of matter depend very much on the scale at which we look at them. Properties of matter at the large scale (macroscopic) depend on its properties at the small scale (microscopic). • Chemistry is the study of the properties and interactions of atoms and the study of groups of atoms as they combine to form compounds and mixtures.

Appendix C:
The Big Ideas of the *California Science Framework*

Physical Science

P-1.   There are an almost inifinite number of kinds of matter. Matter is sometimes found as a pure substance, but more often found as a mixture of different substances or as a composite structure composed of different substances. Substances are distinguishable by properties that can be observed, described, and measured.

P-2.   Conditions external to a sample of matter affect many of its properties. Examples are temperature and pressure.

P-3.   All matter is composed of the same fundamental building blocks.

P-4.   Matter changes as it interacts with other matter, but the total amount of matter remains constant. Energy is transferred to and from systems of matter as they interact, but if all interactions are taken into account, any energy lost by one system is gained by some other system(s).

P-5.   Chemical interactions are affected by the conditions under which they occur.

P-6.   Forces cause changes in motion.

P-7.   Energy can produce work.

P-8.   Heat flows between regions/objects.

P-9.   Temperature and heat are related.

P-10. Forms of energy are interconvertible.

P-11. The properties of electrical charge explain magnetism and electricity.

P-12. Magnetism and electricity are related forces and have many uses.

P-13. Light travels in a straight line except when it is reflected or refracted (bent).

P-14. Light is a form of energy; the eye is sensitive to some light and sees this range as "color."

P-15. Sound results from a vibration of matter.

Earth Science

E-1.   Geological processes explain the evolution of the Earth.

E-2.   The universe has changed through time.

E-3.   The Earth is one of many objects in the universe, all composed of the same basic elements and subject to the same universal laws.

E-4.   The Earth's resources are limited. The Earth is a product of its history, so resources are unevenly distributed on the Earth's crust.

E-5.   All life is dependent on the water cycle.

E-6.   Oceans play a central role in the global ecosystem and in the formation, evolution, and continued support of life on Earth.

E-7.   Energy from the sun, interacting with air, water, and land, produces climate and weather.

Life Science

L-1.   Living things come from other living things and are distinguished from non-living things.

L-2.   All living things have life cycles.

L-3.   Living things demonstrate a structure/function relationship, enabling organisms to survive in an environment.

L-4.   Cells are the basic unit of function in living things. There are different levels of organization in a multicellular organism.

L-5.   Living things, both within and among species, are diverse.

L-6.   Living things evolve through geologic time.

L-7.   Genetic characteristics are passed from one generation to another with modification.

L-8.   Organisms are classified by shared derived characteristics.

L-9.   All living things play distinct roles as they interact with each other and the physical environment.

L-10. Changes in one part of an ecosystem affect other parts of the ecosystem.

L-11. Matter cycles through ecosystems, often in predictable ways.

L-12. Energy flows through ecosystems.

L-13. Humans are responsible for their interactions/effect on ecosystems.

Appendix D: Information for University of California Approval of SS&C Courses

Information for SS&C Schools
Regarding University of California Approval of Science Courses for (a) - (f) pattern

Supplement to attached BOARS Statement on Science Courses prepared in June 1991.

Introduction

School science departments should expect to undertake a serious self evaluation of their entire science programs with respect to UC Admissions Standards. The goals of the school's science program should be much broader than the UC requirements, but it should be possible, in a well designed program, for individual students to meet the UC requirements and to enroll in elective science courses that go beyond the basic science requirements. The UC science requirement for admission in Fall 1994 is as follows:

(d)   Laboratory Science - 2 years, 3 recommended.
Two years of laboratory science providing fundamental knowledge in at least two of these three areas: biology, chemistry, and physics. Laboratory courses in earth/space sciences are acceptable if they have as prerequisites or provide basic knowledge in biology, chemistry, or physics. Not more than one year of grade 9 science can be used to meet this requirement.

Philosophy

The principle intent of UC admissions requirements is to assure students that they can participate fully in the first year program at the University in a very broad variety of fields of study. Thus requirements are written deliberately for the benefit of all students expecting to enter the university and not for preparation for specific majors. Many students change majors during their studies at the university. Students who are decided on specific majors are expected, where possible, to extend their studies in that field beyond the minimum requirements.

The GOAL of the 1994 Science Requirement is to assure that students have a sufficient grounding in basic science to pursue majors which require enrollment in freshman courses in the traditional science disciplines of chemistry, biology or physics. Recognizing the crowded nature of the high school curriculum, the University adopted a minimum two year course requirement to assure preparation in at least two of these areas.

The document used to provide specific guidance about the nature of such preparation is the *Statement on Preparation in Natural Sciences Expected of Entering Freshmen*. While the University supports the principles embodied in the *California Science Framework*, the latter does not explicitly address preparation for college level work in science.

Self Study Objectives

The overall goal of the self study is to identify and assess how students in the school's program will expect to meet the desired standards. The school has several measurements against which this may be judged, including:

- *Statement on Preparation in Natural Sciences Expected of Entering Freshmen*
- School's previously approved traditional courses in biology, chemistry and physics etc.
- Entrance standards expected for Advanced Placement courses in biology, chemistry or physics. [which are designed to be directly comparable to first year university courses]

The variety of SS&C experiments in place suggests that a single review instrument will be inadequate to suit all needs. School faculty are in the best position to judge how and when students have completed the minimum standards and make recommendation accordingly. The following general guidance indicates questions that should probably be addressed.

1. What part of the program can readily be identified as a two unit sequence to meet the UC's (d) requirement? Is a preparation year required which should be certified to qualify for part of the (f) requirement?

2. What preparation should be expected in order that students can fully benefit from the two year program? How will this preparation be assured? (e.g.: by pre-requisites, or other means)

3. What parallel development in other subjects should be anticipated? How will this be assured (by means of co-requisites or otherwise), and how does this mesh with the other (a)-(f) requirements of the University?
    (Since the University requires three years of mathematics and four years of English, science programs to qualify students for admission should be constructed with the expectation that students are also meeting the other requirements.) [Note: parallel development of skills and competency is also expected by the State Department of Education in its review documents].

4. How does the program address the basic conceptual content areas, including the necessary facts and vocabulary to both apply and communicate about these concepts and their application. This content needs to meet the majority of the basic topic sections of at least two of the discipline areas included in the Expectations statement and some of the ideas included as "optional", or "advanced".

5. How does the program address the basic skills expected for beginning University students as identified in the *Expectations* statement in at least two of the three identified discipline areas? How does the laboratory program provide sufficient breadth and depth to assure that students acquire and practice all laboratory related skills?

6. How does the assessment used in the courses assure that students exiting the two year program have accumulated a substantial body of knowledge and skills that they can apply to further work in science?

7. Does the school anticipate allowing students to complete only a portion of the program? If so, how will students be guided as to how to complete the UC (d) requirement.

8. What opportunities will be offered to students to take science courses beyond the minimum? How will these courses be structured to build on the prior experience and be sure not to be repetitive, especially if they are also open to students from "traditional" courses.

[If the school plans to retain the traditional biology, chemistry, physics, and earth science courses, some discussion of how the entire science program at the school will select or steer students through appropriate pathways of science curriculum is needed.]

Submitting Information to the University of California Office of the President

The enclosed documentation describes how to submit a course for UCOP approval. It is important to note that science faculty at the University will review these materials. The process will be expedited greatly if schools prepare the materials so that the answers to the above questions can be ascertained readily, without a great deal of searching. In other words, the school should carry out a self evaluation of its program with the objective of preparing a case for course approval. The documentation submitted should summarize the case and provide explicit evidence to support the case rather than simply be a set of course materials to be reviewed for compliance by the University. Attached is a checklist to assist in this task.

In the early days of the SS&C project, course aprovals were given to a few schools on the basis of their course outlines. This process was very slow, as many questions could not be answered from the items submitted. Since a large number of schools may now be applying for SS&C course approval it is vital for schools to follow the practice recommended above in order that courses can be reviewed in a timely manner.

Prepared by Geoge E. Miller, UC. Irvine, in February 1993 for guidance only. This document has not received official endorsement by the University of California.

Directions for submitting a request for approval are provided on next page.

Submission of SS&C course(s) for UC Approval.

To aid in faculty review, please provide a summary <u>statement</u> and supportive <u>evidence</u> for each of the following items, where applicable.

Course sequence planned and proposed UC admissions category for each course [(d) or (f)].

List of prerequisites and corequisites.

How the course contents asusre completion of the 1994 UC (d) requirement for students taking the entire sequence[1,2]. (Content equivalent to preparation in fundamental knowledge in two of the three discipline areas.)

How the course activities will assure students acquire the necessary skills and attitudes expected for the University's requirements[3] and the State's goal of 40% laboratory work (*Science Framework*). This should include some review of assessment practices and standards to be expected for passing grades.

How the course(s) will expect students to utilize parallel development of skills in other areas, but particularly in mathematics.

The opportunities that will exist for students to study further science beyond the minimum program.[4] Will UC certification be sought for new extension courses?

Provisions made (if any) for students to come in and out of the program. If the school is to continue to offer competitive "traditional" science courses, describe restrictions (if any) on students switching or taking a mix of courses.

1. *Statement on Preparation in Natural Sciences Expected of Entering Freshmen* pages 15-48.

2. Some of the concept understanding and skill acquisition should probably have been covered in earlier grades. This should be so stated, but detailed curricula for previous work are not expected to be included.
For fully integrated courses, these will generally be acceptable if a student will have covered about 2/3rds of the fundamentals of <u>all three</u> discipline areas provided that some optional advanced topics are also included.

3. *Statement on Preparation in Natural Sciences Expected of Entering Freshmen* pages 7-10: Inquisitiveness, objectivity, open-mindedness, skepticism, perseverance, reading comprehension, writing, listening and speaking, memory, mathematics, analysis, reasoning, generalization, classification, application, scientific investigation including defining a problem, proposing a hypothesis, testing the hypothesis, analyzing results; and laboratory work including planning, measurement, use of statistics, safety, reporting).

4. These could include AP courses, specialized topical courses, integrated "capstone" courses, etc.

Appendix E
Golden State Examination Science Program

- Relevance
- Content connected meaningfully
  — not isolated facts
- Conceptually based science
  content
- Constructed meaning

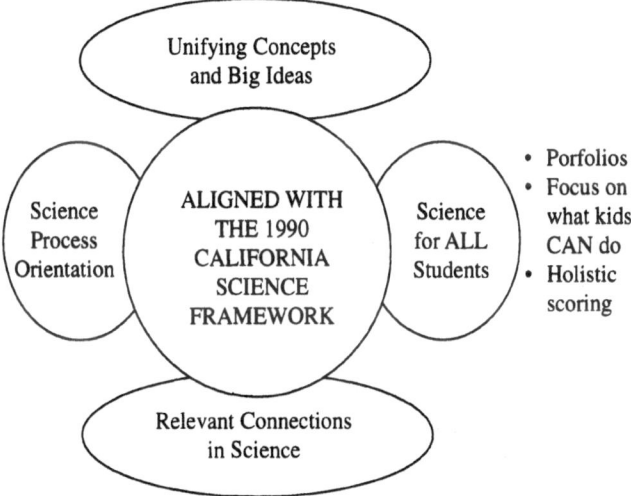

- Experimental
  design
- Hands-on
- Problem
  solving
- Methods of
  science
- Performance-
  based

Unifying Concepts
and Big Ideas

Science
Process
Orientation

ALIGNED WITH
THE 1990
CALIFORNIA
SCIENCE
FRAMEWORK

Science
for ALL
Students

- Porfolios
- Focus on
  what kids
  CAN do
- Holistic
  scoring

Relevant Connections
in Science

- Interdisciplinary
- Applications
- STS-Science Technology Society
- Relate to kids' real experience
- Environmental and ecological
  relationships

Chapter 2

The Different Worlds of Project 2061

J. Myron Atkin
Julie A. Bianchini
Nicole I. Holthuis

School of Education
Stanford University

Contents

The Different Worlds of Project 2061

Why a Case Study of Project 2061?

Project 2061: Education for a Changing Future is a long-term reform initiative designed to transform science, mathematics, and technology education from kindergarten through grade 12. Launched by the American Association for the Advancement of Science (AAAS) in 1985, the goal of Project 2061 is scientific literacy for *all* citizens. Project 2061 defines scientific literacy as embracing the natural and social sciences, mathematics, statistics, engineering, and technology. An adult who is literate in science understands a common core of scientific knowledge; is aware that science, mathematics, and technology are interdependent human enterprises; and uses the habits of mind associated with these fields for individual and social purposes (AAAS 1990 and AAAS 1993). Project 2061 sees science literacy as "essential to the education of today's children for tomorrow's world." "The terms and circumstances of human existence can be expected to change radically during the next human life span. Science, mathematics, and technology will be at the center of that change—causing it, shaping it, responding to it" (AAAS 1989, cover page).[1]

From its inception, Project 2061 has stood apart from other educational reform efforts in its ambition, scale, and inclusiveness; it boasts several distinguishing characteristics. First, the project targets all of science education, from kindergarten through grade 12. It starts with a vision of adult science literacy and works backward to determine the scientific knowledge, skills, and attitudes appropriate for lower grade levels. Moreover, it advances a broad conception of the science to be taught and learned in schools, one that includes not only the biological and physical sciences, but also the behavioral and social sciences, mathematics, and technology.

Second, from the outset, project participants have viewed the undertaking as a long-term commitment. They have expected the job of reforming science education to take decades, maybe longer: science literacy needs to be defined;

---

[1]*Science for All Americans* was published twice: in 1989 and in 1990. For the most part, we have cited the 1990 edition. However, some information from the 1989 edition was dropped in the later publication. In those few instances, we have cited the 1989 edition.

curriculum needs to be developed for all levels of schooling; and thousands of school districts and tens of thousands of teachers have to undergo extensive preparation for new goals and methods.

Third, Project 2061's approach to reform is both national and systemic; it is arguably the first national education reform initiative with a reasonable claim to the systemic label. That is, Project 2061 considers the education system, and its reform, to involve more than students, teachers, and school administrators. The organizational structures within which these people work and the educational and political policies to which they adhere must be changed as well. Furthermore, the efforts of parents, business leaders, academics, textbook and test developers, industrial scientists, and many others must be solicited and skillfully integrated if educational change is to take place at the scale and depth desired.

Fourth, the project has captured the interest and support of diverse bodies of scientists, educators, and policymakers. It has the imprimatur of AAAS, the nation's largest organization of practicing scientists. It has significantly altered the terms of debate about science education in policy circles, specifically, the substance of the national science standards. Private foundations and public agencies have spent large sums of money—over $15 million—to bring Project 2061 to its present level. The project is featured regularly in the education and general press. It is prominent at key education conferences and is known well not only in the United States, but serves as a model in dozens of countries around the world. Virtually all informed commentary is commendatory.

For all these reasons, the authors of this case study deemed Project 2061 an educational innovation worth studying in some detail. We are eager to share our findings with you, the reader, because we believe there are important lessons about the substance and process of educational change to be learned. Project participants have attempted to grapple with issues currently debated within the science education arena: Who decides what science should be taught? How is science to be presented by teachers? What does it mean to try to reach *all* students? What judgments about teachers, schools, and educational change are necessary to effect significant, lasting reform? The project also provides insight into questions surrounding the process of educational reform: The project is now approximately 10 years old. What has it achieved? What changes, if any, have taken place in American science education that reasonably might be attributed to the vision and strategy of Project 2061? How effective were the approaches used? How, if at all, has the project modified its goals and style of operation in the light of experience and changing circumstances? What are some of its continuing problems, disappointments, and challenges? And is the project likely to play an abiding and influential role into the next century, which is one of its most important aims?

In other words, an analysis of Project 2061 can be consequential in planning new efforts to improve science education as well as useful to Project 2061 itself. In the mid-1990s, policymakers continue to express dissatisfaction with schools.

They seriously and persistently search for educational change strategies that work and regularly appropriate scarce funds to make schools better. If sensible educational improvement strategies are to be developed in the coming decades, it is important to understand Project 2061's successes, failures, and continuing challenges. At the very least, the concerned public, as well as those professionally involved in science education, needs to know whether to fund ventures that, like Project 2061, are intended to endure—ventures that have time lines that extend over decades rather than years.

The authors began research for this case study in the fall of 1992. At that time, we established contact with Project 2061 staff members, with teachers and administrators at the six school district centers located across the country, and with several Project 2061 consultants. Since then, we have collected Project 2061 memos and reports; reviewed over 50 articles about Project 2061 published in scientific, education, and popular journals; completed text analysis of Project 2061 publications; conducted 35 interviews with Washington staff members, school district center directors, project teachers, and consultants; attended approximately 20 meetings, workshops, and presentations at national conferences; conducted site visits at three of the six school district centers; and received both verbal and written feedback from participants about preliminary case study findings (see Project 2061 Comments on Case Study).

The research team was headed by J. Myron Atkin, professor of education at Stanford University. He codirected the National Science Foundation (NSF)-supported University of Illinois Astronomy Project in the 1960s and was a close advisor to the Stake and Easley case study project in science, mathematics, and social studies education at the University of Illinois in the late 1970s. Julie Bianchini and Nicole Holthuis were doctoral students at Stanford at the time the research was undertaken. Both have been high school science teachers, and Bianchini is now an assistant professor at California State University, Long Beach.[2]

Our aim in telling the story of Project 2061 is not to present a report card—that would be presumptuous, unnecessary, and misleading. Rather, our goal is to deepen understanding of what happens when a long-term, national, systemic effort is mounted to change science education. While our own analyses and interpretations are not hidden, the major technique employed in this case study is to amplify and juxtapose the voices of project participants. Where possible, we use the language of the participants themselves, in their writing and in their spoken comments. Their voices are often consistent with one another. Sometimes they are not. In addition, we inject perspectives of knowledgeable people who are not intimately involved in Project 2061 at the national or local

[2]Vicky Webber, a graduate student and research assistant at Stanford University, also participated in collecting data the first two years of the study. She provided a unique analytic perspective to the reform and contributed to some of the ideas found in this report. We are grateful for her efforts.

level, but who have a stake in the future of science education. Inevitably, our own predispositions and biases (for example, all the case researchers have deep interests in the role of the teacher in effecting educational change) are also evident, if only in the aspects of the project we have chosen to study and those we largely ignore.

As for what is omitted, the reader should note that we have chosen largely a historical approach. Although we recognize that the project continuously revises its course of action and plans to extend well into the next century, we have, on occasion, focused on past issues or events at the expense of current or future happenings. In doing so, we have produced a document that already is dated and that thus has displeased several key participants in Project 2061. In reacting to a draft version of this report, they reminded us that several of the problems highlighted here may have been troublesome at one time, but now have been ameliorated. Contentiousness between the six sites and the Washington office, for example, is not as prominent now as in the early 1990s. We agree (and we have noted that fact in the document). Nevertheless, we who prepared this case study believe that description and analysis of some of these difficult matters are important, even if the situations have improved. We have chosen to probe several problems that may no longer exist because we think they carry important lessons. As Project 2061 learns from its experiences and builds on its strengths, it is necessary for those interested in its successes to understand those experiences. Accomplishment is seldom easy, and *problems* do not equate with *mistakes*. Educational improvement is not a straightforward or simple process. Examining some of Project 2061's rough spots may lead to a deeper understanding of how they might be avoided in the future.

To complement our aim and approach, we have organized this case study into four sections. We begin with a historical overview of Project 2061. We then present three stories of the project: we describe the development, substance, and purpose of *Science for All Americans*, or *SFAA* (AAAS 1990), Project 2061's first major publication; the development and dissemination of *Benchmarks for Science Literacy* (AAAS 1993), its second major publication; and the organization, goals, and activities of three of the six school district centers. These stories are intended to familiarize readers with important aspects of the Project 2061 enterprise as well as to provide necessary context for subsequent analysis. The next three sections explore issues identified by the case study writers as both central to the project and salient to our audience: Project 2061's conception of science, its conception of the reform process, and its approach to the challenge of equity in science education. We end our case study with a discussion of important lessons learned. Throughout these sections, we have also attempted to underscore the benefits and challenges of a reform that is flexible in approach yet consistent in its vision of science education, that recognizes the autonomy of individual sites yet expects coordinated action across them, and that hopes to effect change at both the national and local levels.

More specifically, our tale of Project 2061 is one of accomplishment. Through its major publications, *Science for All Americans* and *Benchmarks for Science Literacy*, the project has helped determine the course and pace of science education reform at the local, state, and national levels. These documents already have influenced several states and local school districts that are developing guidelines for improving science education. The popular *California Science Framework*, a state document published in 1990, is a case in point. The project's dictum, "less is more" (or, more recently, "less is better"), now frames much of the current agenda about science education. Most informed observers believe that the curriculum found in schools spans too many topics and covers them superficially. Fewer topics should be selected, they argue, with the aim of developing deeper understanding. In addition, the emerging national standards for science education would be different if not for Project 2061. *Benchmarks*, released by Project 2061 in 1993, was promptly acclaimed by much of the press—often on the front page—as representing the long-awaited national curriculum standards for science education.

Another salient effect of the project has been to give energetic and accomplished people at the school, district, state, and national levels a set of principles and an attractive banner around which to rally. These educators have effectively employed the various project documents and, more importantly, their affiliation with a prestigious and highly visible national effort to articulate and legitimate a new vision of science education. As a result of their involvement in the project, these educators also have enhanced their own leadership positions in science education.

A third accomplishment is the translation of a conception of educational change into operational form. Project 2061 is in the process of carefully building a model of how a national vision for a school subject might be translated into action at the school level. It is attempting to create a reform that will effect lasting change, one that is systemic, national in scope but responsive to local needs, long term, and goal oriented.

We have also woven into our story of Project 2061 examples of unexpected events, of as-yet unmet goals, of apparent contradictions, of opportunities lost, of not always successful adjustments to changing circumstances. The Washington office, for example, has not enjoyed a smooth relationship with its six geographically dispersed centers. Quite the contrary. It is still trying to figure out how best to work with teachers and local school districts. The project has had limited success in persuading funding agencies to move toward longer cycles of support for innovation. It has yet to meet fully the challenge of translating the learning goals identified in *Science for All Americans* into cogent pieces of curriculum. In addition, it has voiced a strong commitment to reaching all students, including those not usually served well by the public schools, but has failed to clearly articulate a coherent policy or to make marked progress toward that end.

And throughout, we have attempted to present the story of 2061 as a multi-faceted, ever-changing, and not entirely consistent tale—as dependent, in part, on time, context, and informant. Definitions, descriptions, and opinions about Project 2061 are almost as numerous and varied as those who voice them: staff members at the Washington office; directors, administrators, and teachers at the school district centers; outside consultants; and interested observers. Many people active in the six centers, for example, have seen issues surrounding curriculum development, assessment, and implementation differently from those at the project's Washington office. Not everyone has agreed that the picture of science presented by Project 2061 is completely justifiable, or that it is the most appropriate science for all children. There also have been different viewpoints about how the science disciplines are to relate to one another in the classroom and to the lives of children.

Those involved in Project 2061 have been extraordinarily generous in permitting us access to their meetings, documents, and ideas. They are proud of their accomplishments, justifiably. They have reviewed our drafts, meticulously. Clearly, they would have written a report different from this one. Our fondest hope, nevertheless, is that the dedicated and able group associated with Project 2061—as well as the science education community at large—sees this case study as accurate, fair, and, for the most part, useful.

◆

Project 2061: From Halley's Comet to the Present

As stated in the introductory section, Project 2061 was launched in 1985—its goal, to transform K-12 science education. Why was the project initiated? What is wrong with current science, mathematics, and technology education? According to Andrew Ahlgren, associate director of Project 2061, research persistently shows students learn far less science than we think they do. As explained more fully in *Science for All Americans*, there are problems with access: women and minorities remain underrepresented in scientific fields. There are problems with teacher preparation and in-service education: many elementary school teachers fail to understand even the most fundamental concepts in science and mathematics; many middle and high school teachers of science and mathematics are not adequately prepared. There are problems with science curriculum materials and instructional strategies.

> The present science textbooks and methods of instruction, far from helping, often actually impede progress toward scientific literacy. They emphasize the learning of answers more than the exploration of questions, memory at the expense of critical thought, bits and pieces of information instead of understandings in context, recitation over argument, reading in lieu of doing" (AAAS 1990, p. xvi).

There are also problems with the larger educational system: science and mathematics teachers must juggle too many students in too little time, often with insufficient equipment, minimal funds, and inadequate support from other faculty and staff.

Project 2061's solution to the challenges faced in science education is deceptively simple: "We must attempt to teach less so that important ideas can be learned well," (Ahlgren notes). To operationalize this notion of "less is more"—or now, "less is better"—2061 has engineered a reform that is systemic, national, and long term; one centered on all children, all grades, all subjects, and all aspects of the educational system. Project 2061 began its reform effort with the identification of learning goals. *Science for All Americans*, the first Project 2061 publication, "answers the question of what constitutes adult science literacy, recommending what all students should know and be able to do in science, mathematics, and technology by the time they graduate from high school" (AAAS 1993, p. xi). The learning goals identified in *SFAA* were then translated into a form suitable for curriculum development in *Benchmarks for Science Literacy* (AAAS 1993). *Benchmarks*, published four years after *SFAA*, "specifies how students should progress toward science literacy, recommending what they should know and be able to do by the time they reach certain grade levels" (AAAS 1993, p. xi). In time, these two publications will be joined by additional tools and resources for use in designing curriculum materials and in changing the educational system. Examples include: *Designs for Science Literacy, Blueprints for Reform*, curriculum blocks and models, and *Resources for Science Literacy*, RSL. Ultimately, Ahlgren explains, 2061 is "less interested in fixing up the current system, whatever ails it, than in designing a new system to serve well-specified goals."

In the remainder of this section we describe the project's organizational structure, funding sources, and key events and products. Our intent here is neither to critique nor evaluate various aspects of the Project 2061 reform effort; rather, it is to provide needed context within which to situate subsequent detailed discussions of project activities.

The Project's Structure

Project 2061 is a multifaceted reform effort, operating simultaneously within three organizational spheres as well as across seven physical sites. Reform participants include AAAS; Project 2061 staff; school district center directors, administrators, and teachers; and educational consultants. Sites comprise the following: Georgia; McFarland, Wisconsin; San Antonio; San Diego; San Francisco; Philadelphia; and Washington, D.C. Clearly, participants' views and actions are shaped by their locations within these different spheres.

AAAS. AAAS, the sponsor of Project 2061, is the largest membership-based scientific society in the United States. It has many affiliated scientific groups and carries out numerous programs to advance science and science education. It also publishes *Science*, a weekly professional journal, and *Science Books & Films*, a review magazine for schools and libraries (AAAS 1989).

AAAS created the National Council on Science and Technology Education to oversee and advise Project 2061. Originally, members of the National Council were charged with keeping the public abreast of needed reforms in science and technology education; providing a forum to examine, critique, and improve the strategies and products of Project 2061; and acting as spokespersons for Project 2061 in other appropriate arenas. Ahlgren has cautioned that this description of the roles and responsibilities of the National Council is ambitious and may no longer be on target. At the very least, the council has furnished feedback on such 2061 tools as *SFAA, Benchmarks*, and *Blueprints for Reform*, as well as offered "the inspiration and guidance that was needed to keep this ambitious, long-term project on track" (AAAS 1993, p. vii).

Members of the National Council are appointed by AAAS and include educators, scientists, business leaders, state legislators, governors, school board members, parents, and classroom teachers. At its inception, the council was co-chaired by William Baker, chairman of the board at AT&T Bell Laboratories, and Margaret MacVicar, dean for undergraduate education at Massachusetts Institute of Technology. It had 26 members (AAAS 1990). The National Council was reconfigured in 1990, after the publication of *SFAA*, to better reflect the nature of the work ahead. Currently, the council is chaired by the chancellor of the University of Maryland system, Don Langenberg, and has approximately 30 members (AAAS 1993 and AAAS 1994a).

Project 2061 Staff. At present, the Project 2061 staff, or Washington office, includes approximately 20 members and is housed within the AAAS building in Washington, D.C. (AAAS 1993). Key staff personnel include F. James Rutherford, project director; Andrew "Chick" Ahlgren, associate director; Jo Ellen Roseman, curriculum director; Mary Ann Brearton, field services coordinator; James Oglesby, dissemination director; and Lawrence Rogers, deputy director.

Size of and membership within this sphere has remained fluid. The staff has grown from 5 members in 1989 to over 20 in 1995. Several people have changed positions: Roseman, for example, moved from field services coordinator to curriculum director; Oglesby, from National Council member to dissemination director (AAAS 1993). Several staff members have left: Walter Gillespie, deputy director of Project 2061, retired in 1994; and Carol Muscara resigned from her post as technology systems director. Challenges faced in ensuring continuity and coherence of the project's vision in the midst of personnel turnover and organizational expansion are addressed later in this paper.

School District Centers. Working collaboratively with the Project 2061 staff are six teams, or school district centers, comprised of administrators, teachers, and curriculum specialists.

Rather than one centralized team, we [the Project 2061 staff] wanted teams from several school districts across the country: we believed that conditions at a variety of sites could inspire the development of a range of alternative curriculum models that would suggest possibilities for mathematics, science, and technology curricula nationwide" (Brearton 1994, p. 1).

A number of criteria guided selection of these school district centers: (1) collectively, the centers represent urban, suburban, and rural school districts; (2) ethnic minorities and other traditionally underserved groups are well-represented in these districts; (3) local teachers and administrators are enthusiastic and talented; (4) school district officials agree to give the team of teachers considerable release time, specifically, 40 days during the school year for each of two years, plus two summers; (5) state departments and school districts agree to provide both time and funding for the project; (6) states agree to commit to review their own curriculum guidelines; and (7) a local university is in close proximity to provide academic resources (Brearton 1994, and Lynch and Britton 1992). The following six centers were selected to participate in the Project 2061 reform effort:

1. *Georgia*—A rural center near Athens that includes two school districts. Candido Munumer is center director.

2. *McFarland, Wisconsin*—A suburban center near Madison that includes one small school district. The center director is Deb Larson.

3. *Philadelphia, Pennsylvania*—An urban center that includes one large school district with significant African-American and Hispanic student populations. Marlene Hilkowitz is center director.

4. *San Antonio, Texas*—An urban, suburban, and rural center that originally included 16 districts with a large Hispanic student population. Currently, the center is being restructured. The new director is Joan Drennan-Taylor.

5. *San Diego, California*—An urban/suburban center that includes one school district with students from a variety of cultures and ethnicities. The center director is Gary Oden.

6. *San Francisco, California*—An urban center that includes one school district with students from a variety of cultures and ethnicities. Bernard Farges is center director.

Originally, the model was that there should be 25 members at each center: 5 elementary school teachers, 5 middle school teachers, 10 high school teachers, 1 principal from each level, and 2 curriculum specialists. Team members were selected to represent a variety of disciplines, including the life and physical sciences, social studies, mathematics, technology, and the humanities (Brearton 1994). They also were expected to reflect the gender and ethnic composition of the district teaching staff, to have a distinguished teaching history, and to believe that all children can achieve the learning goals outlined in *SFAA* (Lynch and Britton 1992). Each team was guided by one to three team leaders (now called center directors) as well as a steering committee.

In turn, Project 2061 pledged to provide each team with four kinds of support; two of these four support mechanisms were included in the criteria for site selection:

- From their school districts, team members received four days per month release time for two years (Georgia districts would only agree to two, so teachers spent one weekend a month working on the project; California districts would only agree to three) as well as six weeks for each of two summers. (For most team members, this was extended for two additional summers.)

- Each team was provided with academic resources from one or more nearby universities. (This rarely happened as planned, despite considerable investment of time and money.)

- Each center received computer equipment supplied free by IBM. The computers proved quite helpful for work within a given site but were seldom used to bridge the communication gap between the centers and the Washington office.

- Each team had a dedicated workplace. San Antonio and San Diego, for example, had comfortable quarters in unused classrooms, while Georgia was originally housed at the University of Georgia.

Today, as 2061 staff members note, despite the fact that the collaborative relationship between Project 2061 staff and site participants was intended to last only a few years, the sites remain active in the reform process. However, their list of participants, kinds of organizational support, and range of responsibilities have changed. At the San Francisco site, for example, the number of teachers involved in the reform process has grown to span eight schools in the district; the entire staff at three of these eight schools participates. Farges reports that four district-level science resource teachers have also been hired on a part-time basis to work with him to support teachers' implementation of Project 2061-based curriculum materials. In contrast, the financial support provided by the Washington office has dwindled over time. School districts have been expected to assume more and more of their center's financial burden. Centers, in order to sustain and expand their implementation activities, have begun to solicit outside agencies for additional funds. In 1995, center directors were funded only 50 percent time by the project. In addition, Project 2061 staff members have voiced a desire to work with other schools and school districts. They have not pledged to continue funding all six centers.

At the intersection between these regional centers and Project 2061 staff is found much of the project's strength as well as many of its tensions and dilemmas. The relationships among centers, and between centers and Project 2061 staff, are explored in three later sections: "Entering the National Reform Fray: *Benchmarks for Science Literacy*"; "The School District Centers"; and "The Conception of Reform in Project 2061."

Other Participants. In addition to the Washington office and center personnel, Project 2061 has invited a large number of scientists and educators to participate in various phases of the reform process. Over 300 professionals, for example, contributed to the creation of *Science for All Americans*: 8 to 10 scientists on each of five scientific panels, 5 to 22 consultants to each panel, 85 advisors to Project 2061 staff, and over 240 reviewers. More recently, Project 2061 has commissioned 13 experts or teams of experts to write Blueprint reports—concept papers on aspects of the educational system that must be changed to accommodate curriculum reform. For example, Robert Floden, James Gallagher, and Mary Kennedy at Michigan State University, agreed to coordinate the Teacher Education Blueprint; Robert Donmoyer, at Ohio State University, the School Organization Blueprint; Ronald Good, at Louisiana State University, the Research Blueprint; Sharon Lynch, at George Washington University, and Cora Marrett, at the National Science Foundation, the Equity Blueprint; and Wayne Welch, at the University of Minnesota, the Assessment Blueprint (Project 2061 1991b).

Key Events and Products

Since its inception in 1985, Project 2061 staff, center participants, and educational consultants have worked to advance the project's reform agenda: to develop a set of tools and resources to help schools and districts design their own science curricula, and ultimately, reform the educational system (AAAS 1993). Below, we provide a brief overview of key events and products in 2061's history: the publication of *SFAA*, the adoption of the six school district centers and subsequent summer institutes, the creation of *Benchmarks*, and the ongoing development of *Blueprints for Reform* and *Designs for Science Literacy*. In so doing, we also attempt to highlight ways in which the project has grown and changed over time; shifts in the project's time line, priorities, strategies, and structures. Figure 1 provides a summary of Project 2061's tool and resource materials.

Science for All Americans. *SFAA* is a set of recommendations "on what understandings and ways of thinking are essential for all citizens in a world shaped by science and technology" (AAAS 1990, p. xiii). The document

> is the result of a three-year collaboration involving several hundred scientists, mathematicians, engineers, physicians, philosophers, historians, and educators. It is, we [the Project staff] believe, as close as it is possible to come to a valid expression of the view of the science community on what constitutes literacy in science, mathematics, and technology" (AAAS 1990, p. x).

Figure 1.   Project 2061 Vision of Systemic Reform

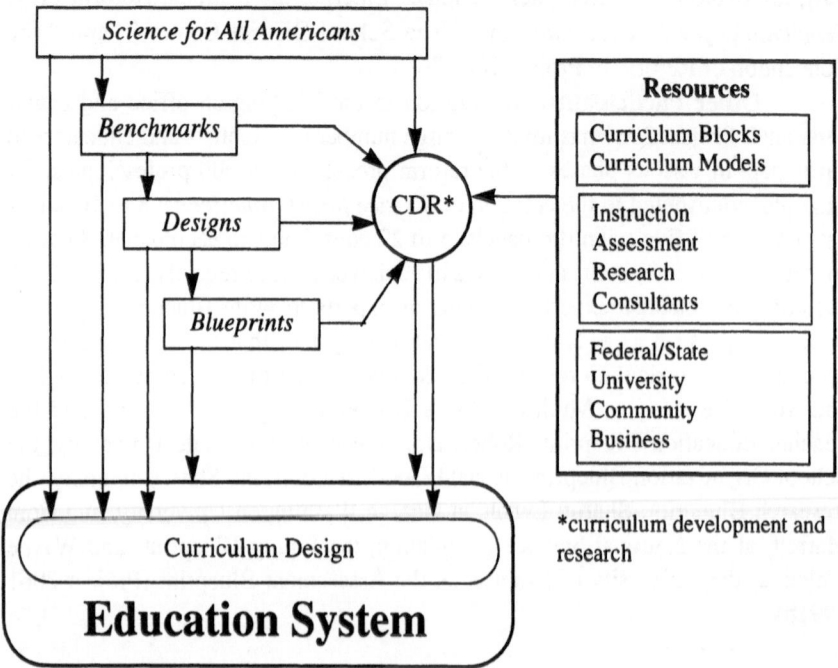

*curriculum development and research

First published in 1989, *SFAA* remains a bestseller in science education circles. We provide further details about the development and content of this seminal document later in this paper.

SFAA also presents Project 2061's original time line for educational reform; it offers a vantage point from which to view changes in the project's direction and pace. As first conceived, Project 2061 was divided into three phases of reform. Phase I was to define the substance of scientific literacy, to culminate in the publication of *SFAA*, and to take two years. Phase II was to translate the literacy goals specified in *SFAA* into curriculum models and *Blueprints for Action* (later renamed *Blueprints for Reform*). Phase II was expected to take an additional three years. Finally, Phase III was to be a widespread collaborative effort in which scientists and educators would use the resources of phases I and II to change educational practice, and thus, to move the nation toward scientific literacy. Phase III was projected to take a decade or more.

As will soon become evident, Project 2061 found this original time line impossible to follow. Project participants discovered each task was more extensive and difficult than originally envisioned. In addition, they concluded that phases II and III of the reform process could not be separated. Indeed, Ahlgren notes that today, the project no longer divides its time line into phases. And, although already at work for a decade, participants expect to continue their efforts for a minimum of five more years.

Summer Institutes. After publication of *SFAA*, Project 2061 staff brought six school district teams and approximately 150 team members on board. During four consecutive summers from 1989 to 1992, these team members met with each other, the Project 2061 staff, and project consultants. These four four-week summer institutes are noteworthy for several reasons. One, they offered Project 2061 staff the opportunity to deepen team members' understanding of the vision, progress, and products of the project. Two, to a lesser extent, they furnished team members with uninterrupted blocks of time to develop resource materials and implementation strategies. Three, they provided the setting for certain tensions between the Washington office and center personnel to be played out. And four, it was during these summer institutes that the Washington office staff learned much about teachers and school districts.

At the first four-week summer institute, held at the University of Colorado at Boulder, team members attended seminars on a wide array of topics in science, mathematics, and technology. Each week of presentations was organized around one of the following themes: microscopic explanations of macroscopic phenomena; evolution as a unifying theme in biology; risk and probability; and the interaction of science, mathematics, and technology (Lynch and Britton 1992). Team members were also divided into cross-team working groups. One of these working groups, the "strand" group, was introduced to the idea of backmapping (the precursor activity to development of benchmarks). In backmapping, one takes a statement, a science literacy goal from *SFAA*, and tries to determine what kinds of information or experiences students need to have in K-12 to achieve it.

Some team members found this first summer institute a valuable use of their time and effort. Others did not. Andrea Foster, a teacher from the San Antonio team, thought the weeks at Boulder worthwhile. For her, "it marked the beginning of a great collaboration between scientists, researchers, and educators." Linda Nott, also from San Antonio, grew frustrated with the kind and tenor of interactions between teachers and Washington office personnel. She explained:

> The Washington staff was very conceited in their approach to "training" us . . . Somehow, they were going to "bring us up to speed on science content and pedagogy" so that we could be useful to them in the long run. The misconception they had then, and continue to have now, is that we are putty that they must mold into something that is worthwhile to them.

The San Francisco team members were not entirely satisfied with their Boulder experience either. They had begun working on their curriculum model—an outline of a K-12 science, mathematics, and technology curriculum—prior to Boulder. They had intended to continue work on their model that summer.

The second institute was held in the summer of 1990 at the University of Wisconsin at Madison. Here, all teams were given the assignment of backmapping. Ahlgren and Rutherford explain this process in greater detail:

[In backmapping,] educators work in multigrade-level teams to think through the likely growth of understanding over K-12. Mapping involves identifying a plausible sequence of levels of understanding for every major idea in *SFAA*, including necessary "precursor" ideas that do not appear explicitly in its post-graduation recommendations. Then the process involves discerning connections among the ideas and establishing approximate grade levels for every step. Backmapping requires both logical structure of science and an understanding of learning, gleaned from teachers' experience and from research into how children learn (Ahlgren and Rutherford 1993, p. 20).

The six teams rarely met together during this summer institute. Instead, each team worked on its own set of backmaps at different locations around the university. Again, some team members thought that backmapping was an important exercise, ultimately leading to the development of benchmarks.

Approximately 120 team members attended the third summer institute at the University of Washington in July 1991. Here, each team presented a draft of its K-12 curriculum model—a draft members had been working on for two years (AAAS 1991c). Presentation and discussion of these models proved a source of considerable disagreement: Project 2061 staff thought the team models conceptually incomplete; team members felt their work not thoroughly appreciated. A more detailed description of these differences in perspective is presented later in this paper.

The fourth and final institute was held at Cornell in the summer of 1992. Team members split into four groups to work on one of the following: benchmarks and curriculum blocks, a computerized design and resource system (the precursor of *RSL*), a communications strategy, and an implementation plan. To some extent, this summer institute served to structure later work at the sites. The computerized design and resource system group, or resource and technology group, remained particularly active at the site level for several years: members of this group received separate funding, conducted technology training sessions for other participants, and met nationally three times a year. (This group no longer exists; technology training has been integrated throughout each center.) The project's sizable documentation effort also arose from interest expressed by teachers at this summer institute.[3]

---

[3]In 1993, Project 2061 centers began systematic and rigorous documentation of reform activities. This documentation effort arose, Brearton notes, from center participants' requests for assistance in capturing what they were learning as a result of curriculum development, dissemination of *SFAA* and *Benchmarks*, and implementation of the curriculum models. There were a number of additional reasons for pursuing documentation: to inform Project 2061 and other educational reform efforts, to build credibility for more funding, and to enhance teachers' ability to be reflective practitioners. Thus, Mary Jo McGee-Brown, of the University of Georgia, with help from Brearton and the centers, translated these participants' interest in qualitative research into the design of a documentation workshop and a procedure for submitting information gathered at the centers for compilation and analysis.

Benchmarks for Science Literacy. *Benchmarks for Science Literacy* serves as a companion document to *SFAA* as a tool to be used in curriculum development. Published in 1993, *Benchmarks* identifies what all students should know and be able to do in science, mathematics, and technology by the end of grades 2, 5, 8, and 12. The 400-page report includes over 830 goal statements, or benchmarks, for student learning. Individual benchmarks are supported by introductory essays and informed by educational research (AAAS 1993). Impetus for and development of the *Benchmarks* document is described later in this paper.

The creation of *Benchmarks* marked a major shift in Project 2061's priorities. After publication of *SFAA* in 1989 and subsequent adoption of the six school district centers, project participants first turned their attention to development of alternative K-12 curriculum models. Development of these models was expected to be "the main creative activity of Phase II" (AAAS 1990, p. 221). A curriculum model is defined below:

> A curriculum model for a set of subjects and grade levels is most obviously a design for how students spend their time. But in addition, any curriculum model should contain some other basic features: goals to be attained, purposes of education and principles of teaching and learning that it assumes, and requirements that all these model features place on the educational system (Project 2061 1992).

However, after the six sites presented drafts of their curriculum models during the 1991 summer institute, Project 2061 staff appeared to lose interest in them. Instead, the staff's attention was drawn to talk of national science standards. Refinement and implementation of the curriculum models were placed on hold. The teams were directed to use their knowledge of *SFAA* and their backmaps to develop precursor learning goals; they stopped discussion and refinement of curriculum models to generate lists of benchmarks. Indeed, the process of backmapping and the development of the *Benchmarks* document became the focus of project work.

Blueprints for Reform. *Blueprints for Reform* is one of several Project 2061 tools under development at the time this case study went to print. When finished, *Blueprints* will identify aspects of the educational system in need of reform; it will recommend changes in views, practices, and structures to better meet the goals of 2061. In making its recommendations, *Blueprints* will synthesize the ideas and arguments of 13 individual Blueprint reports. As stated in a previous section, these reports were commissioned by Project 2061 staff and written by individuals or teams of educational experts. Topics covered in these Blueprint reports include the following: teacher education, assessment, materials and technology, curriculum connections, school organization, higher education, equity, business and industry, policy, parents and community, research, and finance (Project 2061 1991b).

The development of *Blueprints for Reform* is noteworthy for two reasons. First, it provides a specific example of the project's shifting time lines and priorities.

Originally, the Blueprints were to be a cornerstone of phase II efforts, closely tied to the development and refinement of the curriculum models. Blueprint writers were expected to begin writing in the fall of 1991. Drafts were to be reviewed and revised in 1992 (Lynch and Britton 1992). And the Blueprint reports were slated for publication in the spring of 1993 (AAAS 1991b). Instead, 2061 deferred work on the Blueprints to immerse itself first in the development of curriculum models, and then in the writing of *Benchmarks for Science Literacy*. The 1993 publication deadline came and went.

The synthesis document, *Blueprints for Reform*, is targeted for publication in 1997. (Separate release of at least some of the Blueprint reports is being considered.) Blueprint authors began submitting their first drafts in 1994. These drafts varied in length and imagination. Their close ties to the curriculum models were never found. Review and revision of these drafts remain in progress. Each draft Blueprint report has been reviewed by approximately a dozen experts in the field, deliberately selected to include some with divergent views, and then revised in response to reviewers' suggestions. Project 2061 has sponsored two Blueprint conferences—one on teacher education and one on materials development—to spark further discussion. Roseman has raised the possibility of holding such conferences for each document. In addition, the project has commissioned Michael Kirst of Stanford University and Steve Schneider of Far West Labs to synthesize each Blueprint report into a standard document of approximately 10 pages.

Second, the development of *Blueprints* underscores important questions about the process of systemic reform: From where does knowledge about changing the educational system come? Who are the experts in this arena? Ahlgren and other members of the Washington office have not been entirely satisfied with the depth and breadth of recommendations for systemic change made by Blueprint authors. Ahlgren notes:

> There's not a long history of changing anything about the educational system—other than school lunches, perhaps. So that's a much harder job and for that we need to invite more people still to help us with the array of Blueprints. The Blueprints, if you've seen them, are turning out to be not bad at all as a way of sketching what the current situation is and its problems, but are much less successful, so far, in advising on how to solve all those problems. That's not a surprise, but it certainly makes them more like site surveys than blueprints for houses to be built.

Ahlgren readily admits that Project 2061 staff and center personnel do not possess the requisite understanding of systemic change either. Project staff, he explains, are experts in science education, not in educational change:

> We are . . . sort of at the pinnacle of what we know we know how to do well. We're science educators. We knew we were as competent as anyone to make a sketch of science literacy, what it would look like, and then to break it down by

grade levels—through our acquaintance with the research and researchers and our ability to think about these strands of connected thoughts—to write *Benchmarks*. Experts in changing systems, we are not.

At present, then, many of the questions planned to be addressed in *Blueprints* remain unanswered.

Designs for Science Literacy. *Designs for Science Literacy* is another Project 2061 tool under development. "While *Benchmarks* and *SFAA* defined goals for science literacy, *Designs* will provide a systematic process for planning curricula to reach those goals" (AAAS 1994b). Couched in the language and philosophy of engineers and architects, *Designs* is expected to examine decisions central to planning a curriculum; consider opportunities, constraints, and tradeoffs implicit in curriculum design; describe the kinds of activities to include in a curriculum; and suggest different ways of configuring those activities (AAAS 1994a and AAAS 1994b). The nature and substance of *Designs* has been revised and refined over time. A recent outline, for example, was modified in response to "cries for help" from the field, Roseman says: the topic of "Getting Started" was expanded from a single chapter to an entire section of the book. At last check, the document was slated for publication sometime in 1996.

Designs marked a shift in Project 2061 strategy, a different process of tool development. In creating the other 2061 documents—*SFAA, Benchmarks*, and *Blueprints*—project staff drew from the work of a body of experts—scientists, teachers, or researchers—brought together for a clear purpose: to offer a conception of scientific literacy, to construct a road map to achieve literacy goals, or to identify needed reforms in the educational system. In writing *Designs*, however, the Project 2061 staff members have chosen a different route. Rather than convene a new group of experts, they have decided to synthesize work already completed as part of their reform effort. As they explain in an early outline:

> The intellectual basis for the content of *Designs* comes from four sources: the experience, extending over a period of more than four years, shared by Project 2061 school-based team members, expert consultants, and Project staff; the draft reports of the 12 [now 13] teams of *Blueprints for Reform* authors; commentary in the literature by thoughtful, informed writers; and the research literature (Project 2061 1994a).

Over this last year, the project staff has begun to draw on additional resources. Roseman identified some of these sources in a recent letter:

> I think the R&D [research and development] coming out of the workshop effort is contributing significantly to *Designs*. Of course we will "mine" the team models and what they are learning as they attempt to implement them . . . But there are many other schools and districts who have been using *SFAA* and *Benchmarks* to guide their reform. We would be remiss if we looked only at six sites, particularly since they had four years of special experiences and resources.

Shifting Time Lines, Priorities, Strategies, and Structures

As we hope is now evident, Project 2061 has experienced major shifts in its time lines, priorities, strategies, and organizational structure. In attempting to highlight these various shifts, we are in no way criticizing project participants; we interpret these changes in mid-course as signs of strength and self-reflection, not of weakness or indecision. Indeed, we think it important to highlight shifts in Project 2061's reform effort to drive home three points. First, the project is in a continual state of evolution. The description of Project 2061 presented here in this case study does not quite match descriptions of it yesterday and certainly will not fit such descriptions of it tomorrow.

Second, the willingness to change, to keep as many options open for as long as possible, is a trademark of Project 2061. Ahlgren has explained that the project staff consciously keeps its options open. Even decisions labeled as final are never treated as such.

Third, and most importantly, a degree of flexibility is necessary in any long-term effort, if that effort intends to survive and grow. Foster, a teacher from San Antonio, provided an eloquent explanation of the balance Project 2061 has achieved between following a prescribed course and needing to change over time:

> I think [the ability to change over time] is the basis of truly understanding reform in science education. Project 2061 is dynamic by the mere nature of how it was designed. So why then do we feel compelled to adhere to phases and stages? Have we really fallen far behind? Or is this the path that reform inevitably takes? We set deadlines, they go by, we set more, they go by . . . this is how it should be, no?

◆

Launching the Reform: *Science for All Americans*

The Washington office staff of Project 2061 considers the goal-oriented nature of the project to be one of its most important and distinguishing characteristics. This feature of the project is epitomized in the first of its major publications, *Science for All Americans* (AAAS 1989), which describes the goals—or learning outcomes—toward which science education in the United States should be directed. The task of preparing *SFAA* was seen from the outset as exclusively a responsibility of the practicing scientists, not teachers.

From Panel Reports to *SFAA*: The Role of the Scientific Community

As part of phase I of the project, groups were convened in 1985 to represent five areas of science: biology and health, physical science, mathematics, social and behavioral science, and technology. Each group, or panel, consisted of 8 to 10 scientists who, for the most part, were located in roughly the same region of the country (for example, the mathematics panel consisted of mathematicians from the greater San Francisco Bay Area). The panelists were recognized and respected figures within their fields. Most were actively and productively engaged in research in academic institutions. A few were from industry.

These experts were charged with the task of outlining the fundamental concepts that underlie their respective fields. To help the experts in this task, the project provided some guidelines (though it is unclear exactly when these were instituted). First, the panels were to identify only those concepts and skills that are scientifically significant and that can serve as a foundation for a lifetime of individual growth. Second, the knowledge and skills selected were to be those most likely "to prepare students to live interesting and responsible lives" (Appley and Appley 1989, p. viii). Individual growth and satisfaction were to be considered as well. Third, members of the panel were to begin with a clean slate. Specifically, a recommendation could not be justified solely because it was currently being taught. Fourth, the panelists were to ignore the limitations of the existing education system. This "blue sky" approach allowed the panelists to assume that money, time, materials, and other resources would not restrict educators from achieving the desired learning outcomes. Fifth, the project asked panelists to take a "less is more" approach: they were to identify only a small core of essential knowledge and skills rather than cover the entire breath of the field. Finally, the recommendations made by the panels were to serve all students, "regardless of sex, race, academic talent, or life goals" (Appley and Appley 1989, p. ix).

This assignment was challenging. It is far from obvious to the expert what concepts in one's field should be known to the nonspecialists. Particularly for the cutting-edge researchers steeped in the most advanced research in a field, a significant conceptual reorientation is required to identify the material that everyone should know.

The panelists met for over a year, usually convening on a monthly basis. From accounts of these meetings, the participants were said to be engaged deeply in their tasks. Deliberations were diligent and intellectually challenging. The resulting five documents—now known as the panel reports—were to make up five of the seven chapters of *SFAA*. Ahlgren and Rutherford planned initially to add introductory and concluding chapters to complete the book. However, this plan changed as work on the reports progressed, as Ahlgren explains:

> It turned out that . . . what the panels produced was not directly usable. We kept it completely open. We said, "Write whatever you have to. The point is: what should adults be like in their understanding of science? We want you to describe that. You've got 15 pages. It can be a poem. It can be a song. It can be a test. It can be whatever you like. We're not going to constrain you." . . . As they started to actually produce some things, we started to say, "Well, this isn't quite as helpful as it could be," and began gradually to form an opinion about what it should be like, as we saw [it]. I think they thought that we really knew what we wanted all along and were keeping it behind our back. But the simple fact was we thought we would use their reports and sandwich them between an introduction and a summary and publish them. It was only as we got into it a

couple of years, we saw that what they were producing took very different degrees of relevance among the different panels for different reasons . . . By the time we got around to trying to constrain them—"do this, don't do that"—they were to into their own thing to be able to respond.

Ahlgren goes on to provide an example:

We wanted their reports to be about what adults would know, not about how people would learn things initially. We considered the scientists experts [who] had really something to tell us about what their fields were like and what it would take to understand that knowledge and skill, but really were not very knowledgeable about how people learn . . . We didn't want them to tell us about how to teach. [But] they could not stop themselves.

As a result, the panel reports provided 2061 with varying degrees of insight and ideas—some of the reports were more closely aligned with 2061's vision than others. For example, according to Ahlgren:

One of the panel reports was almost right on target. We reshuffled a bit and rewrote it a bit for the style, but the language and the topics were very much what got in the report. [For] another panel report, [the authors] never figured out what we wanted. It was almost all *sales* for the importance of the subject and almost *nothing* about "what would kids actually learn?". Finally, at the end of that, I said, "Well where are we going to get this stuff about what kids would actually learn?" "Oh, you can go to the library and look that up. We just had to establish the need."

Thus, Rutherford and Ahlgren recast each of the reports to varying degrees. After several rounds of review and editing, the book was approved by the AAAS board. According to Rutherford, this was a rather painful process. "We went through all that external review, [then] through [the Project 2061] council, through the board . . . it took another six months . . . It was a long, hard thing, but I think it was very valuable."

Difficult though it was, getting the approval of the AAAS Board gave *SFAA* the imprimatur of the scientific community. However, whether *SFAA* represents the opinions and ideas of most scientists is debatable. For example, Ahlgren reports that members of the social sciences panel objected strongly to the project's suggestion that the chapter be *primarily* about methods rather than about what the social sciences have found out about the world. "To our surprise," Ahlgren recalls, "the social science people took great umbrage at this because it was suggesting that the social sciences were different from the natural sciences in some way. They didn't want to be treated differently." In addition, some panel members publicly opposed some parts of *SFAA* at the 2061 National Council meeting. They argued that some of the writing no longer represented their views and was not a completely accurate portrayal of the knowledge and skills of their field. In the end, all five panels signed off on the final draft.

In 1989, the five panel reports were published as independent, stand-alone documents. *SFAA* was also first published in 1989. Today, it is in its eighth printing.

The Content of *SFAA*

According to Project 2061, *Science for All Americans*

> recommends a coherent set of learning goals in science, mathematics, and technology for all high school graduates. *SFAA* recommends not only specific, interconnected understandings in these three domains, but also habits of mind essential to science literacy and some principles to guide teaching and learning for science literacy (AAAS 1994d, p. 8).

To satisfy these multiple purposes, the authors created a 15-chapter book. In *SFAA*, chapters 1, 2, and 3 focus on the nature of science, mathematics, and technology, respectively. The authors describe similarities and differences among these three enterprises, the reliance on evidence and use of imagination, and ways they identify and avoid bias. Two examples, provided below, offer some insight into the substance and tone found in these three introductory chapters.

For example, in chapter 1, "The Nature of Science," the authors describe the nature of scientific inquiry. They do so in a discursive form which preserves the rich and complicated character of inquiry:

> Scientific inquiry is not easily described apart from the context of particular investigations. There simply is no fixed set of steps that scientists always follow, no one path that leads them unerringly to scientific knowledge. There are, however, certain features of science that give it its distinctive character as a mode of inquiry. Although those features are especially characteristic of the work of professional scientists, everyone can exercise them in thinking scientifically about many matters of interest in everyday life (AAAS 1990, p. 4).

In chapter 2, "The Nature of Mathematics," the authors present a description of mathematics. Here, we get a flavor for the authoritative tone of much of the text: there is little question, according to the authors, as to what is or is not mathematics. One is told:

> Mathematics is the science of patterns and relationships. As a theoretical discipline, mathematics explores the possible relationships among abstractions without concern for whether those abstractions have counterparts to the real world. The abstractions can be anything from strings of numbers to geometric figures to sets of equations. In addressing, say, "Does the interval between prime numbers form a pattern?" as a theoretical question, mathematicians are interested only in finding a pattern or proving that there is none, but not in what use such knowledge might have (AAAS 1990, p. 16).

Mathematics is more than just a theoretical abstraction, however. The authors go on to describe mathematics as an applied science—one in which "mathematicians focus their attention on solving problems that originate in the world of experience" (AAAS 1990, p. 16).

Chapters 4 through 9 "present the picture that science currently paints of how the world works" (AAAS 1994d, p. 8). The first of these five chapters, "The Physical Setting," describes basic knowledge about the overall content and structure of the universe on astronomical, terrestrial, and sub-microscopic levels, as well as the physical principles on which the universe seems to run. Chapter 5, "The Living Environment," considers the knowledge we have of living things, how they function and interact with one another and their environment. The chapter focuses on six major subjects:

> The diversity of life, as reflected in biological characteristics from one genera-
> tion to the next; the structure and functioning of cells, the basic building blocks
> of all organisms; the interdependence of all organisms and their environment;
> the flow of matter and energy through the grand-scale cycles of life; and how
> biological evolution explains the similarity and diversity of life (AAAS 1990,
> p. 59).

People are the topic of chapters 6 and 7. "The Human Organism," chapter 6, presents our species as one that is both similar to and different from other living things. Again, a passage from the chapter gives the tenor of the wording and presentation. "Within a few hours of conception, the fertilized egg divides into two identical cells, each of which soon divides again, and so on, until there are enough to form a small sphere. Within a few days, this sphere embeds itself in the wall of the uterus, where the placenta nourishes the embryo" (AAAS 1990, p. 74). Thus, the science is described with minimal vocabulary, technical language, or detail.

Chapter 7, "Human Society," discusses individual and group behavior, social organizations, and the process of social change. In the following chapter, "The Designed World," the authors present the way people attempt to shape and control the world through technology. Then, in chapter 9, "The Mathematical World," we read about the basic mathematical ideas that play a key role in almost all human endeavors.

The final five chapters are of a slightly different nature. In chapter 10, "Historical Perspectives," the authors present 10 historical examples that illustrate how the scientific enterprise works, and how and why science progresses. Examples include moving the continents (the theory of plate tectonics), uniting the heavens and earth (the Newtonian revolution), splitting the atom (radioactivity and nuclear fission), and explaining the diversity of life (Darwin's theory of evolution). Each of these, according to the authors, is of exceptional significance to our cultural heritage.

Chapter 11 presents four important themes that pervade the disciplines and can serve as tools for thinking about phenomena as diverse as ancient civilizations, comets, and the human body. These cross-cutting themes are said to be outside the content of any particular field of study. They include: systems;

models; stability, constancy, and change; and scale. The concept of systems, for example, is defined as:

> Any collection of things that have some influence on one another . . . The things can be almost anything, including objects, organisms, machines, processes, ideas, numbers, or organizations. Thinking of a collection of things as a system draws our attention to what needs to be included among the parts to make sense of it, to how its parts interact with one another, and to how the system as a whole relates to other systems. Thinking in terms of a system implies that each part is fully understandable only in relation to the rest of the system (AAAS 1990, p. 166).

With regard to the theme of change, *SFAA* authors state, "Descriptions of change are important for predicting what will happen; analysis of change is essential to understanding what is going on, as well as for predicting what will happen; and control of change is essential for the design of technological systems" (AAAS 1990, p. 174). They then continue their discussion of change by distinguishing between trends, cycles, and chaos.

The authors make recommendations in chapter 12, "Habits of Mind," regarding the values, attitudes, and skills associated with science, mathematics, and technology. The first part of the chapter focuses on four specific aspects: the values inherent in science, mathematics, and technology; the social value of science and technology; the role of social values in science; and people's attitudes toward their own ability to understand science and mathematics. The second part of the chapter focuses on skills related to computation, estimation, manipulation, observation, communication, and critical responses to arguments.

A rationale for teaching such scientific values is provided in the introduction to the chapter: "To the degree that schooling concerns itself with values and attitudes . . . it must take scientific values and attitudes into account when preparing young people for life beyond school" (AAAS 1990, p. 183). The authors' description of scientific values provides some sense of the message of the chapter:

> It is also important for people to be aware that science is based upon everyday values even as it questions our understanding of the world and ourselves. Indeed, science is in many respects the systematic application of some highly regarded human values—integrity, diligence, fairness, curiosity, openness to new ideas, skepticism, and imagination. Scientists did not invent any of these values, and they are not the only people who hold them. But the broad field of science does incorporate and emphasize such values and dramatically demonstrates just how important they are for advancing human knowledge and welfare (AAAS 1990, p. 185).

When discussing computation and estimation skills, the authors get slightly more specific as well as behavioristic. That is, they outline what it is students should be able to *do*:

To make full and effective use of calculators, everyone should also be able to do the following: read and follow step-by-step instructions given in calculator manuals when learning new procedures, make up and write out simple algorithms for solving problems that take several steps, figure out what the unit . . . of the answer will be from the inputs to the calculation . . . round off the numbers appearing in the calculator answer to a number of significant figures . . . and judge whether an answer is reasonable by comparing it to an estimated answer (AAAS 1990, pp. 189-90).

In chapter 13, "Effective Learning and Teaching," the authors describe principles of learning and teaching. Using these principles, they then make recommendations for how these learning goals might be achieved. For example, the authors present a theory of learning reflecting a conceptual change perspective. As such, they go on to suggest that students "be encouraged to develop new views by seeing how such views help them make better sense of the world" (AAAS 1990, p. 199). The authors also advocate the use of group work and activities while discouraging memorization of technical vocabulary.

In the final chapter of the book, "Reforming Education," the authors assert that there is a need for reform in science, mathematics, and technology education. They then present the premises that underlie the 2061 approach to reform: it takes a long time, collaboration is essential, teachers are central, it must be comprehensive, it must focus on the science learning needs of all children, and it requires positive conditions.

Scope of the Content. Science, according to Project 2061's definition, includes the social and behavioral sciences, mathematics, the natural and physical sciences, and technology. As Rutherford argues, "Science literacy is concerned less with having students understand the disciplines, as such, than with having them understand the world through the eyes of science. Thus, science literacy draws on all of the natural sciences, the social sciences, and mathematics, and statistics" (Rutherford 1993, p. 12). This notion appealed to teachers who were moving toward integrated or interdisciplinary curricula, as Foster explains: "Schools impose artificial boundaries on curriculum and subject matter. *SFAA* is a document that makes us think about dissolving those boundaries in an effort to help us come to know our world and how it works. Math, science, technology . . . who cares? It's a thoughtful society that we're after." For center director Bernard Farges, this appealed to ideas he had had for years:

The world out there is messy. Problems don't come to you as math, science, or technology problems. You build a dam—well, of course there is some math involved, some physics involved, there are also some social issues involved, some environmental issues involved. You have a whole set of problems in the challenge of building that dam. So it is the way the world comes to us. That is why we want to prepare the students to deal with that. We should be modeling that in the way we are teaching.

Unfortunately, as explored in chapter 6, there were those who did care; this sweeping definition of science literacy ruffled more than a few feathers. Some felt the project was trespassing into their territory. Ahlgren reports: "The technology people, by and large, weren't unhappy about being tucked in under science. I think they saw it as a helpful way of designating things, as an interim anyway . . . The math education people certainly felt distanced by that, although they haven't welcomed our opinions anyway, so it's hard to know."

Breadth of the Content. The "less is better" approach of the project required authors to make decisions about what content was worth teaching. Such decisions were made by considering the cost effectiveness of teaching certain topics and concepts. *SFAA* authors asked themselves, "How much do you learn from this compared to how much time it takes to learn it?" For example, the authors decided to eliminate discussion of the periodic table. As Ahlgren explains:

> The periodic table: I remember thinking about it and arguing with myself about it . . . To get something out of the periodic table takes a tremendous amount of input and unless you know something about exclusion of the electrons from the same orbit, why should orbitals fill up and so why should you have to start a new series and why that's now causing new properties to appear again. The judgment was, "There's too much. To get something out of it, there's too much input and it's too hard" . . . It wasn't going to be worth it.

Breadth does not only include what to teach. Authors also needed to determine the level at which these concepts selected for inclusion should be taught and learned. With regard to the level of content in *SFAA*, Roseman argued that it is appropriate to aim high.

> I think *Science for All Americans* is a very optimistic set of goals, but it's *only* when we see how real kids do in *good* schools, with *good* teachers and *good* materials over 13 years, that we'll learn what's really possible and what isn't. I'm not willing to back off on *SFAA* now. I'm not willing to say, at this point in time, "Well, we should only do part of this for some kids."

The Educational Community Responds

How did the educational community receive *SFAA*? With more than 100,000 copies in circulation, *SFAA* is the best selling book in the history of AAAS (AAAS 1994c). In addition, many state science frameworks, including California's (Science Curriculum Framework and Criteria Committee 1990), cite *SFAA* as an influential document. It is also being used in some teacher education programs, as well as in numerous schools where it serves as a tool to help organize teachers' efforts. In addition, in the *Draft of National Science Education Standards* developed by the National Research Council (NRC), the authors acknowledge the influence of *SFAA* on the standards:

> The many individuals who have developed the content standards sections of the National Science Education Standards have drawn extensively on and have made independent use and interpretation of the statements of what all students

should know and be able to do that are published in *Science for All Americans* and *Benchmarks for Science Literacy*. The National Research Council of the National Academy of Sciences gratefully acknowledges its indebtedness to that seminal work by the American Association for the Advancement of Science's Project 2061 (NRC 1994, p. I-2).

From Skepticism to Acceptance. Rutherford acknowledged that designing *SFAA* without reference to schools was an unpopular decision initially. At present, however, the book has been well-received by both science educators and reformers. Ahlgren explains: "I think there were science educators, early on, who were very worried about our procedure of using scientists first, but once they saw the product they got to feel much better about it."

Why was the book so popular? First, it appears readers were attracted to the content of the book: it covered the vital and interesting topics students need to understand the natural world around them. Sue Matthews, a teacher, expressed her high regard for the book:

> I think that there is a certain level [of scientific knowledge] that everyone needs to have. I don't know whether it is as in depth as *Science for All Americans* is, to be quite honest with you, but I do think that a lot of what is in *Science for All Americans* is science for *all* Americans—that they really need an understanding of that. One of the things that drove me crazy, the first time I even read through it (and I started with one chapter and I read all the way through it), I kept saying, "Why don't they just come out and say, instead of going through this long explanation of what this process is? Why don't they just say photosynthesis? Why don't they just say it? Over and done with it. This is photosynthesis." And then I began to realize and appreciate the carefulness with which the book was written so that it could *not* be turned into a multiple choice test. The concept of photosynthesis is so incredibly complex that you just can't write down the equation and have the kids learn it . . . [They need to] understand the photosynthetic process, [without] which none of us would be here.

Second, in many ways, the book was cutting edge without being radical. For example, the nature of science described in the book is progressive—including imagination and creativity as part of the scientific method is not something found in many science texts—but not outside of the mainstream. The authors did not overwhelm readers with overly complex or revolutionary ideas about what science is.

Third, the book's appeal may be due not only to its content, but presentation: it's a simple text lacking the technical jargon found in so many science books. Ahlgren explains: "The biggest contribution of *Science for All Americans* was to demonstrate ideas that were valuable, worth knowing, and didn't involve all the detail that typically is associated with them in science education. A real simplified scientific view of the world." A teacher echoes this point of view:

I took a copy on the airplane with me one day and I was reading it and it just crystallized everything that I had been learning about the reform effort. The chapter, I think it's called "The Nature of Math and the Nature of Science," [was] just so easy to understand and so clearly talked about what mathematics was and what science was and what the relationship between the two are. It just really got me excited because it was so easy to understand. It gave the essence of the entire reform movement to me. I think anybody could have [SFAA] . . . and know everything they need to know about math and science.

The ease and comfort with which many educators read and understood this document may be attributed to its discursive style. In fact, the writing style distinguishes this publication from other documents both within and without the project. Unlike *Benchmarks* or many state frameworks, the book is not a list of bulleted items that students should know. Rather, it reads like a narrative which, for many, is interesting, easy to understand, and engaging.

Finally, the book attracted many teachers with whom we spoke because it legitimated their work. That is, it provides a rationale for teaching certain topics and leaving others out, for using cooperative learning, and for focusing on concepts over vocabulary. One teacher explains,

I've had to justify the way I do things differently from other people in my school, other people in my school system. And I've used *Science for All Americans* to do that, to a certain extent. But now I feel as though I have more support to go back and say, "There is a reason that I'm doing it this way and this is that reason."

In many cases, teachers did not consider the book's ideas new or earth-shattering. Rather, the opinions expressed in the book comported with their own ideas about schools and teaching. Finally, they had company. Finally, there was someone out there backing them up. Nott states, "Those who did sign on, and especially those who stayed the course, . . . saw their own philosophy of public school and literacy articulated in the text of *Science for All Americans*."

From Acceptance to Understanding (and Misunderstanding). While the document has been enthusiastically embraced by many educators, it has not always been well-understood. Ironically, the discursive style of presentation may have been the source of some problems. For example, Roseman explains: "*Science for All Americans* . . . was elegant prose but misled people . . . in the sense that it was such an easy read for many, and it was also easy for people to read into it—so that if the word DNA appeared, people imagined all that they would normally teach and they would say, 'Yes. I'm doing that.'" While it may be an easy read, Roseman maintains, "It's a hard think—and not until you dig into it over many years and engage in discussions with the authors that you appreciate not only what's there but what's *not* there and why those decisions were made."

"The narrative style has become very popular," according to Ahlgren. "I think it also leaves a lot of room for people to read in their own ideas. 'Oh, you have something about the cell. Yes, this is nice.' Well, that means we can do all the old stuff we used to do about the cell because it's the same topic." Ahlgren continues:

> People will come up at a meeting or a presentation and say, "We based every-thing we did on *Science for All Americans*. It was wonderful! We're so glad to meet you and have you here. Here, look at our curriculum." And here's floating and sinking and series and parallel circuits and we think, "My God. How can they think that this was consistent with what we tried to sell as an idea of what science education is?" It's minutiae, it's hands-on with no reflection, it's end-less details of things with no large concepts or connections.

Of all the chapters in *SFAA*, chapter 11, or the common themes, may be the most widely misconstrued and misused. The themes, 2061 staff argued, are not meant to be used to organize curriculum and they cannot be taught until after the students have had a wealth of experience with topics through which the themes can later be explored. Yet, some educators and curriculum developers have inter-preted the themes as ideas around which curriculum can be designed. Thus, a unit devoted to the theme of scale or models is, according to Project 2061 staff, an example of how the ideas of *SFAA* have been unadvisedly applied.

Another particular dilemma faced by the book's authors and users alike is understanding what is meant by goals. Brearton explains, "Clearly *Science for All Americans* is about goals. But that can be said and be lost, because . . . so often when you say to teachers, "We have to write a goal and rationale for our goals,' they respond, "Well, let's do that quickly, because that won't mean any-thing.'"

Ahlgren suggested another misunderstanding that often arises as educators attempt to understand the book: "The distinction between the set of goals and what the curriculum should be like is very important . . . and not at all self-evi-dent. We have many people who will look at *Science for All Americans* and say, 'I don't think you should separate learning about *Habits of Mind* from learning about physical science.' But that's just in the book! We have to order them in *some* way. We can't have *random* access to miscellaneous goals."

From Understanding to Use. While *SFAA* became increasingly popu-lar, some educators experienced frustration in using the book to inform class-room practices or educational reform efforts. As Sue Matthews of the Georgia site explains, "[The Washington office] handed us the book and they said, 'Here it is. Go to it.' It was not a usable tool. When it's accompanied by *Benchmarks* and *Benchmarks* on disk, it becomes very useful." The Washington office, in reflecting back, realized the mistake it made.

> It's hard now to imagine how we could have been so optimistic back then; how much work there [was] to do . . . Initially we gave them *Science for All Americans* and said, "Invent K-12 curriculum. We'll give you three or four years." Just distributing the goals over time, which is to say *Benchmarks*, was a work of several years.

Roseman similarly reflects on this problem:

> There's been a lot of optimism that the project could produce goals and then people would understand how to use them. When I came to the project with the publication of *Science for All Americans* and I asked, "How can anyone who wasn't part of the development of these goals design curriculum?" And that is not to say that people would have to start from scratch on the goals. But I think the amount of time that it takes to make *sense* of the goals is something that is easily underestimated.

It was this struggle to use *SFAA* to inform instruction and curriculum that began the long process of backmapping—a process that eventually led to the publication of *Benchmarks for Science Literacy* discussed in the next section.

◆

Entering the National Reform Fray: *Benchmarks*

Presented to the public in October 1993, *Benchmarks for Science Literacy* is one of Project 2061's most ambitious undertakings, representing years of work and labor by countless individuals. And though *Benchmarks*[4] required a major commitment of time, energy, and money, it is not mentioned in early Project 2061 documents. Rather, upon completion of phase I, the project intended to complete *Blueprints for Action* and the curriculum models. What happened?

Benchmarks for Science Literacy evolved slowly: what was originally considered a process for moving site-level teams toward curriculum models eventually became a major publication with a release accompanied by all the fanfare the project could muster. This transformation from an internal product to a national reform tool provides an example of the project's eagerness and ability to help frame the national educational agenda and make significant, substantive contributions. In addition, the 855 benchmarks represent 2061's attempt to convert *SFAA* into something useful and practical for teachers. The *Benchmarks* story also illuminates the project's conception of science, teaching, learning, and curriculum.

From *SFAA* to Backmapping: The Role of Teachers

With the publication of *SFAA* in 1989, groups of teachers were constituted and asked to "design curriculum models that could be used by school districts to plan curricula that serve local needs and meet the goals of *SFAA*" (AAAS 1993).

---

[4]In this report, "*Benchmarks*" refers to the official publication of Project 2061, whereas "benchmarks" refers to the individual items that are today found in the document. The distinction is necessary because benchmarks were being developed before *Benchmarks* was ever considered.

It did not take long for teachers and staff alike to realize that *SFAA* could not be translated directly into curriculum. Roseman explained,

> I realized that until a [conceptual] task analysis [of the goals described in *SFAA*] . . . was done, then there would be no way to develop goals for the earlier grades. There's no way you can, in a goal-driven project, think about what the activities ought to be, what the curriculum ought to look like, unless you know right where it is you're headed. But I had a lot fuzzier sense of that at the time. My background in biochemistry provided me with a helpful analogy. I likened the K-12 conceptual maps to metabolic pathways. Metabolic pathways represent the conversion of S1 to S2, of S2 to S3, and so forth, by an arrow that connects them: S1 ➤ S2. The name of the enzyme that catalyzes this reaction is frequently written above the arrow. Activities that help students move from a more primitive to a more sophisticated concept were like the enzymes in the metabolic pathway.

Thus, in 1989, 2061 staff introduced the idea of 'backmapping,' or simply mapping, to one of the cross-team working groups at the Boulder summer institute. The 'strand group,' as they were called, attempted to develop a map for the water cycle. Subsequently, in spring of 1990, more teams began the process of backmapping. That is, the ultimate learning outcomes described in *SFAA* were analyzed to identify the sequence and approximate grade levels at which those ideas should be taught. See figure 2 for a partial backmap of the concept *structure of matter.*

Though the process of mapping was arduous, Roseman began to view it as essential:

> What I came to see was that the mapping did far more than just providing us with a set of benchmarks; that it actually helped the team members to understand the science. It was very powerful. I remember a discussion in Philadelphia when a K-12 group was attempting to map the *SFAA* section on heredity that deals with the passing of the genetic code contained in the DNA molecule. The biology teachers on the team were saying, "Oh, well, yes. So, if we're going to map this, then they need to know about the nucleotide bases and they need to know about hydrogen bonding and . . . " I said, "Wait a minute. It seems to me as if what this is all about is continuity of information and codes." And so, in the earliest grades, the map is about codes and then the DNA becomes nature's solution to this problem of how do you transfer information. That's a very different way of looking at it than, "What are all the terms that the kids have to learn before they can learn that?"

> That was very labor-intensive work and what I realized is that to map requires somebody who knows the science and someone who can get it down to the simplest manifestation of that. See, I didn't know little kids, but once I said codes, then the teachers of young children could talk about "what we do" and they helped us to write the benchmarks in that way. But you can get lots of university scientists who will go on and on about all the gory details and that wasn't the point. The point was to help get the real important ideas out.

Figure 2.   **Structure of Matter**

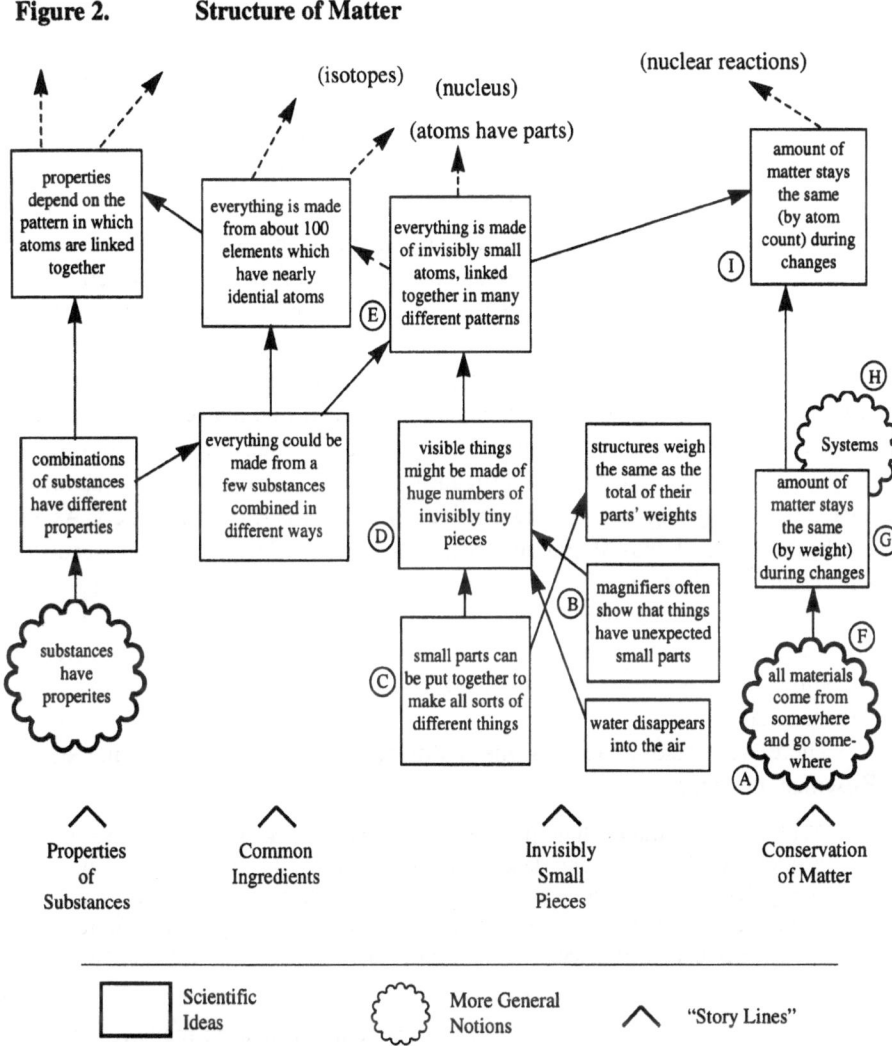

While Roseman believed strongly in the value of backmapping, there was no consensus among the teams, team members, or Washington office staff. As Roseman recalls, "By the fall of 1990, Jim [Rutherford] didn't believe in devoting so much time to backmapping. Chick [Ahlgren] was committed to the importance of mapping but was reluctant to push it on unwilling team members. [But], I wondered how sites would link models to *SFAA* if they hadn't engaged in backmapping."

Likewise, some sites engaged in the mapping process sooner than others. Roseman recalls:

The Texas team, as a group, bought into mapping long before the other sites. The five San Antonio members who were part of the Boulder summer institute had appreciated the importance of thinking about how students' ideas would grow towards science literacy. When I visited the site and showed the whole team how to map, they decided, "That's what we should do!" . . . The mapping notion was reinforced by Mary Budd Rowe during a subsequent visit to the site . . . the Texas team spent a lot of time mapping.

Some teachers found mapping, although arduous, rewarding and useful—even enjoyable. However, Farges, the center director of the San Francisco site, while valuing the concept of backmapping, called the backmapping process incredibly frustrating. According to Farges, no one at the Washington office could give him a model of a backmap, yet none of the backmaps his team constructed were thought to be good enough. His team was on the verge of a major protest to the exercise when Rutherford decided that San Francisco and San Diego—a team that had been significantly engaged in backmapping—should abandon the mapping process and continue with their curriculum models. As a result, the Philadelphia, McFarland, San Antonio, and Georgia sites proceeded with mapping while San Francisco and San Diego focused on developing and editing their curriculum models. Roseman is critical of this turn of events, arguing that, as a result, the San Diego and San Francisco curriculum models were developed in the absence of a framework about what students should know and be able to do.

There were also some tensions within teams. For example, at one site, Roseman remembers,

[The] leadership was pushing [teachers] to do model work and I was pushing them to map. The elementary teachers began to get into the mapping and they were good at it. I wrote a memo saying to the high school teachers, "Cut them loose. Let them do this." That legitimized what they were doing and they were then permitted, at their meetings in the future, to spend half of the day working as an elementary group, mapping.

After a painstaking process, the teams eventually produced maps for many of the learning outcomes. In a proposal to NSF, the project reported that "while sites worked independently on backmapping tasks, analysis of their work showed remarkable agreement on outcomes, or 'benchmarks' as they have come to be called" (AAAS 1992, p. 4). These maps became the basis for the benchmarks (though the two are not identical) which, notably, were initially conceived as an internal product.

This internal product eventually became a nationally published and disseminated document touted often as a product for teachers, by teachers. In fact, it is characterized by the project as a tool "largely the result of the vision, commitment, and inventiveness of . . . in-the-trenches educators" (AAAS 1993, p. vii). However, it is evident from this history that control of the document was rarely, if ever, in teachers' hands. Teachers did much of the work, they offered suggestions; but they did not have the power to make decisions, to guide the reform, and to make substantive changes.[5] As one 2061 consultant stated, "Teachers were asked to work on *Benchmarks*. But it was clear that [2061 Washington office staff] controlled the writing of *Benchmarks*." Also, the Washington office staff claimed teachers "demanded" backmaps and benchmarks. However, as was pointed out, some teachers and sites felt the backmapping process did not contribute to the more important and immediate goals of reforming science curriculum and instruction—it was seen as overly time consuming and distracting.

From Backmaps to *Benchmarks*:
The Role of the National Standards Movement

The decision to publish *Benchmarks* was made, to some degree, as a response to the national science standards movement. In 1991, *America 2000: An Education Strategy* was issued by the secretary of Education. This report declares that "standards will be developed . . . for each of the five core subjects [that] represent what young Americans need to know and be able to do if they are to live and work successfully in today's world." With this manifesto, the need to establish national educational standards gained nearly unanimous political support (Atkin 1994). (More recently, incidentally, this support has begun to wane as political and monetary struggles have ensued.) The mathematics community, specifically the National Council of Teachers of Mathematics (NCTM), already had published K-12 standards in mathematics in 1989. With the favorable reception of the NCTM standards, Project 2061 staff began considering the connection between its work with the benchmarks and the soon-to-be developed national standards in science. Ahlgren explains, "I guess *Benchmarks* were always part of the plan . . . but [they were] an internal product. It was something we would generate on the way to building alternative models of the K-12 curriculum. As the scene shifted in the world, it became evident that there was a big market for just that internal product."

However, the Department of Education and NSF, the federal funders of the development of science standards, decided that the task of developing the science standards would be given to the National Research Council (NRC) of the National Academy of Sciences, rather than Project 2061. The general belief at the time was that the funding agencies did not assign the task to 2061 because it was a preexisting project—the Department of Education and NSF wanted a fresh, unbiased approach. As a result, Project 2061 made the conscious decision to influence NRC's development of standards. Ahlgren notes:

> There were going to be standards out there . . . whether we had anything to do with it or not, so we had better divert our energy into being sure we had something to say . . . When the NRC got its mandate that somebody actually was

---

[5]On this point, the Washington office staff states that teams did have a noteworthy level of influence, though indeed the power was not theirs, nor should it have been.

going to go out and do this—get money to do it and make a lot out of the results—we said we had better have an influence on this. At the very least, we have to pipe our internal product to them.

The fall 1991 issue of the project's newsletter, *2061 Today*, describes the purpose of this move toward benchmarks for external publication and use:

[*SFAA* benchmarks] will contribute to the formulation of national standards for science education, an enterprise to be orchestrated by the National Research Council and funded by the U.S. Department of Education. This welcome opportunity for working together will produce a much-needed resource for the nation (AAAS 1991c, p. 3).

It is important to note, however, that the decision to publish *Benchmarks* was not made solely in response to the standards movement. Roseman underscores another factor contributing to the decision to publish the benchmarks: teachers and curriculum specialists. Project participants in the field were unable to use *SFAA* to develop curriculum and assess students' work. An intermediary document was necessary. In addition, in Roseman's words, "curriculum specialists in districts were wild about the maps;" they wanted the maps, rather than the models.

According to the historical account presented in a concluding chapter of *Benchmarks*, the draft benchmarks were distributed for review to thousands of teachers, administrators, scientists, publishers, researchers, district curriculum specialists, state curriculum groups, and other project affiliates. As a result of the more than 1,300 responses received, the benchmarks were revised. In October 1993, a press conference was held in Washington, D.C., announcing the release of the publication *Benchmarks*. Newspapers nationwide heralded the book as the arrival of the science standards. In fact, the benchmarks and national standards were considered to be synonymous in most of the press reports (Atkin 1994).

The decision to publish *Benchmarks* was not without cost. First, it had a significant impact on other work in progress in the project. Several products, specifically the curriculum models, *Blueprints*, and *Designs*, were put on hold to produce the benchmarks. Second, according to one 2061 staff member, NSF denied the project funding because it was competing with the NRC standards.

Despite some negative consequences, Roseman expresses her confidence and pride in the product 2061 has produced: "I'm really glad we did it . . . It's a service to the nation. The NRC document wouldn't have been nearly as good as it is without benchmarks as a starting point. So that's good."

Benchmarks and Standards: An Inevitable Comparison

At the time of this study, while *Benchmarks* is enjoying considerable attention, NRC's standards are still in development. Not surprisingly, there is ambiguity and some tension associated with sorting out the roles of Project 2061 and NRC in developing national science standards. One way of clarifying responsibilities was articulated by Jim Rutherford in an article in the winter 1992 issue of *2061 Today*. In it, he argues that

> *SFAA* has been widely accepted by scientists and educators as a statement of what it means to be literate in science. This vision, I trust, will serve as the starting point for the creation of national standards for science . . . So let us work together to articulate forward-looking standards for science literacy that are truly worth achieving, and then do what it takes to enable all schools to meet them" (Rutherford 1992).

The project staff takes pains to point out that NRC and 2061 have worked and are continuing to work collaboratively. As Jim Oglesby told participants at a recent *Benchmarks* workshop, "There is no competition." The benchmarks, the staff argued, may serve to inform NRC's Content Standards, but are independent of the Opportunity to Learn Standards, Curriculum Standards, Program Standards, and other types of standards NRC may produce. In addition, Project 2061 contends that the benchmarks satisfy reformers who need standards now, rather than later. Roseman states:

> I think more people are coming to appreciate *Benchmarks*, so that there is a greater cry from states [because] the standards aren't ready [while] the benchmarks are . . . [States] are developing frameworks now. Standards aren't going to be out at least until the fall and likely later, and that's a reviewable draft. So what appears to be happening . . . is that the standards will turn out to be more general. [The] benchmarks will be a very useful tool in support of standards. Standards will do the work of integrating assessment, teaching, and learning (as they said they were and we're expecting them to do), and the documents will be perfectly compatible, with the exception that standards [from NRC] will not bite off mathematics and the social sciences.

In addition, project staff point out differences between the NRC standards and the benchmarks. The 2061 benchmarks, according to Ahlgren, start with the question: What do adults need to know to be scientifically literate? The standards, on the other hand, start with the question: What should constitute school science?

Benchmarks also overlaps with the standards developed by NCTM. Some mathematics educators take issue with the 2061 mathematics benchmarks; they consider them to be redundant with their standards. Other mathematics educators consider the benchmarks to be incomplete or disjointed. However, Ahlgren sees the NCTM standards as being very different from the benchmarks. They are "both very vague and very specific compared to ours. That is, the telegraphic standards statements themselves are very vague, whereas the examples that follow are highly specific, virtually lesson plans. In addition, Rutherford argues that while *Benchmarks* is not committed to any particular instructional strategy, the NCTM standards describe what mathematics instruction should look like.

The Content of *Benchmarks*

The development of *Benchmarks* is only half of the story. Just as crucial is examination of the substance of the *Benchmarks* document. In brief, for every idea expressed in *SFAA*, *Benchmarks* contains recommended knowledge goals for the end of grades 2, 5, 8, and 12. *Benchmarks* is organized much like *SFAA*. In fact, there is a deliberate correspondence between the chapters of both documents so people can easily use them together and understand the relationship between the two.

Benchmarks begins with background information on Project 2061 and *SFAA*. The authors then outline distinguishing characteristics of *Benchmarks* and address some of the misconceptions early reviewers of the document had. For example, *Benchmarks* is said to be "different from a curriculum, a curriculum framework, a curriculum design, or a plan for curriculum" (AAAS 1993, p. xii). Also, the authors assert that *Benchmarks* is a report from a cross section of practicing educators working collaboratively to create, critique, and edit the document.

The benchmarks in the document demarcate thresholds rather than average or advanced performances. Thus, *all* students are expected to reach the levels of knowledge described. Operationally, 2061 staff consider "all" to equal 90 percent of the population. Some participants voiced skepticism about this. They argue that some of these goals can't possibly be met with particular populations such as mentally handicapped kids, or in extremely heterogeneous classes.

In the introduction, the authors also include information about using *Benchmarks*. The authors suggest it could be used by study groups "to explore the concept of science literacy in relation to instruction in the early elementary, upper elementary, middle-, and high-school grades" (AAAS 1993, p. xv). It is also suggested for use by teachers and curriculum specialists to evaluate their current curriculum or framework and guide the creation of new curriculum.

In chapters 1 through 12, the authors lay out the actual statements students should know at the end of grades 2, 5, 8, and 12 in order to achieve the outcomes described in *SFAA*. As stated above, the chapters correspond to chapters 1 through 12 in *SFAA*. Indeed, each of these 12 chapters begins with a few paragraphs taken directly from the parallel chapter in *SFAA*. The excerpt is followed by some summary comments about what students should learn and how that learning can be fostered. The chapter is then divided into sub-sections, each of which also begins with an overview. Lastly, the actual benchmarks are listed for each grade span. The list is prefaced by two or three paragraphs detailing some of the specific information for that grade span.

For example, chapter 5: "The Living Environment" begins, as they all do, with a quote taken from the first page of chapter 5 in *SFAA*. The authors then link the general ideas expressed in *SFAA* to education and children:

This sense of wonder at the rich diversity and complexity of life is easily fostered in children. They spontaneously respond to nature. However, attempts to give them explanations for that diversity before they are able to handle the abstractions, or before they see the need for explanations, can dampen their natural curiosity (AAAS 1993, p. 100).

As is evident, these brief essays differ significantly from *SFAA* in their tone and content—greater attention is placed on children and education rather than just the science itself.

The chapter is then subdivided into six sections: diversity of life, heredity, cells, interdependence of life, flow of matter and energy, and evolution of life (again, these correspond to the subsections of the *SFAA* chapter). In the last of these subsections, for example, the authors provide insight into teaching and learning evolution:

It is important to distinguish between *evolution*, the historical changes in life forms that are well substantiated and generally accepted as fact by scientists, and *natural selection*, the proposed mechanism for these changes. Students should first be familiar with the evidence of evolution so that they will have an informed basis for judging different explanations (AAAS 1993, p. 122).

Once readers get a greater sense of the difficulties an educator may face when teaching evolution, they then can read grade band essays which begin to explore *how* evolution should or might be taught. For example, in grades 3-5, the authors suggest:

Students can begin to look for ways in which organisms in one habitat differ from those in another and consider how some of those differences are helpful to survival. The focus should be on the consequences of different features of organisms for their survival and reproduction. The study of fossils that preserve plant and animal structures is one approach to looking at characteristics of organisms (AAAS 1993, p. 123).

It is important to note the purpose of these grade band essays: while they are designed to put the benchmark's statement into the context of schooling, they are not offered in any great detail or number. Rather, they are intended to provide and provoke ideas.

Following the brief essays are the benchmarks themselves. For example, in the K-2 grade band, the authors argue that students should know that "different plants and animals have external features that help them thrive in different kinds of places" (AAAS 1993, p. 123). This idea is expanded upon in the 3-5 grade band: "individuals of the same kind differ in their characteristics, and sometimes the differences give individuals an advantage in surviving and reproducing" (AAAS 1993, p. 123). By the end of eighth grade, according to the authors, students should know that "individual organisms with certain traits are more likely than others to survive and have offspring. Changes in environmental conditions affect the survival of individual organisms and entire species." Finally, this idea culminates in the grades 9-12 band as:

Natural selection provides the following mechanism for evolution: Some variation in heritable characteristics exists within every species; some of the characteristics give individuals an advantage over others in surviving and reproducing, and the advantaged offspring, in turn, are more likely than others to survive and reproduce. The proportion of individuals that have advantageous characteristics will increase (AAAS 1993, p. 125).

The natural selection benchmarks exemplify some important characteristics of the benchmarks in general. First, the grade bands themselves are significant. They are said to be based on children's cognitive development, though they are only approximations. (The K-2 band, for instance, might be considered to be the "primary" or "early elementary" band, according to the authors.) The authors rejected the more common convention of cutting along grades 4, 8, and 12 on the grounds that the developmental difference between kindergarten and fourth grade was too significant to aggregate them.

Second, the benchmarks are presented as knowledge statements rather than more commonly used behavioral objectives. According to the authors, knowing "implies that students can explain ideas in their own words, relate the ideas to the benchmark, and apply the ideas in novel contexts" (AAAS 1993, p. xviii).

Third, they are written in a language that approximates the language students of that age would use and know. For example, the words "evolution" and "natural selection" are not used in the first three grade bands. They are used for the first time in the 9-12 grade band. The statements also avoid technical language for its own sake: the concepts are taught before the vocabulary.

Fourth, the benchmarks are not exhaustive. They spell out a common core of learning without covering everything one could know about evolution. The authors acknowledge that "most students have interests, abilities, and ambitions that extend beyond these core studies" (AAAS 1993, p. xiii).

Fifth, the beginning of each section contains a reference to research notes. For example, the reader is referred to page 343 of the document for research notes regarding evolution. Turning to this page, the reader finds a summary of the research regarding the teaching and learning of natural selection, adaptation, and evolution and reasoning ability. Approximately 10 representative references are cited.

Sixth, included in each section is an "also see" box. Here, the interconnected nature of the benchmarks is evident. For example, those sections of chapters 1, 4, 6, 8, 9, 10, and 11 that relate to evolution are listed along with, when necessary, a brief explanation in parentheses. Thus, section 4c, "Processes that Shape the Earth," is referenced because it includes a discussion of fossils.

Following the benchmarks are three concluding and supplemental chapters "written to put the benchmarks statements into the context of education issues and the Project 2061 vision, as well as to make explicit the research basis for the substance and grade level placement of benchmarks" (AAAS 1994d, p. 13). Chapter 13, "The Origin of Benchmarks," provides a history of the document. This history related here is a bit simplified and does not include many of the behind-the-scenes struggles. Nevertheless, it does serve to help readers better understand how *Benchmarks* came to be, with the intention of increasing their grasp of its purpose and possibilities. In chapter 14, "Issues and Language," the authors outline the rationale behind some of their decisions. For example, they justify their use of knowledge outcomes over behavioral, or action, outcomes. Finally, as explained above, the research base is presented in chapter 15. This chapter underscores the project's emphasis on research. Decisions about whether a benchmark should be included, where, and when, were said to be informed by the research provided.

Dissemination and Use of *Benchmarks*

Since the release of *Benchmarks* in the fall of 1993, Project 2061 has been busy disseminating the *Benchmarks* book; as well as the more recently produced Benchmarks on Disk (a computer program available in either IBM or Mac format), to schools, teachers, administrators, and many others. In 1994, a workshop was held in Maryland to train participants in how to lead *Benchmarks* workshops. The 38 participants, most of whom were teachers, serve as a cadre of disseminators who are enlisted to do presentations. Project 2061 staff in Washington act as brokers for those people who request training.

During *Benchmarks'* dissemination, the project has experienced difficulties similar to those that arose after the publication of *SFAA*. First, some educators have misunderstood the document. For example, some teachers have thought the order of the benchmarks to be purposeful and significant. As a result, they have viewed the document as a linear and chronological representation of how they may organize their school year. At the other end of the spectrum, some educators have pulled out particular benchmarks without considering the connections to benchmarks in other grade bands or other chapters.

Second, the project is finding that it is not directly obvious how to use the benchmarks as a reform tool. For example, how can the benchmarks be used to create or evaluate curriculum? How can one assess whether a benchmark has not just been taught, but learned? How can a teacher use it to answer questions of worth (e.g., if a unit or activity addresses x number of benchmarks, should it be taught?)? What does it mean to target a benchmark?

Thus, teachers at the sites have spent considerable time and energy developing curriculum that targets the benchmarks. In San Antonio, for example, many teachers have designed new units and coded them for the benchmarks addressed. In addition to curriculum, teachers have found that assessment tools are necessary—tools that allow teachers to evaluate students' progress with respect to benchmarks. In San Francisco, for example, the team members have focused on developing alternative forms of assessment and corresponding scoring rubrics.

Other uses of *Benchmarks* vary. Some teachers find certain parts more useful than others. One teacher notes:

I particularly like the passages at the beginning of each of the benchmark sections. It's an overview as well as a preview for me and I refer to the passages a lot. And I know some people don't feel they need to do that, but for me, that works. Especially when I'm figuring out what I need to use for a unit and what types of things I want to cover and what resources to bring in."

Other teachers use them to evaluate materials and curricula they already implement: "Using [*Benchmarks*], I can sit down and I can document my units a lot better and say, 'This is supported here. This is where we're going. This is going to take care of a small portion of it,' and if they're successful, then in the future they can do some other things."

Some teachers with whom we spoke sang the praises of Benchmarks on Disk. As a result of its sorting, searching, and sequencing capabilities, the computer program is said to ease resource analysis and curriculum development. As one teacher states, "*Benchmarks*, the book, is wonderful. But Benchmarks on Disk is heavenly . . . Being able to sit down and with a few clicks pull up what you know is there is really great . . . Benchmarks on Disk is what's going to be what makes *Benchmarks* usable." This teacher also notes, she went on to state, "Without a doubt *SFAA* has changed my life as a person and teacher. It has influenced all my students for the last 6 years and now with *Benchmarks* available, I really can't wait to see how much of what I do is really as close as I think it is to *SFAA* goals and philosophy."

◆

The School District Centers

As described earlier, six school district centers joined the 2061 reform effort in 1989. These sites differ considerably. Each serves a very different population (inner city, rural, and/or immigrant). Each has a unique relationship with the Washington office and with the participating schools. Each has focused on very different aspects of Project 2061 with different results. And each has overcome different and unique obstacles. Nonetheless, these sites also have had some common experiences. They have faced some of the same struggles and dilemmas, have commiserated when things got tough, and have banded together when necessary.

In this section, we detail the structure, activities, and priorities of three of the six centers: Georgia, San Francisco, and San Antonio. Our decision to feature just three of the six sites was a difficult one. First, there was the question of *how many* sites to feature. We decided to focus on three sites to provide an adequate picture of the scope of activities at the site level while still providing the time and attention required for individual analysis. Second, there was the question of *which* sites to feature; all have contributed significantly to the national effort, each in unique ways. We chose to feature these three sites not because their work surpassed that of the others, but because they exemplify both some of the important differences and similarities among the sites. Each provided us with a different story as well as a different perspective. From Georgia, we heard the story of a center whose participants have learned to navigate the uncharted waters of a national, long-term reform. In San Francisco, center director Bernard Farges provided us with insight into the work of a center focused on implementation of the 2061 model in schools throughout the district. And from San Antonio, we heard from the teachers—their praise, their criticisms, their successes, and their challenges. Their voice is evident throughout the San Antonio story.

Our decision to focus on three of the six sites also elicited varying responses from the 2061 participants. On the one hand, some center directors felt we did not pay enough attention to all the sites: a focus on just three of the sites could be construed as slighting the others and ignoring the unique aspects of each. On the other hand, a Washington office staff member felt a chapter on the sites paid *too much* attention to their work: the centers are no more important to the work of the project, we were told, than other participants such as outside consultants and panel members.

The sites represent a unique and distinctive aspect of the reform. Attention to site-level activities is important to understanding the role of the centers within the Project 2061 reform. As is evident, we have presented these activities primarily from the perspective of those at the sites. We have used the words of the teachers and center directors—rather than the Washington staff—to describe the centers and the work being done there.

The Georgia Center

The two-hour drive from the metropolis of Atlanta to rural Elberton, Georgia, took us through a number of small, and not so small, towns. An abundance of churches dot the highway. Entering Elberton, we drove down the main street. Lining it are a WalMart, a McDonalds, a roller rink, hamburger stands, hardware stores, and some restaurants. The town's primary industry, granite, was evident as we passed a number of large blocks engraved with the names of local mills. In the McDonald's, there were granite plaques and tiles paying tribute to the industry that employs the majority of the town. It is clear that the lifeblood of Elberton, Georgia, is in the granite.

The lifeblood of the Georgia 2061 center, however, is in its teachers and center leaders. In the local high school, they met on their day off to discuss their current efforts at resource analysis and to receive updates on the work of the center. Sue Matthews, the center's curriculum director, led the session.

History and Structure of the Center. In 1989, Director Jim Rutherford agreed to place a 2061 center in Georgia, using the University of Georgia as the academic liaison. Three counties were chosen to be involved—Elbert, Greene, and Oglethorpe—because of their proximity to each other and to the University of Georgia. Leadership was divided among the three counties with each providing a leader or key representative.

After fulfilling its initial two-year commitment to 2061, Oglethorpe County withdrew from the project. The time requirements began to be too much for the small county to handle: the conferences, work sessions, and travel meant teachers spent a lot of time away from the classroom—time that the county could not afford. In 1991, involvement by Greene County teachers also waned. For a time, Elbert County teachers, with Candido Munumer as their leader, remained the only strong participants. Greene County became actively involved again in the summer of 1993.

As the focus of activities at the Georgia center moved to the one county, Munumer became the official center director. He remains in the position, sharing his leadership duties with Sue Matthews, the curriculum director for the center. Richard Matthews serves as the technical support person. Currently, approximately 20 teachers from seven schools from Greene and Elbert are directly and consistently involved in the 2061 reform in Georgia. Most come from elementary and middle schools, only a few from high schools.

There are no full-time employees on the project—the staff all have teaching or administrative positions as well. Munumer is a principal at Falling Creek Elementary School. His official contract for the project is for 27 days a year. Sue and Richard Matthews are both half-time employees, teaching at the middle school the other half of the time. Notably, working without any full-time personnel has been difficult, according to the Georgia staff. Fortunately, the Washington office funded a full-time curriculum director position for the 1995-96 school year. The position is to be shared by Sue Matthews and Joan Jordan.

As with all the centers, the Georgia team has undergone many changes and some growth pains. Notably, the relationship between the center and both the Washington office and the University of Georgia has changed over time. These changes were brought about by different forces. Some of the changes reveal the growth and learning of the center leaders as a result of their local involvement in this national effort. Others reflect the rapidly shifting landscape of the reform in both Georgia and the nation.

As at many of the centers, tensions arose between the Georgia center and the Washington office, particularly in the early stages of the project. However, the relationship with the Washington office has improved over time. As discussed later, the Georgia center leaders have become more knowledgeable about the differing worlds in which they and the Washington office operate. They expect change, while previously they resented it.

Since its inception, the Georgia center has maintained a relationship with the University of Georgia in Athens. Initially, this relationship was an intimate one. Project 2061 computers were located at the university, meetings were held there, and university faculty members were directly involved in the project. However, during the period when only Elbert County was involved in the project, it was decided it would be more practical to house the computers and office space in the county rather than at the university an hour away. Today, a small office in the town of Elberton houses the Project 2061 computers, an administrator, resource materials, and Sue Matthews's office. A more informal relationship is maintained with the university.

The Georgia Curriculum Model. Occupying a significant portion of their time in the early years, the Georgia curriculum model and blocks represent the collective work of many Georgia teachers and center personnel. In the *Project 2061 Prospectus* for 1994, the Georgia curriculum model is described as one that "connects the curriculum to the real-life experiences of students and calls on teachers to analyze their practice" (AAAS 1994c, p. E6). The model begins by delineating the nature of children and how they learn. For example, the authors state, "Because children are naturally curious, they will learn if given appropriate learning opportunities . . . There is not one way or one rate at which all children learn" (AAAS 1994c, p. 9). In addition, echoes of the conceptual change perspective can be heard: "Children form their own ideas, beliefs, and explanations about the world around them and these often differ from what is subsequently taught to them in school" (AAAS 1994c, p. 9). Based on these beliefs, suggestions for curriculum and instruction are then outlined. For example, the model suggests that "curriculum should emphasize conceptual understanding of fewer major ideas in science and how they interact, at the expense of details and isolated facts" (AAAS 1994c, p. 9).

Included in the curriculum model are a number of what the Georgia center calls "schemes"—each of which is "based on a topic in which major scientific, mathematical, and/or technological ideas are addressed" (Project 2061 1991a, p. 3). Examples include forces, communication, and scale. Each scheme is, in turn, broken down into various underlying concepts and treated in an interdisciplinary manner. These schemes are taught through a set of experiences, or blocks.

None of the blocks included in the original Georgia model have been implemented. One block, written more recently by Matthews, has been implemented and entered into the *RSL* database. As Matthews explains, Georgia's blocks were written to take to the Cornell summer institute and to satisfy the requirements of the Washington office. The blocks had not been field tested. In contrast, the one block that has been implemented was written as Matthews taught the unit.

Though only one curriculum block has been implemented, teachers involved in the project report a change in their teaching. Some teachers told us the exposure to new ideas and people has given them the opportunity to teach in new ways. For example, one teacher reports placing less emphasis on textbooks and including more hands-on activities in her teaching. Others report a greater self-confidence to tackle science in the early grades—something they were not well-prepared to do in the past.

Georgia: The Rural Site. The Georgia site was chosen initially to represent rural schools. Project 2061 participants in Georgia have struggled on both political and intellectual levels with this label. In an interview, Munumer reflected on the center's designation as rural:

At first, we fought the "rural" label. We don't see any differences in how people deal with the children or teach them, but we recognize how radically different the environment is. Much more conservative view. Didn't do something as radical as vistas—we did things slowly and closely with parents. We're aware of the need to be slow, careful, and cautious . . . We're always aware of the religious right.

Rural is environmental, sociological, and attitudinal but not pedagogical. Initially, we didn't accept the rural designation but then found it might be politically savvy. Rural Georgia represents mainstream America more closely. Our kids don't know what cows are . . . Rural does not mean agriculture. When you hear rural, you think pig slop.

The team's ambivalence about the rural designation is evident in a section of its curriculum model: "Rural Model?" Here the framers of the document argue that the rural condition is unique—the students' experiences, the community, and teaching conditions are different from those in suburban or urban districts. However, they contend that these differences should not mean that 'rural-ness' should affect a school's curriculum or instructional practices for science, mathematics, and technology. We believe, in fact, our model will apply to almost *all* students in *all* schools. We do not feel that rural students should be dealt with differently (Project 2061 1991a, p. 11).

Current Activities and Priorities. The Georgia center spent the 1994-95 school year on primarily three activities: outreach, the Consortium of Schools for Science Literacy, and curriculum analysis. With respect to outreach, the center hosted week-long summer in-services for those interested in finding out about 2061. According to Matthews, "The first summer was not successful in terms of bringing people on board with the project." Only 3 of the 32 teachers who attended are still involved. The second workshop was more successful: 10 of the 15 people who participated are still active on the project. Matthews meets with them regularly to discuss their plans and experiences in the classroom. This extra contact with and support for the participants is vital, according to Matthews, drastically lowering the rate of attrition.

The Consortium of Schools for Science Literacy is another major project of the center. Created in 1993, the consortium is cosponsored by the Georgia center and the Washington office. It is comprised of school teachers, administrators, and curriculum developers from Georgia. The consortium serves a number of purposes. First, it provides a vehicle for dissemination of information pertaining to science, mathematics, and technology education. Second, the consortium serves to network educators across the state. Third, unlike state and national science teacher associations, this group explicitly focuses on reform. Lastly, the consortium represents the center's efforts to make itself self-sustaining in the event that the 2061 Washington office cuts off funding to the sites—a move the Georgia site sees as imminent, according to Matthews.

The first annual conference of the consortium was held in 1994 in Macon, Georgia. This two-day conference, attended by 69 participants, focused on the

Project 2061 tools: *SFAA* and *Benchmarks*. Also included was networking time, presentations by several reform projects (e.g., Project LIFE, the State Systemic Initiative in Louisiana), and break-out sessions on various topics including action research. More recently, the consortium held a day-long workshop on *Benchmarks*.

Lastly, the center has spent considerable energies analyzing curriculum resources. A Project 2061 conference led by the Washington office staff in September 1994 marked a shift from a focus on curriculum blocks to resource analysis. Attended by Matthews, the meeting provided a systematic way for analyzing curriculum materials with respect to the benchmarks. Since then, Matthews has trained Georgia participants in the resource analysis process, which, according to Matthews, is a more manageable and concrete activity than curriculum development.

The San Francisco Center

While the Georgia center may be best characterized as specializing in outreach, dissemination, and resource analysis, the San Francisco center identifies itself as an implementation center. According to Bernard Farges, the San Francisco center director, the implementation and assessment of curriculum blocks, what San Francisco participants call "challenge-based learning experiences," is a major priority of the center.

According to Farges, this focus on implementation is reflected in the center's twofold mission: (1) to serve as a research and development center for designing ideas, tools, and resources that contribute to the national effort; and (2) to foster reform of K-12 education in the San Francisco Unified School District.

History and Structure of the Center. Center Director Bernard Farges has worked with the center since its inception and is a powerful force in the district as well as the national project. He currently works with nine schools that joined the project at three different points in its history. In 1992, three schools (now referred to as the cluster A schools) began their involvement: Cesar Chavez Elementary School (grades K-5), San Francisco Community School (K-8), and Horace Mann Middle School (6-8). These schools have continued their strong and active presence in the reform. In 1993, four cluster B schools joined the project: Buena Vista Elementary School (K-5), Spring Valley Elementary School (K-5), James Lick Middle School (6-8), and Balboa High School (9-12). And more recently, the center has been involved in the creation of two new cluster C schools: Thurgood Marshall Academic High School (presently, 9-10) and Gloria Davis Academic Middle School (presently, 6).

A large number of teachers at each of these schools (the totality of the faculty at six of them) have committed themselves to active participation in the project. They have attended the regular center meetings; designed, developed, implemented, and documented the implementation of challenge-based learning experiences targeting *Benchmarks*; disseminated project information to other teachers and schools; developed assessment tools; and worked on the San Francisco model. "They are truly collaborators in a democratic reform process," according to Farges.

While Georgia is said to represent rural America, the San Francisco center is thought to represent the inner city. The center serves a very ethnically and linguistically heterogeneous population: it is a "microcosm of the world," says Farges. The diversity within the site is consistently reflected in its products. For example, the San Francisco model pays particular attention to issues of diversity and equity, and recognizes the "unique student population" attending its district schools (Project 2061 1994b, p. 6). Similarly, in meetings we have observed, participants consistently asked each other if their work addressed the needs of *all* students. In one instance, as teachers attempted to create a new curriculum challenge, they struggled to find a topic that was not biased toward a socioeconomic or ethnic group.

The San Francisco Model. The San Francisco model "proposes to create a learning environment and learning experiences to maximize the opportunities for all students to make connections both within each, and between all, disciplines, including social science and the humanities as well as science, mathematics, and technology" (Project 2061 1994b, p. vi). These words were chosen carefully: they highlight the importance of making connections among ideas without mentioning the word *integration*—a word that, according to Farges, means too many different things to too many people.

The fundamental tenets of the San Francisco model emphasize content-specific learning that is meaningful from the *student's* perspective:

- Students must find the experiences meaningful to them at the time they are engaged in them.

- Learning occurs best when the object of learning is contextualized in a way that makes sense to us as individuals living in specific societal contexts.

- Learning is an organic process of construction realized through necessarily subjective experiences rather than a mechanical process of accumulation of canonical knowledge (Project 2061 1994b, p. vi).

Responding to the proverbial student question, "Why do we have to learn this?," the San Francisco model suggests teachers create learning experiences prompted by *challenges*. Challenges, according to Farges (1995)

> are multi-faceted tasks which are meaningful from a student's perspective . . . Students are engaged in investigating and responding to environmental and social issues, in making decisions and solving problems of local and global concern, in designing and creating products and performances, and/or in inquiring into "How do we know what we know?

As such, teachers have designed learning experiences, or tasks, prompted by the challenges.

For example, Debbie Faigenbaum, a teacher at San Francisco Community School, designed a three-week-long challenge to build a kayak. There were two requirements: (1) the boat should carry at least one student, and (2) students could use only cardboard and plastic tape. This challenge-based learning experience was then approached from all of the disciplines. Students studied the history and culture of the Aleut and Inuit, Native American tribes that used and depended on kayaks for survival. They used mathematics to design, measure, and create models of their boat, and science to formulate and test their hypotheses about buoyancy. They heard folktales, read poetry, and wrote letters and creative stories—all pertaining to kayaks, their use, history, design, and significance. The class then culminated the learning experience by trying the boats out on the San Francisco Bay.

Often, students pursue these challenges with others not in their same grade. In the case of the cardboard kayak challenge, students ranged in grade from fifth to eighth. In addition, they are usually provided a choice of challenges and given the opportunity to frame the challenge, set the goals, and define the assessment criteria and standards to assess the quality of their products and performances.

The convictions underlying the San Francisco model and challenges are also reflected in the ways in which the participants work with one another. For example, Farges has placed particular weight on teachers' activities in the reform being "authentic." Activities have not been undertaken solely for the sake of team building or learning the *Benchmarks*. Rather, the team has worked on concrete and immediate goals that, in the process, have built camaraderie and increased the teachers' conceptual understanding of the benchmarks. "Professional development of teachers takes place," according to Farges, "in the context of, and for the purpose of, planning and implementing learning experiences."

In the process of refining their implementation efforts over the past three years, the teachers have developed 12 models addressing areas such as instructional strategies, assessment, partnerships, professional development, and organizational structures, notes Farges.

Current Activities and Priorities. As indicated in figure 3, the center is currently involved in several different activities, financial support for which has been provided by the Washington office and the San Francisco school district, as well as private foundations, such as the Clarence E. Heller Foundation and the Robert N. Noyce Foundation.

First, as a result of the center's commitment to implementation, participants have spent considerable time creating, implementing, and evaluating challenge-based learning experiences. In an effort to facilitate this process, center participants developed two documents: the *Teacher's Challenge Handbook* and the *Teacher's Challenge Organizer*. Each of these is intended to improve the quality of implementation of the learning experiences.

Figure 3.   **San Francisco Center's Activities**

Creation of a New Middle
School and a New High School

Implementation of
K-12 Schools

Refining the
San Francisco Model

Documentation

Presentations

Dissemination of
Benchmarks for
Science Literacy

SAN FRANCISCO
PROJECT 2061

Teacher Preparation
Programs

Resources and
Technology

After-School
Enrichment Programs

Fundraising

Partnerships

Source: Project 2061 San Francisco Center, August 25, 1995.

In the process of implementing their model, participants encountered dilemmas as well as many successes. One dilemma, Farges reports, was incorporating all subject areas into a challenge. He, as well as some of the teachers involved in the project, has expressed his concerns about including enough mathematics and science in the challenges:

> When you start with a real-world issue, or the real world problems, the social science is always very strong. And it is true that we must be very conscious of the benchmarks—science-based benchmarks—making sure that the science and math are used to provide evidence that then can be the basis for argumentation.

Center participants have also identified the tremendous need for assessment tools and strategies to evaluate students' understanding. Thus, they have spent considerable time and energy creating assessment tools to answer their question: How do we know what students have learned?

While there have been some struggles and dilemmas, the curriculum challenges implemented have produced satisfying results. For example, a recent article in a neighborhood San Francisco paper highlighted one six-week learning experience at the new Thurgood Marshall Academic High School. Students were asked to study "housing and transportation issues in Bay Area neighborhoods and communities and, based on their findings, make recommendations to improve the quality of life in these areas by 2020" (Perrigan 1995, p. 8). The student products ranged from land-use plans for the Presidio to mass transit recommendations.

Second, in addition to curriculum and assessment work, the center also established two new "2061" schools in the district. Unlike the cluster A and B schools, these cluster C schools (Thurgood Marshall Academic High School and the Gloria Davis Academic Middle School) are a product of the project. That is, they were designed specifically with the goals and agenda of Project 2061 in mind: from the day they opened their doors, teachers and administrators were working toward the implementation of 2061 tools and resources such as *Benchmarks* and the San Francisco model.

While the schools are considered by the public to be 2061 schools, their identity within school walls is slightly more complicated. As one teacher at Thurgood Marshall told us, there is an uneasy mix of visions for the school—visions that don't always peacefully coexist. For example, in addition to being "a 2061 school," it is also a technology magnet school. Politically, it must satisfy the needs of both the local African-American community who live in the neighborhood as well as the Asian-American community who are looking for an alternative to the prestigious, but over-crowded, Lowell High School. Nonetheless, Thurgood Marshall opened for the 1994-95 school year with 300 ninth graders in attendance. In the 1995-96 school year, a new class of ninth graders was added, bringing enrollment to 600. Gloria Davis opened its doors to 200 sixth graders for the 1995-96 school year.

Third, participants have continued to document their reform efforts. Farges notes that teachers in cluster A schools "have been provided in-services on teaching as a process of systematic inquiry to document the impact that Project 2061 tools are having on science literacy education for students and professional development of teachers." As a result of outside financial support, some teachers have been provided release time and/or compensation in order to engage in proper professional development and act as documenters.

Farges considers documentation to be an important step in changing what happens in the classroom. He argues that for teachers to change their instructional strategies, they need evidence of what works and what doesn't work. In addition, documentation also serves a communicative function. By putting their ideas, trials, and errors on paper, teachers can provide others with important insight and expertise.

Fourth, dissemination and outreach remain a major priority of the project. Thus, participants continue to disseminate *Benchmarks* to teachers in the district via presentations and in-service workshops. Farges has also initiated and coordinated a collaborative effort between the center and local colleges and universities to, as he explains, "create and implement a model of teacher preparation programs based on *Benchmarks* and San Francisco Project 2061's challenge-based approach to learning and teaching."

The relationship between the San Francisco center and the Washington office has evolved over time, as it has for all the centers. There have been periods, particularly between 1989 and 1991, when, Farges says, San Francisco partici-pants felt they were receiving conflicting directions and little feedback from the Washington office. The San Francisco team felt that the Washington office staff could have been more responsive to the needs of those in the schools. For exam-ple, after the publication of *Benchmarks*, Farges expressed a need for assessment guidelines. Because of the urgent teachers' need for alternative forms of assess-ment, he has taken on the assessment issue at the center level.

More recently, according to Farges, the Washington office has focused greater attention on the work of the sites. For instance, the Washington office has provided support for the creation of site-based leadership teams in San Francisco 2061 schools during the 1995-96 school year. In Farges's words, "I think there is an increasing appreciation of some of the work which is being done [at the sites]. The work which is happening at a site may help the national agenda, so I think that members of the Washington office see increasing value in what is being done at the San Francisco center."

The San Antonio Center

The San Antonio center is in flux. It was an active and significant participant in the reform beginning in 1989, but recently underwent dramatic changes. Peggy Carnahan, the original center director, left the position in August 1995. As of late 1995, the Washington office restructured the site so as to include only one San Antonio school district, rather than the 16 that had been involved either directly or indirectly. This new center is working in collaboration with the Texas Urban Systemic Initiative. Meanwhile, as the following description notes, other participants have reconfigured their activities under a new name. Their work preserves much of the language of Project 2061 and builds on years of formal affiliation with the project. They believe their work preserves the project's spirit.

History and Structure of the Center. In 1989, when Project 2061 established a center in San Antonio, Texas, four San Antonio districts were involved (a number that eventually grew to 16). Initially, the center was housed in a building owned by one of the participating school districts. When the group lost this space in 1994, it moved to an office in San Antonio College. Through all these changes, Peggy Carnahan served as the center director.

The work of the center initially focused on the Texas curriculum model, known as the "Designed World Model." According to a San Antonio center brochure, the original model proposed

> a unique and dynamic system for educating our children based on the desire to provide science literacy for all Americans. The Designed World Model calls for the revitalization and restructuring of the initial educational system so that stu-dents will become life-long learners, informed decision makers, and confident problem solvers.

More specifically, the brochure continues, the Designed World Model suggests an interdisciplinary curriculum organized around "six major areas of human endeavor: information processing and communication, materials and manufacturing, health technology, energy, transportation and agriculture." In addition to this work on the model, the San Antonio center, as noted in section 4, was also heavily involved in backmapping *Science for All Americans*, eventually mapping the entire document.

The Teachers' Perspective. While teachers at the center were introduced to the project in various ways—either by chance, by involvement in other professional development groups, or through the teacher grapevine—their reasons for involvement were similar. The project's vision, teachers reported, fit well with what they felt teaching should be all about. As one teacher told us, "It fit with my experiences as a teacher, while also providing me insight into where to go." "It fit my gut. I felt like what was being offered as a philosophy—less is more—fit into what I had always done as a teacher," reported another.

As such, the project helped to legitimate the teachers' work. In the words of one teacher, it "provided the structure and the scaffolding to make decisions [about what to teach and what to leave out] and a great deal of support and research to do it." In addition, as one teacher explains, the 2061 tools gave their teaching a stamp of approval from a national, well-regarded organization:

> *Benchmarks*, and particularly *Benchmarks on Disk*, gives us a tool, a point of validity . . . If you have this, you can say, "This is where it's coming from. It's based on research." This gets people to listen. So when parents want to know, "Why are you doing this?," you can trot this stuff out and say, "You wanna know why? Okay, here's why." You gain some credibility with them. And that's been a big bonus.

The project not only confirmed participants' own beliefs about teaching but challenged them to try new strategies and explore new avenues, while affording them vital professional development opportunities. One teacher notes:

> It's been great fun being on the cutting edge of what's happening. It's the kind of place that you feel like all teachers need to be. We need to be right out there experiencing the world instead of being shut up in a little classroom . . . This project just opens you up, lets you experience all that there is out there. It's changed my life tremendously.

And finally, participation in the center provided teachers with the chance to share ideas with other professionals. "This was a place to express my views and get feedback, to share my experiences," one science teacher reports. The acclamations continue: "I was looking for a support group and I found one . . . I can come to this group and get energy, information, and guidance. And I can go back to the classroom and hopefully give the types of science lessons that interest kids and catch their attention. It allows me to do what I need to do in, hopefully, a better way." For many teachers, particularly those in rural and small schools where there is tremendous isolation, this interaction was considered vital.

Current Priorities and Activities. During the 1994-95 school year, the activities and priorities of the center were varied. While some participants concentrated on providing technical assistance to educators around the state, others focused on redesigning the center's curriculum model, and still others on outreach and networking.

A subset of San Antonio center teachers focused their efforts on redesigning and updating the center's original curriculum model which had received little attention since the 1991 summer institute in Seattle. Based on criticisms that the model was filled with too much "education-ese," teachers focused their efforts on rewriting it for a more general audience. They also developed and field tested curriculum blocks and reviewed print and technology resources for use with the Designed World Model.

Another group of participants focused on technology. The group explored ways of increasing students' and teachers' access to technology as well as strategies for using technology more effectively. As a result of this work, some teachers reported feeling a renewed interest in teaching and an increased level of skill and expertise. Many used these new forms of technology in their own classrooms.

Still other center participants designed activities and curricular units of their own. For example, teachers at Judson High School developed two new science units on the topics biosphere and wellness. Students address each topic across six themes: human presence, environmental interactions, changes over time, systems, energy, and structure. These themes were taken directly from the state's science curriculum development model developed by the Texas Education Agency. In this manner, the curriculum units fulfill the goals of both the state and Project 2061.

Carnahan reports that the center has always worked carefully with state agencies to integrate the objectives of the national project with local and state initiatives. Toward this end, teachers from the center participated in the development of the Texas Elementary Science In-service Program, a state training module for elementary teachers. Others were involved in the development of the state's *Coordinated-Thematic Science II*, a science framework for teachers. Still others developed training guides and videos for the Texas Assessment of Academic Skills test.

While project participants did not often list implementation as a major priority of the center, all of the teachers with whom we spoke reported having changed their classroom practices as a result of their participation in this project. For some, the changes were dramatic; for others, more subtle. For example, in one classroom we visited, the teacher was implementing new curriculum materials that required students to design their own car. Introduced to the materials by other teachers on the project, the teacher had tremendous praise for the activities, "It's incredible to see kids so engaged in science."

As participants undertook these tasks, they faced challenges and frustrations. For some, support from school and district administrators was less than ideal. District personnel, they note, can act as impediments to this type of professional development because they don't encourage participation in such groups, nor do they provide the necessary support. Others feel that their administrator supported the project superficially but did not really buy into it or provide the necessary support to do it 100 percent. These administrators, members argue, "just want the 2061 name." However, there was not consensus on this point. As Joan Drennan-Taylor points out, the school districts gave teachers considerable leave time to work on the project and funding to attend and present at state and national education conferences. According to Drennan-Taylor, "We were supported and still are supported by our district."

Monetary support also has been difficult. While funding has come from a variety of sources, including Project 2061, the Texas Education Agency, and districts, it has not always been constant. Most significantly, Carnahan reports, support from the Washington office was reduced in 1994. As a result, Carnahan lost some administrative staff members and half of her salary.

Carnahan attributes some of the center's challenges to being a multi-district center. As such, there is a diversity of schools including small, rural, suburban, and urban. In addition, Carnahan notes, "[there are] multiple politics and multiple agendas, so one has to facilitate the teachers more so that [they] can work and be involved." Some center participants feel the Washington office never understood or appreciated the unique challenges facing centers that work with more than one district.

A New Direction. In 1995, the relationship between the center and the Washington office deteriorated. As a result, there was been a bifurcation of activities. Some participants established a new official 2061 center. Others continued their work under the new name of Texas Center for Science and Math—a group that has no *official* affiliation with Project 2061.

The reasons for restructuring the San Antonio center were not entirely clear to us, the case study researchers. However, some points of contention between the Washington office and the San Antonio center were evident. Issues regarding the levels of documentation, the budget, and control over the center's activities and priorities appeared to be sources of disagreement. Some of these issues may have been a result of 2061's model of reform and the challenges and dilemmas it has invoked. That possibility is elaborated later in this paper.

The restructured 2061 center is under the guidance of Center Director Joan Drennan-Taylor. Some of the participants include original members of the San Antonio center. According to Brearton,

> The Washington office is working with Dr. Diana Lam, Superintendent of the San Antonio Independent School District (SAISD), San Antonio central office staff, and members of the Project 2061 Center to restructure the Center. This includes physically relocating the center to a school facility and expanding its membership. As Principal Investigator of the San Antonio Urban Systemic Initiative (USI), Dr. Lam is eager to use the personnel and material resources of the Project 2061 Center in the professional development program envisioned

for teachers in the nine school districts identified in the USI. The program for the Center . . . includes plans for [an] extensive, long-term professional development program using Center members as leaders of this effort. The Center would also contribute to the development of an infrastructure for professional development that is the focus of a Rockefeller grant awarded to the SAISD.

Thus, the center has begun working with the USI mentor teachers. The budget calls for the training of approximately 450 teachers in the construction and implementation of curriculum based on the *Benchmarks for Science Literacy* and the review of resources currently being used and those being considered for purchase. "Work on these programs has already begun with a two-day professional development workshop by members of the Washington staff and center members for 40 USI mentor teachers," Brearton notes.

Another group of teachers, led by former Center Director Peggy Carnahan, has continued its work on the project under new auspices. These teachers had little doubt about their endeavor: they would continue to disseminate the ideas of Project 2061 to a population that is, according to participants, in desperate need of assistance. Too many teachers in Texas, the teachers told us, were teaching as they had 20 years ago, but not because they were necessarily lazy or resistant to change. Rather, many teachers simply had not heard about exciting reform initiatives such as 2061. Not enough Texas teachers had the opportunity to be a part of such a fulfilling and challenging professional development experience. The problem is exacerbated in the small towns of Texas where teachers are in tiny districts without a science coordinator or even another science teacher. Thus, participants say they hope "to get the word out and provide teachers with the support to do things that might be risky in their classrooms." Efforts thus far to disseminate have been well-received in many areas. Teachers around the state have learned about *Benchmarks* and voiced interest in getting involved in the group.

Carnahan would also like to see greater emphasis on "teacher as reflective practitioner":

> The inconsistency of documentation of the change process has been the project's greatest loss . . . In four years, Project 2061 teachers were associated with the greatest people and curriculum resources in the nation, causing unbelievable changes in [teachers'] ability to help students become science literate. Their paradigm change toward teaching and student understanding is a process that should help other teachers and reform efforts. Through intense professional development on reflective research, the center could be recording changes in teacher instruction and student achievement. There should have been hundreds of publications by teachers on how Project 2061 has influenced instruction in the classroom.

However, while some support for documentation was provided (the San Antonio center had two days of documentation training by Mary Jo McGee-Brown of the University of Georgia), no follow-up training was funded. As a result, some participants, in Carnahan's opinion, were unable to document their work adequately. Carnahan hopes greater training in reflective, or action, research will provide teachers with the necessary skills for documentation.

These priorities for the future will be undertaken under the name Texas Center for Science and Technology. Though no longer an official 2061 center, the center's new mission remains wedded to the ideals of the project. The language is familiar; Carnahan notes:

> The Texas Center for Science and Technology is a resource center whose mission is to facilitate the systemic reform in public education that leads to science literacy for all Americans. Through collaboration with school districts, colleges, universities, and the private sector, the Center has helped to implement changes in materials and practices that are supported by the research on how children learn.

For this group, severing ties with Washington was difficult. However, most of the participants see the dissolution of their formal relationship as a move in the right direction. Teachers told us they felt increased freedom to accomplish the goals important to them without the politics and bureaucracy associated with the Washington office. In the past, some had felt frustrated that the center was made to engage in activities to advance the work of the Washington office, at the expense of helping its own schools and districts. As one member states, "I don't think I ever lost the vision of 2061—it sure gets tough slogging around in the quicksand. I still see the mountain top, it isn't obscured yet. I'm still hanging in there."

Perhaps, while rocky, the experiences of the San Antonio center may provide the Washington office with important insight into productive ways to work with schools around the country. The situation suggests the possibility of a modified dissemination model—a model that would take two forms: (1) official Project 2061 centers that are funded and overseen by the Washington office, and (2) independent bodies that are not under the direct control or coordination of Washington, but still provide the national effort with useful information about school-level activities that are guided by local interpretations of the national 2061 vision.

◆

The Conception of Science in Project 2061

The starting point of science curriculum reform in the 1960s was to enlist scientists in universities to identify the content that students should learn in elementary and secondary schools. The impetus and much of the financial support for this approach came from the National Science Foundation. The agency had been created by Congress in 1950 to improve both science and science education, but its hallmark and its fundamental reason for existence was to support basic research. So it was to this group that the foundation turned for guidance about the directions for science education in the public schools. About a score of projects were initiated to create new curricula; each one had close links to the scientific research community in universities.

The scientists usually began by identifying the concepts they considered basic in their respective disciplines. Textbooks of the time tended to emphasize some applications of science, but there was seldom a coherent approach to the underlying scientific principles; furthermore, the basic science that did get included was often outdated. What, in the opinion of the country's best scientific researchers, should a modern student know about physics? Or biology? Or earth science? Which ideas have the most intellectual mileage? On completion of the task, the next steps usually were to prepare texts and other materials to convey that content to students, and lay plans to bring teachers up to date in the newly identified science. With an exception or two at the middle and elementary school levels, the approach was discipline by discipline: a course in chemistry, a course in earth science, a course in physics, and (because consensus about what constitutes the fundamentals of biology apparently was harder to achieve among biologists) three separate texts for high school biology.

One project, Harvard Project Physics, incorporated an additional perspective: understanding the principles of modern science is the essential goal for science education, but it is also important for those who would have deep understanding to know how certain scientific ideas have changed over the centuries, and why. Gerald Holton, a historian of science, was a key figure in Harvard Project Physics. So were James Rutherford and Andrew Ahlgren.

Decades later, when Rutherford conceived of the science curriculum-reform effort that was to be called Project 2061, he brought with him the same general orientation, both with respect to primacy of the academic scientist in identifying essential ideas of science and to the need for students to understand something of the origin and development of those ideas. By then, he had held major positions in science education in academia and also in the federal government (both at the Department of Health, Education, and Welfare and at NSF). He had also come to be recognized as the most prominent national figure in American science education. A proven and effective advocate, he was able to secure the financial and other support necessary to launch the new effort. Rutherford brought many novel features to Project 2061. But one similarity to curriculum reform of the 1960s was unambiguous: first go to outstanding researchers at the university to get the science right, then figure out how to get it into the schools.

The amount of content that conceivably might be covered in science, however, is potentially enormous. The field generates new information at a dizzying rate; scientists themselves find it challenging even to keep abreast of developments in their own fields. What should everyone who graduates from high school know? What are the core concepts for *all* Americans? Or, as put by Project 2061 staff, what must one know to be science "literate"?

A fundamental premise of Project 2061 is that the schools do not need to be asked to teach more and more content, but rather to focus on what is essential to science literacy and to teach it more effectively.

Science for All Americans is based on the belief that the science-literate person is one who is aware that science, mathematics, and technology are interdependent human enterprises with strengths and limitations; understands key concepts and principles of science; is familiar with the natural world and recognizes both its diversity and unity; and uses scientific knowledge and scientific ways of thinking for individual and social purposes (AAAS 1990, pp. xvi-xvii).

Use of the word "literacy" by Project 2061 is noteworthy. The term was not generally employed during the 1960s round of science curriculum reform. In the 1970s, however, there was a move to emphasize the "basics" in elementary schools. The schools were seen as tolerating what many people saw as "frills." The main purpose of public education, it was said, is to stress core skills, traditionally associated with subjects like reading, computation, and communication. Those educators interested in the arts, physical education, and even social studies and science, began to become concerned about the place of their subjects in a back-to-the-basics era. Would they be minimized, or even crowded out? Many of those interested in promoting and improving science education began to liken their field to the generally accepted essential competencies: it is just as important, they said, as the 3 Rs. To underscore the point, they began to talk about scientific *literacy*, with the implication that it is just as necessary for functioning in the modern world to know about science as to know how to read, compute, and communicate.

The concept of scientific literacy advanced in Project 2061 is novel in several interrelated aspects. First, the project chose to define science as encompassing not only the customary disciplines of the natural sciences as they usually are organized in schools and colleges (chemistry, biology, physics, and earth sciences), but also mathematics, technology, and social sciences. Second, science literacy is characterized by "balance" across the science disciplines; it is not enough to know only one, or even a few, of the special fields within science; a person who is science literate must have a certain level of understanding about all of them. Third, the "essentials" can be taught well only if the coverage of science topics is reduced: less can be better. Finally, the project is neutral on the matter of which curriculum structures or methods of teaching best advance the goal of achieving scientific literacy. Unlike almost all other attempts at reform in science education over the decades, Project 2061 posits that science literacy can be achieved by teaching within traditional subject matter boundaries, or as cross-disciplinary courses, or in connection with examination of social issues, or even—presumably—outside formal course structures altogether. Furthermore, there is no one best pedagogical approach. The important criteria for judging whether scientific literacy is achieved lie in the nature of the content and the quality of the student's understanding, not in curricular organization or mode of presentation.

These core beliefs, most of them articulated early in the project's history, lie near the heart of the reform in science education that Project 2061 is promoting.

Each of them has both predictable and surprising implications. And, as a set of principles, there are some apparent internal contradictions.

What Disciplines Constitute Science Literacy?

Science, according to the Project 2061 definition, includes mathematics, technology, and social and behavioral science—as well as chemistry, physical science, and biology. It was a bold and unconventional step, even a brash one, to extend the boundaries of scientific literacy in so unorthodox a fashion. Why was it done? Says Ahlgren:

> The rationale was a mixed one. One is that these things [mathematics, science, technology, and the social sciences] are closely related and that if you split these out you're losing an opportunity for coherence. The second point, not entirely independent of the first, is that the AAAS has constituencies in mathematics, and in engineering, and the social sciences—as well as the natural sciences . . . And [an additional reason for including the social sciences] is that they tell you something about the world that was based on scientific research or a scientific way of looking at things.

The idea of one's discipline being encompassed within a broad conception of scientific literacy has proved troublesome for some mathematicians, technologists, and social scientists, however. Rutherford remembers that, even by 1993, "We already [had] problems with mathematicians."

Why the resistance? Mathematics is a well-established school subject in its own right. Mathematics teachers form a strong group in every high school, and the subject is a centerpiece in all elementary schools. Furthermore, the professional association of mathematics teachers, the National Council of Teachers of Mathematics, had created an independent and large-scale enterprise to identify core mathematics content for the schools. NCTM's standards project, in fact, has attracted—and continues to attract—considerable attention, almost all of it favorable. It is widely hailed as the first successful attempt to articulate standards for a school subject and is often cited as a model that should be emulated for all subjects. To include mathematics as part of an inclusive subject called "science" has the potential for weakening the well-established and distinctive position of mathematics in the curriculum. If it is one of many scientific fields to be studied, might its large and prominent place in the present lives of students be lost?

More substantively and conceptually, perhaps, might the subject of mathematics be redefined as something different from the discipline that has evolved over centuries if it is incorporated within a broad category called science? While certain subfields of mathematics are influenced by disciplines like physics and chemistry, as well as other sciences, and it indeed may be the "language of science," as is often asserted, it has its own history, its own challenges, its own rhythms. Mathematics need not relate to the world in the same sense that the sciences do: much more than the natural and social sciences, and certainly more than technology, mathematics can be the study of a disembodied set of patterns and relationships, unconnected to real objects and events. Might that defining feature of mathematics be compromised if the subject is categorized as it is in Project 2061?

Inclusion of the social sciences presented a somewhat different kind of challenge: the kinds of explanations generated by social scientists may not paint the same kind of picture of the world that the physical and biological sciences do. Perhaps social scientists provide a way of talking about the world, rather than explaining and predicting in the same sense as the natural sciences. But, as noted earlier, social scientists involved in Project 2061 took umbrage at this distinction. In the end, those social scientists involved in developing the Phase I panel reports seemed satisfied with inclusion of their field in the project.

Indeed, by early 1995, the consolidation of social sciences into Project 2061 seemed a minor and uncontroversial matter—if not for reasons centering on the principle of whether the field is sufficiently related to the natural sciences to be part of Project 2061. Like the mathematicians, perhaps, social science educators by then saw no particular threat from Project 2061 to the special and established position of their subject in the curriculum. On the contrary, one might speculate that social scientists welcomed inclusion in the territory defined by Project 2061 because "social studies" was being challenged as a distinctive curriculum subject. Strong advocates began to advance the view that there should be separate courses in history, economics, geography, and civics, not a generalized "social studies." The challenge was real: the federal government supported identification of standards in civics, history, and geography, but not in social studies. The National Council for the Social Studies, in fact, deemed it necessary to launch its own effort to delineate standards, without governmental support.

Ahlgren notes that there is a high degree of conceptual consistency between Project 2061's view of the social sciences and the vision for the subject projected by leaders within the social studies education community:

> Just a few weeks ago [autumn 1994] I saw the National Council for the Social Studies recommendations, their standards. I marked all those that were comparable [to ours] and where we had something to say about the same topic. They had about 80 bulleted items. Half of those, 40, I thought were consonant with what we had done. That's a tremendous overlap. What they had but we didn't was very largely civics types of things: how your government works . . . There was very little there that I felt, "Oh, yes. We should have had that in there, too."

> . . . A worry [for us] would have been the feeling that we're poaching on some other curricular area's territory. But since the social studies people just got their standards out now, five years later [than publication of *Science for All Americans*], that hadn't been much of an issue. Right now, I think, the feeling is pretty ecumenical.

Professionals in technology education are a more diffuse group. Take the engineers. Unlike mathematics and social studies, engineering has never enjoyed a strong position in elementary and secondary school education, so there is no established interest to defend. There are intermittent moves to introduce more technology education in American schools, but that pressure seems to wax and wane depending on the sense of crisis in the body politic about economic productivity and issues of balance of world trade. (In at least one of the Project 2061 centers, San Antonio, there has been considerable interest in devising a design-based technology curriculum.)

There are also the teachers of vocational and technical education. But their subjects have never been considered central to precollege education. Perhaps, rather than seeing a potential threat during the early days of Project 2061, these teachers saw the possibility of an enhanced place for their subject with the increased attention to technology from Project 2061. Additionally, in late 1994, the National Science Foundation and the National Aeronautics and Space Administration jointly funded an effort to define standards for technology education in elementary and secondary schools.

In our view, there remains a serious conceptual difficulty for those who would embrace the broad definition of scientific literacy proffered by Project 2061. Scientific, social, and technological capabilities are not the same. Communities of practice in these fields are driven by different sets of purposes and, sometimes, different methods and criteria of quality. Deep scientific knowledge does not necessarily lead to wise use of that understanding. Technology and social science, to a greater degree than the "natural sciences, encompass considerations of what is prudent, value laden, timely, and ethical. Modes of deliberation and decisionmaking differ. The relationships between science, on the one hand, and beliefs, values, and other nonscientific knowledge, on the other, would seem to require more serious attention than has so far been accorded the matter in Project 2061.

"Balance"

"Less may be better" ("less is more," in the early years) is a slogan often associated with Project 2061, both by those within the project and those outside. The dictum is appealing to many people, not least to the science teachers who constantly feel enormous pressures for content coverage, and who believe that such directions are counterproductive educationally because they easily become superficial. The pressures stem both from textbooks, which almost always aim to be inclusive so as not to omit someone's favorite topic, and from standardized examinations, which usually claim to be comprehensive.

But it is by no means clear that Project 2061 has reduced the pressure for coverage. Project staff members insist that for a person to be scientifically literate there must be a "balance" of subject matter in the science curriculum. To know primarily about essential concepts in biology, for example, or about earth sciences is not enough. Literacy, according to Project 2061, implies that a person has a nontrivial familiarity with the full array of major ideas identified so articulately in *Science for All Americans*.

A curriculum consonant with Project 2061 would certainly consist of ideas worth knowing. The concepts it would emphasize would have considerable explanatory power. Similarly, the major ideas outlined in *Benchmarks* are worthy. But to understand *Science for All Americans* and the concepts in *Benchmarks* well, to begin to comprehend some of the evidence that supports the essential ideas of science (as Project 2061 also advocates), almost certainly requires more time than most school districts are prepared to devote to science, probably more time than they devote to the subject now.

"By the end of the 5th grade," says *Benchmarks*, "students should know that a great variety of kinds of living things can be sorted into groups in many ways using various features to decide which things belong to which group [and that] features used for grouping depend on the purpose of the grouping" (AAAS 1993, p. 103). The goal is a significant one. And Project 2061 holds that students should learn such ideas in ways that have meaning for them, not solely by rote. A considerable amount of classroom time is therefore necessary, so that students try out and ultimately comprehend the underlying rationale for each of several different criteria for sorting. Teachers would want students to learn important subtleties within and across different classification schemes. A major point of the exercise would be to help children understand that there is no method that serves all purposes; there is no "right" way to classify living things. Instruction toward such goals is time consuming.

There are, however, well over 800 benchmarks, about 350 of them in the natural sciences. They are seldom simple: "Viruses, bacteria, fungi, and parasites may infect the human body and interfere with normal body functions. A person can catch cold many times because there are many varieties of cold viruses that cause similar symptoms" (grades 6-9). Or, "The forces that hold the nucleus of an atom together are much stronger than the electromagnetic force. That is why such great amounts of energy are released from the nuclear reactions in the sun and stars" (grades 9-12). Virtually all the benchmarks are packed with concepts at least as complex as these. Is it necessary that *all* of them be taught in a manner that leads to the understandings that define 2061's conception of science literacy? And, if so, where is the time to be found? It is still very early days in Project 2061 terms, but it is not difficult to anticipate the need to cut back on some of them to achieve the levels of understanding for each one desired by Project 2061.

It remains to be seen, then, whether the coverage demanded by Project 2061 in the hundreds of benchmarks is sustainable, or whether in most situations it will be necessary to temper the conception of balance to which all members of the Project 2061 head office staff are committed.

Whose Essentials?

Several issues that bear on the conception of science reflected in Project 2061, and that still reverberate within the project, go back to phase I. At the outset of

Project 2061, project leadership turned to the established scientific community—and particularly to those engaged in academic research—to make the initial effort to identify the essential components of scientific literacy. The technology panel had significant representation from industry, but not from any of the other groups. Science teachers of children in elementary and secondary schools were not included. Nor were those at colleges and universities whose primary responsibility it is to prepare science teachers and conduct research in science education. (The plan was to involve these groups centrally in phase II of the project as attention turned to curriculum implications.) There are both conceptual and political problems associated with the decision about whom to include. The two sets of problems are not unconnected.

Some members of the science education community, particularly those in universities, saw the early reliance on scientists as possibly repeating what were seen by many of them as strategic errors associated with the science education curriculum reforms of the 1960s. Scientists then, too, took the lead. However, in the end, the reform was judged by many to be less effective than it might have been because teachers and other science educators were enlisted late, and then not to assist in the identification of content, but primarily to implement the ideas that already had been identified by the academic scientists.

According to Ahlgren, whatever problems may have been associated with the decision not to include university-based science education specialists in phase I activities have moderated markedly.

> I think that prominent science educators may have felt that the procedure that we used in phase I didn't give enough ear to professional science educators, and, to that extent, we lacked a political mandate for what we had done, [but] when the results came out, I've heard nothing but good things from science educators.

The problems associated with identification of essential content by scientists, however, seem to be more persistent in the case of teachers. While most of the tensions between teachers at the six centers and staff at the Washington office revolve around expectations and prerogatives (and are elaborated in the next section), there is an underlying uneasiness about some of the content and how it is best identified.

Do research scientists based in universities accurately understand the special quality of today's students? A minority of children in school intend to enroll in a university, and an even smaller number plan on majoring in science. Are the science essentials identified by scientists sufficiently related to the lives of students to engage them in worthwhile classroom activities?

Many teachers generally believe that for students without apparent academic interests, it is necessary to emphasize the science that has an impact on their lives and on their communities. They often talk about "applications": a child will be better motivated to learn about statistical probability, for example, if she understands its relationship to a decision she is making about whether to start smoking. Selection of appropriate content that has immediate meaning to students is no less essential to many teachers than the science deemed fundamental by university-level researchers who are operating at the scientific frontiers. Furthermore, they claim, it is they, the people closest to the students on a daily basis, who are best positioned to make the decisions about the personal and social needs of their students and therefore the science that should be taught.

Sometimes the issue is couched as pedagogy as contrasted with content. A key figure at the Georgia center says, "There is a tension between content and pedagogy [in Project 2061]. Teachers are concerned about what goes on in the classroom, with pedagogy. Washington is concerned with content." But underlying such statements is the possibility of a mismatch between the content that is to be emphasized in Project 2061 and the needs of the students for certain kinds of scientific information. How do I protect myself from disease? What do I need to know to get a good job? Teachers often see the curriculum as needing to emphasize more practical matters associated with science than do university professors, who—they assume—know little about challenges in today's classrooms.

It is not that scientists do not concern themselves with the proximate needs of children, which many teachers deem important. Many of them do. But a sizable number of those scientists do not work primarily in university settings. And they do not engage in basic research. Rather they are employed in industry or in government laboratories where, because of internal and external pressures and the resulting expectations held for these institutions, they often work on issues of greater immediacy, and more personal and social impact, than do university-based scholars. The matters investigated by such scientists and engineers—developing better fire-retardant material, figuring out the relationship between exercise and good health, designing structures that are more earthquake-resistant—might be of greater interest to students, too, because of their closer link to everyday lives.

Curriculum Integration

An issue receiving considerable attention in the science education community in the mid-1990s is that of blurring the conventional disciplinary distinctions among earth sciences, biology, chemistry, and physics at the high school level (and eliminating them altogether, where they exist, at the elementary school level). The Scope, Sequence, and Coordination project of the National Science Teachers Association places this issue at the center of its effort to change high school science teaching. It attempts to coordinate the teaching of the separate science disciplines, year by year, occasionally even to integrate them. The state of California is moving in this direction; several high schools in the state have created courses with titles like Integrated Science I, Integrated Science II, and Integrated Science III. Project 2061 itself introduces the importance of cross-disciplinary "themes" in *Science for All Americans*, and suggests four as illustrative: systems, models, constancy and change, and scale.

The intention of the themes in *Science for All Americans* is to accent the fact that, while the world one wants to understand and explain is usefully and manageably analyzed by creating disciplinary distinctions, people do not encounter a naturally compartmentalized world. Scientists, too, find it important to recognize that there are various unities in science. One of them is that certain over-arching themes have valuable explanatory power and that they transcend the separate disciplines. Stressing the interconnectedness of science is a priority for many scientists and science educators.

Project 2061 staff members, however, are insistently neutral on the matter of how the curriculum should be organized. Roseman notes:

> I don't think Project 2061 is arguing for integrated curriculum. What I think *SFAA* says is that understanding concepts well depends on understanding other concepts, whether it be the importance of the conservation of matter for understanding how ecosystems cycle resources, or whether it's understanding systems in one context helps you understand systems in another context. [On integration] the jury is still out. Most of the literate people that I know started in a single discipline, and that sophisticated understanding of one discipline made it easier for them to broaden their knowledge in others.

It may be of more than passing interest, however, that at almost every one of the six centers affiliated with Project 2061, models of curriculum were devised that featured integration of the science disciplines. These models were met with no particular enthusiasm by the Washington office, leading some teachers at the site level and several of the center directors to assert that the Washington office is opposed to subject integration.

Neutrality on an issue like curriculum integration can certainly be interpreted as opposition by proponents. At the very least, however, there is ambiguity about this issue. Says one head office staff member: "I see no incompatibility [in organizing *SFAA* along well-understood disciplinary boundaries] when it comes to topics like environmental studies. When you write a book [of goals], you have to organize it in some way. It doesn't mean you have to teach [toward those goals] that way."

Promoting Inquiry

As with the matter of curriculum structure, Project 2061 appears to be neutral on the matter of exactly how one actually instructs children in science, at least within very broad boundaries—though it advocates a variety of approaches. One can employ hands-on techniques. Presumably one can be entirely expository, as well. The important criterion is that students acquire the science concepts associated with scientific literacy.

Project 2061's position on this point again differs substantially from that of other important and knowledgeable groups. NRC's National Committee on Science Education Standards and Assessment, which is the governmentally sanctioned effort to develop science standards, placed science inquiry as the first content goal in its November 1994 draft standards.

> Science as inquiry is a basic and controlling principle in the ultimate organization of and activities in science education. The standards on inquiry highlight the ability to conduct inquiry and develop understandings about scientific inquiry. Students at all grade levels and in every domain of science should have the opportunity to use scientific inquiry and develop the ability to think and act in ways associated with the processes of inquiry, including asking questions, planning and conducting an investigation, using appropriate tools and techniques, thinking critically and logically about the relationships between evidence and explanation . . . The science as inquiry standards are described in terms of activities resulting in student development of certain abilities and understandings.

An argument sometimes advanced for involving students in actual scientific investigations is that such activity is highly motivating and that active engagement improves the quality of learning. The draft NRC standards clearly go further, however, in suggesting that a person does not meet standards for scientific literacy without having been personally involved in scientific investigations. Project 2061 head office staff members reject this position. Requiring all students to conduct inquiries is fine, but not central. The objective is for students to acquire an understanding of fundamental concepts in biology, physical sciences, mathematics, social sciences, and technology—and, in the process, begin to understand how scientists work. For this understanding, it may or may not be necessary to involve children in scientific investigations, as long as they understand as adults how scientists work.

A person also should possess certain "habits of mind" associated with science. These include "knowledge of the values inherent in science," openness to new ideas, and certain attitudes. Furthermore, "teaching should be consistent with the nature of scientific inquiry," and teachers "should engage students actively" in such activities as collecting, sorting, observing, using instruments, and measuring. There is an emphasis on "careful observation and thoughtful analysis" throughout the major documents of Project 2061. But it is not essential for adults (or students, presumably) actually to conduct investigations. Commenting on the draft NRC standards, Ahlgren says,

> [The NRC standards say] get kids good at inquiry. They [NRC] don't mean understanding inquiry so you can make sense of articles in the newspaper . . . They mean you should end up as an adult able to conduct scientific investigations. They emphasize it as much as they do, in part, through fear that unless they emphasize it that much, it's going to be neglected.

> [But] can you expect all adults to be able to design and conduct scientific investigations, or can't you? We'd be delighted if adults could do that. We think we have a humbler goal, which is that they should be able to read the newspaper

and say, "Hey, that seems to be not quite right." It isn't a difference in value [with the NRC standards]. I think it's a difference in prudence, perhaps, in making it part of a minimal set of goals.

Specialists in science for elementary schools are often particularly insistent that children investigate the world around them, that they cannot know (or like!) science in any deep sense unless they begin to experience science in some of the ways that science researchers do. Ensuring the quality of that science experience is of high priority, both for understanding the science and for ensuring engagement.

"But," asks Ahlgren,

What do you learn, other than to think that science time in school was fun? Are they going to say that they learned about series and parallel circuits? Given that there's not much around these days that's engaging for kids, I'll go with the priority that we [figure out ways] to avoid turning them off. But let's imagine that we're 10 years down the road, and people have come up with a greater variety of engaging things on science for kids to do, now wouldn't it make some sense to select from among those activities the ones that also taught important ideas? Right now we're in a psychology of poverty. If you haven't got any food at all, you'll eat whatever it is. You don't worry about excess fats or not enough vitamin E.

Reflecting on these issues, Rutherford also couched the place of teaching through inquiry as a matter of priority.

I think when you see the work of Doug Lapp [a national and respected figure in elementary science education] and listen to him talk, there is a high premium on the quality of [children's] experiences. If that is high, other things will follow in due course.

At any one time, we have a notion of what's the worst trouble [in the schools]. What's the danger? Doug says, "Right now, the schools are bookish. Teaching focuses on kids learning words." So I say, "Well, Doug, we go to classrooms and the kids are doing marvelous experiments, but it doesn't appear they're learning anything."

Sofia Kesidou, a project research associate, at the Washington office, notes:

Many people would want [us at Project 2061] to have a more inquiry orientation, to have students be able to conduct scientific investigations and be able to do controlled experiments, rather than more on content kinds of things. The project does not say we should not include such investigations in the curriculum, but we see these investigations more as a means to understanding the nature of science and some habits of mind, rather than as the goal to be able to conduct these investigations . . . Of course it would be a wonderful thing to find activities that both engage kids and teach them worthwhile goals.

Ahlgren is less ecumenical: "I call [placing priority on children's inquiry] aerobic science education. It doesn't matter what you study as long as you're breathing hard."

We leave the last word on these matters of subject integration and teaching by engaging students in active investigations to Rutherford:

> We say, sure reflection, sure hands-on, sure integration. We've got goals. If you teach them through the disciplines, fine. Through integration [of the disciplines], fine. Through hands-on teaching, fine. By and large it doesn't happen, though, when people create one form or another of an interdisciplinary course . . . We're trying to say that if people have tools and instruments that promote thought and analysis and creative efforts and clear goals, then people can look at those issues in the context of what they want to do.
>
> It's the sense of balance between kinds of subjects and contexts for learning that we're looking for. Whether it occurs in something labeled a science course or social science course is of less concern to us.
>
> I think the next edition will say, "Yes, not only does there have to be some balance in content, but a 2061 curriculum would have a balance of modes, styles, and kinds of experiences, as opposed to finding the right, single best way of doing something, then doing it over and over".

◆

The Conception of Reform in Project 2061

Introduction: The Worlds of 2061

From the beginning, four key features of Project 2061's model of reform have distinguished it from other educational reform efforts. First, in recognition of the fact that the task is unprecedented, complex, and costly, the undertaking was conceived from the outset as a long-term commitment. The reform is expected to take decades—maybe longer. Second, as a national reform effort, the project's scope reaches far beyond the district or state level—project participants are attempting to change science education nationwide. Third, in today's educational parlance, the project is systemic. That is, to effect change in science education, Project 2061 targets not only science curriculum but "policy, teacher education, the design of learning materials, assessment practices, and much more" (AAAS 1993, p. 323). Fourth, and finally, the project is goal oriented in nature: participants have outlined the knowledge a scientifically literate adult should know and based subsequent recommendations on those goals.

Project 2061's conception of reform has required the support and participation of many groups, including teachers, administrators, scientists, engineers, mathematicians, and university educators. As a result, the project operates simultaneously within three organizational spheres as well as across different physical sites: a Washington, D.C., national office, six school district centers, and academia. Because each world has its own demands, cultures, and concerns, the interplay among them is complex and, not infrequently, bumpy.

In this section, we explore how this conception of reform—a conception that is long term, national, systemic, and goal oriented—is articulated across the worlds of Washington, the centers, and academia. The following questions frame our analysis: How is Project 2061's conception of reform understood, articulated, and implemented across these worlds? How do the distinct goals, pressures, and conditions of each world affect its ability to accept and implement the features of this reform? And, as the features of the reform are operationalized, what difficulties arise at the intersection of these different worlds? In presenting our findings, we hope both to provide insight into the successes, challenges, and limitations of this large-scale project, and to inform present and future reform efforts.

The Reform as Long Term

What's in a name? For Project 2061, its name, the calendar year 2061, symbolically frames both the project's goal and time line: its commitment to effect lasting change, and its conviction that lasting change takes time (Rutherford 1991). First, the year 2061 represents the project's attempt to reform—fundamentally and permanently—science education. As explained in *SFAA*, Project 2061 "was started in 1985, a year when Comet Halley happened to be in Earth's vicinity. That coincidence prompted the project's name, for it was realized that the children who would live to see the return of the comet in 2061 would soon be starting their school years" (AAAS 1989, p. 11). Second, the year 2061 suggests that the project's time line for design, implementation, and dissemination of its reform agenda is long-range. (It does not imply, however, that the project expects to work toward its reform goals until 2061.) When first introduced, the reform was expected to span a decade or longer (AAAS 1989). More recently, Rutherford has projected the effort to last two to three decades. It is at the intersection of these two definitions of 2061—2061 as a long-term reform effort designed to effect lasting change—that we begin our discussion of the reform process.

      Why Make Project 2061 Long Term? As suggested above, the name 2061 not only frames the project's goal and time line, it underscores the connection between them as well. Participants across the worlds of 2061 agreed that, to effect significant, lasting change, a reform must be long term. Roseman expresses her conviction that change must be long term if it is to be lasting in this way:

> If you want the change to make sense to people, then people need time for sense making. I guess an analogy is psychotherapy. If you really want people to change—to feel comfortable with it and incorporate it into their personal lives—then there have to be time and support for the change. Anything short of that, and they will "say" what they think you want to hear but it won't affect what they really think or do. And it's not just the time, it's knowing that you'll have the time. This was an important factor at the sites, at least I think so. When we started working with them, not only did they have release time and summers but they also *knew* they were going to have it. And I think knowing that you're going to have it makes you willing to take risks.

Gary Oden, a center director, echoes Roseman's sentiments:

The battle of illiteracy will not be won in my lifetime . . . A reform project's place is where we are right now: to provide the intellectual material for the future. That's the basis of our efforts. By accepting that fact, it gives me hope that one day it may make a difference . . . It won't happen in our lifetime, but the fact is that these things are going to happen eventually. They help shape and guide things. If we can do that, I think that's a great contribution and that keeps me going actually. So even though I may sound negative, the fact that we may be contributing to something that will help direct change—not necessarily provide the soil, minerals, and resources for that change, but help direct that change—that, to me, is something to keep working for. That's my positive note.

And teachers explain that the long-term nature of Project 2061 served as an incentive to join the reform effort, as reassurance that they would have the time they needed to make lasting, significant changes. A teacher from San Antonio underscores the importance of a sustained, long-term commitment to reform: Project 2061

is a long-term effort. So that, once we got into it—once we committed emotionally, philosophically—[we knew] they wouldn't pull the rug out from underneath us. I think that was why a lot of us bought into it and bought into it so profoundly . . . I think that's what sets this project apart from most of the ones I have seen.

A colleague from Georgia expresses her hope that changes instituted by the project will last: "[I look at] a lot of school classrooms now, and they look the same as they did 50 years, 25 years ago. And I'm hoping that in 25, 30 years, we will see that we have made changes in the way a classroom looks and is organized."

Challenges Faced. Although participants across the worlds of 2061 have proudly billed themselves as part of a long-term effort, they do not always agree on how to translate their long-term agenda into effective, coherent action. In this section, we focus on three challenges participants have faced in trying to work out what it means to be long term: the tension between producing immediate results and planning long-range strategy; the difficulty in ensuring continuity of ideas, personnel, and funding over time; and the desire to remain responsive to the ever-changing context of science education reform.

Immediate Results Versus Long-Range Planning. One challenge has been trying to achieve a balance between addressing immediate needs and planning long-range policy. As a long-term reform effort spread across several physical locations, project participants have, at times, disagreed over reform priorities and opportunities for action. During a meeting between Washington office personnel and center directors, for example, Roseman and Ahlgren proposed discussing an outline of *Designs for Science Literacy*, the third 2061 tool slated for publication. Center directors raised what they perceived to be more immediate needs: an action handbook for reform and a how-to-use-the-*Benchmarks* guide to assess or revise curriculum. One center director, Deb Larson of Wisconsin, planned to use the *Benchmarks* that summer to detect gaps or duplications in her district's K-12 curriculum. A second, Oden of San Diego, hoped to convince the eighth grade teachers in his district to wait one more year to buy textbooks. As he explained: "The people have the *Benchmarks* but we need to give information about curriculum blocks and how they interact with *Benchmarks*. We need a short document to buy time . . . People are planning for next year." A third, Carnahan of San Antonio, noted that groups of teachers in her state were coming together that summer to discuss curriculum. They met only once every five to seven years. She made this point: "We need a how-to book. Those topics in the *Designs* can't wait for the year. How do you get implementation? How do you get things going? We're not at 2061 now but . . . when you're trying to get things started, what do [you] do?"

Project 2061 staff members attempted to respond to center directors' concerns. They drafted a survey for use in finding other center members willing to contribute to an action handbook. There was never any follow-up by center directors after the meeting in San Francisco where this discussion took place. The staff also changed the outline of *Designs* to satisfy various center director requests and altered the configuration of *RSL* in response to comments by Larson and Oden. More recently, staff developed a whole new (and then unanticipated) product—*Resources for Science Literacy*—which has greatly displaced the work on *Designs*.

The Washington office staff and teachers have clashed over this subject as well: the staff has steadfastly declined to move toward the broad-scale implementation of its ideas before it perceives itself ready, while some teachers have been eager to begin revising and refining their classroom practice. One teacher, Foster of San Antonio, voiced her growing frustration with 2061's reluctance to demonstrate more than a passing interest in classroom implementation. She has decided to shift her energies toward work that will have a more immediate impact on her students:

> AAAS has not provided feedback in years to those of us still plugging away trying to make sense of science literacy and how it can be achieved. They select a few individuals to receive some training and forget that there is [a] whole group out there who needs to be revitalized and encouraged. [For me,] the motivation to spend time and energy on products that AAAS doesn't appreciate is not there any more. I'd rather spend my time on more meaningful work, work that directly impacts my students and my own education.

As a long-term reform effort on a short-term funding schedule, Project 2061 has also been forced to balance the production of periodic products with its efforts to develop a comprehensive reform plan. For some project participants— including several project teachers—the production of concrete, immediate results to satisfy the demands of funding agencies has served as a distraction. A teacher attending a *Benchmarks* training workshop in Washington, D.C., complained that 2061 funders' insistence on products prevented them from appreciating the full scope of 2061: "NSF and others do not know how much they're getting for their money. All they can see is two books." Matthews, a 2061 teacher in Georgia, described the production of products as an unfortunate but necessary evil in working on a long-term project:

> [A] very clear disadvantage is that people expect to have something very tangible. After all, we've been working on this for five or six years—"What do you have to show for it?"—kind of thing. "Where is your product?" "Well, there's *Benchmarks*." "Well, is that all you've got? Just one book?" . . . There are milestones, there are benchmarks, mile markers, but I don't think it will ever be finished. And, if it is ever finished, then it's not 2061 because I think . . . [Project 2061 represents] change over time.

Rutherford, the project director, offers a more expansive view of funders' demands for short-term products. He agrees that individual products inadequately represent Project 2061's long-range, systemic agenda. However, he also views these products as a way to sustain and refine the reform's vision: to attract and satisfy funders, to motivate participants in the reform, and to keep the reform focused. He elaborates:

> When something is very long and where the things, no matter how hard you run, seem to be out on the horizon, you've got to figure out ways to have frequent enough payoffs for people to have a sense of accomplishment. And that's one of the reasons we have to get products out every couple of years. Something out there. Do things like the newsletter . . . Got to have some concrete things so that the horizons aren't always too far away. And then not only for funders, but for the sake of the people doing their work . . . So that's, I think, something that has to be dealt with in a long-term project of this sort. I think failure to do that would lead to the difficulty of keeping it focused.

Continuity. A second challenge faced by project participants has been putting organizational structures in place to ensure continuity of personnel and funding. Several Washington office staff members have recognized the need to infuse new blood into Project 2061, at both the Washington office and site level, in order for the project to survive and grow over time. Ahlgren expresses concern at the inability of the 2061 Washington office to recruit young intellectuals into the fold. Roseman explains her interest in creating continuity between old and new participants at the site level: "This question of turnover of human lives is something that concerns me a great deal . . . I'm concerned a lot about the continuity. I think this is an issue in school districts: the turnover of teaching staff and certainly a turnover of superintendents." And Rutherford notes that, although he initiated the reform effort, he did not plan to see it to completion. He has begun to consider both who would become the new project director and how to embed or distribute some of his responsibilities across the project's structure: "One of the things I have to solve, besides the funding thing, is succession. After all, other long-term enterprises don't depend on an individual."

Similarly, 2061's attempts to ensure continual funding over long periods of time have not met with much success: most funding agencies have been reluctant to extend their funding cycles beyond three to five years. Oglesby describes the project's funding problem in this way:

> I've consistently said and continue to say that we have a long-term, systemic, science education reform plan, but we don't have a long-term financing plan. So it could be the politics of lack of resources [that stalls this project]. For whatever reason, the effort could die from a lack of support. Not that it's not a good idea, but it's the fact that people tend to tire of long-term efforts.

Rutherford agrees with Oglesby: procuring adequate funds for project activities has been and remains difficult. Recently, he suggested enmeshing 2061 more firmly into the AAAS network, placing AAAS in a role similar to that of a trustee. Such an approach, he continues, would ensure a continual source of money. "One of the things I'm hoping for is AAAS [becoming] part of a system . . . [Then,] we could build a revenue base that would take a lot of the pressure off having to constantly scramble every year for money to survive."

Keeping Up With the Changing Times. A third challenge of being long term has been to keep up with and remain responsive to new ideas and movements within the project and within the science education reform arena. Washington office personnel have attempted to remain at the forefront of science education reform by being flexible in approach, by frequently revising or reversing their course of action in midstream. They have deemed flexibility of approach both a benefit to being a long-term effort and a way to make the reform better. As Roseman explains:

> [2061 as a long-term effort] makes us willing at the end of a year, when we find we've been heading in the wrong direction, to say, "Let's regroup. Let's rethink. Let's start all over" . . . This project can change after four years of investment in something. That isn't to say—"Throw it all out"—but to really rethink . . . [One example:] The notion of models. Where are we on models? Another example: The resource base that we've been developing. How useful is it to people? Our original notion is that we would find resources that would help people get started. The more we looked into it, the more we didn't want to name resources because we could see intellectual dead-ends in using those resources. So, we're not sure where we are on a resource database of approved resources. Perhaps an analytic tool would be more useful to people. We'll do that after investing two years. We won't do it lightly.

However, in responding to new currents in the sea of science education reform, project staff has sometimes created ripples in its relationships with centers and funding agencies. Staff members have, on several occasions, asked center participants to stop, restart, and/or refocus project work. For example, in the summer of 1990, the Washington office asked all center personnel to put aside development of the curriculum models to engage in the process of backmapping. Then, that same September, at a joint retreat of San Diego and San Francisco,

Rutherford changed his mind: He decided that these two teams should abandon the backmapping process and continue work on their respective curriculum models. The project has also failed to meet deadlines promised to funding agencies: the development of *Blueprints* and *Designs* remains behind schedule.

Because Washington office staff has not followed a firm and detailed set of plans, changes in project priorities have not always made sense to or been readily accepted by site participants. Some teachers have lost their patience with the Washington office. As one teacher explains: "The head office has changed their directions so frequently that I suspect that when center directors were hearing something that was questionable, they simply waited for a shift in the wind before stirring up problems." Rutherford recognizes that staff members' willingness to change their minds has harmed their relationships with sites: "We're like academics. We change our minds often and think that's okay. In their world, it's not okay."

By not following a firm and detailed set of plans, 2061 also has walked a fine line with funding agencies. It has had to negotiate between promises to produce products and freedom to respond to changes in the national education reform scene. Ahlgren explains this tension between reliability and flexibility well: "So it becomes a little tougher to, at the same time, assure funders that we have something definite in mind that they can count on and yet keep enough freedom to change our minds as we find out what goes on."

Roseman provides a specific example:

> I think it was a wise decision on our part [to engage in the national science standards debate] but I think there has been an enormous cost in terms of our energies that have gone into analyzing [NRC's] work, have gone into advising them. I think we estimated maybe 1.5 staff years over the past three years. That's a lot! . . . Some of the things that have not been done as a result are *Designs,* because we have not turned our attention to it in as focused a way as we turned our attention to *Benchmarks* . . . Our funders have expectations about time lines. They haven't paid us to be advisors to the nation.

Before moving on, it is important to note that 2061 has attempted to keep up with the changing times by altering its *approach* to reform, not by revising its vision of science education. One could argue that a consistency of vision is necessary: How can participants be expected to work toward a better science education if what is considered better continually changes? Changes in approach have caused enough turmoil and resentment; how would project participants react to additional changes in message as well? Moreover, the intent of 2061 was to draw on principles and positions that were already common in the enlightened science education world, to synthesize rather than invent ideas. By continually revising and updating its vision, wouldn't the project quickly leave most of the science education community behind?

However, one could also argue that, because 2061 has held fast to its original vision of science education, it is already dated. A teacher from San Antonio,

Marilyn Brien, expresses her concern that 2061 has remained entrenched in 1989, that it is no longer poised on the cutting edge of science education. She explains:

> I suspect that Project 2061 has lived a full life. I have not seen any growth in vision. I have not seen any growth in direction. We are a different world than we were five or six years ago . . . [but Washington staff members] are still in 1989. I have not seen any vision, any change, any focus to anything else.
>
> The Texas model [for example] may have appeared radical in 1991. The model has been validated by subsequent publications and initiatives such as SCANS, Prisoners of Time, Goals 2000, and School to Work. We were told to be vision-ary and unhindered by current restraints. Today, when I read our latest revision of our model, I think that it was cutting edge four years ago, but is almost main-stream thinking today. We still have hardly made a dent in putting theory into practice. That is our next challenge.

The Reform as National

"Project 2061 is a long-term reform initiative whose mission is nationwide sci-ence literacy. Being 'long-term' may be the most distinguishing feature of the Project. Nevertheless, its national character is an important one" (Rutherford 1993, p. 12). From the outset, Project 2061 has distinguished itself from many other reforms in science education by attempting to transform not only the edu-cational system of any one school, district, or state, but that of the entire nation. That is, as a project brochure describes, the project has set its sights on imple-mentation "in school districts across the country— in all 50 states, in all 80,000 schools, for all 50 million students." Rutherford suggests that this may be accomplished "not by offering a standard curriculum to be adopted locally but by providing educators in every state and school district with a powerful tool to use in fashioning their own curricula" (Rutherford 1993, p. vii).

In some respects, Project 2061 participants are in uncharted waters: national reforms in science education have been few and far between. Thus, project par-ticipants have had to blaze their own trail, design their own system of dissemina-tion, and deal with the unique dilemmas and challenges of such an effort. For example, while state-level reforms often have an organizational and political structure already in place to assist in delivery (e.g., state educational mandates or policies and government agencies), national reforms have no functional equivalent. Consequently, as a national project without a federal support system, 2061 has had to design a unique delivery system—one that would provide the project with a means of disseminating its ideas and agenda to schools around the country while simultaneously engaging and coordinating the many players of 2061: scientists, university educators, administrators, teachers, center directors, and Washington office personnel.

This goal to affect schools across the nation has several implications for the structure, agenda, and politics of the project. Thus, in this section we explore the ways the reform has managed to be *national*: the particular dilemmas, challenges, and accomplishments that are associated with coordinating efforts across the worlds of 2061 in order to improve science education nationwide.

Division of Labor: Teachers, Center Directors, Washington Office Staff, and Consultants. As with any reform, the presence and involvement of many different groups require some division of the labor. However, at the national level, where those divisions are drawn is not always immediately self-evident. Moreover, the process is particularly complex and arduous when the groups involved are both geographically separate and culturally distinct. Communication and clear articulation are essential.

Responsibilities have been divided among the groups involved: teachers, center directors, the Washington office staff, and external consultants. Each has been assigned quite different tasks. Below, we provide descriptions of the responsibilities assigned to or appropriated by each group.

The role of the teacher in 2061 is at best ambiguous, and, at times, contradictory. On the one hand, Project 2061, as represented in its official documents, places teachers at the core of its reform movement: Project 2061 is described as a reform "for teachers by teachers." *Benchmarks*, in particular, is said to be *for* teachers, resulting from teachers' demands to move forward in changing curriculum. In addition, the publication is presented as a product *by* teachers much as *Science for All Americans* is by scientists.

On the other hand, Project 2061, as embodied in the Washington office, has a less expansive vision of teachers' abilities, and thus, their role in reform. Teachers are seen not as collaborators, co-reformers, or even reformers, but as consultants to the project. As Ahlgren argues, "We still see [the teachers] essentially as advisors to us rather than people in whose hands we would put the project."

This is, in fact, only one of many different descriptions given by the Washington office of the teachers' role. Rutherford, in a speech to teacher-reformers, characterized the teachers as "working with us to help create the national capacity for reform." In addition, the teachers have been labeled the technical assistants to the project as well as the leaders.

While the Washington office tends to position teachers somewhat at the periphery of the reform, at the six sites, the teachers' role is that of equal and active participant. There, Project 2061 is seen as an avenue of professional development, as a way to transform teachers into initiatory and reflective practitioners. It is also seen as an opportunity for expert teachers, teachers already on the cutting edge of instruction, to extend and refine their own work. Teacher after teacher with whom we spoke felt they had experienced "tremendous professional growth" and gained "extraordinary experience."

In addition, teachers see themselves as a vital part of the collaboration process necessary to really effect classroom change. As one teacher says:

> This project offered the promise of teacher collaboration, teacher enhancement, lifelong learning, and having teachers involved. Until a project can affect the people who are in direct contact with children, we're not going to have a real reform . . . [To] get reform, there [needs to be] a support group for the teachers. Teachers need to be lifelong learners and this was an opportunity for teachers to have that kind of collaboration. That was the promise that this project had for me.

The center directors are situated in an intermediate position requiring negotiation of the two distinct and separate spheres: the schools and the Washington office. Each of the center directors carries out this role in a slightly different manner. Two of the center directors attempt to insulate their teachers from the politics of the national movement and the Washington office. In particular, these center directors, seeing themselves as most in touch with teachers and schools, act as filters of information to the sites; information that is perceived as detrimental to the work of the teachers is not passed on.

Some teachers appreciated this buffering role assumed by the center directors. As one teacher states, "I'm not sure that I really cared what Washington did as long as [our center director] protected us from it." However, one teacher says she need not be insulated from the politics and struggles in Washington:

> As a teacher who has worked with Project 2061 since 1989, I can honestly say that I have resented being treated like a child that needs protection from the big bad world of politics. If only the [Washington] office and the center directors could see that to achieve the Project 2061 vision, a level of collaboration and trust must be established between all stakeholders. For years I have wanted to tell Washington to get it together, we don't need information that has been filtered and diluted. Give us the meat. We can handle it.

Other center directors, such as Munumer in Georgia, do not attempt to insulate their teams in any way. In any case, the Washington office does not see the center directors as adequate communicators of Project 2061's ideas. As Roseman explains:

> I think one of the things that was not terribly successful was counting on the leaders to communicate our [the Washington office's] ideas . . . So we would have somewhat effective meetings with the leaders and then the ideas would never get passed on . . . By and large, there was a lot lost in the translation and it was diluted so far that it just didn't get through to a lot of team members.

For many center directors, building a community of practitioners and fostering a culture that is amenable to reform is essential. While they accomplish this task in different ways, each has spent considerable time and energy creating a community of teachers who are agents of change in their schools or districts. In addition, some center directors consider themselves facilitators: rather than set the goals and agenda for the center, they help teachers do so by providing the tools and environment conducive to such work.

The role of the Washington office personnel is distinct as well. The Washington office determines project priorities and, alone, decides on reform tools. In addition, the Washington office staff describes its role as one of bringing people together, listening to their ideas, and pulling from their sometimes disparate views what they think is useful to the 2061 cause. There are, however, some internal disagreements as to the role of the Washington office: "Some prefer to believe we are a think-tank. Some believe that we should be an outreach organization," Oglesby notes.

A large number of external consultants have been hired by the project at one time or another to lend their expertise. The responsibilities assigned to these consultants have varied widely: some have worked closely with the teachers at the summer institutes; others have written or reviewed documents, provided input to the Washington office staff, consulted with center directors and teachers, or served on the National Council of Science and Technology Education (the project's overseeing body). Scores of scientists, mathematicians, engineers, and historians were involved in writing the panel reports.

The Washington office has not always agreed with or approved of the work of these consultants. However, their participation serves multiple purposes. They not only perform some very necessary tasks that other project participants cannot for lack of time, money, or expertise, but they serve a legitimizing purpose as well. That is, the project can refer to these experts to lend credibility to publications such as *Science for All Americans*.

Turf Wars. As a national effort, Project 2061 has encountered problems in the process of defining and articulating the roles of the various players involved in the reform as well as the organizational bodies which they represent: the sites, districts, universities, and the Washington office. In particular, the role of the teachers and sites has been a thorny issue. The Washington staff has not always known what to do with the sites or how to handle them. First, there have been debates regarding the suitability of tasks initially accorded the teachers. With regard to the curriculum models, one external consultant argues,

> Reliance on people from local sites is an innovation. But they didn't provide the right kind of support or give them the right kind of task . . . Teachers, when asked to think about education, often think of it in idealistic terms. Teachers don't have to think about budgets and constraints . . . They "blue sky" it all the time. You take a bunch of people predisposed to ignore those realities and tell them explicitly not to worry and the result is something so far from reality that it is not useful . . . It would have been fine to give blue sky for what kinds of activities, what kinds of resources, what kinds of assessment of learning. Those kinds of tasks that teachers really know about, teacher-pupil interaction . . . end results. And then have some other group, or have the teachers collaborate with some other group, to figure out how much can we make happen in some feasible context. I think things would have gone much better. But they had them plan a whole school program.

The criticism goes both ways: some teachers are critical of the way the Washington staff has executed its responsibilities. Particularly, teachers voiced some dissatisfaction with the staff's role in managing the project and its participants. As one teacher notes: "The Washington staff was placed in the position of managing a project without having the 'management' skills to accomplish the task well . . . [While they] have many and varied talents, managing people is not one of them." She goes on to state, "The Washington staff still does not have a clue about how to interact with such a large and talented resource. I suppose they must see themselves as the sole proprietors of the keys to science education reform. It appears that their roles are essentially as 'filters' of reform ideas, not "facilitators' of reform."

Second, there have been difficulties in assigning the teachers and the sites legitimate tasks as the project has evolved. With the publication of *Benchmarks*, the need has dwindled for tasks to be performed that engage sites and contribute to the project on a national level. However, Brearton recognizes that the partnership can not be maintained if the sites are not given some legitimate tasks— defining what those might be and how they might serve both the local and national interests is the current challenge. Says Brearton, "The centers, with the national office, have begun to do that in terms of study, as well as use, of the tools."

Third, in addition to issues of suitability, problems regarding power also have surfaced. That is, assigning groups of individuals certain responsibilities is not a neutral exercise. Particular tasks appear to hold more prestige and power than others. Creation of *resources* (e.g., local, working documents, or raw materials, such as the curriculum blocks, instructional strategies, and assessment tools) is considered suitable for teachers, while the Washington office has jurisdiction over the nationally disseminated *tools* (e.g., the official, published project statements such as *Science for All Americans*, *Benchmarks*, and *Designs for Science Literacy*). Center directors and the Washington office staff clashed over this point at a meeting held in Washington. A center director argued vehemently for the legitimacy of the sites to produce tools, while the Washington office maintained its ownership over the tools and the sole right to create them.

The assignment of tasks is rationalized using notions of experience and expertise. On the one hand, teachers are said to possess expertise, or a "craft knowledge" as Brearton calls it, of the children. As a result, they are assigned to tasks that utilize such expertise. On the other hand, Ahlgren describes the center directors as having "experience" in school reform but not "expertise" (another point that has been contested by center directors and teachers alike). He thus argues that center directors should not be placed in charge of writing *Designs*. A teacher notes:

> There exists an inherent mistrust between AAAS staff and all others who, to them, don't have the "appropriate" understandings of the goals of the Project, and therefore, are incapable of producing anything of substance and worth . . . this sort of mistrust ultimately undermines the creativity and openness to new ideas which is what Project 2061 is all about.

Finally, tensions between what teachers need or want and what the Washington office needs or wants are evident. For example, the need for teachers and center directors to address local and immediate pressures is, at times, in opposition to the Washington office staff's desire to forward the national (and long-term) effort. Roseman explains:

> Some of them [the center directors] are more interested in implementing at their own sites than they are in contributing to another book . . . There are some aspects of what the book is meant to include . . . that some of them are not interested in contributing to. They are interested in having something that can advise them on implementation . . . I don't think that's what Project 2061 central office should be doing . . . But what we are good at doing is provoking people, listening to what they say and capturing in simple terms messages that we can all pay attention to.

Changes Over Time. The tasks and responsibilities assigned to or assumed by the various groups in the project have certainly changed over time. In particular, the role of the teachers has evolved. Oglesby explains:

> It's an evolutionary process. First of all, when these teachers at these centers were selected, we knew that we had very talented people, but we didn't know what their potentials or capabilities were at the time. We knew that, given the right environment for them to grow—and they've grown significantly—that they would be able to take on much more complicated tasks.

He goes on to express the tremendous changes he has seen in some of the participating teachers: "These teachers now think of education in different ways. They relate to others in different ways." Thus, the tasks assumed by the teachers have changed as they have gained experience in and knowledge about reform.

Changes in the responsibilities assigned to the teachers are not due solely to growth in knowledge. The Washington office has also learned from experience what tasks are most appropriate given the teachers' abilities, constraints, and demands. For example, initially, curriculum development was thought to be one of the appropriate domains of the teachers. It was envisioned that curriculum development would not be the *sole* responsibility of teachers, but, as Roseman states, "that they would be in the driver's seat and they would come to understand the goals, and their knowledge of the classrooms would help them to design real curriculum models and they would commandeer whatever resources they needed to get the help along the way."

More recently, the Washington office appears to have reconsidered this role. According to Brearton, "I would rather envision [the teachers] as a group that is able to say to materials developers and publishers, 'This is what we want,' and is able to review that. To be very honest, I think expecting sites to come up with exemplary blocks is challenging and to some extent unrealistic."

Roseman concurs with Brearton:

We believe that our products will not be as good as they can be or as useful unless they pass muster with teachers. I say "pass muster" because that's different than saying teachers should develop them from scratch, and I know my own thinking about that has changed a little bit. I think in the beginning we really believed that teachers, aided by universities . . . could design real curriculum models.[6]

As the role of teachers in the reform has evolved, so too has the purpose of the sites and their relationship to the Washington office. Roseman states,

The original plan was to engage the sites for two years in developing curriculum models for the nation. We didn't ask them to commit, in advance, to implementing them. Why should school districts commit to implementing something sight unseen? There was another reason for discouraging them from thinking about implementation too soon: we didn't want today's realities to constrain their vision. But by the end of the second year, we encouraged them to start planning for implementation. Unfortunately, the financial scene at the sites had deteriorated considerably since teams began their work. School districts were unable to pick up the tab for the release time (30 to 40 days per year) and the six weeks of summer work that our NSF grant had funded.

Subsequently, the emphasis began to shift from model-and-block-development sites to places where approaches to 2061-type reform could be studied, and where one could glean insights valuable to the nation and certainly to the project. The change of names for the sites (from "sites" to "R&D centers") reflects this evolution. The Philadelphia center director, for example, considers this change to be very significant: "I believe that the change in sites from teams to R&D centers was a turning point for Philadelphia. It allows us to develop a cadre of professional development leaders to help the school district move toward change." A second center director, Oden from San Diego, does not think changing from site to centers was anything more than a semantic exercise.

Currently, the Washington office staff hopes the sites will attain some degree of specialization. Roseman says, "[Rutherford] has identified maybe four or five areas that [the sites] may want to concentrate in, rather than [all] the sites doing everything." As discussed previously, Farges echoes this idea of specialization: he considers San Francisco to be the *implementation* site as the San Francisco schools have focused their time and attention on implementing and assessing curriculum blocks.

It is not only the Washington office that views the role of the sites differently today from in the past. The sites themselves do as well. Roseman reports that initially few sites saw themselves as doing a task for the nation. In Seattle, however, a number of them came to see that they were contributing to a national project rather than simply doing a task for their school districts.

---

[6]On this point, Ahlgren notes that "there is little evidence that anyone, Washington office staff included, knows how to produce high-quality blocks. Coherent, option-rich blocks may still be quite a ways in the future."

The Reform as Systemic

> Reform must be comprehensive and long-term, if it is to be significant and lasting. It must center on all children, all grades, and all subjects. In addition, it must deal interactively with all aspects of the system—curriculum, teacher education, the organization of instruction, assessment, materials and technology, policy, and more. All of which takes time (AAAS 1993, p. xii).

2061 is not merely a curriculum development project; it is an attempt to effect change in all arenas inside and out of the educational system, including business, higher education, curriculum, and pedagogy. Although most of 2061's efforts to date have centered on curriculum, instruction, and the use of technology, there have been a few forays into the systemic arena. For example, the Washington office has asked various experts in the field to craft 13 *Blueprint* reports, each designed to address a different aspect of the educational system. Both Roseman, from the Washington office, and Farges, from the San Francisco center, have sponsored forums on teacher education. Moreover, some 2061 schools have experimented with changes in their organizational structure, thus more closely approaching the goal of systemic reform.

Besides the crafting of tangible products, the arranging of meetings, and the dividing of roles and responsibilities, 2061 has been engaged in an ongoing debate over how to orchestrate a systemic reform. Rutherford admits that the project has yet to arrive at a definitive answer to the systemic question:

> I think one other element is trying to take systemic seriously beyond most of what I see, which is a lot of meetings and people saying, "Get the stakeholders together. Let's talk. Let's share ideas. Let's try to work together. Camaraderie. Partnerships." I think that's all very important, but I take systemic to mean something fundamentally deeper, and so our blueprint thing. The academy is trying to do part of that, but [it's] very superficial. Ours is superficial right now, but here we are struggling . . . We've got to find in here some small number of ideas or something that runs through that we can, and with all the help we can get, sort out some distinctive ways of thinking about systemic and [its various] elements. It isn't enough to say the assessment has to match the goals . . . So we may or may not come out with something that will be useful, but I think [our] taking a run at it is pretty different.

Engaging, Enabling, and Coordinating Different Worlds. Although the project continues to search for a satisfactory answer to the systemic challenge, it has forged ahead with attempts to change fundamentally all aspects of K-12 education. One of the challenges encountered along the way to systemic reform has been effectively engaging, enabling, and coordinating the various project participants. Toward this end, the project instituted a two-pronged approach to reform, one that can best be described as a combination of top down and bottom up, of a centralized vision and decentralized implementation.

On the one hand, Project 2061 has portrayed itself as a national, high-profile project emanating from Washington, D.C., under the auspices of AAAS. From this perspective, it is an effort poised to shape the terms of the national debate regarding the meaning of science literacy and the means of achieving a scientifically literate populace. It draws upon the expertise and experience of scientists, policymakers, and educators across the nation to offer a national vision and approach to science education reform that are both broad in scope and exceptionally ambitious. Specifically, the project outlined in *SFAA* has goals for science literacy that reach well beyond the present boundaries of school science. It offered *Benchmarks* to educators as a road map for reaching these goals. In addition, the project has been working to construct a framework for systemic change in the form of *Blueprints* that, if realized, would significantly alter the face of American education.

On the other hand, Project 2061 has been portrayed as an invitation to teachers and administrators to make creative use of a central storehouse of tools, to fashion a grassroots response to local and changing conditions. This perspective emphasizes the diversity of undertakings that Project 2061 supports by the use of carefully crafted and broadly applicable guideposts like *SFAA* and *Benchmarks for Science Literacy*. Tools like these have provided teachers, schools, districts, and the nation with a way of building on their own talents and opportunities.

Much of our data support this notion of 2061 as a reform with a centralized vision and decentralized implementation. In a 1994 speech at the California Science Teachers Association, for example, Rutherford described the project's approach to reform as both top down and bottom up. Members of the Washington office, he explained, originated, control, and guide the project; they have provided a strong vision. "In that respect, we [the members of the Washington office] are enablers. We're not doers. We're not going to reform the schools. We haven't the authority to do so. [However,] we can create some things which will enable the only people who can do it, to do it, and that's the people all up and down the line." The teachers, he continued, do not lead the project; rather, they have been intimately involved in the training of other teachers, the writing of curriculum materials, and the implementing of new approaches to curriculum and instruction in their classrooms.

> People make a certain virtue out of grassroots. People always talk about grassroots; how important that is. It is important; it's just that it doesn't make it reform. However nice it is in 110,000 schools in America to believe that, if you just let each [school] go and give teachers a little encouragement, they could collectively create a national set of experiences for children, it's expecting too much.

In short, Rutherford concluded,

> we have to have *both*. Another word for top down is leadership, vision, pulling things together, creating awherewithal with policies to make it possible for people at the working level to put in changes. We have to have grassroots because the only place in the end that matters is where there is the interaction between young people and teachers . . . Only they can bring about reform.

Our characterization of 2061 as having a centralized vision with decentralized implementation, it is important to note, has met with some objections. Some center members argues that the Washington office has never offered a central vision; rather, they have sent a confusing series of mixed messages. As one site member explains: "A vision implies a clearly articulated goal which everyone moves toward. It's a moving target with 2061." Other participants agree that this is the current model of systemic reform held by the project; however, they wonder whether this approach effectively engaged, enabled, and coordinated participants in different worlds. In our next section, we take up this second issue: Who should control the direction and pace of Project 2061?

Who Should Control the Project? As suggested above, Project 2061's approach to the coordination of participants engaged in systemic reform—a centralized vision with decentralized implementation—has raised issues of ownership and control. Some teachers have expressed frustration at their inability to direct the reform effort, to shift the center of gravity away from the Washington office. One teacher explains:

> Teachers claimed ownership of the project. The staff [however] did not have sufficient experience with K-12 teachers to anticipate what a powerful constituency they were guiding. They created a "monster" that they could not control, so by default, they refused to offer feedback that would encourage the teachers to pursue goals other than those narrowly defined parameters which would lead to a publishable document or a report for funding sources.

After all, why shouldn't the teachers be in control? It is they who operate in the world of schools. Munumer points out, "[The school] is the real world— it's where the rubber hits the road."

Washington office members have countered teachers' responses by insisting that they are and rightfully should be the leaders of Project 2061. Ahlgren, in commenting on a preliminary draft of this case study report, made the following point: "I'm reflecting off a turn of phrase in your preliminary report, and I forget what it was now, but it seemed to suggest that somehow we had sinned by taking control of the reform away from the teachers. We don't see the teachers as having control of the reform. We see them as essential helpers in it." In fact, he explained during a second interview, the Washington office has come to view the sites as somewhat peripheral to the reform movement. In the last few years, "the sites have pursued their own agenda, and that has caused us to lose interest [in some things they are doing]."

The Need for Understanding. Another challenge participants have faced in attempting to effect systemic change is learning to understand and accept the worlds of fellow project participants. Presumably, project staff has been frustrated by teachers' lack of understanding of its world—of the pressures, responsibilities, and interests of those housed in the Washington office. The teachers, for their part, have been very vocal about project staff's perceived lack of understanding of *their* world. A teacher from Georgia, for example, says that the Washington office failed to adequately appreciate or make accommodations for the world of schools. Staff members did not understand

> what [teachers] have to deal with. But because they don't deal with it, they don't think about it. When you're up on H Street in Washington in an office and you've got five or six secretaries, you're sort of in a think-tank. And you're thinking up all these wonderful things, and yeah, they are wonderful for some people but you're not, you're not exposed to all of the diversities that [teachers] have to deal with. You don't collect lunch money. You don't monitor a bathroom break. You don't take care of all the nuts and bolts . . . I would like to say to head office, "You're not dealing with the real world." I would like them to come teach my boys.

A teacher from San Antonio, Brien, agrees that project staff has demonstrated little understanding of the world of schools. Members of the Washington office, she continues, would be well-served to listen to and draw from teachers' experiences:

> I have not felt that [project staff members] have really had any inkling of what it's like to be in the classroom, what it's like to be in real schools, what the political situation is, what the whole political framework of this is. I don't know that they can be knowledgeable for each different area of the country, but they need to have confidence in the people they have put in those positions in that area. And I think that is the biggest thing, right there. Also, they need to have teachers and people in real schools making some fundamental decisions, not people who are so far removed that they really don't know what occurs.

Whose Perspective Is the Systemic Perspective? In its drive for comprehensiveness and coherence in educational reform, Project 2061 has experienced several challenges. One way of understanding and perhaps avoiding such challenges in the future is to identify from what perspective 2061 has decided to view the educational system; in other words, to recognize the importance of perspective. In attempting to effect systemic reform, Project 2061 has brought together participants from different, sometimes conflicting, vantage points in the educational system. If one understands systemic reform as requiring comprehensive change of the entire education system, from whose vantage point does one decide to conceptualize the system? Whose perspective—policymakers', administrators', teachers'—is the most useful for making decisions? How should what is considered most useful depend on the kinds of actions needed to be taken?

Like most systemic reform efforts, the perspective Project 2061 has emphasized is the broadest one, the one that comes from viewing an object or event at a distance, the one held by inhabitants of the world of Washington. And, to a large degree, such a broad perspective makes sense. After all, it is natural when striving for perspective to seek distance. To get a comprehensive view of a university campus, climb the clock tower. To view a city, scale a nearby hill. To get a sense of the flora and fauna in a state, fly a plane. But those inside a classroom encounter a different university from those at the top of a tower. A commuter stalled on the freeway experiences a traffic jam differently from a newscaster. A park ranger has a different view of her national park from those who fly over it. In truth, then, a broad perspective and the insights and experiences that accompany it are no better than a local view; the perspectives are simply different.

What are the consequences of according priority to a view from afar? One consequence evident in the 2061 reform effort has been the privileging of national, broad-scale efforts over local implementation, an issue already discussed in this section. A second consequence is that such a stance tends to fortify the power and prerogatives of those whose responsibilities are at the most general level, and to weaken the policymaking prerogatives of those closest to the sites at which educational services are actually provided: the teachers. Such a stance is unfortunate. Teachers have a different view of the education system from state governors or national policymakers. They are unlikely even to use the same vocabulary. More importantly, one of the clearest lessons of successful reforms is the importance of according considerable weight to the insights and initiatives of those closest to the point of provision of educational services. Teacher networks, for example, have proved a major vehicle not only for sustaining reform but initiating it.

The Reform as Goal Oriented

Participants in Project 2061 place great emphasis on the project's goal-oriented nature. In fact, according to some, this is one of the project's most distinguishing characteristics. In this regard, Roseman says, "To my knowledge, in the history of education, there's been no other project that attempts to systematically take a look at all 13 years of schooling, in several related disciplines, and derive what kids should be doing in school from what you want adults to be like." And Kesidou notes, "The commitment to goals is, I think, one of the major characteristics of the project . . . A lot of our efforts in education should be driven by a set of coherent goals . . . the need for a set of coherent goals which will drive all our efforts." However, as with the project's efforts toward being long term, national, and systemic, the goal-oriented characteristic of the reform has brought with it some difficulties as it is articulated across the different worlds.

First, not all of the players involved in the reform place equal weight on being goal oriented. When queried, some participants were unable to see this characteristic of the reform as unique or defining—not because they saw it as unimportant, but because they seemed to be unable to imagine doing it any other way. They realized the goals themselves might be debatable— that others would place greater emphasis on inquiry or wouldn't include such-and-such—but starting with something other than a predefined set of content goals was inconceivable. Second, not all participants agree with placing content goals at the forefront of the reform both conceptually and temporally. And lastly, the various participants have often interpreted and applied the notion of goal-oriented reform to their work in very different ways. This is particularly evident when we look at the events that transpired in the summer of 1991.

One Summer, Three Worlds: The Curriculum Models of 1991. In the summer of 1991, Project 2061 participants met in Seattle for the project's third summer institute. At this meeting, the six site teams unveiled their curriculum models, which were to provide "an alternative approach for K-12 curriculum that would produce the outcomes defined in *Science for All Americans*" (AAAS 1991a, p. 1). The models presented, however, were not met with enthusiasm by the Washington, D.C., office. As participants reflect back on the events of this summer, explanations about what happened differ.

On the one hand, the Washington office's response to the models angered most of the teachers. Matthews explains, "We were frustrated. We were hurt to a certain degree because there was no feedback. We were confused . . . That was an extreme amount of work and hours represented by many, many, different people and there was just no response. It was very peculiar." In the eyes of the team members, they had done what they had been told to do—now that wasn't good enough. Some participants felt that Washington was using them, that the Washington office had had the models all figured out and were simply looking for the teachers' stamp of approval. In hindsight, however, the summer of 1991 looks different to some of the teachers. Matthews states that she now realizes that the Washington office really had no clearer conception of what a curriculum model should look like than the teachers did.

While the teachers speculated on why the Washington office staff members had not asked for what they wanted, the staff members wondered why the teachers had not done what they had requested. Ahlgren explains:

> A lot of [the models] were good ideas, but they didn't grow out of having thought through how students would learn. They would have come up with those whether it was Project 2061 or any other group who said, "How shall we improve schools?" The way you want to schedule a class or organize a school wasn't derived from what you wanted students to end up like. They already had these ideas waiting. So, that doesn't mean that they're wrong at all, but it means that it wasn't the job that we wanted to have done.

An external consultant to the project recollects that

> The teams presented radical stuff, [but they were] quite naive about a whole bunch of realities. Too complex and cumbersome . . . Outrageously unrealistic. [Project staff members] were very disappointed in the results of the teachers' work. That was my reading. They didn't say it to the teachers. But the teachers

felt they were disappointed, too. The teachers were upset. They had worked years on this thing. They poured their best ideas and thought they were extremely visionary.[7]

In response to the models submitted by the teachers, Rutherford and the editorial board suggested the models be distinguished on the basis of how each model organizes large blocks of *SFAA* content. These model *types* were said to have been met with disfavor by the teams. However, in the fall 1991 issue of *2061 Today*, a quarterly project newsletter, two of these types—the inquiry model and design model—were profiled by the Washington office. In the end, the models were reassigned to the category of "raw materials" rather than their initial position as "published products." They are being used, to some extent at three of the sites—the other three sites put them aside after the Seattle institute to work on projects that they felt were more immediately useful and necessary. According to Brearton, for some of the centers,

> the model is scarcely referenced in their dissemination of *Benchmarks*, analysis of resources, exploration of curriculum development, and study of the project tools. Those centers who have explored implementation of their models have realized considerable revisions are necessary if the models are to be useful in their own work and to others engaged in Project 2061 reform.

In addition, the models have not been distributed nationally as was the initial intention. However, they have not been forgotten. As Ahlgren states, "In spite of the severe sag in attention to models, we have often described the team models in speeches, and recommended to many reformers that they obtain copies of [some of them]."

Why the Friction? When it comes to interpreting and applying this ideal of a goal-oriented educational system, it is evident that tensions exist between teachers and the Washington office. We have identified a number of reasons for these tensions. First, the goals defined by the Washington office are content goals rather than instructional or curricular goals. Yet many teachers, as described earlier, place greater emphasis on pedagogy rather than content. Thus, for teachers, the content goals are secondary to the ways in which they are taught and the organization of the schools. Second, communication among the

---

[7]Upon reading a draft of this report, the Washington office staff added the following:

> Originally, we could only say that a model was supposed to be "a sketch of what the whole curriculum would be like." It was to be sketchy, but also complete—and, when filled in with local design details, would lead to the *SFAA* picture of science literacy. Like everything else, we learned about design slowly, and were not clear ourselves about what kind of detail would be necessary for "models." Having learned more about design since then, we would now say that what we wanted from teams were "design concepts," a more sketchy and less complete ambition than we had then. Teams did produce interesting design concepts, and they will take that role in the upcoming *Designs for Science Literacy*. It's just that they leave a lot more to be done than we had imagined."

Washington office, academics, and centers has been, at times, less than optimal. Clearly, the lines of communication broke down during the Seattle summer institute. Teachers were told to "blue sky," yet criticized for not starting with the content goals of what scientifically literate adults should know. This perceived contradiction frustrated teachers and staff alike.

Again, as with the other characteristics of the reform, being goal oriented had implications for the work and organization of the project. For example, the decision to focus first on the content goals rather than on instructional strategies or organizational structure has meant that some issues have been sidelined. Equity, for instance, has not yet been fully addressed in the 2061 tools because it was thought that equity would be achieved not through the content goals, which are said to be appropriate for all students, but through the instruction. However, implementation has been continuously delayed while the goals and ways to work toward them have been considered and reconsidered.

Conclusion: Implications of Different Worlds

The worlds of 2061 have and continue to differ in physical location, pressures, and foci. Across these worlds, participants' conceptions of their roles and responsibilities, of 2061's reform agenda, and of the reform process have not always been identical. At times, they have been contradictory. As a result, when participants attempt to interact across spheres in order to develop and implement the project's multifaceted conception of reforms (1) communication becomes vital, (2) issues of status arise, and (3) opportunities for professional development and personal growth emerge.

The Crucial Role of Communication. Because the worlds of 2061 are physically separate, communication among participants has been crucial for smooth and efficient progress. For many project participants, however, communication across worlds has been inadequate. Teachers at the San Antonio site, for example, express the desire for more frequent and substantive conversations with Washington office staff. They think site visits were too few and far between: Washington office members had never come into their schools or seen their classrooms. In addition, they point out that communication from sites had been sometimes ignored or lost by Washington.

When the flow of information among Project 2061 staff, consultants, center directors, and center participants has been sporadic, misunderstandings have arisen. Brearton, for example, expresses frustration over the failure of center directors to execute their role of liaison adequately between the Washington office and teachers:

> Often there are long periods when you don't communicate, and I made the mistake of assuming that my communication to the center director was then communicated to the teams, and it wasn't. So often you were talking one on one, and you weren't making any headway, if indeed what you said to the center director never went anywhere. So, for example, all the work you were doing on

the roles and responsibilities document, or all the work you were doing on the proposals—I had this assumption in my head that this was being dialoged about at the site. But when I visited a site, I would hear, "We don't know why Washington is making us do such and such." My response to this was to explain that the programs defined in the center's proposal were intended to represent activities the center members wanted to do; that the programs matched the interests, talents, and time availability of members. Naturally, our funding of the programs was linked to their potential for producing products and information useful to those engaged in the reform efforts of the project. I think those misunderstandings were sometimes my fault, and I think that sometimes they were the fault of the center directors. I don't think that the center directors are the communicators that they need to be.

Lack of regular communication among project participants across spheres makes it all the more difficult to appreciate each others' roles in the reform effort. The Washington office is situated in the national education policy arena and concerned, to a greater degree, with issues of scientific content. Roseman, as noted previously, saw participants at the centers as focusing on local matters of interest rather than on national reform priorities.

Teachers, in contrast, operate within the world of schools; they are concerned with issues of implementation. One teacher expresses his frustration with the Washington office in this way: "They don't really understand what it is like for the classroom teacher." He thought that 2061 staff members espoused a great deal of empty rhetoric, that they continued forward without reaching consensus on basic issues of curriculum and institutional structure. A second teacher dismisses the Washington office's ruminations on national policy as somewhat superfluous to the task at hand: "I think that it is a lot easier to sit down and write about it as they do in Washington than it is to actually try it out to see whether it would work or not."

A center director summarizes this lack of appreciation among worlds in the following manner:

> There is a tension between content and pedagogy. Teachers are concerned with what goes on in the classroom, the pedagogy. Washington is concerned with content. There is a tension between these two goals. The last team meeting, I heard from the team that their primary interest in CRIS [now called RSL] was to go to it and get quick information about things that play out in the classroom. The [Washington] office is concerned with how benchmarks get addressed. There is a tension between the two. Teachers are ultimately concerned with their classroom—what it will look like, what he or she will do. People like Sue [Matthews] can envision both goals. Others, at the two ends of the spectrum, they can't appreciate how difficult the other end is.

As with most aspects of 2061, this lack of articulation across worlds and across participants' roles in the reform has changed over time. Over the years, it appears that participants have learned to accept the existence of these different worlds, have become more sophisticated in their negotiations among them, and have developed a greater appreciation for the work of others. A center director, Munumer of Georgia, describes how his understanding of the role of the Washington office evolved:

> We expect change now. Previously, we resented it. We take things in stride, expect a fluidity in vision. We're going to do what we're going to do in Georgia, but it's within the confines of the national vision. The consortium was not envisioned as national but it's supported. We don't just take things from the top. There's dialog. We want to further the national goals and be a reality check. Three years ago, when they didn't listen to us, we called them names. Now, we accept it. Three years ago, we were so naive. We thought we were all in the same world and that the project was beyond politics.

A colleague of Munumer's, Sue Matthews, provides similar reflections on what she has learned over the years:

> We had preconceived notions about a lot of things. We talk about the two different worlds: the world of public school versus the world of Washington. We didn't understand in the beginning and still don't fully understand what it would be like to have all those other things to deal with. And I think that, if anything, what has helped the most is . . . we have developed an appreciation for the different types of circumstances they operate under.

Issues of Status. As participants from these various worlds have interacted, differences in perceived status—differences in the power and prestige awarded various project participants—have emerged. In academic arenas, scientists and professors traditionally have enjoyed higher status than practitioners. Similarly, in Project 2061, Washington office members have been expected to know more, to exert greater influence over decisions, and to be afforded greater respect than center directors or teachers.

Issues of status have created friction among the worlds of 2061. At meetings between the Washington office and center directors, for example, it was the Washington office that set the agenda, moved the discussion along, and made decisions. As a result, center directors voiced frustrations with their perceived lack of expertise, the dismissal of their ideas, and their inability to affect the national agenda. One center director asked us, the case study researchers, if we had caught certain statements made by Ahlgren during a meeting, incisive statements that illustrated his low regard for the center directors. A second director had read about changes made to the name and purpose of *Designs* in a Project 2061 calendar. Why had the Washington office failed to notify the center directors about such a major shift in the reform agenda?

Like the center directors, some teachers have felt their ideas stifled by those with perceived higher status. They viewed themselves as at the bottom of the hierarchy, as underneath both the Washington office staff and center directors. One teacher notes that staff members gave more weight to ideas expressed by educational consultants than those offered by teachers: "The Washington staff does give more credibility to people with Ph.D. attached to their names than those without, even if the ideas expressed are identical (a specific case in point was the math discussion group at the Cornell institute)." A second admits that she was tired of being at the bottom of the decisionmaking hierarchy:

As a teacher who has worked with Project 2061 since 1989, I can honestly say that I have resented being treated like a child that needs protection from the big bad world of politics. If only the [Washington] office and the center directors could see that, to achieve the Project 2061 vision, a level of collaboration and trust must be established between all stakeholders.[8]

Issues of status have arisen not only between the worlds of 2061 but within worlds as well—for example, between secondary science teachers and elementary school teachers. Elementary school teachers at the Georgia site described the existence of "a status hierarchy with the grade levels and what you teach." Especially in the early years of the project, many felt intimidated by the high school science teachers because they did not know as much science. As one elementary teacher explains: "I felt like this little thing that knew nothing. I just knew my little third grade stuff and a little stuff I had acquired." They also felt discouraged from teaching science. As a second elementary teacher explains: "It was, 'Elementary teachers should not teach science. They don't know how. They don't have the background.' And that is what we heard for two years. For two years, [the high school teachers] told us that. We didn't have the background. We shouldn't be teaching." Notably, this was not an issue at all of the sites. At the San Antonio center, a teacher notes, "elementary teachers have been considered the expert on what is appropriate for young children, how children learn, and what foundation is necessary [to teach] a particular concept."

As with issues of communication, there has been some amelioration of status problems over time. In part, participants' expectations for competence have changed. Matthews, a teacher in Georgia, has come to see teachers and the Washington office personnel as operating at the same level of expertise:

At first . . . [I thought], they already have all of this worked out. They're just waiting for us to come up with it. And then, they'll say, "Yes that's it." And they'll put the teacher endorsement on it and off we'll go. You know what we finally realized? They were flying by the seats of their pants right along with the rest of us. And you know how comforting that was when we finally figured that out? That we were all, even two different worlds, we were all in the same boat of taking it a piece at a time and figuring it out as we went along.

In part, status problems have become less noticeable because all participants, whatever their status, have proved valuable to the reform endeavor. For example, in attempting to backmap scientific concepts, the high school science

---

[8]On this point, the Washington office staff wish to point out that difficulties such as these "occur regularly in every office, in every human enterprise;" they are certainly not unique to Project 2061.

teachers knew "the stuff that high schoolers need to know, but they could not break it down. They could not think below that." It was the elementary school teachers that "could listen to them and then say, okay, this is what you do in kindergarten and first grade." Through the process of backmapping, the secondary science teachers finally realized that their colleagues were "just as important."

Benefits of Collaboration. Because participants are situated in different physical and professional locations, 2061 has served as a platform for personal agendas, as a vehicle for professional development, and as a context for collaboration among diverse groups. Participants across spheres have seen 2061 as a way to further personal goals. In fact, Ahlgren considers working at the Washington office to be a golden opportunity for young intellectuals: "If nothing else, this is a good vehicle for their own ideas . . . It doesn't have to be loyalty to what we've done necessarily." Similarly, schools have used 2061 to legitimate and enhance their efforts to advance educational reform at the local level. Teachers at an elementary school in San Francisco, for example, decided to join 2061 out of a pool of possible reform efforts because it seemed most inline with their school's goals. It "provided an open-ended model" that allowed for trying out ideas that went far beyond teaming across grade levels and emphasis on whole language. Moreover, individual teachers joined the 2061 reform effort to improve their own classroom instruction. "For me," explains one teacher, "it was something newer. A new way of doing things. An interesting way that I thought would challenge my children. That was it for me."

Participants across spheres also have recognized the many and varied opportunities for professional development afforded them by their involvement in 2061. Roseman marvels at what she has learned since leaving Johns Hopkins to become, first, the field site coordinator and, then, the curriculum director of 2061:

> I came to the project because I literally fell in love with a draft of *Science for All Americans*. It presented science as relatively simple stories that most people could understand. I must admit that when I came to the project [in early 1989] I thought I knew how to prepare scientists, but I hadn't a clue how to prepare people to be science literate. I think I now have a few clues.

Similarly, almost every teacher we spoke with praised Project 2061 for providing frequent opportunities for personal and professional growth. They were both excited and grateful to meet new people; to expand their knowledge of science, mathematics, and technology; and to enhance their understanding of teaching and learning. Some relished their interactions with educators from across the country. One teacher, reflecting on her experience at a *Benchmarks* training workshop, expresses her enthusiasm in this way: "I think something like this gives me, I don't know, a refresher in the adult world. You can speak in something other than simple sentences for the summer and then go back. It's intellectually stimulating." Others were able to further their careers in education. Clara Tolbert and Rita Rice, for example, are both original and active team members of the Philadelphia 2061 center. They were recently hired to direct the Philadelphia Urban Systemic Initiative. Others thought their presentations at national conferences or their affiliations with national committees, such as the NRC Science Standards Committee or the National Board for Professional Teaching Standards, to be valuable learning experiences. And still others expressed excitement at having access to and training in educational technology.

Moreover, participants in a given world have sought out the experience or expertise of others to accomplish a particular goal or to provide new insight into a persistent problem. Teachers in Georgia, for example, think they have learned much through their involvement with teachers from across the country as well as with educational leaders such as Rutherford. They remain convinced that they will be able to achieve something they could not have attempted alone: scientific literacy for all. More specifically, one day soon, they expect to provide the children of Georgia with a science, mathematics, and technology education equal to that of children in other states.

In turn, the Washington office has recognized the benefits in consulting with center personnel to help write 2061 publications, such as *Benchmarks* and *Designs*. Oglesby describes how the experiences and insights of the sites have contributed to the development of *Designs*:

> Jim's [Rutherford] working on it. We're struggling, because there are some pieces in there that have not been thoroughly worked out. We have a situation here where the intellectual integrity and capability are here [in Washington], but the experience and practicalness of that is out in our centers. We have worked with people out in the centers now for five or six years, over the summer and regularly, working with them in terms of developing a certain amount of capacity for them to provide us a certain amount of support. And they have come along very well. But now we have a problem of how do we extract from them what we need in order to buttress the intellectual integrity.

Linda Nott perhaps best summarizes why most educators became involved in Project 2061 and why many have decided to stay the course. And although she speaks from her world, from the world of schools, her sentiments could easily be applied to all those involved in the reform project:

> To commit to working at least three years on such a visionary project with such ambiguous definition of tasks was no insignificant decision. Those who did sign on, and especially those who have stayed the course, are over-achievers who saw their own philosophy of public school and literacy articulated in the text of *Science for All Americans* and in the speeches so eloquently delivered by Jim Rutherford. They saw the opportunity to make a difference in the lives of thousands of children, and subsequently, in the future of our country. Here, at last, was someone with some measure of political clout who was establishing a forum in which teachers could affect the structure of the whole education sys-

tem, instead of being relegated to working within the current system's confines. There was tremendous personal buy-in from the teachers, emotionally and philosophically.

◆

The Conception of Equity in Project 2061

Most American innovations in science and mathematics education claim a concern for equity in official documents and public presentations. Commitment to equity, however, does not translate in obvious or easy ways into policy or practice. In this section, we examine two conceptions of equity that seem embedded in Project 2061. The first is the idea of inclusiveness, signified by the emphasis placed on the word "all" in *Science for All Americans*. The second is the idea of diversity, with its particular focus on differences among students rather than sameness. We consider how these two conceptions of equity (including their strengths and limitations) seem to manifest themselves in the operations of Project 2061. We conclude with a brief discussion of the challenges associated with achieving the goal of equity in Project 2061.

Equity is a term used liberally and often interchangeably with equality. In fact, there is confusion among educators as to the definition of both terms. Grant (1989), for example, defines equity as "fairness" and "justice." Sanders views equity as "what we do to achieve equality of outcomes" (Sanders 1989, p. 160). Fennema and Meyer (1989) describe equity in terms of equal opportunity, equal treatment, and equal outcomes.

Similarly, Secada (1989) identifies at least six conceptions of equity that can be identified in the work and talk of school personnel. They include such notions as equity as the same treatment, equity as a concern for the whole child, and equity as a safety net for students with individual differences. It is from his work that we borrow the two conceptions of equity that provide the framework for this section.

Inclusiveness: A Focus on All

Equity is conceptualized by Project 2061 partly in terms of inclusiveness. That is, the project has targeted all students, not only those who have traditionally succeeded or who are traditionally underrepresented in science and mathematics. For example, *Science for All Americans* states the conviction that "all children need and deserve a basic education in science, mathematics and technology that prepares them to live interesting, productive lives" (AAAS 1989, p. 11). In their brief to the panel members (from whose work *SFAA* was generated), 2061 staff required that the learning goals they developed be "modest enough to make sense for all students (including those who do not ordinarily perform well academically) but . . . nevertheless ambitious enough to raise the sights of students and teachers" (AAAS 1989, p. 19).

In addition, the notion of inclusivity goes beyond *who* learns to *what* they learn. Project 2061 suggests that all students learn the same content, no matter what their background, gender, race, or socioeconomic status. The project stresses that its recommendations constitute a common core of learning " . . . for all young people, regardless of their social circumstances and career aspirations. In particular, the recommendations pertain to those who in the past have largely been bypassed in science and mathematics education: ethnic and language minorities and girls" (AAAS 1990, p. xviii). As such, the authors of *SFAA* and *Benchmarks* recommend one unified, undifferentiated set of goals for science literacy.

None of the principal documents produced by the Washington office addresses its goal of inclusiveness in any greater depth than may be inferred from the quotes cited above. Notably absent is an extended consideration of issues of inclusion pertaining to historically underrepresented groups. There are several possible reasons. One is that the project will develop such statements in the future: priorities were elsewhere during the early years. Indeed, a blueprint has been commissioned on equity. There may be additional attention to the issue. A second is that 2061 headquarters staff perceive equity to be more appropriately the domain of the sites, whose local knowledge may be viewed as essential to any meaningful attempt at inclusion. A third reason for not giving underrepresented groups greater explicit attention is the belief that, at present, science education underserves virtually the entire population. Why encourage public resistance to extensive reforms by couching them as a "minority" issue? Perhaps it is more strategic to emphasize the idea of a rich core of concepts, skills, and habits of mind to which all must have access: everyone will benefit from these changes.

Science for All Americans moves away from a minimum competency ideology toward the idea that "given clear goals, the right resources, and good teaching . . . essentially all students will be able to reach all of the recommended learning goals" (AAAS 1989, p. 20). However, especially when considering the high level of resources required, a focus primarily on defining what constitutes good science, mathematics, and technology knowledge for *everyone* is problematic in a number of ways.

First, as a concept grounded in theories of justice, equity usually reaches beyond notions of participation in a given enterprise. In particular, it may be flawed to assume that a commitment to inclusiveness necessarily implies a commitment to "equity" according to other criteria for evaluating the achievement of such a goal. One pronounced argument for inclusiveness, for example, is linked to the belief that a scientifically literate populace will enhance national competitiveness. One center director, however, described the shortcomings of a "pipeline"-driven notion of inclusion as a basis for satisfying equity-related concerns. He argues:

> We talk about this pipeline business . . . yet the people who always control the
> gate are still the people who are, number one, advocating the pipeline, the need
> for change . . . and the flow . . . So I really question just what does this mean?
> . . . Are we just creating all this great vision and dreams for security for some-
> one else? Not really security of all but for a very few.

Evaluating the inclusiveness of the word "all" from a standpoint of equity may require more concerted attention to the purposes that are to be achieved by the desired inclusion. Increasing the flow in the pipeline seems to assume that economic and social benefits consequently will flow to all segments of society. Such an outcome is not assured.

Second is the concern that what is cast as "good for everybody" rarely serves everybody equally well. Shirley Malcom, in a 1982 AAAS report addressing the issue of increasing access to and achievement in science, cautions: "Unless programs 'for all' specifically assess the status of, articulate goals for and directly target educational access problems of females and minorities (and also disabled youth), they are unlikely to be effective with these populations."

Lastly, epistemological problems arise from defining knowledge that all students must know. More specifically, there are issues surrounding the process by which particular ideas are constituted and legitimized as scientific knowledge, and then selected as the "official knowledge" to be apprehended by all Americans. Who decides? Whose knowledge is recognized? Whose knowledge is not considered scientific or valid? These are questions we explored earlier in greater detail.

Diversity: Issues of Difference and Representation

Project 2061 staff and participants also envision equity in terms of diversity of both the participants and the students. More specifically, this focus on diversity is found in the sites, the draft of the Equity Blueprint, and the curriculum models.

First, as Roseman notes, diversity within and among the six school centers was a criterion for selection:

> The project's commitment to equity has certainly been in the forefront since I
> joined the project nearly six years ago . . . We have pursued it in involving all
> kinds of people from various teams, having diverse team sites. So, there was a
> lot of input that we sought to at least provide safeguards that we weren't just
> going to be dealing with white suburban teachers in white suburban schools,
> addressing upper-middle-class kids.

These sentiments are echoed in *Benchmarks*. The teams were "very different from one another by virtue of locale, demographics, and available resources, so that they might collectively represent the nation" (AAAS 1993, p. 304). Specifically, the two principal signifying criteria are (1) the rural/suburban/urban distinction, and (2) the ethnic makeup of the student population. This, coupled with a desire to include educators on the teams who represent diverse perspectives, is considered to strengthen the project nationally.

Second, the draft of the equity *Blueprint* reveals an emphasis on diversity. The draft report was divided into several sections, each written by a different author and focusing on a different group. There is a section on girls in science, the gifted, Asian-Americans, African-Americans, and students with mental or physical disabilities. Each section outlines the current status of this group and makes recommendations for policy and instruction.

Third, several of the six models developed by the sites focus on issues of diversity as well. They suggest using instructional strategies that appeal to different learning styles, choosing locally or culturally relevant topics, and—in some cases—paying explicit attention to multiculturalism. The models also address, to varying degrees, the idea that an individual's prior experience, knowledge, interests, and talents are socially or culturally influenced. "The nature of learners includes not only their cognitive and affective characteristics, but also their social and cultural experiences. Curriculum and instruction should reflect the experiences of its students, its teachers, and its community" (Project 2061 1991a, p. 9).

Some sites have begun to translate the equity goal into practice in a variety of ways. For instance, in San Francisco, student support systems (including mentorships, internships, home computer programs, tutoring, and after school programs) are provided in an attempt to guarantee all students access to a quality education. While focusing on the uniqueness of individual students as learners, the teachers are very sensitive to the need to engage students in learning experiences that reflect their diversity and areas of interest. Students are exposed to a variety of positive role models in the educational program, including teachers, community leaders, scientists, and other professionals, all of whom vary by gender, language, race, ethnicity, culture, disability, sexual orientation, belief system, and life style. The contributions from historically underrepresented groups are infused into the curriculum, and the students' cultures and languages are incorporated in the education program.

Conceptualizing equity in terms of diversity can be problematic, however, because sites that consider themselves as lacking in diversity may then consider equity to be someone else's problem, or, at best, a goal already achieved. In addition, as with the inclusiveness conception, epistemological questions arise as a result of the project's attempts at defining the knowledge everyone should know. For example, we might ask whether more diverse views of the nature and content of science need to be included if the project intends to address diverse populations. In what sense should there be the same knowledge base for everyone? Are there pluralistic forms of knowing in science?

Operationalizing These Conceptions

The focus on equity through instruction is most often considered in terms of (1) student-centered instruction, or (2) multiple abilities/intelligences. In the first case, teachers, like those in San Francisco, argued that if a curriculum begins with the concerns of the student and at his or her cognitive level, then issues of equity will be addressed automatically. In the second case, teachers advocated using a variety of teaching styles and modes of expression to address the needs of more students.

We conclude by asking about the dilemmas or struggles that may arise as the project proceeds with its goal of equity. Looking solely at conceptualizations of inclusiveness and diversity, which represent only a portion of what might be encompassed under general concern about equity, we have tried to point out that there is confusion and ambiguity. In addition, structural barriers may exist that impede the project's progress on this front. The reform has not, as yet, entered the stage of full implementation; so there probably are many factors related to equity that are as yet unidentified. What about Project 2061 in schools that serve predominantly minority cities or neighborhoods? How are they different from others in the curriculum they create? What about magnet schools? What about districts with (or without) large numbers of teachers who are themselves members of ethnic or racial minorities? Since poverty is correlated with race, what about Project 2061 schools in impoverished neighborhoods?

Perhaps the key question, as with so many in Project 2061, is to ask the degree to which equity will be considered a matter for national attention in the project and the extent to which it will be addressed primarily at the local and classroom level. So far, the matter has claimed much less attention at headquarters level than issues directly related to the creation of curriculum. This fact is not surprising, given the prior range of tasks necessary to attain plausibility as a serious effort at reform of science education.

Project 2061, however, from the beginning claimed strong and central commitment to dealing with matters of equity. So, even apart from trying to meet a worthy goal, there is a credibility issue. It remains to be seen how seriously equity issues will be addressed, at what level, and to what effect. The challenge well may be the most difficult one in American education, but a project that continually emphasizes its commitment to all Americans has created expectations that seem to suggest the necessity for turning significant resources toward the issue in the near future.

◆

Lessons From Project 2061

In 10 years, Project 2061 has become the single most visible attempt at science education reform in American history. We have outlined its purposes, depicted its methods of operation, and noted several of its major accomplishments. We have also focused on problems the project encountered as it moved from initial conception toward implementation of a curriculum. Can we say more? Are there lessons to be learned from this bold and visionary attempt to change science education in American schools? We believe so.

First, however, a very brief summary of the project's major features and accomplishments.

- Project 2061 produced a clear and comprehensive vision of what everyone should know about science. *Science for All Americans* persuades its readers that virtually everything the nonspecialist adult should know about science is interesting and worth learning. It offers a portrayal of the field that is fresh without being radical, accessible without being trivial. Above all, it looks achievable.

- In operationalizing its conception of educational change, the project has addressed at some level almost all the many factors that are believed to affect education reform. In that sense, it exemplifies and pioneers a comprehensive approach to change that has come to be called "systemic."

- The project is national and inclusive. It offers a professionally certified version of what every school should strive for in science, wherever its location and whatever the composition of its student body. In doing so, it has generated an example of what nationally driven curriculum reform might look like. As the country began to commit itself to the creation of national standards for various subjects in the curriculum, Project 2061 was already in a position to offer an illustration, even a prototype, to demonstrate how such standards might play out in practice.

- Project 2061 has enhanced leadership in science education. The professional influence of people who have been associated with the project has been strengthened; whether at one of the six sites or at Washington headquarters, affiliation with Project 2061 has accorded an extra level of legitimacy to the changes in science education that these people have tried to promote.

What, then, are the possible lessons from all this? Project 2061 reflects largely the educational assumptions and wisdom of a period in educational reform that flourished about 40 years ago. Can its approach to curriculum reform succeed as the country reaches the 21st century? More precisely, what seem to be the strengths, shortcomings, and challenges of an orientation to science education that was forged when many aspects of American society, including the schools, were dramatically different from what they are today—when teachers, academic and industrial scientists, textbook publishers, and governmental policymakers were expected to play somewhat different roles in the improvement of science education than they do now?

Who Owns Science?

The perspectives of academically based research scientists are central to Project 2061's conception of desirable science education. Flush from their impressive achievements in helping to win World War II, academic scientists in the 1950s had the standing, influence, and desire to define their fields for all those who would learn about them in the public schools. They were considered the experts: science was theirs, and many of them believed they had the right and obligation to determine what it is about their field that was to be taught to everyone who would study it.

Today, however, the standing and influence of these academic scientists, while still central in the education world, are not unquestioned. Exactly what constitutes "literacy" in science is less certain and more contested; several groups are becoming assertive about their view of the science to be taught in schools. Subtly, and not so subtly, they are challenging the influence of academic researchers. Some of this erosion of the authority of university professors in determining science content is associated with changes in the nature of science itself. Science is what scientists do: increasingly, scientific activity is being directed toward what the public decides as useful. And what is perceived as useful is not the same as what scientists consider basic.

In the mid-1990s, support for fundamental research is more difficult to obtain, particularly as its cost increases. Despite its priority position for theoretical physicists, for example, Congress has sharply curtailed funding for the super-conducting super collider. This shift away from the university researcher's marked preference for fundamental inquiry is evident even at the National Science Foundation. Priorities of that agency have moved toward more practical matters, and "applied" fields receive greater support.

And science teachers themselves are among the contesting groups today. Teachers are claiming prerogatives in identification of the subject matter for their students that they apparently were content to leave to university professors 40 years ago. Often they want to emphasize topics that they believe to be more closely connected to their students' lives, for example. They claim that their personal knowledge of their own students gives them the warrant not only to make pedagogical decisions—an arena traditionally left to teachers—but content choices as well.

The net result is that it is no longer as clear today as it was several decades ago just who "owns" science. It is even less clear who owns school science. Teachers at some of the Project 2061 sites frame this issue as according priority either to "content" (for which scientists are presumably the authority) or to "pedagogy" (for which they see themselves as the experts). Their new-found assertiveness, however, spills easily into their making more decisions about what should be taught, as well as how.

It might be noted in passing that the tasks that university professors assumed in determining curriculum for elementary and secondary schools after World

War II were virtually unprecedented. During earlier periods in science education, the goal was not primarily to teach science as it was conceived and understood by researchers, but, successively, to improve students' thinking abilities (into the early years of the 20th century) and to help them appreciate how science affects their daily lives (up to and including the World War II years). Sometimes research-oriented university faculties played a role in spelling out the content, but usually they did not. It is possible that the country is moving into a period more like most of its past, in which it is less clear who has the legitimacy to decide curriculum. In such a period, different groups try to enhance their influence, consensus is harder to find, and even the aims of science education become more fluid. Project 2061, priding itself on its long-term vision, takes quite a different view, of course: whatever else changes, the goals of science education do not.

Standards and the Classroom Teacher

A second set of challenges is associated with how new subject matter standards find their way at some acceptable level into American schools. If developers of content standards in the various subjects believe that their work is done when the written standards are released, they would do well to look at Project 2061. It is far from an easy job to translate content goals into usable curriculum materials. Teachers at the six 2061 sites labored long and hard with both *Science for All Americans* and with *Benchmarks*, but, 10 years into the project, there is still no Project 2061 curriculum. It is unclear if or when there will be one. Teachers need time to discuss what the standards mean. They need help in figuring out how key concepts might be taught. They need guidance and financial assistance in the formidable task of designing assessment tools and appropriate scoring rubrics. Most of all, they need recognition that the task is extraordinarily complex. Not least because consensus about the actual curriculum is difficult to generate, the process is contentious and risky for teachers, more so than the leadership of Project 2061 recognized.

What, then, is an appropriate role for teachers in standards-driven curriculum development? Are they the ones to translate standards into action? If not teachers, who? In the 1950s and 1960s, it was believed by some that instructional materials could be crafted cleverly enough that teachers would almost certainly use them the way the developers intended. The goal was sometimes stated as one of designing "teacher-proof" texts and laboratory investigations. Project 2061 personnel never made that assumption, but the project's Washington staff has not always been satisfied with curriculum efforts at the site level, either. At the outset, the Washington leadership of Project 2061 believed that curriculum indeed could be crafted by teachers at the six sites. Now they are not so sure. They certainly were not satisfied with many of the attempts at the sites to build curriculum "models." The teachers, on the other hand, were never sure why their efforts were considered unsatisfactory.

Flexibility and Consistency

On certain matters, Project 2061 is highly improvisational. Activities within the project are shifted to take advantage of new strategic opportunities, as when attention turned dramatically and expensively to development of *Benchmarks*. The development of subject matter standards became a national priority, and Project 2061 has something important to say about the subject. On the other hand, because the staff prides itself on the steadiness of its long-term purposes, it is steadfast in the conception of science that it promotes.

Project 2061 leadership has chosen to influence the entire process of science education reform, from articulation of goals for science education to adoption of curricular materials. This determination to maintain a high level of consistency has caused problems. In the terminology of one of the conflict flash points in Project 2061, for example, do teachers create "tools" for crafting actual curriculum, or do they produce "resources" that then are drawn upon (or not) by those who decide upon the official Project 2061 designs that will be published and nationally disseminated? "Tools," say many of the teachers, refer to the official products issued by the project. "Resources," say those at project headquarters, meaning raw materials. Even in a highly selected group of school districts, perhaps especially in such districts, teachers seek and demand significant levels of latitude and influence—particularly as the reform gets closer to their own classrooms.

Tensions about issues like this one, ever since the Seattle summer of 1991, have colored relationships within the project, negatively and disruptively. The core question is the degree to which the initiators of a project should try to maintain full control of the project's development. Keeping a tight rein helps to ensure conceptual and educational coherence, but it also invites serious and debilitating friction. Project 2061 is still trying to get the balance right.

Science for *All* Americans?

A major ambition of Project 2061 from the outset has been that it reach all students. Indeed, in a bold effort to reform American science education—and Project 2061 certainly is one of the boldest and most farsighted—one cannot long fail to come to grips with the profoundly difficult matter of ensuring a sound educational experience for everyone. However worthy the goal, few people have been successful. Matters of access and equity are both conceptually difficult and politically charged, probably more so than any other set of challenges in American society. It is one thing to say that all students should receive a high level of science education. But in what sense should the education of all students be the same? And how is it to be provided? How can all students be held to the same standard, yet keep the standard meaningful? The access/equity issue continues to command project attention, to be sure, but it is not yet clear that the project has addressed this issue in depth or with much effect. Hardly anyone else has either, of course. Project 2061, however, declared the goal of equity to be central to its purposes. It even embedded the commitment in the title of its first major publication. So expectations were raised, and the education world hopes for some answers.

Different Worlds

Project 2061 was created to change American science education in fundamental ways. It has started to do so. *Science for All Americans* was proffered as a coherent vision of what the schools in the United States should strive for, one that has been accepted, even embraced, by large segments of the education policy world. Individual states and school districts make frequent reference to the key 2061 documents in developing their science programs and will probably continue to do so for several years to come.

Creating actual curriculum is another matter. It is still very early to say how well the science education of tomorrow's students will comport with the vision of Project 2061. Will it be possible to develop the depth of understanding portrayed in *Science for All Americans* for a significant proportion of America's students, let alone all of them? Students, and their teachers, live in different worlds from those who wrote Project 2061's major documents. Certainly the six sites initially enlisted to help with the process are not yet close to the kind of complete and balanced program that Project 2061 envisions—and they have worked intensively within the project, with extra levels of support, from near the beginning.

It could well be that Project 2061's lasting influence will not be found at any of the six sites that so far have been officially part of the project, and that these sites actually will recede in importance in overall project planning in the years immediately ahead. Rather, judging from the pronounced degree to which discussions about science education already have been shaped by Project 2061 (and not solely in the United States), it may be that the project's major effects will be indirect and scattered. State education departments will take inspiration from Project 2061 documents. School districts here, there, and maybe in many places will hammer out curricular interpretations that they will link publicly and proudly to the overall vision of Project 2061. And science education will improve as a result.

But the classroom manifestations of the project may well be strikingly different from one another, and the relationship of those attempts at implementation to the original guiding vision of Project 2061 will be difficult to trace and often questionable. With the country's traditions of local autonomy, the increased assertiveness of teachers, and the changes taking place in science itself, it would be surprising in American science education reform if any central agency—even one as strong, prestigious, and able as Project 2061—will be able to maintain strong and direct influence over actual classroom practices for very long.

◆

Project 2061 Comments on Case Study

Editor's note: We include the following comments by Andrew Ahlgren, Associate Director of Project 2061, at his request.

The account of the project is generally very good, but the authors have not got the story quite right in emphasizing the project's top-down character and omit some very important developments since 1993, developments with greater significance for reform of science education than the earlier project features on which the case study focused.

(1) The "scientist-driven," "top-down" portrayal emphasized in the report, which may derive in good part from themes of prior interest to the investigators, is misleading in two respects: overstating the control by university scientists and disregarding the project's central strategy of supporting diverse local construction of curricula. With regard to the role of university scientists, the report first claims that *Science for All Americans* "began with a long period (lasting several years) during which *only* scientists were involved;" then that Project 2061's view "has been defined primarily by university researchers in various science disciplines and some serious challenges result from this." While we see no need to apologize for asking experts about what the most important ideas in their fields would be, the report's claims needs to be put in some perspective. An adequate perspective would have to take account of the following points:

a. The panel reports were produced in about 18 months, after which the drafting of *SFAA* by AAAS staff—educators—began. Thereafter, panelists were invited to react to drafts of *SFAA*.

b. There were many non-university experts among the panelists, consultants, and reviewers for *Science for All Americans*. This is evident in the extended lists of contributors that appeared in the first, AAAS-published version of SFAA. Non-"research scientists" make up: over half the Advisory Council, half the technology panel, about half the reviewers of SFAA, and almost all the 80 or so advisors to project 2061 staff.

c. *Science for All Americans* and *Benchmarks* have been very well received by science teachers, who the report claims would have done it differently. The validity of *SFAA* and *Benchmarks* is confirmed by their similarity to the content recommendations in the subsequent *National Science Education Standards*, whose development involved large numbers of teachers and teacher organizations. The case report offers no examples of inappropriate Project 2061 recommendations and no suggestions for what would be better alternatives.

(2) The central Project 2061 strategy has been *not* to produce top-down curriculum, but rather to help local educators (state, district, school, or teacher level) design their own, suited to their own circumstances and tastes. The strategy assumed that there were existing curriculum materials from which a set could be selected that would lead students to all Benchmarks. We thought we could, by generating a database of which Benchmarks were served by what materials,

help educators to assemble a great variety of alternative K-12 curricula that would achieve the entire set of Benchmarks.

An early setback to that strategy, as the report accounts, was recognizing how much we had underestimated the difficulty of even just sketching out a K-12 curriculum. Our site teams came up with some interesting design concepts, but had nowhere near the resources that would be required for designing complete curricula.

A more serious setback to the strategy of facilitating local assembly of curriculum occurred when our examination of existing materials showed very few of them to be both (a) focused on Benchmark ideas as specific learning goals and (b) likely to help students learn. One response has been to concentrate on training educators to judge materials more carefully. This has proved to be unexpectedly difficult, because educators are not accustomed to studying specific learning goals closely, tending instead to translate them into loose topical areas (into which many other specific ideas could be fit as well—or instead). A second response has been to try to influence the professional developers of curriculum materials to attend more conscientiously to Benchmarks, so as to increase the pool from which suitable materials can be selected. A third response is to formulate tactics for how to modify some of the most promising materials to serve Benchmarks better.

References

Ahlgren, A., and F. J. Rutherford. 1993. Where is Project 2061 today? *Educational Leadership* 50(8): 19-22.

American Association for the Advancement of Science (AAAS). 1989. *Science for all Americans: A Project 2061 report on literacy goals in science, mathematics, and technology.* Washington, DC.

———. 1990. *Science for all Americans: A Project 2061 report on literacy goals in science, mathematics, and technology.* New York: Oxford University Press.

———. 1991a. Curriculum models coming into focus. *2061 Today* 1(3): 1-3.

———. 1991b. Designing blueprints. *2061 Today* 1(2): 1-3.

———. 1991c. Developing Standards. *2061 Today* 1(3):2-3.

———. 1991d. Project 2061's own classroom. *2061 Today* 1(1): 2.

———. 1992. Proposal to the National Science Foundation. Unpublished manuscript.

———. 1993. *Benchmarks for science literacy.* New York: Oxford University Press.

———. 1994a. AAAS Project 2061 1994 calendar. Washington, DC.

———. 1994b. *Designs for Science Literacy* under way. *2061 Today* 4(1): 4.

———. 1994c. Prospectus to NSF 1995-99. Unpublished manuscript.

———. 1994d. *Update.* Washington, DC.

Appley, Mortimer, and Maher Appley. 1989. *Social and behavioral sciences.* Washington, DC: AAAS.

Atkin, J. M. 1994. Developing world class education standards: Some conceptual and political dilemmas. In *The future of education: Perspectives on national standards in America,* ed. N. Cobb. New York: College Entrance Examination Board.

Brearton, M. A. 1994. *Systematic reflective teacher research in educational reform in science: The Project 2061 school district centers.* Paper presented at National Association for Research in Science Teaching Conference, Anaheim, CA.

Farges, B. 1995. San Francisco. *California Classroom* SCIENCE 7(1): 3.

Fennema, E., and M. R. Meyer. 1989. Gender, equity, and mathematics. In *Equity in education,* ed. W. Secada. New York: Falmer Press.

Lynch, M., and E. Britton. 1992. *Project 2061: A project of the American Association for the Advancement of Science.* Washington, DC: National Center for Improving Science Education.

National Research Council (NRC). 1994. *Draft of the National Science Education Standards.* Washington, DC: National Academy Press.

Perrigan, Dana. 1995. Students rise to meet the challenge in San Francisco schools. *The City Voice* (May 5): 11.

Project 2061. Undated. Blueprints for systemic reform. Unpublished manuscript.

——. Undated. Overhead: Tools and resources. Unpublished manuscript.

——. 1991a. Draft: Georgia team curriculum model. Unpublished manuscript.

——. 1991b. Status of blueprint assignments. Unpublished manuscript.

——. 1992. Draft: Current thinking about curriculum models. Unpublished manuscript, 24 May.

——. 1994a. Draft outline: *Designs for science literacy.* Unpublished manuscript, 10 February.

——. 1994b. Draft: San Francisco model. Unpublished manuscript.

Rutherford, F. J. 1991. What's in a name? *2061 Today* (1)1: 5.

——. 1992. Standards can bite. *2061 Today* 2(1): 5.

——. 1993. Project 2061 from the national perspective. *San Francisco State University School of Education Review* (Spring): 12-16.

Sanders, J. 1989. Equity and technology: An applied researcher talks to the politicians. In *Equity in Education*, ed. W. Secada. New York: Falmer Press.

Science Curriculum Framework and Criteria Committee. 1990. *Science framework for California public schools kindergarten through grade twelve.* Sacramento, CA: State Department of Education.

Secada, W., ed. 1989. *Equity in education.* New York: Falmer Press.

Appendix A: Case Study Team and Research Activities

Case Study Team

J. Myron Atkin is a professor of education at Stanford University and served as Dean of Education from 1979 to 1986. He has chaired the Education Section of the American Association for the Advancement of Science and is a consultant to the Organization for Economic Cooperation and Development in Paris. He co-edited Changing the Subject, OECD's 1996 report on member nations' case studies of innovations in science, mathematics, and technology education. He was a member of the National Committee on Science Education Standards and Assessment and chaired the Committee on Science and Engineering Education of Sgima Xi. His research interests and publications focus on: identification of the science content to be taught in elementary and secondary schools; teacher-initiated inquiry, especially action research; practical reasoning by teachers and children; case methods in educational research; evaluation of educational pro-grams; science education in museums; and development of policies that accord classroom teachers greater influence in determining the educational research agenda.

Julie Bianchini is an assistant professor of science education at California State University, Long Beach. At the time of this study, she was a doctoral student in curriculum and teacher education at Stanford University. Before earning her graduate degree, she taught high school biology and physics in San Francisco. Her dissertation research, completed in 1995, explored how students attending an urban middle school learned science in small groups.

Nicole Holthuis is a research assistant and doctoral student in science education at Stanford University. She has a master's degree in Curriculum and Teacher Education and has taught high school biology, chemistry, and physics. Her research inter-ests include constructivism; history, sociology, and philosophy of science; and equity issues in curriculum and instruction. She expects to complete her disserta-tion in 1997.

Research Activities

Formal Interviews

- Andrew Ahlgren, Project 2061 associate director, Washington D.C. (October 1993; December 1994)
- Mary Ann Brearton, Project 2061 field services coordinator (February 1994; June 1994; December 1994)
- Marilyn Brien, participant, Project 2061 Texas Center (September 1995)
- Karrie Buckles, participant, Project 2061 Texas Center (September 1995)
- Peggy Carnahan, Project 2061 Texas Center, former center director (September 1995)
- Marcia Denton, participant, Project 2061 Texas Center (September, 1995)
- Robert Donmoyer, project consultant (May 1994)
- Bernard Farges, center director, Project 2061 San Francisco Center (February 1994; June 1995)
- Sophia Kesidou, Project 2061 research associate (December 1994)
- Mary Koppal, Project 2061 communication manager (December 1994)
- Sue Matthews, curriculum director, Project 2061 Georgia Center (January 1995)
- Candido Munumer, center director , Project 2061 Georgia Center (January 1995)
- Carol Muscara, Project 2061 technology systems director (September 1994)
- Gary Oden, Project 2061 San Diego Center, director (October 1995)
- James Oglesby, Project 2061 dissemination director (December 1994)
- Larry Rogers, Project 2061 deputy director (December 1994)
- Jo Ellen Roseman, Project 2061 curriculum director (February 1994; June 1994; December 1994)
- Mary Budd Rowe, professor at Stanford University and Advisory Board Member (June 1994)
- James Rutherford, Project 2061 director (December 1994)
- Gil Sanchez, resource specialist, Project 2061 San Francisco Center (February 1994)
- Decker Walker, professor at Stanford University and former consultant to Project 2061 (April 1994)
- Ellen White-Volk, participant, (September 1995)
- David Wong, professor at Michigan State University and author of the *Teacher Education Blueprint Report* (March 1994)
- Head Teacher, San Francisco Community School (November 1993)
- Participants (3) at the Project 2061 Leadership Workshop (June 1994)
- Participants from the Project 2061 Georgia Center, group interview (January 1995)
- Participants from the Project 2061 Texas Center, group interview (September 1995)

Conferences and Workshops Observed

- National Science Teachers Association, Annual Meeting, Kansas City, Missouri (March 1993)
- American Association for the Advancement of Science, Annual Meeting, San Francisco, California (February 1994)
- National Association for Research on Science Teaching, Annual Meeting, Anaheim, California (March 1994)

- National Science Teachers Association, Annual Meeting, Anaheim, California (March 1994)
- American Educational Research Association, Annual Meeting, New Orleans, Louisiana (April 1994)
- Project 2061 Leadership (Benchmarks) Workshop, Columbia, Maryland (June 1994)
- Project 2061 Resource Evaluation Workshop, Washington, D.C. (September 1994)
- California Science Teaching Association, Project Day (October 1994)

Meetings Attended

- Teacher In-Service at Horace Mann Middle School, San Francisco, California (September1993)
- Center Directors' Meeting, Washington, D.C. (October, 1993)
- Meeting regarding *Designs for Science Literacy* with consultants from Stanford University and Project 2061 staff, Stanford, California (February 1994)
- Teacher Education Meeting with Project 2061 staff, representatives from Mills College and San Francisco State University, San Francisco, California (February 1994)
- Center Directors' Meeting, San Francisco, California (February 1994)
- Open-enrollment request meeting for New Academic High School, McAteer High School, San Francisco, California (March 1994)
- San Francisco Curriculum Model Revision Steering Committee Meeting, San Francisco, California (June 1994)
- Teacher In-Service, Georgia Center (January 1995)

Site Visits

- Horace Mann Middle School, San Francisco, California (September 1992)
- San Francisco Community School, San Francisco, California (November 1993)
- Buena Vista Elementary School, San Francisco Community School, and Cesar Chavez Middle School, San Francisco, California (February 1994)
- Project 2061 Georgia Center
- Project 2061 (former) Texas Center

Individuals Who Provided Comments on Draft Reports

Andrew Ahlgren, Project 2061 associate director
Mary Ann Brearton, Project 2061 field services coordinator
Marilyn Brien, Project 2061 Texas Center,participant
Peggy Carnahan, Project 2061 Texas Center, former center director
Chris Castillo-Comer, Project 2061 Texas Center, participant
Marcia Denton, Project 2061 Texas Cente, participant
Joan Drennan-Taylor, Project 2061 Texas Center, participant
Bernard Farges, Project 2061 San Francisco Center, center director
Andrea Foster, Project 2061 Texas Center, participant
Marlene Hilkowitz, Project 2061 Philadelphia Center, center director
Sophia Kesidou, Project 2061 research associate
Sue Matthews, Project 2061 Georgia Center, curriculum director
Linda Nott, Project 2061 Texas Center, participant

James Oglesby, Project 2061 dissemination director
Jo Ellen Roseman, Project 2061 curriculum director
James Rutherford, Project 2061 director
Ellen White-Volk, Project 2061 Texas Center, participant

Chapter 3

The Challenges of Bringing the Kids Network to the Classroom

James W. Karlan
Michael Huberman
Sally H. Middlebrooks

Harvard University, and
National Center for Improving
Science Education

Contents

3

The Challenges of Bringing the Kids Network to the Classroom

Purpose and Focus

For several years, both North American and Organisation for Economic Co-operation and Development (OECD) countries have labored over the stimulation and improvement of work in science, mathematics, and technology. In the American context, one area of special interest has been the middle grades, 5-8. Promising projects have been designed, field tested, and carried out, usually leaving few tracks for us to study closely afterward. Some of these projects are still in operation, but the memory of their adoption and the keys to explain the success of some and the demise of others are missing.

In creating such projects, international educators have known what they were looking for. In the U.S. context, there was a quest for a redefined science content, one that placed less emphasis on quantities of facts and more on replicating basic scientific structure through actual exercises and experiments such as scientists themselves might conduct—expeditions, transformations of materials, predictions of effects from the manipulation of variables. In mathematics, programs were scouted out that emphasized more of a problem-solving or hands-on approach—an application of mathematical knowledge and skills to real-life dilemmas, whose meaning could be probed at any point in the experimentation process.

At the same time, there was a concern that these approaches be applied by teachers with requisite subject matter knowledge specialized for teaching, including the choice of appropriate strategies and representations, the anticipation of children's insights, and the interpretation of student conceptualizations. Finally, there was a particular interest in the use of new learning technologies—from video to computers—to make the environment more authentic, expand interaction and collaboration with others through networks, promote laboratory-like investigations, and emulate the tools experts use to produce artifacts. Such technology also allows students to manipulate, construct, and revise their own representations and artifacts easily and in several media, including text, graphics, video, and audio.

251

Given its objectives and program (described in the next section), it was small wonder that the National Geographic Society (NGS) Kids Network was one of the eight U.S. innovations selected for closer study as part of the OECD case studies in innovations project. This project aimed to illustrate new or recent science/mathematics programs in the upper elementary grades that provided learning experiences built around powerful ideas (basic understandings and principles rooted in the disciplines).

In preparing this case study, we had a threefold focus. The first was to describe closely the enactment of a program that, potentially at least, had students actively and thoughtfully engaged with important issues of science and mathematics. What did such an enactment look like, when observed closely and dynamically? How could such a program be carried out in garden-variety schools—as opposed to hothouses of experimentation—since that is where future executions would take place?

Second, it was crucial to find out how such programs had been discovered, adopted, implemented, and stabilized. What was the real story behind the Kids Network, in multiple—and therefore different—settings? How had the personnel been prepared and assisted on an ongoing basis? What kinds of organizational shifts were required? How difficult was the process of mastery by which teachers reached the richer parts of the program, and how long did that take? Although studies of the innovation implementation process (an area of specialization of the authors) have addressed many of these questions, there was evidence that work in science education had underestimated the complexities and difficulties of implementing inquiry-centered programs.

Finally, how could the promising parts of this enactment of the Kids Network be replicated elsewhere? The study could not answer this question, but close descriptions of the ecology of local use and renditions of the process by which a set of schools implemented the program would give other school districts the information they needed for deciding whether the match between their context and these settings warranted their engagement to the Kids Network and curricula like it.

Our purpose is to describe these experiences, including their technological component and their intersections with science and mathematics instruction. Most important, we seek to track their life history through to the opportunities, experiences, activities, and learning that children derive from them.

◆

The Program

This section describes the main components of the NGS Kids Network curriculum for grades 4-6 as designed by developers and publishers, but not necessarily as enacted by teachers. Most of the remainder of this report focuses on the latter.

According to the developers, all the Kids Network units are guided by three principles:

> That students should deal with real and engaging scientific problems, problems that have an important social context. That kids can and should be scientists (students are working as scientists on real science problems). That telecommunications is an important vehicle for showing children that science is a cooperative venture in which they can participate (TERC 1987, p. 1).

Bob Tinker, director of the Technology Education Resource Center (TERC) and one of the originators of the Kids Network, said:

> We were literally thinking, asking ourselves even then, what kind of science research could students actually be doing? What could they contribute to the scientific community? It was pretty obvious . . . that kids couldn't do anything individually; but could we explore the possibility that [with] the power of numbers, that together they could really make serious contributions, interesting contributions, to scientific understanding?

According to the developers, their program accentuates the concepts that "Measurement is central . . . science is cooperative, and the results of . . . inquiry matter" (TERC 1987, p. 9). All the units involve a

> series of cooperative science experiments in which students use the telecommunications network to send results of their local experiments to a central computer which pools their data and then sends back the combined national results. Participating classes analyze trends and patterns in the national data, examining how their findings contribute to the overall picture. And students discuss their questions and observations with their colleagues, and with practicing scientists, via the network (TERC 1989, p. 1).

All the units are intended to "encourage integration of science with other curriculum areas" (TERC 1987, p. 12). The developers "expect children to learn about the scientific content areas they are studying [e.g., acid rain]" and they claim that "the content is embedded in the overall goal of empowering children to do science" (TERC 1987, p. 2).

The Units

There are seven six-week units designed for fourth through sixth graders: Hello!, Acid Rain, Weather in Action, What's in Our Water?, Too Much Trash?, What Are We Eating?, and Solar Energy.[1] Each unit is on-line several times a year for

---

[1]This study only concerns the first series of programs designed for grades 4-6. Four new units for grades 6-9 were made available in spring 1995:

- Soil: What is it good for?
- Sound: How loud is too loud?
- Oxygen: How do our bodies get the oxygen we need?
- Surface water: How polluted is our local surface water?

an eight-week period; thus, teachers have some flexibility in choosing when to implement a program.

All units include: "a *Teacher's Guide*, with background information, lesson plans for core activities and extensions, and a software manual; the *Kids Handbook*, a richly illustrated discussion of the unit concepts; two software disks; and lab materials, which include science equipment, maps and activity sheets" (TERC 1989, p. 3).

In 1994, the cost for each unit was $375 (Hello! was $350). These materials can be used during other sessions or by other teachers in the school who decide to share one registration. All units also require a per-unit-use tuition and telecommunications fee of $115. A per-unit-use fee allows two or more teachers to come up with their own ways of sharing a registration. This fee covers 120 minutes of "off-line" telecommunications time,[2] access to a toll-free computer assistance hotline, teaming with other schools, the compiling and transmission of data, and two electronic mail messages from a unit scientist.

The telecommunications software "features are automatic, from dialing to uploading and downloading of data and letters. To keep telecommunications costs down, students and teachers can prepare and read electronic mail letters off-line. Similarly, all data entry and analysis is done off-line" (TERC 1989, p. 4).

Schools are permitted to make copies of the software disks and student activity sheets. Teachers who want to participate in a unit must register about three months in advance of a session and pay the telecommunications and registration fee.

Each of the units includes the following activities:

- Read about the topic in the student handbook.

- Write a class-constructed letter describing preselected features about the students' community.

- Use a computer to identify the location of teammates according to their longitude and latitude and plot their location on a wall map.

- Telecommunicate in order to pick up and send letters and data.

- Follow a set of directions requiring the manipulation of materials in order to collect data (e.g., pH of rainwater, nitrate levels, solar gain).

- Collate and insert data into preheaded charts.

- Use computers to create graphs and charts representing the data.

- Look for and discuss patterns in class data as well as other teams' data.

---

[2]"Off-line" means students must complete their messages (e.g., letters, surveys, data) before telecommunicating to NGS.

- Present what one learned to audiences beyond classroom community.

- Read a letter from a scientist who interprets the entire network's data.

The following is a brief overview of the focus and some of the activities of each of the seven units:

- **Hello!** This unit introduces the underpinnings of all the units: "student-as-scientist, use of telecommunications, use of computer tools, and the centrality of data analysis and discussion in science learning" (TERC 1989, p. 3). Students analyze data about the pets owned by them and their teammates on the network. A Spanish translation of student materials is also available for this unit.

- **Acid Rain.** Students take pH measurements of rain water and [share] their results with others across the country. Students also examine the effects of acids on various materials, calculate the weekly amount of nitrogen oxides emitted through their families' car use, make predictions about pH measurements at other sites, . . . and discuss the societal consequences of different approaches for dealing with the acid rain problem.

- **Weather in Action.** Students investigate the major elements that make up weather: temperature, moisture, wind, and air pressure. They survey adults about memorable weather events and about how weather affects jobs. They measure and record the microtemperatures in their classroom; monitor daily temperatures, cloud types, and precipitation on their school grounds; and analyze the results of their research teammates' memorable weather surveys and weather patterns.

- **What's in Our Water?** Students learn about watersheds and determine where their school's tap water comes from. They examine how substances get into water and determine which substances might be considered pollutants. They test tap water for nitrates, do experiments to model the effects of chlorine on microorganisms, and analyze their own and teammates' data on water use and quality.

- **Too Much Trash?** Students collect trash and calculate the average weight of trash produced per student. Students devise and implement a plan to reduce their classroom trash. Trash activities are shared among research teammates across the networks and analyzed for patterns according to geography and grade levels.

- **What Are We Eating?** Students experiment to determine the nutrient levels of various foods. They compare the nutritional value of one of their personal lunches with a lunch that is ideally balanced, design a nutritionally balanced lunch, survey their relatives about their eating habits, and analyze data from their teammates about their nutrition investigations.

- **Solar Energy.** Students build solar collectors, measure solar radiation levels, and examine how solar energy intensity varies geographically. Students also design, build, and test solar ovens. Solar data from teammates are analyzed for geographic patterns.

The Teacher's Guide

The *Teacher's Guide* for the Acid Rain unit is representative of most of the teacher's guides in the other units. It is a 48-page book that provides specific activity suggestions for doing the Kids Network on a weekly basis for six weeks. It begins with a 14-page introduction and overview, divided into seven sections and subsections as follows:

1. What Is the National Geographic Kids Network?
    - What makes the Kids Network special?
        - Investigation, collaboration, geography, computer skills, interdisciplinary approach, cooperative learning

2. Who Participates in the NGS Kids Network?
    - Your class, your research team, the NGS Kids Network, the unit scientist, your computer specialist, the NGS Kids Network hotline staff

3. What Is Telecommunications?
    - How do letters, data, and maps travel through the telephone lines? How are data shared by the classes on the NGS Kids Network?

4. What Is the Acid Rain Unit?

5. Geography in the NGS Kids Network
    - Location, place, human-environment interactions, movement (how are we connected?) regions (what binds us together?)

6. Advanced Preparation and Classroom Management: Frequently Asked Questions
    - Includes answers to 16 questions

7. Overview
    - Organized by weekly themes and their corresponding session objectives:
        - Week 1: Learning About Acids
        - Week 2: Investigating Our Rain
        - Week 3: Acid Rain and Our Environment
        - Week 4: Considering Our Data
        - Week 5: Geographic Distribution of Acid Rain
        - Week 6: What Should We Do About Acid Rain?

Weekly lesson plans are divided into one or more sessions. Each session includes objectives, preparation and materials, activities, homework, and teaching tips and background.

Teacher Training Kit

Also available from NGS is a $100 teacher training kit intended to provide teachers, principals, and technology and science specialists with resources and ideas for presenting the Kids Network to colleagues within their own schools or districts. It includes a 43-minute videotape, sample agendas for training sessions, worksheets for computer activities, a software guide and demo disk, a sample data set, a list of "tips and tricks" suggested by veteran Kids Network teachers, a set of activity sheets, and suggestions for how to use the kit's materials.

Spread

NGS estimates that, as of fall 1991, some 250,000 students in total in approximately 8,000 schools had used the Kids Network. In the winter-spring session alone of that year, there were about 3,200 subscribers. As for international usage, "since the beginning of time through the '93-'94 school year, the Kids Network has had 34,941 sites representing 47 different countries" (Cowley 1994).

From 1989 through 1994, NGS compiled data about the number of classes participating in the Kids Network by state. Classrooms in every state are using Kids Network. Overall, the number of these classrooms is proportional to state population. Large population states such as California, New York, and Pennsylvania had 2,000 to 3,500 classrooms participating; while some less populated states such as Arkansas, New Mexico, and Wyoming had fewer than 100 classrooms participating. These data, however, do not indicate the extent to which classrooms may be using more than one unit per year or the numbers of teachers who share registrations. To the best of our knowledge, no other geographical information has been compiled. Therefore, there is no indication of the spread of the Kids Network according to more local geographic conditions and socioeconomic factors (urban, suburban, rural; percentage of students receiving free lunch; per pupil expenditure; student multicultural composition; etc.). We could not determine whether this study's New England sample, which consisted of a majority of Caucasian students in middle- to upper-middle-class communities, is representative of the users at large. This precludes our understanding the extent to which a program like the Kids Network is being used in a diversity of educational settings and contexts.

◆

Conceptions of Teaching: Paradigms and Time Warps

Curriculum developers, publishers, teachers, and educational researchers are entangled in an assortment of relationships as each contributes to the collective development of science curriculum. Interestingly, however, they do not necessarily operate from the same educational paradigm.[3] Developers are open to new trends but try to anticipate the extent to which teachers will tolerate these new practices. Publishers can be equally sensitive but try first to identify what they think the largest market of teachers wants. Teachers adapt new curricula to fit or possibly extend their existing pedagogy or paradigm; once they've mastered a new program, some are ready to move beyond the paradigm in which it was originally embedded. At the same time, educational researchers, more or less conditioned to regard the latest paradigm as a more powerful or compelling vision of science teaching, need to consider whether and how they might be elevating their preference to some normative status.

The nature of science teaching is in constant transformation. One way to grasp how science teaching has changed is to consider what might have been happening in "typical" fifth grade science classrooms at different points in time. The purpose of making these distinctions is to illustrate how the predominant scenarios or paradigms concerning upper elementary science education are changing and to consider where and how the development of the Kids Network fits along this evolution. This is not to say that the three scenarios highlighted in this discussion do not coexist today, did not exist earlier in some form, or that more scenarios could not be described. The variety and mix have always been there. However, the emphases have shifted.

The following three scenarios each represents a paradigm prominent in upper elementary science education roughly over the past decade,[4] but prominent at different levels: more "constructivist" among researchers, more "experimentalist" among program developers, and probably more "didactic" in prevailing classroom practice. As we shall see, these scenarios provide a backdrop by which to understand better the development and implementation of the Kids Network—which used, by and large, the "experimentalist" perspective, both as a compromise and as an attempt to bring more conventional practice into a more hands-on, group-centered, and dynamic approach to science teaching. Thus we see the inevitable "time warps" between the ripening of new, mostly untried, scenarios; the previous development of state-of-the-art scenarios like the Kids Network; and the hesitancy among sets of teachers to change prevailing practices when the "newest" designs are made available—be they experimentalist like the Kids Network or more openly constructivist like many contemporary programs.

---

[3]Paradigm is "in its most common or generic sense: a basic set of beliefs that guides action, whether of the everyday garden variety or action taken in connection with a disciplined inquiry" (Guba 1990, p. 17). Paradigms refer to "a world view, a general perspective, a way of breaking down the complexity of the real world [into] . . . what is important, what is legitimate, what is reasonable."

[4]We recognize that labeling these three scenarios runs the risk of oversimplifying a paradigm and ignoring the ways in which later paradigms are embedded in their ancestors. Nevertheless, for the reader's convenience, we use short-hand labels to distinguish among the scenarios.

Scenario #1—Didactic

When the development of the Kids Network Acid Rain unit was in its infancy some 10 years ago, a typical class of fifth graders was probably spending 30 to 45 minutes, one to two times per week, doing science. Doing science at that time would have included such activities as watching a film, reading a textbook, filling out worksheets, writing a report, or listening to a lecture by the teacher. In well-equipped classrooms, the teachers might carry out a variety of demonstrations. At the end of a study, students would take paper-and-pencil tests consisting of true/false and multiple-choice questions. The exams typically centered around a single topic and were intended to assess students' understanding of scientific information.

These activities are common to scientists. Like scientists, students memorize information. Through reading and listening, they learn about classical experiments; the value of being precise and systematic; and methods of observation, recording, and analyzing data.

Even though students in this scenario typically experience science vicariously, and less often through replications of experiments, for some pupils such experiences can be interesting. They raise students' sense of possibilities and introduce them to formal and systematic approaches to inquiry as well as to historically important scientific discoveries.

In this scenario, students typically sit in rows, raise their hands to be recognized, and direct their questions and comments to their teacher, who plays the role of "expert" positioned in front of the room. Here, teachers spend a lot of time explaining and elucidating constructs and formulae to students and asking questions like "What are the facts?," as well as preparing and correcting tests to monitor their students' levels of understanding.

Scenario #2—Experimental

In this scenario, we may notice subtle differences. While the overall time spent doing science may not have changed from Scenario #1, what is done during science period has. In place of some of the reading time, students do more hands-on investigations requiring them to physically manipulate and account for the properties and interactions of everyday objects. Typically, these investigations are algorithmic in nature, in that students are provided with everything they need: a list of materials, directions on their use, and a system for recording observations. For example, working in small groups, everyone follows roughly the same set of directions for measuring the pH of various preselected materials. Assuming fidelity to the directions, correct investigations lead to the same results. In Scenario #2, the teacher's predominant line of questioning centers around "What do the directions say to do?," "Where are you going astray here?," and "What were your results?" There is a greater emphasis on the investigation itself, with a minimal amount of analysis and discussion of its results and significance.

These activities are an apprenticeship into experiences common to scientists—carrying out fair tests, being precise and systematic, making hypotheses and observations, recording and analyzing data. The investigations are prestructured, but such exercises can be adventurous. Students are able to generate new ideas and experiences from their participation.

In this scenario, cooperative learning typically centers around the distribution and sharing of tasks. It also, however, shifts the role of expert, allows pupils to organize their work themselves, and allows for scientific problems in which more than one vehicle can produce correct answers. Teachers spend a lot of time managing group activity—for example, suggesting alternative procedures. As in Scenario #1, however, students continue to view the teacher as the expert and the ultimate purveyor of knowledge, although peers may play part of that role. Scenario #2 teachers are not just enforcers of student procedures and disseminators of requisite knowledge perceived as lacking in the experiments, but also guides. For example, by monitoring their students' levels of understanding and engagement, teachers look for ways to challenge their students yet help them stay on track.

Occasionally, worksheets and matching exercises complement these hands-on experiences, as do presentations by out-of-school experts, class discussions concerning current events, and student demonstrations to audiences beyond their classrooms. Students' understanding of scientific knowledge and processes are assessed through more open-ended paper-and-pencil tests.

Scenario #3—Constructive

Although elements of this scenario could be observed in some fifth grade science classes 10 years ago, its presence is considered by some developers and researchers to be more prevalent today, although still relatively rare. In Scenario #3, students may be spending up to an hour or more a day participating in science-related activities. Furthermore, the distinction between where science ends and language arts and social studies begins is less clear.

Comparatively, we find greater pupil choice and less structured experimentation. The predominant questions are now "What do you want to find out?," "How do you want to go about finding this out?," "What's your interpretation of the data?," and "What's an alternative explanation?" Although students are told, for example, how to use pH paper, they are now encouraged to use it to answer their own questions: "What is the pH of our local pond, stream, and school drinking water?" "Does the container a liquid is in affect the pH?" Ideally, they are expected to collaborate with a small group of classmates in designing experiments, to articulate their methods, and to consider revisions based on their peers' reviews. Before students can collect their materials and begin an inquiry, their peers need to approve each group's method for recording observations. Thus, students are responsible for creating a chart, with headings they conceived, that allows them to record and compare their data. After the experiments are completed, students present their data and findings to their classmates for review: Do their findings conflict with or corroborate other experiments or interpretations?

Although these types of activities are more open-ended than those in the previous scenario, a teacher's ability to manage them is no less crucial. Here, teachers work simultaneously with groups pursuing different lines of inquiry. They help students explore connections between their seemingly divergent investigations. They help students recognize and negotiate cognitive conflicts within the group, focus on common ideas, and make generalizations.

As in Scenario #2, but to a greater degree, the notion of collaboration and peer review is no longer constrained by one's geography. With the help of on- and off-line telecommunications capabilities, students design research and compare their findings with students from around the globe to a greater degree than in Scenario #1. Less consumers of extant knowledge, fifth graders in Scenario #3 are now considered part-producers of their own knowledge base.

In many respects, this model—much like the work of the scientist—takes Scenario #2 one step further. Hands-on activities are more heuristic and serendipitous in nature, in that they encourage students to discover unknowns. Consequently, these activities consist of many student problems: which materials to use and how to organize them, how to control variables, when and how to observe, how to record observations in formats that can help analysis, how and when to involve their fellow researchers from afar, how to present their findings to their peers. Furthermore, the teachers encourage not only several approaches to the same problem, but several solutions as well. In a more deliberate mode than in Scenario #2, students are provided with opportunities to build on one another's experiences and findings and thus to "socially construct" an understanding of the topic, with an appreciation for a variety of experimental, recording, and analytic methods.

Although the spotlight has shifted considerably from the teacher to the students, the teacher's role is no less important. Whereas in Scenario #2 teachers were primarily sources of knowledge and guides, in Scenario #3 their primary role is to support student-initiated inquiries. The ideal image is that of a guide and fellow investigator searching for answers to unanticipated questions by helping students evaluate the integrity of their processes and share their various approaches and findings. Consequently, there are more occasions for students to reveal their conceptions about the content and process of their investigations.

In this scenario, teachers employ a variety of strategies to elicit and confront students' conceptions about many acid rain-related concepts, such as gases, liquids, solids, mixtures, transformation of matter, the water cycle, food webs, what must happen to sustain life, and ways to change laws. More than in the previous scenario, this approach has teachers' monitoring of students' conceptions as a prerequisite to instruction. In this scenario, teachers are on the lookout for "teachable moments"—opportunities for a cognitive "stretch" just beyond most of the students. They are also concerned with assessing the students' knowledge of both content and process. To achieve this, they negotiate the assessment process itself more than in Scenario #2. Some invite students to reflect on their work, identify skills they need to develop further, and demonstrate what they have learned within and outside their classrooms (Krajcik et al. 1994).

Shifts Among Scenarios

We can construct the argument that these three scenarios derive from different paradigms, which we have tried to illuminate. Each paradigm has its own belief system and dimensions, but some are more compelling at different points in time—not necessarily better, but more effective, at activating pupil cognition, interest, and experimentation. Often, in the field, any one classroom may demonstrate two or possibly all three scenarios simultaneously. Yet it is through one of these perspectives—a predominant paradigm—that innovations in science, mathematics, and technology curricula are meant to be enacted into practice.

Figure 1 illustrates what we have called "time warps" and potential tensions that can occur when multiple paradigms operate simultaneously across three interconnected domains: practice, literature, and curriculum development. When a new paradigm comes along, curriculum developers incorporate some of it into a program that is perceived as ahead of ongoing practice (Moment 1). To accommodate the contexts in which teachers must work (existing programs, schedules, infrastructure), along with teachers' own repertoires; developers may compromise, adapting what they would like to do according to what they think ongoing classroom practice can tolerate (Moment 2). In the case of the Kids Network, the program was met by a generation of teachers of whom many had just assimilated Scenario #2 (Moment 3, Row 1). Consequently, they could execute the units fairly well by being faithful to the structured, hands-on curriculum. However, as teachers' experiences with the program matured, their practices became less embedded in the paradigm in which they originally greeted the program. After a few years, many found themselves intrigued with, but without adequate support for revising, the program so that they could incorporate more components of Scenario #3 (Moment 4). It is important to remember, however, that Scenario #3 emerges out of contemporary constructs that few science programs have as yet been able to codify, and about which, as we shall see, some educators have concerns—notably regarding managing the learning environment, science content coverage, and elevating one process of learning at the expense of others.

To be sure, the prominence of these scenarios at different moments in time is less linear and fixed than figure 1 suggests. Rather, as Worth (1994) explains,

> Scenario #1 is still the major one in schools, whereas Scenario #3 has been around in a variety of forms for a long time, always in very limited settings. Similarly, some form of Scenario #2 has been on and off again since the mid-1960s. In the '60s, we had a little bit more of the "constructivist" movement

Figure 1.   **Multiple Time Warps**

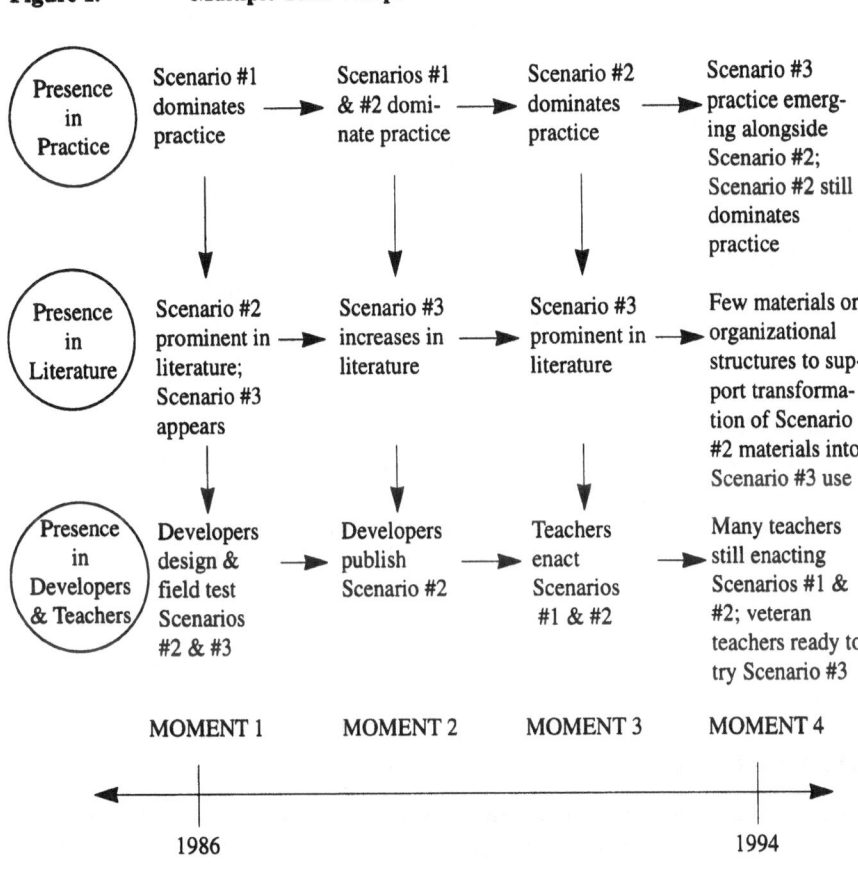

(Scenario #3), reinvented to Scenario #1, then followed by a more "hands-on" approach (as in the 1930s). Then we got a small shift toward Scenario #2—something called hybrid texts—in which there was an attempt to put a bit of hands-on into texts.

Today, Worth continues, there is a real rhetorical shift. Everyone uses the words "hands-on science," "inquiry," etc., with a resulting recent increase again in hands-on materials (Scenario #2); pushed further in terms of teacher role "but still very little activity we can characterize as inquiry teaching (Scenario #3)." Clearly, we are consistently within and between paradigms at all levels of science activity.

Territories of Paradigms

There is another way to represent the tension resulting from the lack of congruence in the state of the art as reflected in educational literature, in ongoing practice, amid developers, and among publishers. This alternative places less emphasis on when these scenarios appear in time and more emphasis on the movement among the scenarios.

Imagine for a moment that these scenarios represent three neighboring territories. What is of concern is not only the nature of each territory, but also the ways in which teachers move among them—the pathways of development that depend, in part, on the territory in which a teacher is host to an innovation. A crucial feature of this landscape is the type of gap between the respective territories. If it appears too large to potential users of an innovation, the passage is impassable; if it appears small enough, there may be some movement. For some teachers, the boundary between Scenarios #1 and #2 is a hurdle—passable, but requiring some concerted effort. It's like moving from reading a textbook to following the directions of an experimental guide. For others, moving from Scenario #2 to Scenario #3 is more like trying to traverse a large chasm, a risky leap from the sure-footedness of an experimental guide to the less certain footing of following students' initiatives. "To shift to Scenario #3," says Worth (1994), "requires a deeper transformation of learning and teaching and a deeper understanding of science than most teachers have experienced."

Some developers have apparently tried to move the field to this extent, but few practitioners adopted their programs; others have catered to predominant practices with small increments of change. The Kids Network seems to fall somewhere in between these extremes.

The Kids Network

The Kids Network is a conglomeration of the three scenarios. All units contain elements of each scenario. To illustrate this, we focus on the design of the Kids Network Acid Rain unit. Our purpose here is to consider how the nature of science teaching portrayed in the Kids Network fits within a broader context of changing practices in science education. To do this, we examine the program's goals, underlying pedagogy, and structure as described by the developers and publishers during interviews; as well as in their annual reports, advertisements, and curricular materials. But it is important to emphasize that the actual experiences of teachers, students, and researchers—included later in this report and reflecting how the Kids Network is actually enacted under normal classroom conditions—are not necessarily faithful to its design.

The Design of Acid Rain. A class of fifth graders participates in the Acid Rain unit for six weeks, perhaps for as long as an hour a day, five days a week—some days even longer. If a teacher is faithful to the published curriculum, students would complete at least the following sequence of specific activities, as illustrated by the student activity sheets:

- **Activity Sheet #2: "Acids in Foods and Drinks."** Record the names of different foods at home and the particular acids they contain.

- **Activity Sheet #3: "Measuring pH."** Follow a set of directions for measuring the pH of various teacher-selected liquids.

- **Activity Sheet #4: "Building a Rain Collector."** Design and build a rain collector that meets four criteria.

- **Activity Sheet #5: "Collecting Rain Samples."** Record the pH of rainwater and explain variations in readings. Teachers explain what to record and how to clean containers with distilled water.

- **Activity Sheet #6: "Acid-Producing Gases."** Examine and discuss a map of acid-producing gases.

- **Activity Sheet #7: "Describing Our Community."** Write a section of a letter describing your community's population, transportation system, sources of electricity, and industry. Answer questions under your assigned topic and telecommunicate the class letter to research teammates.

- **Activity Sheet #8: "Family Use of Cars and Public Transportation."** Record and then analyze family use of private and public transportation.

- **Activity Sheet #9: "Acid and Nonliving Things."** Follow a set of directions for observing what happens to any object immersed in three different liquids representing three levels of pH. Record observations in chart.

- **Activity Sheet #10: "Our Rain."** Collate individual pH readings of rainfall into a chart describing the class's most common, lowest, and highest pH over a three-week period. Telecommunicate the pH data to research teammates.

- **Activity Sheet #11: "Acid Rain on Our Team."** Prepare an oral report summarizing one research team's community letter with answers to questions.

- **Activity Sheet #12: "We Need More Information on Acid Rain"** and **Activity Sheet #13: "We Need to Act Now to Stop Acid Rain."** Discuss two written positions about acid rain: we need to know more versus we need to act now. Telecommunicate the data reflecting the class's position to research teammates.

- **Activity Sheet #14: "Acid Rain Worksheet."** Examine a map of an imaginary island and explain which of three communities will have the most and least acidic rainwater; suggest ways the communities might deal with their acid rain problem. Measure the pH of a mystery sample, compare it to previously collected samples, and predict what would happen to a shell and paper clip that were put in the mystery sample.

The *Teacher's Guide* includes additional activities. For instance, all units begin with a letter from a unit scientist who summarizes the unit and sets the investigation within a personal and contemporary context. Toward the end of all the programs, the unit scientist sends another letter, this time including collated data from the participating team and from all the classes participating in that unit.

Other activities in the Acid Rain *Teacher's Guide* include:

- share ideas about the meaning of the term "acid rain,"

- listen to the teacher's explanations and read in the *Kids Handbook* about the sources and effects of acid rain and how to interpret a pH scale,

- discuss the connection between the effects of various pH on student-selected objects and the effects of pH on our nonliving environment,

- predict and look for patterns in the pH of research teammates' rainwater and network's rainwater,

- write a letter to teammates in other schools describing the results of their vote regarding an acid rain dilemma,

- present what was learned to an audience outside your classroom, and

- consider what you can do about the acid rain problem.

Initial Evaluation Data. The evaluators of the first field tests of the Acid Rain unit discovered early on the discrepancy between the program's design and how it was enacted; they showed in particular that some of the field testers were in fact well-embedded in Scenarios #1 and #2. For example, the evaluators found that "Most teachers rarely took the time to synthesize or discuss [acid rain] findings with their students—doing the investigation was emphasized much more heavily than discussing and analyzing the results" (TERC 1987). The evaluators also commented on the students' responses to the hands-on activities: "The activities themselves were not 'inventive' enough. Students had few choices in constructing, trying out, and making variations in experiments. While students enjoyed many of the hands-on activities, some felt stymied when they were unable to try out their ideas" (TERC 1987). Furthermore, students wanted to communicate more with other schools. "In particular, many wanted more opportunities to write and receive individual letters, rather than only writing letters as a whole class."

The evaluators recommended that the revised unit include more hands-on activities that are "inventive and less prescribed" so "students will have thus more opportunity to take more initiative and will be encouraged to ask 'what if' questions" (TERC 1987). According to Candace Julyan, a main program developer, resolving this tension between the need for replicated procedures and the appeal of more student-centered and -initiated inquiries was challenging: "I tried to keep a balance between that: . . . to make that data set valid, but then as much as possible to have activities within that . . . that gives them the room to explore and to fool around with stuff. And that was a real difficult balance." Julyan also believed that the curriculum might be a district's first "move toward a more constructivist approach . . . a first step at it that they could use to have teachers start to think about kids' understanding." However, Julyan did not hesitate in saying that the Kids Network "is still very prescribed."

Even so, according to TERC director Bob Tinker, the fundamental characteristic of the Kids Network is its experimentalist approach. Tinker believes that content oriented around classification schemes (e.g., six kinds of pulleys and three kinds of levers) is not science. "Practicing scientists don't think in those terms . . . If you really want to understand the way a lever works, you look at it and do some analysis of it and think about forces and what have you. But you don't do classification schemes." In contrast, "We want to focus people on the empirical, on the measurement, on the observation . . . Science . . . isn't some mystical six-step procedure that you have to follow. Go out and open your eyes, measure things, report, and look for patterns."

Experimentalist approaches de-emphasize the value of learning by reading, especially when compared to the prominent role reading plays in Scenario #1. Says Julyan, "Somebody can do the entire Kids Network and never use the *Kids Handbook* and it would be just fine." In her view, "the *Kids Handbook* is primarily a crutch, a security blanket, for teachers who feel there should be some reading material. And for teachers who feel there needs to be more than just explorations . . . If the kids could relate what they read in the handbook with what they did with the hands-on . . . that would be the optimal use of the handbook."

The Kids Network's Fit With Student-Inquiry and Constructivist Approaches. There is an interesting story about the evolution of the Kids Network, particularly in respect to the developers' thoughts about their program and more constructivist approaches. At the onset, the developers were aware that their approach was in some ways incompatible with inquiry-based or constructivist approaches. They attributed this constraint to one of their three guiding principles:

> Using telecommunications can impose a certain rigidity on the curriculum. For the national data collection experiments, students must relay particular predetermined types of data simultaneously at prescheduled times. Not necessarily the most natural route to inquiry-based learning! We have had to counterbalance this need for structure with a more constructivist approach in several ways (TERC 1987).

The developers made two suggestions. The first was to provide students with a bulletin board serving smaller clusters of about eight classes and "e-mail capabilities for more individualized and self-selected communications among students." The other solution was to create a unit called Investigate (TERC 1987). Neither of these suggestions was realized. A closer look at the unpublished Investigate unit shows constructivist approaches that are not included in any of the published Kids Network units.

The Investigate Unit. The Investigate unit was among the first three units designed by TERC. It was intended to be a 10-week "departure from the way science is taught in all but the most progressive of classrooms" (TERC 1989). The

> students in each class choose a topic which interests them and arrive at a research question. Each class then designs a data collection instrument, which they test with another class in their school. Once they have satisfactorily modified their instrument, students ask other schools in their research team to collect data for them. These data are sent via the telecommunications network. Students must then organize and analyze the data they receive from other schools in an attempt to answer their original research questions. They are then called upon to report their findings (TERC 1989).

The publishers decided that Investigate would not go to press, according to the Kids Network project director at NGS. They thought that the unit was too much of an outline, too open-ended, and that it lacked substance.

The Investigate unit includes many Scenario #3 components: collaboration with peers within and outside their respective classrooms, students' creation of new knowledge, and critical reflection about the process of science as well as one's participation in that process. Unlike the other units, Investigate pushes for diversity of perspectives, multiple interpretations of phenomena, new ways of creating knowledge, and new methods of organizing information. Moreover, Investigate, unlike the other units, centers around student-initiated investigations. This component, according to Tinker, would have improved the quality of science in all the units: "If we wanted to make this a very good science experience, there was one thing that we totally left out, which is the kids choosing the science that they do. [What was published] is science that we tell you to do."

Other elements of Scenario #3 appeared in the original field test materials of the Kids Network but did not survive in the published versions. For instance, the original Hello! *Teacher's Guide* presents a constructivist approach to leading discussions. These suggestions do not appear in the published Hello! unit nor in any of the other units. The original Hello! unit also contains several statements about pedagogy that follow a constructivist approach, along with suggestions for brainstorming possible methods for gathering data. These texts also do not appear in the published Hello! unit nor in any of the other upper elementary units.

On the other hand, some components of Scenario #3 have been both developed and published in the newly produced Kids Network materials for junior high students. For example, whereas suggestions of assessment strategies are essentially absent in the teacher's guides for the upper elementary units, the junior high programs include a fairly detailed alternative assessment plan, including portfolios and self-assessments. Julyan explains:

We have a pre-unit questionnaire, not as a test, but as a way to think about what they know [and a] post-unit questionnaire to see if they've learned anything. And then there's a document we give them, called the Before-After Comparison, where they can reflect on what changed for them about each of the 10 questions in the questionnaire. We also built in portfolios and support for teachers who have never used portfolios . . . We tried to make it not terribly intimidating and have the kids take some responsibility in deciding how they did.

Concluding Remarks About Paradigms and Time Warps

The Kids Network has experienced an evolution that reflects some of the predominant shifts and turns in science education from the early 1980s to the mid-1990s. Its evolution also mirrors the differences between and transformation of pedagogies espoused by developers, publishers, teachers, and researchers. As such, it is a story that does not end here. As paradigms shift, so will the reigning beliefs about science education, along with the interpretations of the original program and the ways in which it can be implemented. Time warps may be endemic to the educational enterprise—and a vehicle for understanding the social and historical context giving rise to the development and publication of such programs as the Kids Network.

◆

Research Design and Methods

Case Study as the Method of Choice

Case studies are precisely that: the study of cases, much as a lawyer or forensic scientist studies cases. A case can be a person, a group, or a school or other organization. It can be an event, a type, an instance, or a process.

Cases are studied in their natural habitat to better understand their histories, their contours, their dynamics, their core events and actors, and—in the best instances—the reasons they take the shape they do. When several cases (teachers, schools) are studied, as is true here, each is given somewhat less time and attention; but the core facets of each are studied carefully and over time, and the comparisons bring to light insights that one would have overlooked with a single exemplar. When, for example, one set of teachers works routinely from the *Teacher's Guide*, and a second set turns over the choice and design of an experiment to groups of students while peppering them with questions and apparently incongruous findings, we see significant differences in pedagogy, learning opportunities, cognitive activity, and—most of all—the range of experiences and understandings one curriculum makes available.

Case studies are best conducted as a process of gradual understanding of the core processes and determinants at play. This is achieved through multiple questioning combined with a process of observation of and listening in on different arenas of activity. Progressively, and informed by their research questions, researchers become more locally knowledgeable. Observations can be highly formal—accompanied by instruments or formal notetaking—or they can be more informal. Often, as in this study, observations are mixed with other types of data-gathering devices, especially interviews, typically prepared in advance but shifting as some questions take on more importance and some explanations begin to come clear.

In this study, the research team looked closely at seven exemplars of the Kids Network during the 1993-94 school year. The case study method allowed for a fairly systematic examination of each project's history; a description of the physical, social, and cultural environment of each school site; and, above all, the observation of each project's life under natural conditions, month to month. The methodology allowed researchers to become comfortable with the particulars of local sites. Collectively, researchers and informants could reconstitute the local genesis of the Kids Network; tell the story of its implementation through the eyes of teachers, administrators, students, and other involved parties; document the constraints and supports along the way to daily practice, including many factors still present in the environment; and witness its daily enactment within the mathematics and science curricula.

At the same time, visits by the research team gave the opportunity to local participants to describe their perceptions of the Kids Network from the inside out and to tell their personal history of involvement with its materials and activities. Finally, the research team spent much of its time observing how the Kids Network was acted out: how the teacher's guides were actually used, how teachers modified or otherwise experimented with the curriculum, what kinds of learning opportunities were provided on a daily basis, and—above all—what pupils and teachers actually made of those opportunities as program activities designed to deepen understandings of scientific experimentation and mathematical principles applied to everyday and some exotic settings. To understand the Kids Network was to grasp its local surroundings; understand the strategies of, capacities of, and constraints on its staff; observe the nature, variety, and depth of its practices in different contexts of use; and then compare the seven research sites on meaningful dimensions to tease out important commonalties and differences.

Sample Selection

Initial Design and Criteria for Sites. Because seven sites are not representative of the hundreds that use the Kids Network, our initial criterion was to look for diversity of use and local conditions. We developed a sample matrix divided by the following criteria (shown in table 1): location of the school (urban, suburban, rural); socio-economic status (SES); and racial mix of student population. In addition, we established these other qualifications:

- elementary or middle schools, grades 3-6;

- use of the Kids Network for three or more years;

- self-contained classrooms or teaching within a subject area, for example, science; and

- connections or collaborations with other users within the school or district.

Site Selection Process. Selecting the seven sites was further complicated by the initial difficulty of acquiring lists of Kids Network users. Finally, we obtained a list of about 100 teachers who registered for the Kids Network in the Northeast during 1989 and used this as an initial resource for contacting potential participants by phone. Efforts to meet the minimum criteria for participation described above reduced potential candidates to fewer than 15 schools.

There were few problems in locating suburban and rural sites. The greatest difficulty was in locating urban sites and sites with low SES. We ultimately selected for a diversity of use and users so as to depict how each innovation looks under different conditions. At the same time, although diversity was valued, it was thought that a site should not be so unique that the innovation could not be replicated elsewhere.

Emergence of Promising Configurations. Casual visits to classrooms provoked the need to redefine the terms "exemplary" and "garden variety." These additional interviews and observations showed that teachers strong in some areas were weak in others. For instance, a teacher who helped students reflect on their conflicting ideas had not mastered program components; on the other hand, a teacher who appeared to have a mastery of both program components and content matter did not encourage students' posing and testing conflicting ideas and answers. Based on the sense that sites would exhibit a cluster of attributes rather than being purely exemplary or garden variety, a list of promising configurations was devised as a more accurate way to display the full range of dimensions likely to characterize a given case.

Visits to Potential Candidates. Although conversations with teachers and principals by telephone were helpful, visits to 10 potential sites were key to making the final selection. These visits provided opportunities to observe teachers conduct a Kids Network lesson; to see a school's computer lab; to interview teachers, principals, and specialists; and, in some instances, to listen to children describe their experiences. Finally, these visits revealed those who seemed comfortable as well as interested in the project and those who seemed best able to coordinate access to key players at the site.

Seven sites were selected for the study. It is noteworthy that all sites finally selected, after a wider search, were in the Northeast corridor of the United States. This was done for reasons of convenience, but also after careful examination of alternative sites—which presented very similar characteristics, given the standardized format of the program. The overrepresentation of middle- to upper-middle-class communities participating in the study and the fact that two out of the seven sites were developers' pilot sites are some limitations of the sampling universe.

As table 1 shows, the seven sites are located in three states in the Northeast. Two are in rural areas, three in suburban settings, and two in urban settings. All are elementary schools. Appendix B includes a profile of each participant.

Table 1.   Kids Network Case Sites

Characteristic	Urban		Suburban		Rural	
Site	1	2	3	4**	5	6
State	NH	NY	NH	NY	VT	VT
SES*	Working to middle	43% receive reduced or free lunch	Middle to upper	Middle to upper	Middle to upper	Middle to upper
Racial and ethnic diversity	Primarily Caucasion	50% minority (30% black, 20% Hispanic)	Primarily Caucasian, some international students	Primarily Caucasian, 15% Asian, 5% Pacific Indian	Primarily Caucasian	Primarily Caucasian
School grades	K-6	K-6	K-5	K-5	K-6	K-5
School size	600	600	340	a=271 b=300	600	220
Number of teachers using Kids Network 1993-94	2	12	3	a=2 b=1	6	2

* SES is derived from teachers and principals.
** Site 4 consists of two schools in the same district. They are referred to as Sites 4a and 4b.

Instrumentation

Several versions of the interview and observation schedules were developed. Instrument pilots were also conducted. The final set of instruments included the following:

- interview schedules for developers and school-level and district-level administrators, teachers, pupils, and assistance providers near to—or, in some cases, far from—the site;

- observation schedules and checklists for classroom activity;

- guidelines for nonparticipant observation and participant observation (work in the classroom);

- formats for document analysis and work samples, both from the site and beyond; and

- formats for analytic memos, document summaries, and contact summaries of site visits.

Site Field Visits

During a nine-month period from October 1993 through June 1994, the researchers observed 11 classrooms at seven schools that were implementing one of five NGS Kids Network units. Typically, each school was visited on three occasions for two days at a time, making for a total of approximately 36 days on site. Most visits occurred within specific six-week periods, the length of a Kids Network unit.

During approximately six days at each site, the field researcher carried out between 12 and 20 hours of interviews with teachers, students, principals, technology specialists, and other key supporters. Additionally, there were 4 to 18 hours of classroom observations per site. Since site visits typically lasted all day, the researcher was able to observe both Kids Network and unrelated activities.

Sites varied by the number of teachers using the innovation and the number of support staff. For example, at Site 1, an urban elementary school, only one of the three original teachers who agreed to participate had registered for Kids Network by the time observations began; unexpectedly, a student intern was solely responsible for program implementation. On the other hand, at Site 5, a rural elementary school, we observed and interviewed three teachers and their students, as well as their technology specialist.

In general, we conducted formal and informal interviews during and after school with teachers during each site visit. Interviews typically lasted between 30 minutes and three hours. Whenever possible, we also interviewed school principals, science specialists, technology specialists, key supporters, and students. The researcher's handwritten notes were supplemented by an audio recording. Notes and tapes were transcribed verbatim with the researcher's interpretations set off in brackets.

Researchers carried out an estimated 59 hours of formal interviews with 15 teachers, 54 students, and 5 principals and other support staff; they conducted an estimated 103 hours of observations. In addition, the primary researcher for one case joined classes on a field trip to a district landfill.

Interviews at each school comprised:

- two to three teachers who were actively using the innovation (two to three hours each);

- one or two additional teachers who had either completed using the innovation, were planning on starting, or had discontinued use;

- one principal (one to two hours);

- significant support staff, including technology specialists, special education teachers, and parents (one to two hours); and

- students, individually and in small groups of two to four students, during the second and third round of visits (20 to 40 minutes per session).

Observations at each school entailed:

- 45 minutes to three hours a day in each classroom while the innovation was in practice, and

- 20 minutes to one hour a day of non-innovation-related classroom activities.

Research Questions

The research questions for the study were numerous. They came from previous work on the Kids Network; from studies of the innovation process; and from our desire to capture the history, the implementation, and the patterns underlying the day-to-day enactment of the Kids Network, notably in relation to the opportunities for learning made available to students. The research questions can be compressed into these categories:

- **Origins.** Our concern was with the historical context of program development; the arrangements between funding sources and developers; the dominant paradigm of mathematics, science, and technology; and the initial characteristics of the program. We also wanted to look closely at the main stages in the development of the innovations, including their evaluation, and the ways in which developers contacted potential participants to disseminate the innovation.

- **Background and history.** We wanted to examine the circumstances by which the innovation was adopted at the site, including the chronicle of local adoption—sequences, key events, determinants—as perceived by the main actors and the innovation's prime advocates. We also needed to look into the local context at the time of adoption: the social, economic, political, and historical frame; the characteristics of the school, district, and neighborhood; and the innovation history of the school.

- **Process of local assessment and adoption.** We sought to determine the motivations and incentives to adopt; along with the initial perceptions of the innovation (perceived merits, promises, doubts, limitations) and the expectations invested in it. To this were added questions about the characteristics and orientations of teachers using the practice, their professional profile, and their stance toward science/mathematics teaching—for instance, were they more constructivist or more structured?

- **Organizational fit.** We wanted to know more about the "fit" of the innovation to the organization; that is, its nesting in the organizational structure and working arrangements of the school—for example, with the curriculum in science and mathematics, with the sequence of preceding and following curriculum units, with school/district policy or vision in general, or with respect to science and mathematics.

- **Innovation characteristics.** A series of important questions had to do with innovation characteristics: philosophy of mathematics and science; view and practice of technology; multicultural sensitivity; vision of science expertise and inquiry; epistemology, complexity, and difficulty; portrayal of learning and motivation; components and their links.

- **Principals—role and perceptions; school district personnel—role and perceptions; assistance.** In parallel, we wanted to examine the roles played by the principal and district office personnel, especially their knowledge of the innovation and their behavior in time, their technical and social support, their candid assessment of the Kids Network, and their special role in the crucial assistance network required to sustain such innovations during periods of preparation, uncertainty, and strain.

- **Early implementation (teachers—role and perceptions); later implementation: teacher/classroom variables (teachers—role and perceptions).** Several questions pertained to the innovation's actual implementation through its progressive stages, as it interacted with the local infrastructure to a typically higher level of technological mastery in the classroom. For example, did the project influence core aspects of teachers' work: changes in repertoire, measurable changes of knowledge base in science/mathematics, transposition of the project's content into instructional strategies, knowledge of curriculum and curriculum materials in science/mathematics, mastery of pupils' understandings, and perceptions of or relations with students?

- **Pupils—role and perceptions.** We wanted to know what the innovation looked like on a daily basis from the pupil's view—for example, the typical activities, the materials in use, the tasks undertaken. We questioned the level of task engagement or involvement and looked at task progression assessments made by the students themselves. To the extent possible, we also tried to assess the effect of the innovation on pupils.

- **Organizational outcomes.** Finally, we asked about the extent to which the innovation was settled or stabilized in the school. In which ways were the school organization and yearly routines likely to affect its structure, operations, and substance?

Analytic Procedures

Data analysis in qualitative case study research is largely iterative; that is, it proceeds as the data are collected and reorients the future waves of data-gathering and interim analysis. In this study, interim analyses were driven first by the original research procedures—translation of provisional research questions into codes, creation of provisional data collection instruments, multiple field visits spread throughout the school year, production of audiotaped or handwritten interviews and observations, examination of key documents, photographs (see appendix C for a summary of analysis products). This formed the basis for the successive waves of interim data analysis using a variety of devices. The procedures were dictated, however, mainly by the structure of the site visit, allowing for day-long contact (multiple observations and multiple interviewing of the same and different informants) and for multiple-day visits; between which researchers could review their tapes and notes, identify meaningful trends, devise provisional hypotheses, prepare follow-up questioning and observations, etc.

   Devices. More specifically, analytic devices included the following:

- **Coding** of transcripts and field notes, including the identification of new codes, the discarding of inapplicable codes, and—above all—coding for recurrent themes, which were then discussed at team meetings. Data were coded using HyperRESEARCH (Hesse-Biber et al. 1993), a qualitative analysis software program.

- **Contact summaries**, written after each field visit, each of which corresponds to a set of field notes and highlights potentially significant themes and other ideas the researcher found interesting or illuminating during a visit. The summaries also include a set of questions to be addressed during the next field visit as well as a list of collected documents.

- Regular production of analytic **memos**, elaborating a theme or idea that struck the researcher during an interview or field visit. Memos were also written about readings and off-site observations—for instance, of training events.

- **Document summaries** of innovation-specific materials (teacher and student guides, workbooks, etc.), as well as literature relating to science, mathematics, and technology education.

- Researcher **meetings** at regular intervals, with analytic work explicitly on the agenda.

- **Interim reports**, prepared mainly for overall project coordination and as information for other research teams. The preparation of the reports forced out provisional findings; more thematic or superordinate codes and reviews of memos for recurrent themes; and promising relationships, clusters, and explanations to be explored later.

A focus group consisting of two Kids Network teachers and a districtwide science coordinator convened in Texas during the final analytic phases of this project. During this meeting, participants responded to and confirmed our resumé of principal findings.

Thematic Codes. We worked with 17 functional coding categories, regrouping 32 thematic codes and 173 general codes. Thematic codes included such categories as "algorithmic vs. heuristic," which covered examples and definitions that distinguished these two approaches to teaching and learning. (See appendix D for a sample of the thematic codes.)

The thematic material, keyed to the original and revised research questions, was then subjected to analyses using formats or templates devised by Miles and Huberman (1994). In particular, checklist matrices, event listings, and time- and role-order matrices were used to gather similarly plotted data on one chart. The same was done with conceptually clustered matrices and causal chains. Comparative analysis then allowed us to cluster cases, establish significant similarities and differences across cases, and—from there—identify systematic sources of variation. Here lay the explanatory material we were after, as well as the narrative, contextualized contingencies under which the key determinants were seen to hold. We would then account for what the Kids Network was in seven educational settings, how it functioned, and the reasons for its enactments and effects in specific contexts.

Research Team

The team comprised three working researchers, all based at both Harvard University and the National Center for Improving Science Education. Michael Huberman, a specialist in the area of innovation implementation, learning theory, and qualitative methodologies, directed the study. As research associates, Jimmy Karlan and Sally Middlebrooks were responsible for the science and mathematics components of the study from both developmental and school-related perspectives. Karlan was also instrumental in much of the software development (and debugging) for the project and was especially interested both in the epistemological and pedagogical issues raised by the study. Middlebrooks added significantly to the phenomenological dimension of the study—that is, the perceptions and meanings derived from teachers and students. (See Appendix A for more.)

Limitations of the Study

The research team was handicapped by its orientation toward social-constructivist approaches to teaching mathematics and science—that is, student-centered approaches emphasizing the construction of knowledge by students through the use of open-ended experimentation, and confrontational class discussion, and independent work around problems and solutions that emerge gradually. Aware of this bias, all members deliberately sought to be as descriptive as possible in their observations and accurate in the conduct and transcription of interviews. We realized that this approach constituted a stretch—both conceptually and in the classroom—for some teachers and an incompatible or inappropriate perspective for others, and we tried to reflect that variation in thought and action. The problem was somewhat compounded by the fact that program developers had themselves advocated a less traditional perspective, while providing supporting materials for teachers and students that were highly prescriptive—what more than one of our informants called a "cookbook approach" to group work, experimentation, and individual work.

Other limitations warrant mentioning at the outset:

- restriction of the sample to Northeast area projects;

- over-reliance on middle- to upper-middle-class case sites;

- use of a "process" framework (the implementation of change);

- as in all multiple case studies, noncontinuous presence at the field sites;

- restriction of participants to those with more than three years of experience (to study the project in a minimally consolidated form);

- disproportionate number of more experienced teachers in the sample;

- reliance on self-reported data for earlier events where no documents were available;

- absence of observations of the final series of Kids Network activities in which students share what they learned with an audience outside their classroom; and

- absence of interviews with some principals, but only where these informants were judged not to be knowledgeable (e.g., recently hired, too remote from the project).

The strength of this study emerges indirectly from these limitations: its presence at multiple field sites and—once there—the focus on the process and nature of enactment of the Kids Network, together with the perceptions of its key actors, the teachers and students. When conducted with a nonevaluative perspective, as is the case here, such studies are both unique and highly instructive for understanding how innovative projects get down to the core level of application—the classroom—and what learning opportunities are actually offered and realized.

◆

Resumé of Principal Findings

Many factors and conditions can help explain how a particular curricular innovation is implemented three to four years after its first arrival. What follows is a summary of the principal findings as they are reflected in each of the succeeding sections. In the final section of this paper, "Concluding Remarks," we present some explanations of these factors and their relationship to the nature of the program's implementation. For now, we present some of the principal findings concerning the program's arrival, its adoption, enactments and assessment, sources and types of assistance provided, technological components, student perspectives, effects on teacher practices, program continuation, and equity issues.

Arrival

In our Kids Network sample, a set of sequences of players and events converges on a number of key points. For example, all sites had:

- the financial resources to purchase the original set of materials and pay an annual registration fee;

- the minimal technological requirements (one computer, one modem, and one phone line) and the capacity and interest to upgrade their technological resources;

- an administrator or specialist who learned about the program and then introduced it to a receptive teacher;[5]

- an interested teacher who was willing to try out challenging curricula involving new content and instructional strategies; and

- for those schools in which the program spread to several users, a respected advocate.

Adoption, however, is a complex phenomenon. It is not simply a matter of principals and specialists introducing innovations to receptive teachers. Rather, all the players, when faced with the decision to invest in the Kids Network, had to contend with a variety of ambivalent feelings about it. Our participants described their ambivalence, both anticipated before and realized soon after their first experience with the program.

Incentives and Uncertainties Regarding Adoption

The relevance inherent in the Kids Network was cited most frequently as an incentive to adopt. Teachers identified the unit topics as themes both they and their students could embrace because these topics affected the quality of their lives. The second most frequently mentioned incentive to adopt the program was its ability to create "global classrooms" by way of telecommunications. Finally, most teachers indicated the program's interdisciplinary aspects as one of its original appeals.

---

[5]KL at Site 3 is an exception; she learned about the program from an acquaintance who was a relative of a Kids Network project manager.

After their initial experience with the program, seven participants (32 percent) identified the ability to modify the program as an incentive to stay with it. Six participants indicated that having six unit topics to choose from was appealing. Not only did this provide teachers with some choices, but it also gave them opportunities to vary units from year to year. This helped teachers avoid boredom on a particular topic and work with multigrade classrooms. Five of the participants identified the computer component as an important aspect for adoption; they saw it as a good introduction to a variety of computer applications (e.g., telecommunicating, graphing, word processing, mapping).

More than half of the participants indicated that they adopted the program because it fit with either their curriculum, structure, or personal needs. For example, four teachers from as many schools indicated that one of their incentives for adopting the Kids Network was its fit with the structure of self-contained classrooms. Since the program requires flexible scheduling (i.e., recurring extended periods of activity and intermittent involvement throughout a day), the structure of a self-contained classroom fits well.

A good personal fit was important as well to a few of the teachers, particularly those at Site 1 who highlighted how unit topics satisfied their personal interests or provided them with new experiences that they wanted to learn, such as telecommunications. For one teacher, the program helped her teach a required subject in which she did not feel competent. For another teacher at the same site, the program provided the opportunity to collaborate with a colleague.

Half the participants identified the program's initial appeal to students as a primary incentive for adopting the Kids Network. Unlike most of the other incentives already highlighted (which preceded the participant's first use of the program), the program's appeal to students was instrumental in its continuation. Most participants justified their definitive adoption after their students exhibited a high level of motivation and excitement. With only a little experience, the teachers observed that most of their students expressed tremendous excitement about sending and receiving letters from their teammates.

Often conjoined with the participants' incentives to adopt the Kids Network were a variety of uncertainties and concerns. Most of the participants expressed uncertainties about the program's telecommunications components. These concerns ranged from lacking the necessary equipment to inadequate technological support. Even so, the appeal of the telecommunications component was too strong for even these technological concerns to arrest initial interest in the program. The rest of the concerns and uncertainties emerged after participants' first use of the program. For example, concerns about the program's inherent time constraints were widespread; many of the teachers described the telecommunications program as tedious, inflexible, unsophisticated, or ridden with bugs. Four teachers were initially uncertain about adopting the program in part because they felt the science experiences were not authentic. One teacher said the program did not provide enough interaction with real scientists, the measurement materials were unsophisticated, the unit scientist's results overgeneralized local data, and the reading materials were too general and shallow.

Another concern described by most of our participating teachers was the consistently high drop-out rate of their teammates during the course of a unit's session.

While even our most enthusiastic advocates shared some ambivalence about the program, their experiences convinced them that its adoption had been well worth it. Consequently, their concerns about telecommunications, time constraints, software, and other program components were overshadowed by positive experiences.

Enactments and Assessment

Whether the enactment was a faithful replication of what was prescribed in the *Teacher's Guide* or a spontaneous diversion, whether students worked side by side doing different tasks or went on a unit-related field trip, many of the enactments at the different sites shared some common attributes having to do with learning pace, group work, and assessment.

Most of the Kids Network activities provided in the *Teacher's Guide* and activity sheets are algorithms in which the topics, problems, and strategies are mainly selected by the program and teachers. For instance, the directions for how to create runoff in a cup of soil (What's in Our Water?), and the headings for recording data in all the charts in all units, require students to read and follow a prescribed set of procedures. There were only a few occasions on which we observed students solving problems they themselves generated, using strategies they developed.

In our sample, the nature of cooperative learning was limited to sharing materials. Typically, directions explained how to use the provided materials; consequently, many of the students' actual cooperations with one another centered around getting, using, and returning materials. Given the algorithmic nature of the program in general—as well as of the individual activities—to which most teachers tended to remain faithful, students rarely cooperatively designed their own lines of inquiry, the methods to answer the program's lines of inquiry, or the means to make sense of their data. A corollary to this issue is that teachers tended to spend more time managing the organization and distribution of materials than their students' learning process.

The program was often executed at a breathless pace. There was thus little time for reflecting, revising work or theories, probing students' thinking, or exploring divergent but relevant inquiries. Furthermore, most of the teachers typically paused for no longer than one to two seconds after asking a question before answering it themselves or moving on. Perhaps not surprisingly then, students typically limited their responses to only a few words.

Most teachers were more concerned with students developing science process skills (e.g., theorizing, controlling variables) than understanding scientific concepts (e.g., the effects of pressure on weather conditions). They reported that they made informal assessments around students' levels of engagement and not around specific content knowledge related to the unit topic. Consequently, teachers often were not aware when their students maintained inaccurate explanations and ideas about the topic. In general, the cognitive demands of group members were minimal since most activities required students to follow a prescribed set of directions for solving a preselected problem, rather than opposing and testing different perspectives and representations.

Assistance

Teachers identified a number of sources from outside the school that assisted their implementation of the Kids Network. Each source offered a different type of assistance: the publishers provided computer aid via their toll-free hotline, the developers offered "teacher-friendly" guides, professors and scientists provided technical expertise, parents were benefactors, and a researcher provided an opportunity to reflect. These sources of assistance, external to a school's center of activity, made significant contributions to the teachers' efforts. One source of external assistance that is noticeably absent from our participants' reports is the private sector (i.e., local businesses).

Teachers also identified sources from within their school that assisted implementation. Again, each source offered a different type of assistance: the district administrators used block grants to upgrade their schools' local networking and telecommunications capabilities; school boards shared an educational technology vision with the principal and staff; principals were credited with providing a variety of assistance, including disseminating the program and doing the annual paperwork; and technology specialists assisted teachers as computer troubleshooters. These sources of assistance, internal and external, played vital roles in helping teachers accomplish what they wanted with the program.

The types of assistance reported by our teachers concern issues primarily related to purchasing the program and requisite equipment, and the mechanics of running the program. Teachers were not observed, nor did they talk about, assisting one another in respect to more substantive pedagogical issues. The absence of philosophical and theoretical discussions and debates with colleagues about the pedagogical aspects of the program may explain in part why so many of the veteran users remained faithful to the original program.

Technology

The Kids Network has been promoted by NGS and considered by our participants as a "technology-based science curriculum." Even though using the Kids Network was most teachers' first introduction to using telecommunications with students—and provided others with their first experience in using computers to gather, analyze, and share data—none of the teachers in our sample indicated that they felt intimidated by the program's technological demands. All participants had access to more than the minimal technological resources necessary to enact the program when they started it. Since their first use of the program, their access to not only more computers but also to increasingly more sophisticated hardware and software has grown considerably.

The rapid pace of schooling in general and the role of the computer within the Kids Network's intense time frame in particular put teachers in the position of having to seek quick remedies from other sources (e.g., calling the hotline or seeking help from a technology specialist). Consequently, teachers and students have limited opportunities for solving problems through experimentation. When students are uncertain about what to do, they default to their teacher's assistance. Compound this tendency to rely on others to solve imminent computer problems with an annual 10-month absence from the Kids Network materials, and the difficulty in mastering the computer components becomes even more apparent.

All the teachers—as well as the developers—did indicate, however, that they had one or more concerns about the technological components of the program. Three technological concerns raised by teachers were that the telecommunications demands inhibited student-centered inquiry, the attrition of teammates was too high, and the software was slow and unsophisticated.

A couple of teachers and one of the developers indicated that the program could be as successful either without the students' direct participation with the telecommunications component or without the telecommunications altogether.

Student Perspectives

In general, students' responses to the program were very positive. Most of them considered the Kids Network to be very different from other subjects. They said it involved more "doing," more experimenting, less reading and writing, and—for those doing Acid Rain, What's in Our Water?, or Too Much Trash?—learning about something "real." Most students said telecommunicating was also novel and one of their favorite parts. This view was even shared by a few students who never experienced firsthand a telecommunications session. Some students also identified doing experiments and going on field trips as favorite Kids Network activities. Some students from all sites said they did not like the slowness of the telecommunications sessions and were frustrated by teammates who dropped out. When asked about what they would choose to do next, students had little difficulty suggesting one of four types of activities: experimenting, constructing, communicating, and social action.

Changes in Teacher Practice

The ways that teachers change with the program are reflected in how they changed the innovation itself. Except for a few cases, teachers generally modified the innovation to meet certain technical and practical demands as well as to enrich the program with other adults and excursions. Teachers at three of the sites enhanced the unit with field trips and guest speakers. One teacher modified the program by replacing some of the more hands-on activities with worksheet-driven exercises. A few teachers, however, described their evolution with the program as being pedagogically significant.

Interestingly, none of the teachers modified the student activity sheets, even though the activity sheets represent the majority of students' time with the program. This is a crucial place where one might expect to see pedagogical modifications in the program reflecting teachers' changing conceptions of the nature of science teaching and changing levels of concern and comfort with program execution.

The metaphor of journey does not only concern the ways in which the teachers' relationship to the program changes over time; it also charts the changes in teachers' practice—and, in some cases, their careers. As a result of discovering faulty litmus paper that was provided in the Acid Rain unit, one teacher ended up collaborating with state officials in the development of a curriculum about lake ecology. The student teacher who implemented the Kids Network for her first time during her internship credits the Kids Network with changing her teaching experience.

Continuation

The principals at all of our sites hoped the program spreads among more faculty and have taken various initiatives to support that direction. Furthermore, every participating teacher anticipated using the program for at least the next five years.

Although financial constraints were not an obstacle to continuation for schools in this study, it is certain that the auxiliary costs to run a Kids Network unit are prohibitive for many teachers. First-time users have to invest at least $500 for a unit's start-up costs (*Kids Handbook, Teacher's Guide*, software manual, maps, software, etc.) and the annual telecommunications and registration fee. Add to this the costs of a dedicated phone line, an Apple IIe computer or better, and a modem. Even if these initial expenses are covered, the annual fee and the costs of upgrading and maintaining necessary equipment may undermine a lasting commitment to the program.

Organizational instability can also be an obstacle to continuation. A number of teachers and schools reported that they had discontinued their participation during periods of organizational change. In some instances, a person's retirement or departure from the school led to noncontinuation of the Kids Network. In other situations, a district's restructuring seemed to make the program particularly vulnerable. When schools are better equipped with access to computers, modems, and dedicated phone lines—and as teachers' technological competencies continue to improve—programs like the Kids Network may not be as vulnerable to more tangential changes in the institution.

Equity

The Kids Network touches upon a number of equity issues. All the teachers in our sample indicated that the program appeals to both boys and girls as well as to students with a range of academic abilities. Most of the teachers indicated also that the curriculum materials such as the *Kids Handbook* represent different genders, races, and cultures respectfully. Since NGS has not maintained a data-base on the SES levels of the schools that have purchased the Kids Network, it is difficult to assess the extent to which poorer schools have access to innovations like this. Our unsuccessful attempt at finding poorer schools using the Kids Network may indicate that there might be many schools that are not using the program because they do not have opportunities to do so.

Concluding Remarks About Principal Findings

These are some of the principal findings that have emerged from this study. The sections that follow discuss each of these aspects in more detail. Of course, the issues of adoption, implementation, and continuation are not discrete entities, divorced from one another. In the final section, "Concluding Remarks," we present a theoretical framework to describe some of the relationships among the more salient aspects of the program's journey.

◆

Introduction to the Cases

Below is a quick portrait of each of the participating schools, beginning with the context of each site. This includes some of the features of the school—its physical structure, student composition, the surrounding community, and a summary of Kids Network users. Following the context is a description of one or two observed Kids Network activities.

There is an obvious risk here in presenting one or two isolated activities to introduce a case. These cameos were selected both to illustrate the variations among participants and to disclose some general tendencies in program enactment. We provide far more detail to these and other cameos in the section entitled "Analysis of Enactments."

Site 1

Context. Site 1 is a K-6 school of 600 students in urban New Hampshire. Most of the students are Caucasian and come from middle- and working-class families. The two-story school building is located in a residential district, within a couple of miles of the city's downtown, where the main street is wide, the sidewalks are clean, and the shops and restaurants are elegant. From various urban perches, distant wooded hillsides can be seen rising above the sprawling maze of malls and shopping plazas in the foreground.

Surrounding the school is a narrow wooded band, a hilly playground area topped with a handful of tall pine trees, and a large cement playing area where most children congregate during recess for talk and games. Each school day begins with the principal's ritual broadcast of schoolwide announcements and a student-led broadcast of the Pledge of Allegiance. On the walls outside most classrooms are large bulletin boards exhibiting the class's most recent focus. During six days of observations, researchers noted only one student behavior problem. According to one teacher, the school has a lot of support from parents, many of whom are interested in environmental education. All student report cards state, "Students are evaluated and assessed in relation to their own ability."

Three teachers have used the Kids Network three or more times. MD, a fifth/sixth grade teacher, had used Acid Rain four times before using What's in Our Water? for the first time during this study. HA, a fourth grade teacher, has used the Kids Network the longest: Hello! twice, and Weather in Action three times. JM, also a fifth grade teacher, has used Acid Rain two times and What's in Our Water? and Hello! once. Although all three teachers had planned to participate in this study and use the Kids Network during the fall of 1994, only MD's student-teaching intern, KA, was observed enacting the program. The other intended users, HA and JM, had registration problems. Nevertheless, they both participated in extensive interviews.

MD's undergraduate student-teacher, KA, began her required four-week solo teaching period when MD was to start the Kids Network. During this time, KA had sole responsibility for teaching all subjects to MD's students, including implementation of the Kids Network. Even her supervising teacher was not permitted in the classroom while she was teaching. Occasionally, KA conferred with MD between classes and after school about her Kids Network experience.

KA and MD's Classroom. The 24 fifth and sixth graders' seats are arranged in six homogeneous clusters of two to five boys or girls. The classroom is filled with many large, full-color posters and drawings made by students displaying environmental and ecology-related themes: a huge cross-section of a lake, a poster titled "Water Pollution Solution," and an illustration of a food chain. There is also a student exhibit called "How Does the City Interact with Beavers?," displaying a full-sized stuffed beaver and otter as well as miniature clay models.

Suspended in the center of the room is an inflated globe with continental borders. It hovers over a three-foot-wide model of a watershed. Also sharing center stage are student-constructed models of Roman roads, domes, and a coliseum. Hanging from the ceiling are student models of the solar system made out of colored styrofoam balls. There's an NGS wall map with nine research names and addresses. A guinea pig is in a cage with a running wheel near the teacher's desk. There are six microscopes, numerous fish mobiles, three aquariums, and a MacPlus with an Imagewriter. Since MD does not have a classroom modem, all Kids Network telecomputing takes place in the library.

Some Kids Network Activities. During the field visits, KA generally spends at least an hour a day doing What's in Our Water? It typically fits into a daily schedule of textbook-based mathematics drills, silent reading, independent research and writing about inventions, and studying about Rome.

During the second round of field observations, students begin Activity Sheet #4, "The Grass Experiment." Throughout the Kids Network session, KA holds onto the *Teacher's Guide* and reads aloud many paragraphs from two pages. During this session, KA follows exactly activities A-C: (A) review the watershed demonstration, (B) tell what is nitrate, and (C) set up the grass experiment and make some predictions. For the grass experiment, KA explains to the students that they will do an experiment that will help them decide how much fertilizer is too much, and that "We don't want too much of a good thing." KA collects and organizes all the materials (cups, sand, distilled water) on one table. She calls up one group at a time to collect what's needed according to the directions on the activity sheets. Some groups spend at least five minutes collecting their materials. The groups that are waiting write down their answers to the four questions on the back of the activity sheet:

1. Will most of the seeds sprout in Cup 1? (No fertilizer)

2. Will most of the seeds sprout in Cup 2? (A little fertilizer)

3. Will most of the seeds sprout in Cup 3? (A lot of fertilizer)

4. Which cup will have the tallest grass?

Most of the students predict that the seeds in cup 3, which has the most fertilizer, will grow the fastest and the tallest.

While KA is assisting a group of boys at the materials collection table, she seems frustrated by how much difficulty each group experiences while reading the directions on the activity sheet. She looks up from the group and announces to the class, "This is really a test on how well you read directions." Because it is taking each group at least five minutes to collect their materials, many students have to wait either to begin the experiment or for others to finish it.

The next day, KA begins to modify the program. Although she continues to hold the *Teacher's Guide*, she organizes the materials so that each group of students can begin simultaneously with its own set of materials. Furthermore, she does not hand out the activity sheet; rather, she talks students through the directions, step by step. Unlike the onset of the grass experiment, everyone is performing the same tasks at the same time; and KA asks questions about what they are doing, why, and what the different materials in use represent in real life.

A couple of weeks later, students record their observations of their cups of grass. Their results do not agree with either the *Teacher's Guide* or the students' predictions. The *Teacher's Guide* indicates that the cup with the low amount of fertilizer should grow the most; all the students predicted that the cup with the highest amount of fertilizer would grow the most and fastest. Unexpectedly, the majority of the groups observed the most growth to be in the cup with no fertilizer; the least growth, if any, to be in the cup with the most fertilizer; and the second most growth in the cup with the low fertilizer. After compiling these results, KA reads aloud from the *Teacher's Guide* the reasons why the students might not have gotten the "correct" results.

Students then measure the amount of nitrate in each of their cups' runoff. KA reads and leads them through the directions:

> This is what the nitrate strip is going to look like. It's still white, but you can see the difference. Don't put your fingers on it. This is the detecting source. What you're going to do, take this strip and dip it in the runoff. After one second, take the strip out of the runoff and shake it gently to remove any excess water; then you need to wait one minute and examine it. You're going to be in charge of the nitrate charts. I need you to record the number that you get.

Again, the students' runoff results do not agree with either what the *Teacher's Guide* indicates should happen or what the students predict will happen. The runoff from cup 3 is supposed to have the highest nitrate reading. Instead, students discover that cup 1 is the highest. One student then notices that the sand was contaminated with cat scat—a source of nitrates students read about in the *Kids Handbook*. Apparently their regular teacher, MD, collected the sand from his basement, a favorite haunt of his cat.

During an interview, KA reflected about the grass experiment, offering her thoughts about alternative strategies:

> It would be more of a problem that the kids would need to figure out . . . They would come up with a question like, demonstrate how different amounts of fertilizer on grass seed would affect the grass. And then the kids would need to figure out how much stuff to put in, whereas this is very directed . . . The way it was presented to us at [my college] is, there's a problem, and the kids figure out how to solve it and the teacher is just the facilitator.

Site 2

Context. Site 2 is a K-6 magnet school in urban New York. As 1 of 11 magnet schools in the city, its focus is science, mathematics, and technology. Although the school is surrounded by strip malls and subdivisions, there are a few remaining signs of the city's rural roots, such as a recent sighting of a black bear on the school grounds. The one-story school building surrounds a concrete courtyard.

The 600 students are picked by lottery from a pool of names submitted by parents. According to the principal, all of the magnet schools are completely integrated. They achieve this by regulating the kindergarten registration to reflect the population. Last year, the population was 50 percent majority and 50 percent minority (30 percent black and 20 percent Hispanic). The lotteries then had to allow recruitment from the different schools to reflect those percentages.

Most students are bused in from the inner-city area. All but four staff members are white.

The school receives public grants from city, state, and federal sources. Public and private funding has provided two computer labs—one containing Apple IIGS's, the other Macs—and a well-stocked science and mathematics lab, each with its own specialist. Most classrooms have one or two computers. Within the year, all the computers throughout the entire school are expected to be networked. Since outside funds for magnet schools are anticipated to run out at the end of next year, the future role of the science specialist is uncertain.

A total of seven units were used during the 1993-94 school year (three Hello!, one Weather in Action, two Too Much Trash?, and one What Are We Eating?). AT, a third grade teacher who moved up with her second graders, used Hello! for her first time in the fall and the weather unit in the spring. AT's implementation of the Kids Network is the primary focus of the fieldwork at Site 2.

KW, the science specialist and initiator of the Kids Network at this site, facilitated the scheduling of the Kids Network units and the telecommunications activities, since the modem is located in her science lab. In grades K-3, classroom teachers accompany their classes to the science lab for 30-minute periods; the fifth and sixth grade classes meet for 45 minutes. Unlike the science labs, which are facilitated by KW, teachers conduct their own lessons during computer labs.

AT's Classroom. In addition to 25 students sitting at desks in groups of four or five, this third grade classroom is "stuffed." Every space displays or holds something. Some of the more permanent items include a closet for students' coats; a "book nook" with a round table covered with students' Kids Network folders, a small globe, and a map of the United States indicating teammates' locations; two computers displaying a mathematics program; two other computers displaying the Kids Network desktop; teacher-made charts about language and spelling; a container filled with maps; and a bulletin board covered with mathematics charts. In the fall, there is a chart displaying various student-generated definitions of "pets," five photographs of students with their pets, and a chart showing the results of a survey about numbers of students with and without pets. In the spring, the bulletin board is titled Weather in Action. It lists four terms: "temperature," "moisture," "air pressure," and "wind." In a red holder are the words "scientist," "geography," "geographer," "telecommunications," and "data." Also on this board are two cards: "Weather (Climate)" and "Whether (If)." At least nine books about weather are on display around the room.

Some Kids Network Activities. AT begins the Weather in Action unit a week before its official start. On the first day of the unit, AT modifies the guide. She begins by asking students where they might find information about weather (part of activity B), then facilitates a class discussion about the four weather ingredients (part of activity A); she ends the first day of her first session by introducing students to a chart comparing four weather reporting sources according to their reports on one of four weather ingredients (extension activity). These activities are similar to the ones recommended in the *Teacher's Guide* at the onset of the unit. Most of the recommended activities are in the form of discussions about weather, weather reports, major weather events, and procedures for carrying out a survey on memorable weather events.

AT begins the unit by leading a discussion about "Where can we learn about weather?" Within a few minutes the students make nine suggestions. Two of the nine recommendations include firsthand experiences: observing and using thermometers. The rest involve gathering secondhand information by watching television, reading weather reports, and listening to the radio. Each student makes no more than one brief suggestion. The purpose of this activity is simply to direct students' attention toward a variety of places where they can learn about weather and to introduce a homework assignment—bringing in the weather section from the newspaper.

After AT describes the homework, she asks students to read silently a chapter about weather in their science textbook. Students read for about five minutes, sitting at their desks or lying on the floor. Then AT asks such questions as "What did you find?" and "What else is important about weather?" and leads her students in a discussion of the four ingredients of weather. After a student mentions "water vapor," AT draws a bunch of small dots close together and labels them "cold" and then sketches a bunch of dots spread apart and labels them "warm." She reminds the students of their study of airplanes at a nearby air force base.

Teacher: What did we find out about the molecules? What is the word
            for this?

Student: Air molecules.

[AT then holds a piece of paper to her mouth and blows; it bends downward and flutters. A student exclaims, "Lift. Pressure."]

Teacher: So air pressure. [She writes "air pressure" over the drawings of
            molecules.] Fast moving air.

Student: Light over the top [of the wing].

Teacher: Warm air has less pressure than cold air. So, again, pressure.
            What else is important about weather?

Student: Hot.

Teacher: What is that?

Student: Pressure.

Teacher: No.

Student: Air pressure.

Teacher: We talked about pressure. When we are talking about "hot" and
            "cold," we are talking about—

Student:	Temperature.
Teacher:	Louder. [She adds "temperature" to "pressure."] What else causes our weather?
Student:	Water.
Teacher:	What is that?
Student:	Water . . . moisture. [AT writes "moisture" on chart.]
Teacher:	One other thing.
Student:	I know.
Teacher:	Does it change—in a day?
Student:	Wind. I know what controls the wind, I know what controls the wind—when there is a full moon, there is high tide.
Student:	When there's a halo thing around the moon—it makes snow.
Teacher:	Sounds like we have a lot of info right from the beginning of this.

After lunch, AT begins an extension activity in which students will record later, in a prepared chart, information to compare how different weather reporting sources describe the weather. AT has copied her chart from the *Teacher's Guide*; however, she excludes "telephone reports" from the list of sources.

	Temperature	Moisture	Wind	Air Pressure
Local newspaper				
National newspaper				
Radio				
TV				

AT has deviated slightly from the *Teacher's Guide* on at least two grounds. First, students read about weather from their textbook rather than from the two pages in the *Kids Handbook* recommended in the guide. Second, AT skipped three primary activities recommended in the guide for the first session. Even with these modifications, AT has satisfied one of the four objectives for this session: "To consider weather 'ingredients' and weather reports."

Site 3

Context. Site 3 is a K-6 school located in a small suburban college town. It is surrounded by agricultural and forest land and is about 10 miles from the ocean. Most of the 340 students at Site 3 are Caucasian and come from middle- to upper-middle-class families. Just before the first field visit, a bond issue to build a permanent addition on the elementary school failed by 20 votes.

Consequently, at the time of this study, the district was planning on restructuring the next year so that there would be two K-4 schools and one large 5-8. The K-4 schools were expected to increase their enrollments from about 273 to 450 students. One of the teachers using the Kids Network shared her concerns for the future: "The school's leaving, we won't have a school here, so we don't even know where we'll be."

Two out of the three teachers who used the Kids Network during the 1992-93 school year had planned on using it again in 1993-94. Only one of these teachers, KL, was actually using the program when the field visits began in the fall of 1993. KL teaches a combined, self-contained third/fourth grade class. She is a veteran Kids Network user, and on a number of occasions has presented the program with the NGS marketing manager at state and national conferences on science and technology education. According to NGS, KL is the first teacher to use the Acid Rain unit with students as young as second grade. She has used the Hello! unit for four years, Acid Rain for three years, and What's in Our Water? for two years. During this study, she used the Weather in Action unit for her second time. She was assisted by a student intern from a local undergraduate education program.

JC teaches a self-contained third grade class. During the fall of 1992, he used the Acid Rain unit for his first time with his third graders. JC is the school's computer specialist and has created a school-based local area network (LAN). He is in charge of spending $1,100 per year on software for the entire school. He finds the Kids Network annual cost to be "prohibitive" and that the return is little because the "kids have little connection to computers." JC says the Kids Network computer components need greater "accessibility" and more "interactivity," and that the program in general needs a "science expert hotline." He adds that other word processing and desktop publishing software offer more interactivity and can be used more often by more students. Furthermore, since the school was able to recently access the Internet at no cost, JC feels the original appeal of the Kids Network's telecommunications component has diminished. Nevertheless, JC participated in the Acid Rain unit for his second time in the spring of 1994.

SR is also a third grade teacher; she used the Acid Rain unit for the first time during the spring of 1993. This was her only use of the Kids Network. In a short interview, SR said that she "hates science" and in particular was frustrated by the NGS deadlines. She thought that third graders may be too young to appreciate the program's components and understand its concepts. She claimed that telecommunications and "keeping deadlines and getting information back was not interesting" to her students. SR was excited by the program at first, particularly the computer parts; however, she is quick to point out that few of her students took part in that aspect of the program because the modem and computers were not in her room. She also felt that the topic did not tie in with other thematic units she covers and that there were few hands-on activities provided in the *Teacher's Guide*.

KL's Classroom. KL's class of 24 third and fourth grade students includes five special needs children, some with attention deficit disorder. One boy is from New Zealand. Students sit in six coed clusters of four to five desks per cluster. The 16 children who had not done the Kids Network before are partnered with those who had done so the previous year and matched up with a team from another school. Most of the students enter their classroom in the morning with a smile and enthusiasm.

Throughout the entire Weather in Action unit, the classroom atmosphere is thick with weather. Students are surrounded by walls displaying maps, charts, graphs, definitions, survey results, and poetry—all having to do with weather. On the front wall is a six-foot sign with one-foot-high computer-printed letters saying "Weather in Action." Beneath the sign is a large map of the United States with two strings extending from the map to a card indicating the global addresses of two teams. On the east chalkboard is: "Weather is: moisture (humidity), rain, snow, hail, fog or clouds, sleet, wind (air), temperature, air pressure." On the room's south side, a newsprint stand reads:

Total # interviewed

# of weather events remembered

The weather event remembered most

It took place in (year, season)

Major weather ingredients most noticeable

Effects of that event were:

Of all weather events, the most unusual

Name of Class

Global Address

City, State, Country

At the time of the second field visit about a week later, there is an addition: a map showing temperatures in different places in the classroom. During the last round of visits (about two weeks later), the classroom walls hold many student-made graphs (primarily bar and line graphs) comparing the daily temperatures taken by individual students with their class's average. On the wall, written on a large sheet of newsprint, is a summary of students' interviews with parents about how weather affected their work. There is also a student listing of the most unusual uses of weather data. Added to the U.S. map on the front wall is a statement about each team's most memorable weather event. Also posted on newsprint is the class's fourth week of average temperatures.

Throughout the observations, KL selects pairs of students to work on their one classroom computer, an AppleIIGS connected to a modem and an ImageWriter. Their computer activities include plotting their teammates' schools on the NGS computer map according to each team's longitude and latitude. Students also type in sections of their community letter, as well as send and pick up electronic mail. Some students do the Kids Network computer activity before school, during recess, and while KL's class is doing unrelated activities.

Some Kids Network Activities. Each day during circle time—a morning ritual in this third/fourth grade classroom—a different set of six students settles onto one of six cushions designated for that day's readers. The students read poems they selected from the array of poetry books shelved along the blackboard, from their own collection, or from poems they had written. They then explain why they made their particular selection. During the Weather in Action unit, KL asks students to choose weather-related poems.

Students begin the Weather in Action unit by interviewing their parents or grandparents about memorable weather events (Activity Sheet #3). After students have collected this information, KL asks them for suggestions: "My goal is to get as many of you involved in it. How can we get everyone involved in completing this? How can we do it so more of you can learn the information from the surveys and be part of the letter?"

KL frequently pursues this line of questioning with her students. She invites their input into how they should proceed with a Kids Network activity so that everyone can be involved. After asking for suggestions, KL waits silently for responses and follows each student's ideas with questions of clarification or with, "Does anyone else have another idea?" In this example, each group ends up tallying one piece of the data collectively at a time, then exchanges this information as a class and tallies all of the groups' information.

KL often invites her students to relate their personal strategies for solving mathematics problems in their Kids Network activity. For instance, after each of her six in-class groups indicated that they had interviewed four people respectively, KL asks the students to explain the different ways they figured out the total number of people interviewed by the class. Three students share three distinct strategies. One girl explains that she multiplied 4×6. The boy from New Zealand explains that he rounded $4 + 4$ to 10 and stored in his mind that it should really be 2 less. He did this three times and got 30, and then subtracted 6, the total amount he had stored in his mind. Another girl explains that she knew there were 24 students in the class and that everyone had done one interview. After sharing the various strategies, KL says, "There are lots of ways to get it."

Site 4a

Context. The fifth grade students in AN's and PM's classes shared the Acid Rain registration. They were the only Kids Network users in the school of 280 students at Site 4a in suburban New York. Most of the students are Caucasian and come from middle- to upper-class families. Like Site 4b, the school is one of five elementary schools in a town consisting primarily of two- and three-story single-family homes on landscaped lots ranging from one-half acre to three acres.

Three sides of the school are surrounded by playing areas: to the east are swings, see-saws, and cement tunnels; to the south a large paved area on which box ball, basketball, and handball are played off the exterior wall of the two-story gym. Within a minute's walk from the school is a block of stores including a delicatessen, a toystore, a bank, and a liquor store. As at Site 4b in the same district, the school's financial resources free Kids Network users from any economic or technological constraints. According to the principal, there are no conceivable risks of the program being cut from the budget; in fact, in about 1990, the school district adopted both the Kids Network and the Voyage of the Mimi for fifth grade.

Says AN, "There isn't a child in this room who doesn't want to be in school and learn. The purpose for living in [this town] is to have kids going to good schools." Most of AN's students report that they have access to computers at home, and, says AN, "Over half my children have modems in their homes; they know what telecommunications is about . . . I've never had children as savvy as that before, so I think that speaks more to the technology in 1993 as well as the level of the community."

AN teaches all subjects to her self-contained fifth grade class. After doing the program for four years, AN is sharing one registration with another fifth grade teacher, PM, a second-time user of the Kids Network. They joined their classes to form 11 teams of four to five students per team. All teams comprise students from both classes, and most are coed.

AN's Classroom. AN's class of 26 consists of mostly white students, with four Asian boys, four Asian girls, two black girls, one Brazilian boy, and one Indian boy. AN's classroom is located in one of the "tentacles" of a one-story brick building shaped somewhat like an octopus. The octagonal classroom overlooks a narrow paved path bordered by a tunnel of woods. Several *New York Times* weather maps are suspended from a wire extending between two walls in AN's classroom. Students record on a teacher-made chart the daily high and low temperatures, cloud coverage and type, and precipitation in a city of their choosing. Each student records these data daily on his/her own chart for the respective locations, which extend around the globe. By the second field visit, the acid rain questions that the students made on cards during the first visit are posted on the classroom wall.

AN has a telephone in her classroom from which she can receive and make calls. During the first visit, it rings four times between 8:00 a.m. and noon. Immediately outside AN's and PM's classrooms, in a large common hallway shared with two other classrooms, are 16 LC II computers. They are networked to one Imagewriter and one personal LaserWriter. The one Apple IIGS that is hooked up to a modem is used for all the telecommunications, mapping, and graphing aspects of the program.

Some Kids Network Activities. The second round of field visits began the day after Halloween. A typical snapshot of AN's classroom while doing the Kids Network reveals students involved simultaneously in a variety of activities: Four girls are trying to telecommunicate on the IIGS right outside their classroom; eight students are lined up, waiting to transfer the newspaper's weather data for various cities to a data sheet made by AN; and, on the floor outside the classroom, three boys are creating a huge wall chart comparing the pH of various foodstuffs. Other students are typing and printing in huge letters the name of one of the many foodstuffs that is to be included in the pH chart. After recess, AN's and PM's classes meet in the hallway to look at and discuss the chart. A girl explains what she has learned from measuring the pH of the various liquids: "I'm not sure which has acid, but I think that some of the acids are good. Some of the foods are good, because you add acid to your body and it helps your body and it matches with your body acids to break up foods so your body can digest."

A typical snapshot of AN's classroom on the next day again reveals students involved simultaneously in a variety of activities; this time, however, they're participating in a variety of Kids Network and non-Kids Network activities. A handful of students are working alone on one of five mathematics worksheets; a few students are reading a social studies textbook; a pair of students are talking about how to format, on computer, a letter asking friends to test the pH of their rainwater; and four students are reading aloud an electronic letter they just received from one of the Kids Network research teams.

Meanwhile, PM's students discuss as a class what they think should be included in the letters they are about to write to their friends asking them to measure the pH of their local rainwater. This is an activity created by AN. They then spend part of the morning writing their letters on computer. Some students work alone; others collaborate. Three girls from PM's class, each on her own computer, write about two sentences in approximately 15 minutes. They talk about and assist one another with their letters' appearance. They try to insert a picture into their letters from the word processing program's large picture file.

Another student reads a letter he is writing to a relative in Sweden:

> Dear Louise, how are you? At school for a few weeks we've been learning about acid rain. So far I've learned a lot about it. Some things I never knew about acid rain, I know now. We have been testing a lot of liquids with pH paper and pH scales. Have you been doing anything about acid rain in Sweden? It's easier for you because in Sweden it rains a lot. I know you're wondering what is in the little packet I gave you. That is pH paper, the strips of yellowish paper. The other paper is called the pH scale. The pH paper you use to test rain or other liquids. If you would like, you can make a rain collector by cups, besides glass, and dip the pH paper in the collector for as long as 10 seconds, and then face it to the scale I gave you and see what color it reflects on the scale. Make sure the pH paper does not touch the scale and then record it on sheet.

Site 4b

Context. SE is a fifth grade science teacher at a K-5 school in suburban New York. Most of the school's 271 students are Caucasian, from middle- to upper-class families. She is the only Kids Network user at her school. Huge playing fields, each large enough for at least three soccer fields, surround the front and back of the one- to three-story building. The front area also includes two cement-surfaced basketball courts and a brick wall (the gymnasium's exterior wall) on which a variety of ball games are played. The front field is bordered by a well-traveled two-lane road and is within an earshot of a four-lane highway. The back of the school, in addition to a soccer field with two goals, has a well-maintained baseball field with a 20-foot fence behind home plate and a paved area containing a large swing set, slides, and see-saws. Like Site 4a, the school is one of five elementary schools in a town consisting primarily of two- and three-story single-family homes on landscaped lots ranging from one-half acre to three acres.

As with her colleague AN, who teaches fifth graders at Site 4a in the same district, SE is free from any economic or technological concerns for implementing the Kids Network. The school has the financial resources to maximize the technological components of the Kids Network and to invest in any unit. Most of the students have access to computers at home.

SE has used the Acid Rain unit for three years. Although she is the only teacher who is doing the Kids Network in her school, she communicates with AN at Site 4a, as well as with SF, the former science support person and now principal at another school in the same district. In addition to teaching other subjects to her homeroom students, SE teaches three fifth grade science classes a day.

SE's Classroom. Daily, SE writes a new quote of the day on the blackboard. During the second field visit she quotes Doris Mortimar: "On self-esteem: Until you make peace with who you are, you'll never be content with what you have." On another day an anonymous quote reads: "A politician is a person who can sit on the fence and yet keep both ears to the ground." Above the blackboard are more permanent quotes such as "A good angle to approach any problem is a try-angle."

To the right of the blackboard is a poster on the "Scientific Method" that reads:

Hypothesis—"If . . . then"

Materials—timer, measuring cup, container, etc.

Procedure—step 1, step 2, step 3, etc.

Observation—data tables, graphs, diagrams

On the back wall is a "For the Love of Literature Poster" highlighting a variety of genres such as adventure fantasy, nonfiction, mythology, and short stories.

Students' desks are positioned in two identical arrangements consisting of two rows of six desks facing each other in which boys and girls alternate side by side.

The year before participating in this study, SE registered her three science classes as three different research teams for a simultaneous Acid Rain session. Consequently, SE coordinated her students with approximately 60 teams. By the time she was done, she felt so overwhelmed that she decided that this year all three classes should share the same registration. To ensure that each class has the same opportunities, SE explained, "We'll keep erasing the . . . data so that the kids can have the experience of doing their own mapping. I won't save it, but each kid can print it out and when we're finally ready to save it, we'll send it off."

Although the school has a two-year-old computer lab with at least a half-dozen Macintosh computers, she and her students do all of the Kids Network computer components with the one IIGS and modem located in her classroom.

Some Kids Network Activities. During the first day of the second field visit, SE executes a whole-class computer demonstration around one computer to plot the longitude and latitude of their research teammates. This is her third Kids Network class for the day. This is not the first or the last time SE facilitates a whole-class Kids Network computer demonstration.

SE's approach to doing whole-class computer demonstrations is to provide step-by-step verbal instructions to one or two students who are controlling the mouse and keyboard while everyone else watches. Getting everyone into a comfortable viewing position takes time and is directed by SE.

Teacher: Who can't see now, that I'm in your way, all right just calm down. I need my chair over there, Michael, just wait. Right there, don't get on top of it, Danielle. Brian, can you see Julie from here?

Student: Not really.

Teacher: Switch these seats. Danielle, stop moving. All right. How can't you see, you're right in front. Okay. All right, let's try it again.

Once students have settled into their new positions and SE has selected two students to operate the controls, she tells them how to proceed.

Show me how you got 40. Go over and get the house and we'll find 40. Go to the icons over here and get a house. Excellent. Now let's go back and make the 40, the first numbers. You're going in the wrong direction, you've got to go up and down to get 40. Good. Now, we're all the way over in the middle of the Pacific Ocean. Now, you have to move your, click, west to get to 80. Move it over, go, go, go . . . Over to your east, good, good, good. Don't worry about losing your 40 a little bit, you'll get it back later. You're looking for that 80, keep going, moving across the United States, moving along, 80, 80, 80 . . . go, go, a little more, stop. Now you've got to go which way to get to 40?

After the plotting demonstration, SE selects Maggie and Jenny to telecommunicate, while other students watch to see if they receive any network mail. Maggie smiles when she hears the modem dialing and then exclaims with enthusiasm, "We're connected!"

By the end of the demonstration, SE prints out a global address map with the two schools they have just marked. And at the end of class, when she asks who wants to telecommunicate, all raise their hands.

Site 5

Context. Site 5 is a K-6 school in a small, crafts-filled, historic college town in rural Vermont. Most of the 600 students are Caucasian and come from middle- to upper-class families. Part of the school's grounds consists of the largest and most elaborately constructed playground the researcher had ever seen: a complex of swings, forts, rooftop lookouts, and suspension bridges, all surrounded by long-range views of the Green Mountains and the Adirondacks.

Says the acting principal, "The school board here is tremendously supportive of these kinds of activities [e.g., the Kids Network] . . . Even in difficult economic times, we find ourselves able to preserve our programs, able to preserve, for the most part, our purchases."

Most of the teachers have at least one Macintosh or Apple computer in their classroom networked on a school-based LAN installed and coordinated by a full-time technology specialist (GM) and a full-time technology assistant. In addition to keeping a pulse on the development of educational electronic communications technologies, GM maintains and coordinates schoolwide use of the library's computer lab of 25 Apple IIGS's, advises teachers on the purchase of the school's approximately $4,000 annual investment in computer software, troubleshoots most software-related Kids Network problems, and orders new Kids Network units.

One teacher said the school is currently involved in a restructuring process. Recently, for example, teachers eliminated some extracurricular activities, such as chorus, so that they could gain an extra hour and a half per week of in-class academic contact with their students.

Four of the seven teachers who currently use the Kids Network have used it for three or more years. SL (fifth/sixth grade), the first and longest user of the Kids Network, has been teaching for 22 years and recently received a prestigious science teaching award. She shared the registration for the Acid Rain unit with HM (fifth/sixth grade), a five-year veteran teacher who had already done Acid Rain, Weather in Action, and Too Much Trash?. Although they collaborate regularly in their execution of Acid Rain, SL and HM carry out separately most of the Kids Network activities. The researcher observed both SL and HM as well as KH (fourth grade), who used What Are We Eating? A fifth/sixth grade teacher and a fourth grade teacher used the Weather in Action unit in the spring after the field visits. A new user of the Kids Network, teaching a combined fifth/sixth grade class, participated in the Hello! unit during the time of the field visits but was not observed or interviewed.

SL's Classroom. SL's 21 students sit around five tables of three to five boys and girls per table. Two students sit at their own desks. In the room are a microwave, an electric countertop oven, a small compost bin, and a water cycle chart. The only map indicating the research teammates' global address is posted in HM's room. The classroom also contains one Mac LC with modem, two Mac SE's, one MacPlus, and one Imagewriter. On the front wall is a published colored chart titled "Acid Rain: The Effect on Aquatic Species." Also displayed on the front and back walls are two student posters exhibiting the students' understanding of seasons, density, revolution, convection currents, pressure, condensation, gravity, and evaporation. These topics were studied in SL's and HM's classes just before the onset of the Kids Network unit.

The Kids Network fits within SL's daily class schedule of spelling, reading, and mathematics, and computer lab twice a week. Since SL and HM do not distinguish among science, social studies, or geography, "all of the extra time when we're not doing the other three things [spelling, reading, and mathematics] is Acid Rain time." SL and HM exchange some students between their two classes for mathematics and spelling instruction, according to abilities. Students remain in their respective classes for most of their instruction, regardless of their academic or emotional profile, and were provided special assistance within the regular class.

Some Kids Network Activities. From the onset of the unit, SL and HM deviate from the *Teacher's Guide* because of anticipated teacher absences, school vacations, and a delay in receiving the already overdue Macintosh version of Acid Rain. Without these adaptations, they are concerned that they would not meet the telecommunications deadlines. Consequently, they skip five activities recommended for Week 1, Session 1 (e.g., reading pages 2-5 in the *Kids Handbook*, reading the unit scientist letter, discussing the Acid Rain unit, assigning groups to research teams, and identifying teammates' global addresses).

During the unit, SL and HM arrange for three acid rain-related guest speakers: a college chemistry professor who presents the chemistry of acid rain, a speaker from the local college who explains the college's energy requirements, and a student's parent from the National Oceanic and Atmospheric Administration who talks about prevailing winds. In addition, six students spend a day with a Vermont acid rain specialist who is in charge of collecting and analyzing acid rain data for the state.

During the first field visit, SL hands out Activity Sheet #2, "Acids in Foods and Drinks." She then selects one student to read to the class the homework directions for recording the type of acids found in various products at home. While SL distributes one NGS pH chart to each student, one boy says with all earnestness, "Oh, we're going to test it now—oh, goody, goody." At another table, two boys and two girls begin to talk about the pH chart they have just received. Their conversation, not heard by the teacher, goes something like this:

John:	I wonder how often you can use this.
Corine:	Once . . . [pause] Is this the pH paper or chart?
Elizabeth:	I don't know. I have no idea.
Corine:	It's a chart.
Sebastian:	Air has more acid than water.
John:	Water doesn't have acid.

SL then instructs her students to choose one person from each table to pick up a cup of water, and then to measure its pH before exchanging it for a sample of diluted vinegar. Students dip their pH paper for two seconds in the water and then compare the color on their pH strip to the corresponding colors on their pH chart. In between testing the two samples, SL writes the pH scale on the board and says, "The numbers are like exponents in that each number is 10 times more acidic than the previous higher number." With the testing of their water sample complete, SL explains that each table must first agree on one result before sharing it with the class. SL records the students' results on a chart comparing the pH of the samples taken by each group. All the results except one are identical. One student from each table then exchanges a water sample for a mixture of diluted vinegar. There is no discussion about the one group with deviant results, nor is the data analyzed after it is entered in the chart on the board.

Site 6

Context. TA and CM team-teach 43 fifth graders at Site 6, a K-5 school in a middle- to upper-class ski town in rural Vermont. Most of the students are Caucasian. One of the teachers described the community's socioeconomic background as being a "wide variety." She explained that "some kids come from [the] farming community, and some groups come from highly educated backgrounds and professionals." The houses within a few miles of the school vary from multifamily vacation chalets to 100-year-old classical New England farmhouses—some well-maintained, others showing their age—to more contemporary and modest single-story, single-family homes. During the winter, many of the homes seen on the way to the school had wood piles and a smoking chimney.

The school is surrounded by rolling hills of farmland and the Green Mountains. On one side, a mowed lawn meets a rocky hillside that is bordered by a sprinkling of young pine trees and a forest beyond. On another side is a very large playground area of slides, forts, and lookouts made out of pressure-treated timbers and used tires. On another side is a broad swath of pavement on which the fifth graders play kickball during recess. And on another side, bordering one of two main roads leading to the center of a small village a few miles away, is a soccer field. At the time of this study, some parents and the principal were trying to raise money for equipment to light up the soccer field for evening games. The center of the town boasts a variety of fine restaurants, Vermont craft shops, and an antique store. The town is considered a tourist attraction. On the fringes of town are a large grocery chain, a few gas stations, motels, and ski and outdoor sporting shops. Most of the secondary roads outside the center of town are made of dirt rather than asphalt. Within the town's borders is a popular ski resort.

Near the main entrance and at about the center of the one-story brick school building is the library and resource room. Except for moveable dividers on one side and low bookshelves on the other, it is flanked by hallways connecting the different wings of the school. There, self-playing filmstrips on various topics are exhibited. It also has one computer, an LC 520. Next to the library is the teachers' room, which contains one Apple IIGS with a modem, one Macintosh Classic, and an Imagewriter printer. It is here teachers using the Kids Network do their telecommunicating.

The principal describes his school as having "never been a textbook school . . . It's always been a process-oriented school . . . You don't see workbooks here; we just don't buy them." For instance, in addition to a few of the higher grades doing the Kids Network, some of the first grade and second grade teachers are participating in a telecommunications pilot project in Vermont during which their students, according to the principal, "are growing plants and collecting data and floating that back and forth between schools" on Scholastic's electronic bulletin board.

During lunch in a multipurpose room—which is also used as a gymnasium and auditorium—students sit with their respective classes and are served their meals by a class member on a rotating basis. Outside the lunchroom, the common areas seem relatively quiet, void of any bustling activity. An adult stranger can walk into this school and be greeted by students and teachers with many hellos and smiles.

The technology specialist, who assists teachers with this type of curriculum, also works at the middle and high schools. He is expected to shift from a half-time position to full time within the next two years.

TA and CM are the only fifth grade teachers. This is their first year team-teaching all subjects as well as sharing one Kids Network registration. The previous year was the first time TA had used the Kids Network; she participated in both Hello! and part of What's in Our Water?. When TA went on maternity leave, CM took her place and completed the What's in Our Water? unit; before then, CM had used Acid Rain once.

TA and CM's Classroom. Dividing TA's and CM's two rooms is a collapsible wall that remains open at all times. In the room are an egg incubator, one Macintosh LC 575 displaying comparison maps from a MacUSA program, and four Apple IIe's with color monitors. Other activity-specific materials are displayed or stored in this area as well. The persistent ambient murmuring from one or both of the classes interferred often with the researcher's ability to hear what students were saying.

In both classrooms, students sit around six circular tables, each ringed by six removable drawers. One black girl and boy are in CM's class; otherwise, all the students are white. During the first field visit, a physically and mentally disabled boy lies on a bean bag, his hands held by two boys, one on either side of him. At the end of morning circle during the first field visit, CM and the boy's aide lift him to his wheelchair and leave the room.

Students move back and forth between the two rooms depending on the focus of study. Sometimes they sit in groups according to the teacher's assessment of students' abilities, such as in spelling; at other times—for example, for literature—groups are selected by the students. According to TA and CM, their combined open classroom provides for a "real sense of community."

CM's and TA's classroom walls display more posters than any other classroom in this study. Most are large enough to be read across their respective classrooms. For example, in CM's classroom, to the left of the blackboard, is a laminated jobs list with each student's name written on a clothespin that is moved weekly when classroom jobs are changed. Above the blackboard is a large pull-down map display and posters titled "Proofreaders' Marks," "Before You Turn It In," "Principles of Our Class," and "Problem Solving Strategies." To the left of the blackboard are 10 posters, six up high and four below. The titles to all of the posters are made of very large cutouts of letters which can be read easily from the opposite wall. The top row of posters includes "10 Ways to Get Unstuck," "Top 10 Reasons for Getting Stuck in the First Place," "Quotation Marks," "Titles," "SODAS," and "Cooperative Groups."

On the wall map, under student construction, are five index cards indicating the global addresses of four of their Kids Network teammates and their own. Their teams are from Texas, California, Michigan, and New York. On the opposite wall are bookshelves, a sink, and about 30 continuous feet of windows with insulated window quilts.

At no time does the researcher observe students' misbehaving. Students' interactions with their teachers and one another seem to be relatively free of any explicit conflict.

Some Kids Network Activities. During the Too Much Trash? unit, students went on a field trip to a districtwide landfill and to their local trash distribution center. On the second day of the first field visit, TA works with her class in the computer lab while they plot the location of their teammates' global addresses. At the same time, CM meets with her students to figure out the trash profile of a typical student. Both of these activities are presented in the *Teacher's Guide*. The teachers had hoped to switch their groups after about 30 minutes; however, they modify this plan once they both realize that their respective activities need more time.

In the computer lab, TA explains to students how they will mark their teammates' global addresses on the computer. She holds up a sheet of paper indicating four different locations of their teammates according to their longitudes and latitudes, to the nearest minute. While demonstrating on a computer in front of her entire class, TA explains what she is doing:

> As you move around on the computer the cross bar, wherever it is, tells you what the location is at that particular point, so as you move it, you'll be able to try to pinpoint in as close as possible to where somebody has given their direction to be. Then you'll see it's going to be on a tiny little map. Here's the cross bar and as you move it around, see your latitude and longitude changing? So you want to find these locations, when you find a location you can go to the schoolhouse and get a school and just click it on a spot where you think one of our teammates is located. If you click it in the wrong spot, click on it again, the same schoolhouse, and it will just disappear.

She continues to explain and demonstrate the other mapping tools: a magnifier, a minifier, a hand that shifts the entire image, a boundary marker. When she is done, a student exclaims, "Awesome!" She completes her directions by restating their task and then extending it so that when they are done with the computer part, " . . . then you need to go to a larger map or an atlas to get a clue on what could be the name of the town where that team is located."

After students begin working in pairs on one of the computers, TA circulates around the room to help students troubleshoot problems. A few students ask each other, "Which is latitude?" All of the students ignore the minutes. Once they find the point with the cursor, many are uncertain as to what to do next. One pair of boys say, "I don't know what we're supposed to do."

After students finally locate with the cross-hair tool the location of their teammates, they have to return to the tool bar displaying various icons to pick up the schoolhouse marker and then go through the process again, trying to relocate the point they had just established so they can imprint the schoolhouse in the right location. To an observer, the process appears fastidious. Moreover, even after students have plotted a few schoolhouses, their remarks suggest that they have not fully grasped the relationship between the movement of the mouse and the concepts of latitude and longitude.

Settings and Cameo Wrap-Up

The Kids Network teachers participating in our study work primarily in middle- to upper-middle-class communities. All participants have access to the technological resources necessary to enact much more than the minimum technological requirements of the Kids Network. All the participating teachers have supportive principals and access to other technology expertise. The cameos illustrated above highlight a variety of Kids Network activities in action, all of which are offered in the *Teacher's Guide*. These include algorithmic experiments such as the grass experiment in What's in Our Water? and the pH tests of three water samples in Acid Rain. The cameos also include a reading and discussion activity in the weather unit, a variety of simultaneous Acid Rain and unrelated activities, a whole-class computer demonstration, and a computer mapping activity common to all units. A more complete analysis of the Kids Network activities observed within and across sites appears in the section on "Analysis of Enactments."

◆

The Adoption Process

Typically, innovations are brought to a new setting as the result of tours or visits on the part of local administrators (e.g., curriculum specialists) or by interested teachers. They return enthused and become champions of the new program, urging others to try it out or become its first users (the actual innovators). Table 2 shows the chain of adoption events, with the players and their actions, for each Kids Network site. Following this is a more complete description of how the innovation arrived at each site and then a cross-case analysis of the sequences of key actions and events leading to the adoption of the Kids Network.

The Adoption of the Kids Network at Local Sites

Adoption at Site 1. HA (third grade teacher) was not certain how she first learned about the Kids Network—possibly through an article in a teaching magazine or a flyer. Nevertheless, it was her principal who formally introduced it to her and offered his pedagogical and financial support if she was interested. After being the first Kids Network user, HA teamed up with a colleague for a couple of years. Although overwhelmed by her first time using the Kids Network, HA "loved it" and was "energized by it."

The principal also played a crucial role in introducing JM (fifth/sixth grade teacher) to the Kids Network and in providing the financial support—which did not draw on JM's annual classroom budget of $500. JM first did Hello! and then Acid Rain with fifth graders, and then What's in Our Water? with sixth graders.

MD was introduced to the program the day he interviewed for a teaching position at the school. When the principal took MD to visit HA's classroom, she was having problems completing a Kids Network telecommunication. That's when MD played with the software program for the first time. Soon after MD was hired as a combined fifth/sixth grade teacher, the principal asked him to examine the Kids Network materials and see if he'd like to use the program. Taking his principal's suggestion, MD first attended a Kids Network presentation at a New Hampshire Science Teaching Convention and then ordered the program.

KA's involvement with the Kids Network was different; it was a requirement of her teaching internship. At first, the program "didn't feel like it was mine, and I just kind of felt like I just had to do it," she said. In the end, KA described her experience with the program as the best part of her internship.

Adoption at Site 2. KW, the science specialist, brought the Kids Network into the school in spring 1989. She worked with the program initially with two sixth grade classes. Teachers would bring their students to the science lab, where KW was in charge of doing a Kids Network unit. During the principal's

Table 2.   **Sequence of Key Actions and Events Leading to Kids Network Adoption by Site**

Site	Sequence of Events
1	Principal —▶ Interested Teacher —▶ Collaborating Teacher —▶ Principal —▶ Other Teachers Principal introduces to teacher he thinks would be interested. Teacher first uses alone and then collaborates with new user in year 2. Principal provides funds and support for other teachers. Teachers informally discuss and are enlisted.
2	Science Specialist —▶ Teachers —▶ Principal & Science Specialist —▶ Teachers Science specialist offers to enact program in science lab for third grade teachers. Two years later, principal and specialist decide to shift the program's locus of control to the teachers. Program is then enacted in teachers' classrooms.
3	Teacher's Friend—▶ Teacher —▶ Parents/Principal —▶ Teacher —▶ Other Teachers Teacher's friend informs teacher. Teacher obtains donation from parents and wins principal's support. Teacher introduces program to two other third grade teachers; one discontinues after initial use.
4	District Science Specialist —▶ Distrct Science Committee —▶ Fifth Grade Teachers —▶ District Science Committee —▶ All Fifth Grade Teachers Districtwide science specialist introduces program to district science committee. Committee recommends to fifth grade teachers to try it. Teachers approve of project; committee recommends fifth grade districtwide adoption.
5	Technology Specialist —▶ Special Teacher —▶ Principal —▶ All Sixth Grade Teachers —▶ Other Teachers Technology specialist introduces to a "willing to try anything" teacher. Her stamp of approval motivates principal to provide funds for all interested sixth grade teachers. Gradual spread to lower grades.
6	Principal —▶ Teachers —▶ Other Teachers Principal introduces program to teachers who adopt first independently, then one pair of teachers collaborates.

first year at the school, which began the semester after the arrival of the Kids Network, she gave up her private phone line so that the science lab could have a dedicated line for the Kids Network. During that same year, the principal and science specialist talked about centering the program in the classroom. Over a four-year period, the program spread from the intermediate grades to younger grades—almost every second to fifth grade teacher, and possibly all, will be using one or more of the Kids Network units next year. This past year, KW coordinated its use among 7 to 10 teachers of different grades at any one time and continued to facilitate some of the "messier" activities in her science lab.

According to the principal, GC, the freedom to address the state syllabus "in the way that a particular school or a particular classroom is comfortable with" allows teachers to adapt the Kids Network to fit their particular situations. The principal hopes that every teacher in grades 3-5 will eventually choose a unit. New teachers are introduced to the Kids Network during a required in-service workshop "on using a videocamera, laser disc player, computers, CD-ROMs, all of the technology that was available to us at the time," says GC. Furthermore, the principal evaluates teachers informally and formally for their ability to integrate their science lab experiences into their classrooms and to use an interdisciplinary approach.

Adoption at Site 3. Before KL had heard of the Kids Network, she participated in a statewide initiative that put Apple IIe's into teachers' homes for three consecutive years before they had to return them to their respective classrooms. She remembers attending a workshop for those participating in the home computer program. She recalls, "I went through the workshop that was mandatory for five days, got the computer to my home, and I didn't know how to turn it on." This is when she began to increase her level of comfort and expertise with computers. This laid the groundwork for the Kids Network when she first heard about it from a fellow church member, the uncle of the Kids Network project manager at NGS.

> He dropped a little flyer in my mail box and said, maybe I could start doing Kids Network with his favorite niece . . . We had to get the money to buy the software. A parent volunteered to buy that for me. Then the same parent said she'd buy the modem. Then we had to get the dedicated phone line in—that comes during the AT&T strike. So, my principal went through the files, and one of our parents who works for the company [AT&T] said he would come in and run a line for the after-school program to another place in the school building. It was exciting, and with the principal's support, we managed to get it into the school. We had about four different teachers using it.

The parent who donated two Kids Network software programs and a modem to the school indicated how important it was that the request for help come from KL: "It wasn't just any teacher anywhere speaking to me saying I need this. It was someone whose opinion I had great respect for and whose preparation for class is enormous."

A block grant enabled the district to put dedicated phone lines in all of its schools. "However," said KL, "we [initially] used the after-school program's phone line." Ultimately, a dedicated line made it possible to use a modem without interfering with the school's other telephone needs.

Adoption at Sites 4a and 4b. The story by which the Kids Network was adopted is similar for both Sites 4a and 4b, part of the same district. SF, a districtwide science specialist, was one of the original advocates of the program, as was the assistant superintendent for instruction—who, said SF, was "very enthusiastic about it and played an important role in terms of money." SF was unsure how he was introduced to the program; he thinks it was either at a teacher's convention or through the district's computer coordinator.

According to PS, the principal of Site 4a, there is a district committee for every curriculum area. The science committee consists of representatives from the elementary schools and is chaired by the assistant superintendent, who is in charge of curriculum. This committee learned about the Kids Network and then recommended that it be tried on a voluntary basis. Said SF, "Everybody almost wanted to, which made the next step easier because we'd already purchased modems, and we had made the initial outlay for the package . . . And then the next year we said the same thing, do you want to do it another year? This was an option." Since weather was studied in fifth grade districtwide, the weather unit was particularly appealing because it could be substituted for something that already existed rather than becoming an add-on. After the committee approved the Kids Network, it was adopted. However, said PS, it was "the teacher within the school that has really interested other staff members." Initially, the district tried to save money by having three classes in the same school share one Kids Network registration. This strategy created some problems and was soon abolished in favor of one registration per class.

PS imagines that within a couple of years, some of the third grade teachers might be using the program, and that AN and PM might be participating in other units besides Acid Rain. PS believes that the Kids Network is "here to stay."

Adoption at Site 5. GM, the technology specialist, first heard about the Kids Network about five years ago during a computing conference at Lesley College. While listening to a presentation about the Hello! unit by some of its developers and publishers, he recalls being intrigued by students' definitions of pets. "One of the first things that they had to discover was . . . how do you define a pet? I thought that was real interesting . . . and a neat way for kids to be able to communicate that information with other groups of students that were foreign to them."

When GM bought a unit for the school, he introduced it first to SL, the one teacher he thought was "willing to try anything." As soon as SL indicated that it was a valuable program, "I went to the principal and gave him updates . . . He then asked me to make sure that all of the sixth grades did it, or at least had availability of it . . . So, he quadrupled the budget for that particular line on it and said, 'Make sure they all have it.'"

According to the recently appointed acting principal, "The decision to adopt the program—and the program was not mandated—was really made by teachers. They decided that they wanted to buy into this program. They certainly had my full support to do that." However, according to HM, her initial participation was required.

Adoption at Site 6. The principal, DF, was introduced to the Kids Network at a technology fair at Lesley College, which he had attended a few years back with some of his teachers. The developer, TERC, was demonstrating the Kids Network and seeking field test sites. According to the principal, he and some teachers "pushed to become a pilot school . . . National Geographic provided the software, a IIGS, a bunch of things like that." A third grade teacher who attended the technology fair with the principal piloted Hello!. According to the principal, "We were supposed to use it at a higher grade level, but [the third grade teacher] was here at the time and was an outstanding teacher." In addition, a fifth or sixth grade teacher piloted Acid Rain.

When What's in Our Water? became available, "the sixth grade teacher played with that," said DF:

> Vermont schools were just getting into water quality, lead in the water, and so on. So the timing of that was actually pretty damn nice . . . the kids used to work with the custodian taking water samples and checking out all those kinds of issues, so that did fit right into that thicker unit at the time . . . Acid Rain, when that first came out, we actually had some kids do some research . . . did some sampling on Mount Snow and a look in the area. So that kind of fit in.

According to DF, who has been the building principal for 18 years, there have never been any funding problems or other obstacles to adoption or continuation of the Kids Network.

Cross-Case Analysis of the Arrival Process. The principals at Sites 1 and 6 played vital roles in introducing the program to individual teachers and then providing the pedagogical and financial support for other interested staff. Similarly, the technology specialist at Site 5 introduced the program to a teacher he was certain would test it out. Sites 5 and 6 were also two of the developer's pilot sites. At Site 2, the science specialist introduced the program to her colleagues by enacting the program in her lab and then, with help from the principal, transferred primary responsibility for the program into the teachers' actual classrooms. At Site 3, the program might never have been implemented with third graders without a donation and support from parents and an assertive teacher. And the science specialist at Site 4 played a crucial role in helping the Kids Network win full districtwide teacher support. In all the cases except one (a teacher at Site 5), teachers were able to choose whether to do the Kids Network; none felt they were using the program against their will.

Our Kids Network sample demonstrates a variety of sequences of players and events leading to adoption. However, they all seem to converge on a number of points (figure 2).

Adoption, however, is a complex phenomenon. It is not simply a matter of principals and specialists introducing innovations to receptive teachers. Rather, all the players, when faced with the decision to invest in the Kids Network, had to contend with a variety of ambivalent feelings about it. Our participants described

Figure 2.   **Common Factors Leading to Adoption of Kids Network**

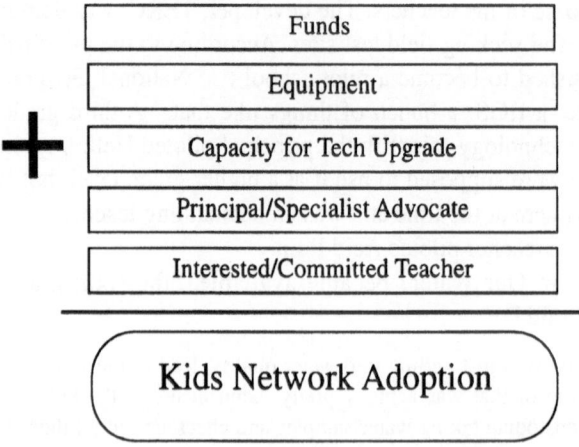

their ambivalence, both anticipated before and realized soon after their first experience with the program.

Incentives and Concerns

Many forces—personal, curricular, and institutional—influence teachers, specialists, and administrators when it comes to adopting an innovation. While most of our participants, individually and collectively, had a variety of concerns about adopting the Kids Network, only one of the participants discontinued the program as a result of unresolvable concerns. On the contrary, for many of our participants, uncertainties were intertwined with their incentives for adoption (table 3).

For instance, a couple of teachers who found the relevance of the Kids Network topics to be appealing were also concerned that the science activities were unsophisticated and not "authentic." Some teachers who found the interdisciplinary components attractive were concerned about whether the program's

Table 3.   **Tensions Between Incentives to and Uncertainties About Adopting Kids Network**

Incentives	Uncertainties
Relevant, real-world focus ————————▶	"Inauthentic" science
Interdisciplinary ————————▶	Inadequate conditions [time]
Student engagement ————————▶	Insufficient computing activity
Hands-on activities ————————▶	"Cookbookish" activities
New technology ————————▶	Not technology savvy

time constraints would allow for interdisciplinary activities. Many teachers who said they felt intimidated by the program's telecommunications component also expressed their appreciation for an innovation that introduced them to new technologies and advanced their own knowledge and experience.

Before turning to an analysis of the tendencies and spread of particular subcategories of incentives and uncertainties, we offer a portrait of three teachers at Site 1 to provide a more coherent discussion of the adoption incentives and concerns at an individual site.

Each teacher at Site 1 was introduced to the Kids Network by the principal, who invited them, at "his" expense, to participate. In general, the adopters were very enthusiastic and receptive to what they perceived initially to be the advantages of the program. Of primary appeal was the idea of a thematic-based, interdisciplinary science program using a new technology that would extend student experiences beyond their own classrooms. In addition, the units fit well with the district's science requirements and teachers' personal interests.

Advocates also expressed some uncertainties about the program following their initial experiences. For instance, one of the teachers thought his students needed more direct interactions with professional scientists using sophisticated scientific equipment. Another teacher, at first enthusiastic about the program, later felt ambivalent about it because she thought one unit needed more hands-on activities and she had difficulty finding her niche with the units. What follows is a more in-depth description highlighting some of these teachers' initial motivations and uncertainties about adopting the Kids Network.

Two Enthusiastic Advocates at Site 1. HA has been both an initiator and advocate of the Kids Network. Other than feeling intimidated initially with the program's telecommunications components, she only had good things to say after using two units a total of five times. After doing the Hello! unit alone, she teamed up to do Hello! again with another fourth grade teacher. Since then, she has been doing the weather unit alone and helping her colleagues with their implementation as problems or questions arise.

When her principal first introduced her to the program, HA was attracted to the professional NGS presentation of materials and intrigued by its novel telecommunications component. Seeking out new material to sustain her personal interests as well as her students', she found the interdisciplinary aspects of the program affirming of her practice and the telecommunications component intriguing, particularly because of its ability to extend her students' experiences beyond their classroom. Also appealing was that the program could satisfy a fourth grade science requirement to teach weather, a topic she found difficult to present.

HA also felt that the program fit her emerging orientation toward a more constructivist pedagogy:

All of a sudden things were making sense to me educationally. It was constructing meaning whether it was reading, writing, doing science, doing math, how
are you constructing meaning? Finding patterns throughout—I mean it sounds
so simplistic, but it was so powerful to me. And I began to become more interested in the constructivist way of learning . . . Whereas a few years before that I
might have thought, how am I going to do that? I don't know how to do that.

HA's colleague, MD, had used Acid Rain four times before participating in
What's in Our Water? for his first time during this study. During his first year
doing Acid Rain he was

thrilled with all the different aspects of curriculum, the multidisciplinary nature
of the unit that kids could run for, in a self-contained classroom. It was wonderful because you could do all different aspects, different times of the day, and it
was rich and we had a lot of students who really jumped off on it and did an
awful lot of side activities . . . It raised a lot of issues as well.

MD viewed the program as an "organizational framework" that "can be
delivered by a teacher much more effectively than if they had to develop it
themselves." It freed him from the more fragmented approaches to experimentation that had dominated his practice. The What's in Our Water? unit also filled in
a conceptual gap that was missing from both the Acid Rain unit and the Lake
Ecology unit MD had developed.

One of the things that I thought was missing was a student's impression of
where this water was coming from and where it was going. Not the natural
water cycle, but the human water cycle, and when I read What's in Our Water?,
the introductory materials for it, I thought it was really a perfect mix, and it
would blend very well with Acid Rain and Lake Ecology.

Incentives Across Sites to Adopt the Kids Network. Incentives to
adopt the Kids Network fell into several categories, including factors intrinsic to
the nature of the program, elements of the program's format, the program's "fit"
with teachers' situations, and appeal to students.

Inherent Incentives. The relevance inherent in the Kids Network unit
topics was cited most frequently as an incentive for adoption (table 4). The
teachers identified the topics as ones both they and their students could commit
to, because they affect the quality of the water, air, and land on which their lives
depend. Because the topics are current, one teacher cited the wealth of resources
available beyond the packaged curriculum. Another teacher hoped the relevance
inherent in the Acid Rain unit would motivate her students to take action, later
on, to reduce the pollution contributed by their parked school buses. Furthermore,
according to this same teacher and her science specialist, the presence of the
acid rain issue keeps it salient throughout the school year, long after the unit has
ended. One of the rural Vermont teachers attributed the program's appeal to the
immediate relevance in the data generated by her students, particularly in light
of the fact that it will be used by other students throughout the country.

The second most frequently mentioned intrinsic incentive to adopt the program was its ability to create "global classrooms"—ones in which the traditional boundaries of interaction are stretched beyond a particular school building. The ability of the telecommunications component to connect classrooms from afar was an important incentive.

Table 4.   **Intrinsic Incentives to Adopt Kids Network**

Incentive	Proportion of Participants (n=22)	Explanation
Relevance, real-world, credible	Two-fifths	Unit topics ranging from acid rain to weather are relevant environmental and social issues
Creates "global" classrooms	A fifth	Nature of inquiry and telecommunications connect students with other students throughout the world

Format Incentives. Participants mentioned eight different appealing features of the program's format (table 5). Many of the teachers indicated the interdisciplinary aspects of the program as one of the its original appeals. The fact that SE, the fifth grade science teacher at Site 4b, was an exception here may be explained, in part, by her being the only Kids Network user in our sample specializing in teaching science to three different fifth grade classes. The rest of the teachers teach multiple subjects in self-contained classrooms. The fact that more teachers did not mention the program's interdisciplinary appeal may thus be more of a problem in the interview design than an indication that this component was not initially appealing to teachers.

After their initial Kids Network experience, seven participants (about a third) identified the ability to modify the program as an incentive to adopt it. For example, the principal at Site 1 indicated that although the program is prescribed, it is not restraining, and that the ability to modify it appeals to his teachers. This was confirmed by one of his teachers who said that he realized more recently that the program provides an organizational framework from which he can pick and choose the appropriate activities and modify others. KL at Site 3, for example, was able to integrate poetry projects about weather. For KA, the teaching intern at Site 1, this meant supporting student-generated inquiries related to the unit topic. For the two teachers at Site 6, it meant taking field trips to local landfills.

Six participants indicated that they liked being able to choose among six unit topics. Not only did this provide teachers with some choices, it also gave them opportunities to vary units from year to year. This provided a way to avoid burnout on a particular topic or to work with combined fifth/sixth grade classrooms.

Five of the participants identified the computer component as an important feature. In particular, they saw it as a good introduction to a variety of computer applications (e.g., telecommunicating, graphing, word processing, mapping) and

Table 5.   **Program Format Incentives to Adopt Kids Network**

Incentive	Proportion of Participants (*n*=22)	Explanation
Interdisciplinary, integrative, thematic	A third	Units integrate science, mathematics, geography, social studies, and language arts around contemporary themes
Modifiable, pliable, open-ended	A third	Program is prescribed but not restrictive; provides a good organizational framework and springboard for other activities
Diversity of unit topics	A quarter	Different units provide teachers with a diversity of topics from which to choose
Computer components	A quarter	Software has variety of tools: word processing, graphing, telecommunications, and mapping
Telecommunications components	A fifth*	Telecommunications is a critical component of an issues-oriented approach
Background materials, guides	A fifth	The noncomputer materials are well-developed, well-laid-out, and teacher-friendly

*Only one of these four participants was a teacher; the other three were a principal, a science specialist, and a technology specialist.

noted that these technological skills were transferable to other subject areas. Four of these same participants identified the telecommunications component as another incentive for adoption, both because it appealed to their own interests and because they thought it was important for students.

"Fit" as an Incentive. More than half of the participants indicated that they adopted the program because it fit with either their curriculum, structural, or personal needs (table 6). Fifty percent of the participants (seven teachers, two principals, and two specialists) indicated that the Kids Network fit well with their curriculum. For instance, two teachers at Site 1 said that What's in Our Water? fills a conceptual hole that exists in the current curriculum. Principals at Site 1 and 4a said the program satisfies the district's fourth grade requirement to study weather. For AN at Site 4a, the program builds on what students studied in earlier grades; whereas for SE at Site 4b, the program lays a foundation of skills that will be developed in higher grades. SF, the science specialist at Site 4, said the teachers have an incentive to adopt the program because it replaces existing curriculum materials rather than adding to an already overcommitted program. Furthermore, both this science specialist and the technology specialist at Site 5 said that the Kids Network is aligned with their school's technology goals.

Table 6.   **"Good Fit" Incentives to Adopt Kids Network**

Incentive	Proportion of Participants ($n=22$)	Explanation
Curriculum fit	Half	Program fits well with prior, current, and future curriculum
Structural fit	A fifth	Program fits well with self-contained classrooms
Personal fit	A fifth	Program satisfies personal needs ranging from interest in topic to providing new experiences for teachers

Four teachers from as many schools indicated that one of their incentives to adopt the Kids Network was its fit with the structure of self-contained classrooms. Since the program requires flexible scheduling (i.e., recurring extended periods of activity and intermittent involvement throughout a day), the structure of a self-contained classroom provided them with the conditions for carrying out the program.

The personal fit was important to a few of the teachers, particularly to those at Site 1, who highlighted how unit topics satisfied their personal interests or provided them with new experiences they wanted to learn, such as telecommunications. For one teacher, the Kids Network helped teach a required subject in which she did not feel especially competent. For another teacher at the same site, the program provided opportunities to collaborate with a colleague.

Student Appeal. Half the participants identified the program's initial appeal to students as a primary incentive for adopting the Kids Network (table 7). Unlike most of the incentives already highlighted, which preceded the participant's first use of the program, the program's appeal to students was instrumental in its continuance. Most participants justified adoption after their students exhibited a high level of motivation and excitement. Even after only a little experience with the program, the teachers observed that most of their students expressed tremendous excitement about sending and receiving letters from their teammates. Even some students who never had a chance to telecommunicate identified this component as one of the program's highlights. It is interesting that only the teaching intern (KA) at Site 1 indicated students having "fun" as one of the appeals of the program. This is the same participant who thought students were most motivated when they pursued their own inquiries—inquiries that diverged from the more algorithmic exercises offered in the *Teacher's Guide.*

Less Frequently Mentioned Incentives to Adopt. Highlighting only the most popular incentives for adopting the Kids Network does not reveal the complete range of reasons that motivated participants to take on the program for the first time (tables 8 and 9). It is possible that these less frequent responses are shared by other teachers but did not come out in the interviews. And, even if

Table 7.   Student Appeal Incentives to Adopt Kids Network

Incentive	Proportion of Participants (*n*=22)	Explanation
Motivated students	Half	Students exhibit a high level of motivation and excitement

these less common explanations are not confirmed by others, they still provide insight into the thinking of experienced teachers who decided to experiment with the Kids Network for the first time.

Table 8.   Less Frequently Mentioned Incentives to Adopt Kids Network— Appeal to Students

Incentive	Proportion of Participants (*n*=22)	Explanation
Appeals to a diversity of students	A tenth (2)	Challenging for students who are strong in science and mathematics and appealing to non-science-oriented students; variety of activities provides something for everyone
Diversity of activities	A tenth (2)	Students like variety of activities: making rain collectors, comparing data from other teammates, writing and receiving community letters
Potential for meaningful student action	(1)	May motivate students when older to take action to reduce pollution at school
Appeals to students' aesthetic and intellectual interests	(1)	Students like the physical presentation of the materials (*Kids Handbook* and software displays) and find information to be interesting

Note: Less frequent responses are those shared by fewer than 14 percent of the participants.

Note that only one participant—HM at Site 5—cited as an incentive to adopt the program the fact that she would not know the answers to the unit's inquiries any more than her students would (table 9). For HM, this made the investigations that much more authentic for her as well as for her class. The fact that she is the only teacher in our sample with formal training in the sciences may help explain her comfort with not knowing the answers to these inquiries.

Uncertainties and Concerns. Often conjoined with the participants' incentives to adopt the Kids Network were a variety of uncertainties and concerns. We first review uncertainties mentioned by a fifth or more of the participants (table 10) before turning to less common concerns. Then we highlight in more detail the ambivalence of two teachers at Site 1.

Table 9. Other Less Frequently Mentioned Incentives

Incentive	Proportion of Participants (*n*=22)	Explanation
Scientific, serious, high quality	9%	Studies provide students with opportunities to work as scientists and researchers
Authentic inquiry for teachers	5%	Answers to inquiries are unknown to teachers as well as students
Promotes conceptual understanding	5%	Program helps students understand basic concepts of units' themes
Good suggestions for activities	5%*	Activities are interactive
Adoption process involved teachers	5%	Teachers participated in adoption process
The rewards of being a pilot site	5%	Developers offered incentives to pilot units (e.g., free use of phone lines, free materials, opportunities for teachers to influence development)

Note: Less frequent responses are those shared by fewer than 14 percent of the participants.

* This was mentioned by one principal.

Telecommunications Components. All of the participants expressed some uncertainties about the program's telecommunications components. Concerns ranged from worries about having the necessary equipment to not having adequate technological support. Unlike the other concerns noted in table 10, which emerged after the participants' first use of the program, the uncertainties about telecommunications technology were present from the start. Even so, the appeal of the telecommunications component was too strong to arrest initial interest in the program. A more detailed analysis of teacher and student use of and ideas about the telecommunications aspects of the program appears in the section "Implementing New Technologies."

Time Constraints. All but two of the teachers expressed concerns about the program's inherent time constraints. Interestingly—and perhaps predictably—however, this concern did not arise among the principals and two of the three specialists who do not actually implement the program.

A variety of issues were raised about the program's data collection and telecommunications deadlines. General school activities, including assemblies, special events, vacations, parent conferences, and teacher workshops, as well as weather-related school closings, created unavoidable interruptions that interfered with the program's continuity and made it difficult to meet deadlines. For one of the rural teachers, these competing demands meant that the six-week program

Table 10.   Frequently Expressed Concerns About Adopting Kids Network

Incentive	Proportion of Participants (n=22)	Explanation
Telecommunications novelty	All	The novelty of telecommunications technology raises a variety of concerns (e.g., equipment, time to learn)
Time constraints	Nearly all	The program's deadlines present intrinsic and extrinsic time constraints that compromise the quality of implementation
High drop-out rate	A third	Too many teammates drop out during unit
Unsophisticated software	A quarter	Telecommunications software cumbersome and inflexible
Insufficient computer experiences	A quarter	Student access to and interaction with computers are very limited
"Cookbookish"	A fifth	*Teacher's Guide* is recipe-ish, as are some student activities which contain too many written instructions
Inauthentic science experiences	A fifth	The experiments are imprecise, and the reading materials are often too simplistic

Note: Frequent responses are those shared by 18 percent or more of the participants.

was in reality a four- or five-week program. As a result, she, like many of the teachers, skipped suggested activities.

AN from Site 4a expressed a common concern about feeling rushed: "We always want more time at the end . . . We are rushed about analyzing community data and doing the debate and getting our votes in . . . "

The science specialist at Site 2, who lends technical support to teachers using the telecommunications in the program, feels this support is very time-consuming. Because a telecommunications session "can run for half an hour" and because she does not want to worry about a student intentionally or inadvertently interrupting the session, she does all the telecommunicating after school hours. Furthermore, since the program requires that each file be printed out separately, "it becomes time-consuming."

AN at Site 4a thinks the software is unbearably slow and thus interferes with her natural teaching pace. She feels too that she should be able to lead her students to "right" answers; however, the nature of teaching presents so many other competing pressures on her time she has not become expert enough about acid rain. CM at Site 6 feels the time demands of the program interfere with her ability to maneuver effectively through the computer and teacher manuals.

For JM at Site 1, the time constraints she experiences when doing the program are specific to the Kids Network and do not occur during other science activities, during which "I can just say, this isn't going well, let's just do it again tomorrow, let's pull these other resources because this wasn't working well . . . And if it's Kids Network and I have to get that information in, well, we have to go with what we have. Because of the time constraint."

Unsophisticated Software. A number of the teachers across the sites think the telecommunications program is tedious, inflexible, unsophisticated, and/or ridden with bugs. In addition to being the third grade teacher at Site 3, JC is his school's technology specialist. He describes the computer components as the program's main constraint. JC thinks other, less expensive, computer programs offer students more accessibility and interactivity and can be used more often by more students. Furthermore, since his school has been able to access the Internet at no cost, JC feels the original appeal of the Kids Network's telecommunications component has diminished. JC and the technology specialist at Site 5 were both concerned that the solar unit, the most recently published of the series, does not work with the school's LAN. HM at Site 5 was frustrated that the telecommunications program only allowed for off-line telecommunications, and was therefore not interactive enough.

"Inauthentic" Science Experiences. Four teachers were initially uncertain about adopting the program in part because they felt the science experiences were not authentic. For example, MD at Site 1 believed the program did not provide enough interaction with real scientists, measurement materials were unsophisticated (e.g., the pH paper was imprecise), the unit scientist's results tended to overgeneralize local data, and the reading materials were too general and shallow. He was disappointed with the "relative inaccuracy of the data that's collected . . . that's perhaps the most important issue I have with the whole process . . . It's extraordinarily inaccurate. It allows wildly inaccurate generalizations that are not founded on anything but wildly inaccurate data results."

For JM at the same site, the Hello! unit lacks content and an overall goal and does not allow for spontaneity. According to HM at Site 5, the experiments in general were shallow and simplistic; consequently, the data generated were not real. CM at Site 6 felt that "their" trash study oversimplified complex issues.

Less Frequently Mentioned Uncertainties. Participants mentioned a number of other concerns and uncertainties with the program, all of which emerged after their first experiences with it. Although each of these concerns is mentioned by only one or two participants, they represent an important part of these teachers' experiences with the program and are weighed with other incentives and uncertainties when these teachers decide each year whether to do the program again.

For example, JM at Site 1 initially felt the pet study in the Hello! unit was "fun and it worked out well and it was a start" even though the topic was not content-rich. However, her experiences with other units left her feeling that for herself, the program was more work than fun.

Two teachers said there were not enough hands-on activities. One teacher indicated that the unit session was too long and tended to drag out. And one teacher was concerned that the Kids Network did not satisfy her students' mathematics requirements.

Two Concerned Users. JM (Site 1) has used three Kids Network units in as many years. Her initial experiences with the program raised a number of concerns; now, she is uncertain about continuing. She feels that two of the units are "cookbookish," which may have contributed to their seeming more fastidious. Furthermore, as noted, JM felt the content in the Hello! unit was unsophisticated and lacked an overall goal and that the Acid Rain unit did not have enough hands-on activities. The absence of the requisite telecommunications equipment in her own classroom did not help either, as it inhibited JM's autonomy and spontaneity. Furthermore, JM feels that she has not found her niche with the units she has tried:

> Some people said that it was cookbookish and maybe that's it, maybe it's just too set that you have to do certain things in order to get the results. To come down to the library to put it in . . . [and] because of where our computer is, it seems really fragmented . . . It takes you out of your room, it requires you to sign up, it's not very mobile, it doesn't allow for spontaneity . . . I don't know, it just seems like work, it doesn't seem fun . . . I know that this is a good program, but I haven't found my particular niche, the one that's really important to me.

SR, also a third grade teacher at Site 3, had so many concerns after her first experience with the Kids Network that she decided to discontinue it. She was frustrated by the NGS deadlines and thought that "keeping deadlines and getting information back was not interesting" to her students. Furthermore, she felt third graders may be too young to appreciate and understand acid rain, the topic did not tie in with other thematic units, not enough students were able to use the computer often enough, and there were few hands-on activities provided in the *Teacher's Guide*.

Concluding Remarks About the Adoption Process and Ideas About Noncontinuation

Throughout this study we came across many teachers who discontinued the program after one or two years. We found them when we were searching for case sites and then again when setting up meetings in other states to validate our preliminary findings. For most of these teachers, their experience with the Kids Network netted more concerns than incentives to continue. This study, however, focuses on veteran users and thus favors those teachers whose incentives outweigh their concerns. Although we did not do a thorough investigation into why teachers and districts discontinued the program, we did hear a variety of reasons for doing so.

One elementary science coordinator in a very large and geographically and culturally diverse district in Texas explained that a number of the schools in her area discontinued the program because they were undergoing systemic changes of one sort or another, from changing buildings to experiencing some other "distracting" restructuring process. A move to another school building not equipped with a modem in itself can stop use of the Kids Network, despite a teacher's desire to continue.

We also met five teachers who had planned on using the Kids Network during their fall semester but missed the mid-August preregistration deadline. They returned from their summer vacations to learn that they could not do the Kids Network until midwinter. For a few of the teachers, the registration paperwork was logjammed on a local administrator's desk. Another teacher did not realize that he had to register over the summer. Perhaps these failed attempts at registration are confined to summer periods, a time during which teachers are least concerned about school matters. Even so, teachers are probably used to deciding more spontaneously when they begin and end a curriculum. Having to sign up for the Kids Network months before its intended use may be an important reason why some teachers stop using the program.

Most of our participating teachers were frustrated by the consistently high drop-out rate of their teammates prior to the submission of a team's final data. HM at Site 5 expressed a common frustration among our participants when she said, "It's very disappointing, when somebody logs on once, and never logs on again. That's the worst part of Kids Network, when they do not follow through with their data." In fact, only some 215 teams out of more than 500 that started one What's in Our Water? unit reported their nitrate data to the NGS network. That means about 57 percent of the teams that made the effort to sign up and telecommunicate their global addresses to the network did not send in their final data—on which the unit's culminating activities depend. Similarly, during a Weather in Action session, only 195 classes out of 426 (46 percent) telecommunicated their final data to the network. The sharp drop in the number of teams participating in the program's final activities enter into this discussion of incentives and concerns on two grounds. First, the frustration of those teachers and students who have been on the receiving end of this decline may contribute to some teachers' decision to discontinue. Second, it would be illuminating to understand why some classes did not send in their data and to test the association between this factor and noncontinuation.

While even the most enthusiastic advocates in our study share some ambivalence about the program, their experiences have convinced them that its adoption has been well worth it. Evidently, their positive experiences overshadowed their concerns about telecommunications, time constraints, software, and other components of the program.

Analysis of Enactments

Every Kids Network unit offers a variety of activities. These include playing games, having discussions, writing letters, telecommunicating, researching, making predictions, doing experiments, recording observations, analyzing data, brainstorming solutions, and presenting findings to an outside audience. We were able to see some, but not all, of these activities in action because logistical constraints made it impossible to coordinate site visits with specific activities. Consequently, our analysis is limited to a relatively small number of enactments and does not include, among a number of things, the final series of activities in which students present what they learn to an outside audience. Reports by teachers and students indicated clearly that they did far more Kids Network activity than what we saw during our approximately six-day period of observations— during which we observed less than a quarter of the program in action.

Our analysis highlights the range and diversity of ways in which the experiment-based activities were enacted. We illustrate examples of students doing experiments by following algorithms; in these cases, teachers were faithfully following the suggestions in the *Teacher's Guide*. We also describe open-ended experiments, occasions when students pursued their own lines of inquiry— sometimes spontaneous and sometimes planned. We then illustrate other configurations to show the range of ways in which this program has been enacted.

These enactments are analyzed according to several criteria such as degrees of prescriptiveness, student self-direction, cognitive engagement, and teacher assessment (Brophy and Alleman 1991). We proceed with caution, however, for breaking an activity into these respective parts may make it appear more fragmented than it was actually experienced by the teachers and students.

Algorithmic Experiments

In general, students did hands-on investigations that were algorithmic in nature. That is to say, most of the experimentation provided everything the students needed: a list of materials, a specific line of inquiry and directions on how to pursue it, and a system for recording observations. During these activities, students worked in small groups to follow a systematic and explicit set of directions to obtain predetermined results.

This is why the pH readings from eight different acid rain collectors from one class should be the same in the Acid Rain unit, as should the results from the fertilizer and nitrate experiments in What's in Our Water?. In experiments such as the pH testing of rain and common household products, the results are expected to be uniform within a class but not necessarily consistent across teams. In activities such as the fertilizer and nitrate/runoff experiments, the results are expected to be uniform among all teams. We observed only one instance in which a teacher provided a student with an opportunity to organize her data in ways that the student thought would help answer her own questions. Overall, most of the observed activities reflect the teachers' tendency to implement faithfully the suggestions provided in the *Teacher's Guide*. It is in the guide itself where teacher-directed, algorithmic approaches to experimentation can be found, and where there is a noticeable absence of suggestions for how to assess student understandings.

Following are illustrations of experiments from the Acid Rain unit enacted in faithful accordance with the program. We then present an analysis of some of the similarities and differences among these enactments.

pH Testing of Household Liquids. SE's (Site 4b) third class for the day meets in the school's science lab. There, four groups of five to six students work around four different tables to test the pH of 12 household liquids (e.g., sodas, household cleaners). When the lab begins Activity Sheet #3, SE assigns one student to be the leader at each table. She explains that leaders are responsible for assigning and coordinating the rotation of roles among the other group members. She points to and reads the six roles, which were already listed on the board: leader, pourer, timer, rinse out cups, test taker, recorder.

SE then explains the lab session, referring to a list of variables written on the blackboard:[6]

> Now, I want to talk to you a little bit about the lab, so I need everybody's pencils down and eyes up here, now. You're testing 12 liquids. You are going to have an opportunity to test all of them, if you can get through with them. The object is just to see how this particular paper records the level of the state of acid in a liquid. And what I want is for you to get the experience of doing some kind of laboratory work where we are controlling most of what we're doing except one thing that we need to find the difference in, and that is the acid liquid. See the labels on the right-hand side? [Referring to the list of variables.] The manipulative variables here are the liquids. We are manipulating, we are changing the liquid . . . Sit correctly on the chair or you're out of here. We are manipulating so that we can see the difference in pH level. However, when a scientist experiments, the most important thing that you have to remember is that you're only going to manipulate one thing at a time. In this case, we're just testing various liquids. But we have to control everything else that we're doing, otherwise we don't know if our results are valid. So, by controlling it, I am making sure that everyone is testing in a plastic cup. Why is that important? Why can't one person do it in a ceramic mug, another person do it in a glass? Why do you think that's important? Maggie?

---

[6]This list of variables reads as follows:

Variables - (Manipulative)
Variables - (Controlled)
- plastic cup
- pH paper
- pH charts
- paper towels

On the blackboard, to the left of the lists of roles and variables, is the following chart which students are asked to copy down and use to record what they think would be the pH of each substance.

	Liquid	Predict	Actual
Seltzer			
Coke (diet)			
Mr. Clean			
Ginger Ale			
Scope			
Ivory			
Milk (low fat)			
Apple Juice			
O.J.			
Water (bottled)			
Lime Juice			
Ammonia			

SE explains what she wants her students to do:

> . . . looking at this chart, would you give it 1 to 2 . . . do you think seltzer is a 6? A 3 or 4 is moderately acidic meaning medium, 5 to 6 weakly acidic . . . Make a prediction. You can put a range, like 1 to 2; or you can say, 2, 4, 7, 9, 11. Do it quickly . . . Your prediction is only personal information for you, what you think.

A student asks if her group has to agree on making a single prediction. SE replies, "No, you predict by yourself." Even so, most of the students share their predictions with others in their group, as in this example from students one table:

Student 1:	How do you know Coke is that?
Student 2:	I know Coke is. The most is lime juice.
Student 3:	Seltzer is going to be the least.
Student 1:	If seltzer is going to be the least, how can . . .
Student 4:	It's going to be ammonia . . . ammonia is very basic.

While students are recording, sharing, and disagreeing about their predictions, SE yells over the rising hum of student activity, "It doesn't matter if you're right or wrong." While the students continue making predictions, SE distributes to each table one wide-range pH chart, a stack of pH paper, and three glasses. Throughout most of the activity, SE is focused on the preparation, distribution, and return of the various liquids. Intermittently, she stops managing the materials to yell across the room at a student she thinks is too loud or disruptive.

After the student leaders assign roles to members of their group, the first pourer from each table picks up a cup of liquid at the front of the room. Upon his or her return, another member of the group sticks the pH paper into the liquid, and then the entire group counts to 10 out loud. SE tells another group that they can only have one "counter" at their table since they're being too noisy.

With five or so minutes left to the lab period, SE tells students to stop testing. She then reads off one substance at a time so students can indicate, by raising their hand, if their prediction matches the actual results. The students express excitement when they're right, and disappointment when they're wrong. SE asks, "How many people feel that they, more than not, were in the range of predicting the actual pH? . . . So that's pretty good. I know that a few of you were thrown off by some of [the liquids]." The activity ends here.

Later, SE says that during the lab she directs and facilitates,

> I'm not necessarily instructing. I'm there to organize them, so that they can self-discover. I'm not there to teach them what they can find . . . I'm there to take the data and look at it in terms of what it meant after we all did it. But, usually in the lab I am not an instructor, I'm a facilitator. I direct them. I sometimes have to end up as being the manager of the whole class in terms of them staying on task.

The Effects of Acidity on Nonliving Objects. During the last two days of field visits, the fifth graders in AN's and PM's classes (Site 4a) join as one group in AN's classroom to carry out an experiment. Their goal is to observe the effects of different levels of acidity on some nonliving objects. This activity is suggested in the *Teacher's Guide* (Activity Sheet #9). AN explains what they are to do:

> You're going to have cups, and what I recommend is that you put the labels on the cups before you fill them up . . . Don't forget how to label it. You're going to need to assign jobs within your group so that somebody's in charge of the labeling, someone is in charge of getting the vinegar, going to use the first vinegar, the straight vinegar for everyone and then if I use diluted vinegar I'll do it with everyone the same. We're going to use beakers to measure; now I'll get one in a minute to show you. One person in each group will get the water. Now when you use the water, you may first go to the white container beside the sink. That is totally purified water. I ran it last night so it would be as fresh as possible. If we need more water, the water in the bucket underneath is also the same thing, as fresh. We will be working off the counter; I will be working at the counter making sure you've got the vinegar. You'll get your sheets, I don't care where your groups end up putting their cups. If there's room on the counter at

the end, fine, they can all stay in here, and we'll cover them all with Saran wrap or tinfoil tomorrow at lunch time, so that there is a tiny evaporation over Thanksgiving. It doesn't matter where the cups are, you make those decisions, just don't put them in front of the chalkboards, because the chalk will get into them.

The students proceed to collect their materials and put their objects into the three different concentrations of vinegar. In some groups, there are as many objects in each glass as there are members of the group. Students then record their first few minutes of observations onto their student activity sheets. Many students have predictions about what will happen. Out of earshot of the teachers, one boy predicts, "I think the only one of our objects that it's going to matter is the crayon. I don't think it's going to make a difference to the penny or to the paper clip." At another table, also out of earshot of the teachers, two students disagree about whether the crayon will dissolve. From another group, an excited exclamation is heard, "Something is peeling off the brick." Then, from another student, "It's turning gray."

A few students in one group disagree about how much more acidic a pH of 5 is compared with a pH of 6. One students thinks it is 10 percent more acidic; another thinks the scale is "1 to 13"; and still another says, "Remember, we just thought it was 1 to 14. I think there's so many different scales." This conversation, too, is not heard by the teachers.

During most of this activity, AN is preparing, distributing, and organizing the materials. Toward the latter part of the activity, she begins to roam around to the different groups. At one table, she asks such questions as "Why do you think different vinegar had different pH readings?" One student thinks it might be due to one of the vinegars being clearer than the other; another student thinks it might be due to the difference in the vinegars' ages.

After students set up their vinegar experiments, one group gets permission to write and send via telecommunications a short note to one of their research teams asking them to send their overdue community letter. A little later, Rebecca reads aloud the letter they plan to send:

> We wrote to say that we were very disappointed that we didn't get your data file yet, and we were wondering if you could send us your data file as soon as possible. We hope you will send us your data file today so we can start communicating with you. You should have received our data file. We would appreciate it when you send us your data file you would include what your delay was and say a little about your city. Sincerely, Fifth Graders, AN's Class, Suburban, NY.

The Effects of Acidity on Living and Nonliving Objects. During the second field visit, HM's fifth graders (Site 5, rural, Vermont) spend about 90 minutes working on two experiments. For the first experiment, students work in groups to observe the effects of different pH levels on materials chosen and brought in by the students, such as graphite, seashells, and Tylenol. Each student puts one of three of the same objects into one of three cups: vinegar, diluted vinegar, and distilled water. For the second experiment, students continue working with their groups to determine the effects of pH on seed germination. One group decides to test the effects of pH on the growth of three plants.

To begin, the teacher explains,

> It doesn't matter how you do it. Then you're going to take your eye droppers and I'm going to come around with three different containers, one with distilled water, one full-strength vinegar, and one with diluted vinegar. You're going to take your medicine droppers and put one or two or whatever you decide, as long as it's consistent and there aren't any variables. On top of the paper towels, soak those seeds, and then we'll put them in a jar and then see if there's any difference in how they germinate.

After HM distributes the liquids for students to saturate the seeds they sandwiched in paper towels, Sarah wonders if they have to give each seed the same amount of liquid. Sarah and her partners decide to make up for yesterday's underwatering of one of the paper towels by providing more water to it today. Their thinking and actions are not known to the teacher.

The teacher also does not overhear Danny and Nathan explaining to the researcher that "We're putting seeds in between paper towels, and seeing if it likes vinegar, diluted vinegar, or distilled water, if it makes a difference in the seeds." Nathan predicts that, "The full-strength vinegar would probably stop the growth of the seeds, or something . . . [because] I think vinegar has acid. So, I think an acid—you're not supposed to put animals and plants . . . "

When everyone is almost done, some students shift into their end-of-the-day, self-selected projects. Two boys choose to continue working on their experiment, dissolving different types of medicine tablets.

The next day, HM's students observe their previous day's experiments, and then spend about 15 minutes recording and exchanging observations. Leah notices that "one of the shells disappeared in full strength." Eugene says he is suspicious of foul play when he can't find any trace of the shell he put in the vinegar the previous day. He thinks the shell he put in the cup of vinegar may have been stolen. Others question his conclusion. HM replies, "Maybe you need to repeat [your experiment]." Tom has been watering three potted plants for a few days with three different levels of pH. He observes, "It looks like the diluted vinegar made 20 leaves fall off. The one that had full strength seems best [healthiest]." Tom did not control for the amount of watering of each plant, nor did he systematically evaluate the condition of the plants before his experiment began. Eric, a student from another group, notices that the letters on the Tylenol in the vinegar are still floating around; he theorizes that the vinegar might help hold the letters together.

Analysis of Algorithmic Experiments. Both SE's lab investigating the pH of different liquids and AN's and PM's classes investigating the effects of different levels of acidity on common objects exemplify enactments that follow closely what is prescribed in the *Teacher's Guide.* In both cases, students were provided with the question of inquiry and the materials and procedures for testing, observing, recording, and analyzing data; the teachers and students spent a considerable amount of time managing materials, and teachers assigned or suggested the roles that the group members should play. Consequently, there were few occasions on which teachers probed into their students' thinking about the content or process of investigation and worked with students' spontaneous dialogues, which often highlighted their areas of disagreement.

Even so, students carried out fair tests, in which they manipulated one or more variables to determine their influence. They experienced inquiry models developed by scientists. They worked at being precise and systematic. They were exposed to testable questions. They made hypotheses and observations, and they recorded and analyzed data. They also shared their data and results offline with students from other parts of the country.

Even though these investigations are prestructured, from a pupil's perspective such exercises can be adventurous. They contain unknowns and obstacles: interpreting directions accurately, setting up the materials, recognizing group or individual errors or misconceptions, and reconciling differences between their classmates' and teammates' data. Furthermore, students often generate new ideas and experiences from their participation with more standard investigations. And, as illustrated in the cases of the student-generated pH experiments, algorithms can lead to more spontaneous and open-ended inquiries.

Student-Initiated Investigations

There were a few instances in which students had opportunities to design their own experiments. The first illustration includes two occasions in which the student intern at the urban New Hampshire site moved away from the *Teacher's Guide* and toward less prescriptiveness and a greater willingness to allow for divergent, yet related, student-initiated inquiries. In the second case, we return to the fifth graders at Site 5. There, students designed and carried out their own experiments.

In both cases, these types of activities were accompanied by an infectious enthusiasm, engagement, and reflection that were not typically observed during other Kids Network activities.

The Making of a "Yeasty-Beasty." A week after carrying out the grass experiment of the What's in Our Water? unit, every student in KA's class at Site 1 is—as part of another activity suggested in the *Teacher's Guide*—following the same set of directions to determine the effects of chlorine on yeast reproduction. At the end of this activity, a sixth grade boy yells out, "Let's make a yeast monster," quickly referred to by some students as a "yeasty-beasty." With their teacher's enthusiastic support, students rise from their chairs, smiling and laughing, while pouring into one container the leftover molasses from each of their experiments. Throughout this spontaneous activity, most of the girls appear animated, laughing, and directly involved in the collective experiment. These behaviors were not noticeable during previous Kids Network activities.

The class gathers around the center table to mix a large batch of yeast, water, and molasses. Many of the students want to mix all of what they had left from the previous activity. KA suggests that they keep the proportions the same as in their original models, which used smaller cups. She asks the students to multiply the amount of molasses they used in one of their chlorine cups, 2 1/4 teaspoons, by 3. One boy calls out immediately that they will need to add 6 3/4 teaspoons. KA asks him how he arrived at that number. He explains that he multiplied the 2 by 3 and got 6, and then multiplied the 1/4 by 3 and got 3/4. Another boy offers an alternative strategy: He converted the mixed number, 2 1/4, into a fraction by multiplying the 4 (the denominator in 1/4) by 2 and then adding one (the numerator) to get 9/4. He then multiplied the 9 by 3 and got 27/4. Before the student finishes his explanation, his teacher indicates that his method is not right. (Actually, 27/4 produces the same result as the first boy's strategy.) While the teacher mixes the ingredients, she is surrounded by a constant buzzing of student suggestions, spontaneous predictions, and the sharing of ideas.

Toilet vs. Tap Water. During another Kids Network experiment in KA's class, students follow the directions on Activity Sheet #9, "Nitrate Tests," to measure the school's tap water for nitrates. As suggested in the *Teacher's Guide*, students are also expected to measure the nitrate from different water sources, such as their teacher's home or their own home, or from surface water. A few students ask the teacher if they can measure the nitrate levels from their classroom's toilet bowl. Not satisfied with their original results indicating higher nitrate levels in their toilet than in their tap water, they interview the principal and the custodian. The principal tells them that all the water pipes in the school originate from the same source, and thus should carry the same amount of nitrate. After the janitor informs them that he cleans the toilet bowl with chlorine bleach, they return to their class to test the toilet water's nitrate levels before and after its first flush. They predict a higher reading before the first flush. Unexpectedly, they observe a higher reading after the flush. A couple of students theorize that the flushing washed and mixed the nitrate from the janitor's cleanser into the bowl, thus creating a higher post-flush reading.

Student-Designed Acid Rain Experiments. After the fifth graders from SL's and HM's classes from Site 5 make their rain collectors and are waiting for it to rain, they are given an opportunity to design and carry out their own acid rain-related investigations. A few days after they begin their experiments, they record their findings onto a chart on the blackboard that was made by their teacher. Once most of the findings are recorded, students describe their observations and findings briefly to their class. Here are a few examples of reports.

Allison and Abbey compared the pH of tap, well, and stream water. They determined that distilled water has a pH of 4.8. Realizing that the students' pH of distilled water is lower than expected, the teacher asks, "The distilled water is 4.8? Do you think that water is very pure or clean? What should distilled be about? What container did you use?" Abbey replies, "It was contaminated." All of their samples were in glass containers—a material the *Teacher's Guide* warns may contaminate pH readings.

Tanna, Ashley, and Luke report that the pH of shampoos and soaps range from 5 to 9. One shampoo was 6.2, and mild soap was 8 or 9. Their teacher asks, "Why do you think that Ivory Soap may be very basic, way up there with 9?" Luke explains that only a little bit of acid is needed to kill germs. Another student speculates, "Maybe the bar soaps, like shampoos, can be acid because it doesn't harm your skin." Another student notes that the shampoos have a pH around 5 and theorizes, "The soap bars are probably basic because acid may harm your hair."

Another student, Ryan, describes his experiment:

> Well, we have three cups with wet pieces of napkin, and have rye seeds on it. One has water, one water and vinegar, and the other one has vinegar, and we want to see which one sprouts the most. And already in water, all of them have sprouted. The water and vinegar one sprouted, and in vinegar one already sprouted.

Ryan did not expect seeds to sprout in the vinegar at all, "Because vinegar," according to Ryan, "is an acid and usually things don't sprout in acid."

In the class next door, HM's students are also designing experiments having to do with acid rain. A few of the students summarize what they are investigating to our field researcher. Leah and Holly are comparing the pH of rainwater in containers made of different materials: tin, styrofoam, and glass. They wanted to do this after learning from their teacher that they could not collect rain in glass containers. Leah explains, "The glass changed something, so we decided to just see how much it changed." A pair of boys were comparing the pH of rain at the beginning and end of a shower. They predict, "I think that the end of the storm will be more acidic than the beginning . . . because [toward the end] when it starts getting a little harder, it makes more acid in the water since there's more of it, just a little more of it."

Some Other Enactments

A Daily Outdoor Activity. During most of the Weather in Action sessions, third and fourth graders from KL's suburban New Hampshire school leave the school building at noon sharp, with thermometers in hand. While Susan, a fourth grader, is taking the temperature in an area she selected, she explains in detail what they are doing:

> Well, the note said it had to be in the shade, so we chose a place in the shade in the woods, and every day of the week we go out to our spot that we chose and then we take our recording paper and a booklet and a pencil and a weather vane and we see what the direction the wind is blowing and then we, after awhile,

we record the temperature Celsius and then we record the cloud type—cumulus, stratus, cirrus, or whatever you call it—and we also record if there was any precipitation—rain, snow.

At the same time, some boys in a shady area of the playground are arguing over exactly where they should position their thermometers, and a group of six girls laugh and then break into a spontaneous group sing-along while waiting for their thermometers to adjust to their new locations.

An Open-Ended Graphing Exercise. During the Weather in Action unit, KL's third and fourth graders at Site 3 make graphs comparing their personal and class's daily average temperatures. As suggested in the *Teacher's Guide*, students are told they can design their own graphs; they can determine their own scale, colors, and arrangements. What is striking about the onset of this exercise is the teacher's effort to meet her students' individual needs. She is like a jazz conductor, helping her class maintain a certain level of harmony while each student plays his or her own variation. This is an example of a more open-ended activity suggested in the *Teacher's Guide*, and demonstrates what might be forthcoming from teachers who are faithful to those activities that give students some latitude in their work.

As soon as KL indicates that they can begin, there is a rapid barrage of questions for clarification, requests for supplies, and personal requests. After a boy approaches her, she says, "Get your temperature chart out." He gets it and brings it to her. She is then greeted by Kate who asks what she can do to make a pie graph instead of a bar or line graph. Her teacher says, "You have to show the percent of students that selected certain degrees. What fraction of 24 got 12° yesterday? And I can show you how to find percentages using your calculator." Kate returns a moment later with a calculator. Before she makes it back, however, Andrew, a fifth grader who has just moved from New Zealand, asks KL for advice on how to make a scale. She suggests that he get a thermometer. When he returns with one, she asks, "So what will you have to do? If it were 1°, where would the alcohol go to?" While Kate is waiting for her teacher, she draws a circle and labels it a pie graph. As soon as Andrew leaves, Louisa approaches KL and asks where there is tape. KL overhears a boy discover aloud that he cannot make the numbers of the temperature scale fit onto his graph paper. She suggests that he get a chart to compare Celsius to Fahrenheit. Brett approaches the teacher with a question and then returns to his desk, looking satisfied. KL then approaches Philip who is sitting next to his classroom aide. He gets his folder out after KL asks him to do. Once he begins working, KL returns to Kate. Kate has already completed the other graphs and now wants to try something else. Even though the rest of the class has not yet studied percentages, KL tutors Kate on how she can turn fractions into percentages and encourages her to make a pie graph. After KL finishes working with two more students, she announces,

We're going to come around and look at those graphs that you've done, and I'd like you, if you've already done your line graph, I'd like you to try a bar graph. As we come around, we'll talk about the bar graph and how they're different. I know that many of the people know what they are, so I'm not going to do a big session on it. I have graph paper. Do you have any questions?

KL tours the classroom, stopping to talk strategically with individual students about what they are doing, examining their graph work, and offering some ideas about bar graphs.

Making Predictions and Justifications. Toward the later part of the unit, HM's fifth graders (Site 5) read aloud, in their respective groups, a community letter received from Maine. Then, by referencing a map indicating where acid-producing gases are emitted in the United States (Activity Sheet #6), each group tries to agree on a prediction of its respective teammates' average acid rain level. HM begins by explaining,

Okay, this is what I want you to do. Everybody remember which team they were hooked onto? Okay. This is what I want you to do. I want you to look at your maps north of the United States, Canada. I want you to think about where it's located, all the information that we've gotten from the guest lecturers that we've had, and make a prediction about how acidic you think the rain might be in those communities. You haven't gotten any information, but if I look underneath the map you can make those. Check where your community is, then look at your pH and then . . .

As the activity begins, HM moves around to each table cluster to listen to the students' ideas; there is a general hum of student assertions and explanations to one another. They then record their prediction on a chart on the board. When the chart is completed, students from each group justify their predictions to their teacher and peers.

Following each explanation, HM asks the whole class, "Do you agree? Disagree?" For example, after looking at the group's prediction regarding the pH of rain in Maine, HM turns to the class and says,

The rest of the class, everybody knows where Maine is. Do you agree with their prediction of 5.0, 5.2, approximately what we've got here [in our town]? Does anybody disagree and if so, why? I'll tell you what I'm curious about. I want you to leave it. This is what I'm curious about. People who went [with the acid rain specialist], what were his pHs? What do we know about our particular area of Vermont? A different feel, he did the research on this. What do we know about our whole area versus the rest of Vermont in terms of acid rain?

Student: It's not that high.

Teacher: Why?

Student: Because a lot of other places don't burn coal.

Teacher: What makes the Valley a little, maybe a little less?

Student:	Limestone.
Teacher:	That's the only thing I'm curious about, so we'll find that out when they come in. You may be absolutely right. That would be the only question that I might have.

Sarah then reports her group's prediction and thinking about its teammates from Lexington:

On the map we saw it's right in the middle or really close to a lot of acid, like producing places that produce acid, so we thought if it flies away as it blows away in a cloud, then it's not going to be very low, because it'll be over some other place. But if it stays there it will be higher. It'll be more acidic than if it just stays there over Lexington, but if it blows to some other place then it'll be more of a base acidic.

HM responds with a question:

Teacher:	What about Lexington? What did you learn from the community letter that you think it might be some type of acid rain? . . . Can anyone remember what else Lexington has? . . . How big is the city of Lexington? What kind of industry do they have? Annie? What are the industries that have that pollution?
Annie:	IBM, Proctor and Gamble.
Teacher:	Anybody disagree with that? I'd like your reasoning, it sounds like good reasoning. Tom.
Tom:	There are a lot of factories and stuff there. It's pretty popular. I guess, they're doing some links in New York City at least, and that will produce a lot of pollution in New York City and Manhattan and all their cities. [Another small city] is right near all those places and that would make it, that weak pressure.
Teacher:	Does anyone disagree with that? . . . It can be lower too. Let's go onto another one so we can talk about . . . Bethlehem, Pennsylvania, which had 4.0. Who had Bethlehem? Okay. Justify.
Student:	Well, in the letter there's that, they're putting things over the smoke stacks to make it clear.
Teacher:	Do you remember the name of those things? Scrubbers.
Student:	First we thought that acid rain is really on the top of them, like right . . . now that we sort of in the middle. So there's like this huge white spot. Do you think that there's a lot of acid rain going in from New Jersey?
Student:	New Jersey is covered with it.
Teacher:	Eugene, do you want to add anything to that?

Eugene:	Yes. Basically, it's in the middle of the industrial area of Pennsylvania and it's right next to New Jersey. I mean that's what happens, it's right next to it and New Jersey is just a lot of pollution. So, that's what we thought. At first we thought it was like 3.5, but then logic started entering. It's definitely not 3.5.
Teacher:	What direction is New Jersey from Pennsylvania?
Student:	East.
Teacher:	Tom?
Student:	It's higher actually.
Student:	Yes, it would be a lot higher.
Teacher:	You don't think it's going to be as acidic as 4.0?
Student:	I think it might be, actually lower. Not lower, it'll. It doesn't seem right somehow, I'm not sure.
Teacher:	Well, you make a guess which way you think it's going to go.
Student:	I think it's going to be lower.
Teacher:	What is the lowest pH that [the acid rain specialist] got in Lincoln, Vermont? 3.0?

The teacher then asks the students to report who predicted the pH in Tucson, Arizona. One of the boys from the group that is matched with a team in Tucson reports: "We think it's 4.5 to 5.0, because on the map it doesn't look like it has any acid producing, sulfur producing, or bad things around it. So we think things like the wind could blow the clouds over there so they would rain there, but . . . we're not sure, all we know, the clouds could blow way over into New Hampshire."

A girl then suggests that their predictions would be more accurate if they knew about the presence of limestone in the respective communities. They had learned already that limestone, in particular in Vermont, has an effect on the level of acidity in lakes and streams. Students were heard earlier explaining that the difference in acid levels in northern and southern Vermont's lakes and streams were due to the difference in the spread of limestone. This girl is applying something she had learned earlier to this activity. (However, no one points out that the acid rain recordings from the other communities were from their rain collectors and not from stream water; therefore, knowing the limestone conditions in their teammate's communities would not have helped them in making predictions about the pH of rainwater collected in their rain collectors.)

Social Action: Reducing One's Trash. During the third field visit, students in both TA's and CM's classes (Site 6) complete, with their respective teachers, the same activity in the *Teacher's Guide*. Students compare two lunch pictures in the *Kids Handbook* according to the amount of grams of waste that are produced in each scenario. Students exchange ideas of how to reduce the amount of trash generated by the lunches. To make the comparisons, students copy and complete an already labeled chart. Some students readily share strategies used by their families. Students also copy the definitions for "reduce," "reuse," and "recycle," as defined in the *Teacher's Guide*. They then brainstorm ways they could reduce their trash during one school day.

On TA's blackboard is a chart presented in the *Teacher's Guide*. It indicates the amount of paper trash produced in three classrooms. It reads:

Paper Trash (in grams per student)		
Class A	Class B	Our Class
120	65	30

TA asks students first to brainstorm why there might be differences between Class A and B and then to explain why their class produces so much less paper waste. Susan suggests that "Class B might have little blackboards to write on rather than paper." Nancy thinks Class B might dry their hands with hand blowers rather than with paper towels.

Teacher:	Total weight of Lunch A for total trash?
Student:	43 grams.
Teacher:	Where is most of that weight coming from?
Student:	Apple core.
Girl:	It weighs more due to . . .
Boy:	You can compost it.
Teacher:	Do Lunch A and Lunch B look like a big difference? [Lunch B weighs 77 grams.]
Boy:	You can feed apple cores to deer and horses.
Teacher:	Which produces the most paper trash?
Student:	Lunch B.
Teacher:	Why would someone have Lunch B instead of Lunch A?
Boy:	Too lazy to carry a lunch box.
Boy:	They couldn't find their lunch box.
Andy:	Paper bag disposal could cost a lot at the dump to dispose.
Teacher:	How can Lunch A be better than they are? What's something they're throwing away that they could use again?
Boy:	They could put carrots in Tupperware containers.

Teacher:	Cloth napkin could be used. What would Mrs. W. say? [She is a teacher who spoke to students earlier in the week indicating that she only throws out three 30-gallon bags of trash in a year.]
Girl:	Stop and think about it.
Jamie:	Instead of a paper bag, they can have lunch box and paper napkins.
Teacher:	Is the fun worth what I'm doing to the environment?
Travis:	Mrs. W. would have a fit if she saw all this trash on this page.
Girl:	The applesauce container and spoon are washable.
Michelle:	If you use Saran wrap there would be less than the plastic container.

In response to why their class's per student paper waste is half that of Class B, Kathryn indicates that her classmates reuse the paper they put in their recycling bin; Andy adds that they are required to take their graded papers home. After a few more minutes of exchanging reasons for producing less trash, TA passes out the *Kids Handbook* and says, "Look at pages 10-11 and record the trash generated during Lunch A and Lunch B." Students record the name of the trash and its corresponding weight in a prelabeled chart (Activity Sheet #10). While students are working, the teacher circulates around the room and answers questions. She suggests that students can use their calculators to check their mathematics.

The teacher then informs her students that they need to make a plan for reducing their trash. "We're going to try really hard to reduce. We produce about 125 grams per person per day of food waste—half liquid and half solids." She asks students to brainstorm a list of ways to reduce, while keeping in mind that recycled paper must be weighed in as trash. One girl suggests, with all seriousness, "We can take stuff home and throw it out there." A boy jokes about eliminating toilet paper.

TA's class then joins CM's class for a joint brainstorming session on ways to reduce their trash in a few days. CM projects overhead: "Fifth Grade Trash Reduction Plan," and then records on the transparency the students' ideas:

Michelle:	If you have chips in a bag, save the uneaten food that won't spoil.
Marlene:	Bring your own cloth napkins.
Teacher:	Do we agree on this? [general consensus]
David:	Instead of bags, use Tupperware and a lunch box.
Teacher:	What else?
Andrea:	Only take as much food as you're going to eat.
Todd:	Bring food to Ms. C's compost.

CM rejects this last idea because, as she reminds the students, their goal is to "generate less" trash. The students continue to make suggestions:

Marlene: Bring in a tape recorder so you don't have to use paper.

Teacher: Is that reasonable?

Hillary: Use a lunch box.

The final Fifth Grade Trash Reduction Plan looks like this:

---

1. Save uneaten food that won't spoil.

2. Bring a cloth napkin.

3. Don't use straws.

4. Use Tupperware instead of paper or plastic.

5. Only take as much food as what you're going to eat.

6. Use a thermos/container that can be reused instead of juice boxes.

7. Use a lunch box.

8. Reuse zip-locks.

Some Other Common Features

Whether the enactment was a faithful replication of what was prescribed in the *Teacher's Guide* or a spontaneous diversion, whether students worked side by side doing different tasks or went on a unit-related field trip, many of the enactments at the different sites shared some common features around such themes as learning pace, group work, and assessment.

Quickness of Pace. The Kids Network was executed a breathless pace. Many teachers were observed rushing students with comments like, "Hurry up and get through it so we can do what's next." Kids Network activity seemed almost frenetic, pauseless. There were a number of times teachers said, "Hurry up, we don't have enough time." Consequently, there was little time for reflecting, revising work or hypotheses, probing into student thinking, or exploring divergent but relevant inquiries. A quick pace also did not help students engage in critical or collaborative activity. Most of the teachers typically paused for no longer than one to two seconds after asking a question before answering it themselves or moving on. Perhaps not surprisingly then, students typically limited their comments to only a few words.

For JM and MD at Site 1, the Kids Network deadlines contribute to a pace that compromises variation and the intrinsic quality of data. MD sees the fast pace of teaching and learning during the Kids Network also as a byproduct of forces unrelated to the Kids Network. He adds, "I know that time is a real factor in this building, because we have a lot of classroom programs that are just districtwide initiatives, that we have to use."

AN at Site 4a expressed a concern shared by many of our participating teachers about the pace of learning:

> We always want more time at the end, and it's not because we haven't adhered to the schedule, I don't mean that, but it seems we get very rushed; but maybe it's because we have half days, parent conferences, and everything else, I don't know. It appears we are rushed about analyzing community data and doing the debate and getting our votes in. By rush, that means that we don't get enough time, we can read the community letters and reflect on them in the class and everything and they usually get all read aloud, etc. But we don't have enough time to write back to a community, you know, maybe we have some questions, maybe we have some comments, maybe some things are really interesting that we would like to know more about. We don't have time to do that, it seems.

Perhaps SE at Site 4b captured best the spirit of the program's pace when she directed her students after printing out some graphs on their teammates' acid rain data. She asked, "Now that is what we have printed based on the three weeks of rainwater that we collected. Who can tell me quickly what they think?"

Group Work: Managing Materials. The introductory materials to all of the units identify cooperative learning as one of seven components that makes the program special. The guides state, "Many of the lessons encourage development of interpersonal and organizational skills necessary for working in a group."

The nature of cooperative learning, in our sample, was limited to sharing materials. Typically, directions explained how to use the provided materials; consequently, many of the students' actual cooperation with one another centered around getting, using, and returning materials. Given the algorithmic nature of the program in general—as well as of the individual activities—to which most teachers tended to remain faithful, students rarely cooperatively designed their own lines of inquiry, the methods to answer the program's lines of inquiry, or the means to make sense of their data. Furthermore, since the program has no concrete suggestions for helping students create, apply, and evaluate strategies for working collaboratively, it is not surprising that these types of activities were not observed.

Informal, Process-Oriented Assessment. Teachers rarely inquired into students' conceptions of the phenomenon being explored. For example, after students plotted their school buildings electronically, one teacher did not know that a few of her students thought that latitude referred to the length of time it takes for a plane to travel in hours and minutes between two points. In another class, the teacher did not know that after doing the grass and nitrate experiments in What's in Our Water?, a couple of her students wanted to send their nitrate results to farmers to inform them that fertilizers do not help plants grow (a conclusion that is discrepant with the *Teacher's Guide*). In that same class, many of her students concluded, after observing the effects of chlorine on yeast, that the bubbles of carbon dioxide released by yeast were the yeast themselves. And in another class studying weather, the teacher did not know that some of her third and fourth graders conceptualized clouds as floating containers that burst when overfilled, and think clouds would feel "very soft, and your hand would just go right through them; you couldn't exactly hold it tight." Another student in the same class explained toward the end of the unit that the temperature in Michigan would be warmer "because they had a heat wave at Christmas once."

In most of these instances, the teachers were not aware that their students were holding these misconceptions. That teachers were not observed probing into their students' conceptual understandings about the unit topic is supported by teachers' self-reports indicating their tendencies to assess student levels of participation and process skills rather than specific content knowledge.

During the field observations, most of the teachers did not ask questions like, "What do you notice?" "What did you find out?" "How do you know for sure?" "What would you like to do next?" "What ideas do you have?" "If that's so, then how do you make sense of this?" In other words, teachers generally did not elicit students' conceptions about phenomena nor the nature of inquiry. As for the few instances in which teachers did elicit their students' ideas, in one case, a teacher did so without ever responding to those ideas that did not match current scientific thinking. In another case, a teacher elicited her students' ideas only during whole-class discussions and not while working with small groups or individual students. And in a number of cases, teachers elicited students' ideas one moment and then did not attend to answers or interrupted them. Too often, teachers seemed to be attending to other demands, such as figuring out whether the modem was connected, telling the students again how to turn on the printer or to operate the line-feed, answering the telephone, or getting resource materials out and accessible.

AN, a 20-year-plus veteran teacher and experienced Kids Network user from Site 4a, expressed her uncertainty about the program's return in terms of content knowledge. AN said, "I'm not sure if the students are getting the content. I have felt that with most of the process learning . . . but at what point do we enforce or lecture the content that you want to make sure they've gotten. Maybe they are getting it. I don't know."

TA at Site 6 described her assessment process of her students during the trash unit:

> It's very informal. It's a lot of observing, as they're doing things like trying to use their latitude, longitude, knowledge. It's watching to see if they're doing anything differently for snack time or lunch time, or if they're at least recognizing if they come in with a glass bottle and say "I'm going to take it home and use it for such and such a purpose; I'm taking my apple core home to feed to my horses." They're at least starting to talk more about trash, and about their own waste. I don't see any kind of a real formal assessment that's going into it.

SL, a fifth grade teacher at Site 5, reflects a common sentiment about assessment that was expressed by almost all the participants in our sample: "I do very little formal assessment. I assess attitude, cooperative learning, and also maybe some of the science process will be assessed when they set up and design experiments. But as to assess their understanding about acid rain, I probably won't make a big deal of it at all."

Informal at best, and process-oriented at least, teachers in our sample employed no formal strategies to assess student mastery of the unit's content knowledge or methods of inquiry.

Concluding Remarks About Enactments

These practices may not be perceived by teachers as being inconsistent or problematic, given the constraints, routines, and competing demands placed on them. It should come as little surprise that the Kids Network activities are executed at an extremely fast pace, that group work emphasizes managerial tasks, that students have few if any opportunities to design and carry out their own experiments, that teachers informally assess their students' engagement and not their content knowledge, and that students' deeper understandings and theories about phenomena and methods of inquiry are not explored. The teacher's guides do not offer suggestions for resolving these concerns.

◆

Assistance

A variety of types of internal and external assistance were available to the teachers in our sample during their implementation and continuation of the program (tables 11 and 12). Depending on the site, various combinations of assistance were provided. All teachers indicated that they received adequate assistance for implementing the program. What follows is our analysis of the sources and types of assistance received and offered by teachers during their implementation of the Kids Network.

External Assistance

Teachers identified publishers, developers, college professors, professional scientists, parents, colleagues from other schools, and a researcher as sources from outside their own schools that assisted them in implementing the program (table 11).

Teachers commended publishers for providing computer aid via a toll-free hotline. They cited developers for providing attractive and well-organized materials (e.g., *Teacher's Guide, Kids Handbook*, software manual, software). College professors and state scientists were guest speakers and, in some cases, loaned or demonstrated sophisticated scientific equipment. In one case, parents donated a number of units and offered technological support to a particular teacher.

Only the two teachers at Sites 4a and 4b talked about their informal collaborations over the years. These collaborations typically focused on fixing com-

Table 11.   **Types of External Assistance Received by Teachers Implementing Kids Network**

Source	Type
Publisher	Free computer hotline
Developer	*Teacher's Guide* and software manual
College professors and state scientists	Guest speaking; loans and demonstrations of sophisticated scientific instruments related to unit topics; benefactor of units
Parents	Benefactors of units and free technological support (e.g., installing phone lines, troubleshooting computers)
Colleagues outside of school	Informal collaboration (problem solving around suspicious data, using the computer, time constraints)
Researcher	Opportunities to reflect

puter "glitches" and, on a couple of occasions, conferring about suspicious-looking data. Finally, a couple of teachers indicated that their participation in this study gave them an opportunity to reflect more deeply about their use of the Kids Network.

Following is additional detail on some of these types of assistance.

Publisher Assistance. The publishers provided a free computer hotline which teachers could call five days a week, eight hours a day, with any questions they had about the program's computer components. It was staffed by four full-time employees and two alternates. According to NGS's Dot Perecca, project manager for the Kids Network, "The hotline is the most expensive part" of the program. Most participants identified the hotline as a positive service that they had used on numerous occasions during their Kids Network career. Says Perecca, "We had analyzed the kind of hotline calls we got and found that the majority of the hotline calls were really simple issues. How to turn your power strip on . . . It was that kind of desperate need." To the teachers, however, the hotline helped them resolve more significant problems. A number of teachers and an extremely well-trained technology specialist reported that they felt "close" to some of the hotline staff. HA from urban New Hampshire voiced the sentiment of the teachers when she said, "I felt that whenever I've had to call the headquarters, they've been very helpful."

College Professors and Scientists. Teachers from both college-town sites (Sites 3 and 5) identified the local college as a resource providing a rich supply of unit-related guest speakers. For example, HM at Site 5 said, "We're in a college town so we can bring in a college chemist, we can bring in the head of the energy plant. An atmospheric research scientist is supposed to come and talk to us next week about prevailing rain, about where our weather comes from." A group of about five students from this same class went on a field trip to measure the acidity of a number of waterways with an acid rain specialist working for the state. Similarly, scientists working for the state were guest speakers in MD's classroom (Site 1). They demonstrated some of the highly precise equipment they use to measure acid rain and other aspects of water quality.

Parents. A few years ago, a student's parents in KL's suburban New Hampshire class played a vital role in the program's initial investment. In addition to buying three units, they assisted KL with getting the computer program up and running. Consequently, KL has done four units with her third and fourth graders in as many years. The student's mother was pleased with its impact. She said, "What we liked about [the Kids Network] is that one gift has proceeded to benefit five years of classes and more than just one classroom, because it's expanded outward too. That's what we wanted . . . a long-range project that I hope will continue to grow."

Internal Assistance

Internal assistance refers to those sources that are directly involved in the daily operations of a school. Teachers identified district-level administrators, school board and community members, principals, teaching colleagues, technology and science specialists, and a school custodian as important sources of internal assistance in implementing the Kids Network (table 12).

District Administration and School Board. According to KL at Site 3, her school district received "block grants to put dedicated lines in all [five] of the schools." Although it is not absolutely necessary to have a dedicated phone line to run the Kids Network, a dedicated line in KL's case ultimately put a modem in her classroom and, consequently, gave her the freedom to telecommunicate at any time during a session. Her other colleagues using the Kids Network did not share this privilege. Site 3 also received state support for purchasing computer equipment which faculty were allowed to take home for extensive periods.

At all of the sites, the school boards have consistently approved financial investment in the Kids Network. The Site 5 principal's comment reflects the kind of support the other sites have received from their boards as well: "The school board here is tremendously supportive of these kinds of activities . . . Even in difficult economic times, we find ourselves able to preserve our programs, able to preserve, for the most part, our purchases." The principal from Site 4a echoed this sentiment when she said: "It's rare that we don't have monies for worthwhile programs. If a teacher comes to me and says this is something new, it's the wonderful part of this; we have a budget . . . In the 11 years, I've never heard of anyone who wanted a program who couldn't get it because of lack of funds."

Table 12.   **Types of Internal Assistance to Teachers Implementing Kids Network**

Source	Type
District-level administration	Win block grants for outfitting schools with computer technologies
School board/ community	Support development of school's computer technology capabilities
Principal	Finds money for unit purchases, promotes program to staff, takes care of annual paperwork, shows sustained interest, troubleshoots computer problems, finds community resources
Colleagues	Collaborate informally, typically around computer-related problems; share resources; provide comfort; teach-teach
Technology specialist	Troubleshoots computer problems; sets up school-based LAN; hotline spokesperson
Science specialist	Performs telecommunications components, completes bookkeeping, teaches teachers, promotes program, provides materials
Custodian	Collects materials, provides expert perspective, and chaperones field trips

Principal Support. Principals were credited with assisting in a greater variety of ways than any other source of assistance. They played key roles in finding the program, getting financial support to purchase some units, supporting teachers who wanted to field test the units, promoting the program to the staff, taking care of annual paperwork, exhibiting sustained interest, troubleshooting computer problems, and finding community resources. JM at Site 1 described how her principal played a key role in enabling her to use the program:

> He found the program itself for us, he found the material; you know, the literature that came to him, he purchased it and did all the paperwork end of it and then we . . . got to pick what time we wanted it done . . . he was additionally supportive in getting the new modem and was supportive when we went to the PTO [Parent-Teacher Organization] to get the computer over there to stay as the telecommunicating center. So he was very supportive in all those hardware problems as well as funding the original year's materials.

JM's colleague MD also felt strongly about the importance of his principal's ongoing support in implementation of the program:

> By asking me how things are going, by being interested in the results, making sure that I know that he feels that it's important by understanding what the program is all about, by knowing [himself] the components of the program, recognizing that it's designed to be certain that kids . . . behave as scientists do . . . Just as students need to know from their teachers what's important, I certainly knew from his actions that the Kids Network was an important source of science teaching education for teachers and for kids.

The principal at Site 6 explained his technology vision for the school. "What we're looking at next year," he said, "everything will be electronic . . . like faculty meeting notices, all that stuff that we put on our own bulletin board here in the building. So, it's going to cause people to have to take it." By networking the school and creating an infrastructure requiring teachers to use their computers, the principal thinks he will enable more staff to take on programs like the Kids Network.

Colleagues. The amount of assistance from fellow Kids Network teachers varied. For example, at three schools (Sites 4a, 5, and 6), pairs of teachers teamed up to combine their classes and share a single registration. For these teachers, meeting during the weekends to discuss their upcoming plans was part of their regular planning cycle. This was very different from SE at Site 4b, who was the only Kids Network user in her school. Her collegial support came from AN, a Kids Network user at another school (Site 4a) in her district. At Sites 1 and 3, there was an informal network of collegial support among two to four Kids Network users, typically focused on troubleshooting computer-related problems. Making sense of the *Teacher's Guide*, generating ideas about field trips and guest speakers, coordinating activities and materials, and troubleshooting computer problems were the most common ways teachers described their colleagues assisting with their efforts with the program.

Many of the teachers at Site 1, for instance, reported that they received significant support from their colleagues. For example, HA, the more experienced Kids Network user, identified her colleague MD as "a troubleshooter; physically his room is near the library and if my kids were in there, and we were running into a problem, then I could say, go tell MD I need help." MD also said that HA played an instrumental role in assisting him with the Kids Network:

> She was the first person that gave me, frankly, all I needed . . . She was the one that introduced me to all the different aspects of it, how they use the menu, how to telecommunicate, what problems we might run into in telecommunications as well as in helping me make sure that everything was done on time and answering any questions that I had at any time.

Only a couple of teachers indicated that a colleague provided pedagogical support. Whereas HA described MD as a technological troubleshooter, MD credited HA with transforming his practice. When she was both teaching at the school and working on her doctorate in education, HA spent a lot of time observing MD's classroom. According to MD, his colleague helped him "do

Kids Network in a more constructivist fashion." He said, "HA has made a real impact on my thinking and it's been about a three-year process . . . I'm not sure that I've said to her . . . how significant the change has been that I feel over my instructional strategies." Before this, MD felt that he had been so consumed by teaching that he spent little time thinking about the philosophical and theoretical aspects of teaching.

Technology Specialists. Across the six case sites, two schools have a full-time in-school technology specialist, two schools share a technology specialist with other schools in their district, and two schools each have a teacher who has earned a reputation as an in-house computer expert (table 13).

Table 13.   Types of Technology Specialists Across Cases

In-School FT TS	Distant TS	S-T Teacher
Site 2, 5	Sites 4a/b, 6*	Sites 1,3

*At the start-up of Kids Network, the school librarian was considered the technology specialist. Later, a more specific position was created to serve the district.

These technology specialists assisted Kids Network teachers in times of need. The technology specialist at Site 5 described a common context:

> Usually when they talk to me, it's about a problem . . . Generally, it's the software isn't working, or we're not getting any connections. "Can you help me set this up, install it," those kinds of things first. It's not always the case, but in general, come see me if it's not working, if there's a problem. That may be unfair where Kids Network is concerned, but it's specific to what my position is, I guess.

TA at Site 6 described how essential it was for her district's technology specialist to "hold her hand" when she began using the Kids Network:

> When I did the first unit, I had somebody holding my hand the whole way. Our technology person was then our librarian, and she had a real interest in the Kids Network and making sure that it was working well. So she devoted as much time as I needed. I would go and say: "I'm not sure how to find where these letters came back." She was there holding my hand through the whole thing.

The science specialist at Site 2 has taken on the primary responsibility for all of the Kids Network teachers' telecommunications components. Consequently, she has provided little assistance as a science specialist. Instead, she assists teachers in a lot of the program's "nitty-gritty" components and paperwork—freeing up teachers from having to be responsible for those elements of the program. She said:

We're going to be starting four [units], so that's one of the problems, because there's so much and it's hard for me just to keep track of whose letters go to whom and who's got what pack for it and so forth. So, I end up, that's what I spend my time on and I don't get a chance to . . . see what the classes have done; just the bookkeeping I'm getting, since this is the place where the modem is and where all the materials are. So I don't get to do as much as I like because I'm doing all the nitty-gritty stuff, the paperwork. I just have to make time. I take all the letters home, and I read them at home.

The districtwide science specialist who first introduced AN and SE to the Kids Network played an essential role in the implementation of the program's technological components. For AN, he did a lot more than just set up her modem. She explains, "I had to have somebody that gave me confidence that what I was doing was good. He would stop by occasionally and I would show him something, and he'd say, 'That's certainly great.' Then he'd ask me to run a workshop on it."

Concluding Remarks About Assistance

Teachers identified a number of sources from both inside and outside their school that facilitated their implementation of the Kids Network. Even so, one source of external assistance noticeably absent from participants' reports is the private sector (i.e., local businesses). Additionally, a type of assistance lacking from most participants' reports is assistance with pedagogical issues. Teachers were not observed, nor did they talk about, assisting one another regarding more substantive pedagogical concerns, such as how to help students ask better questions, ways to make the electronic plotting of teammates' activity more meaningful, how to give students more opportunities to design their own experiments, etc. This is the type of support that Dot Perecca from NGS said will be necessary if more traditional teachers doing the Kids Network are to make the requisite changes in instructional approaches that the program demands. She asked,

How do you train a teacher, who's taught for 15 years from a lecture method, all of a sudden to quit lecturing, and to put the skills of learning into the hands of the children? That's hard, that's an enormous task to do. You've got to have somebody within that school system that's going to say, "This is what we're going to do, this is how we're going to train our teachers." We're going to take them, give them an opportunity to learn and see, and experience how to do what it is that we want.

The general absence of pedagogical assistance may explain in part why so many veteran users remained faithful to the original program.

◆

Implementing New Technologies

A number of factors make the implementation of the Kids Network's computer components particularly interesting. First, the program requires a considerable minimal investment in technology. Second, the program lasts only a fraction of the school year; this in turn affects teachers' abilities to master the computer components. For many of our teachers, this resulted in an underutilization of the program's computer components both during and after a unit. Although teachers expressed a variety of concerns about the program's software and their school's hardware, they generally referred to the program's technological components with high regard. We summarize below key points that have been noted earlier in our case study report.

Minimal Requirements

The Kids Network has been promoted by NGS and considered by our participating teachers as a technological innovation because it introduced them to a way of using off-line telecommunications with their students as well as to a mapping and graphing program. As discussed, at minimum, a teacher needs access to at least one computer (PC, Apple IIGS, or Macintosh); a modem; a printer; and a dedicated phone line. All of our participants had access to more than the minimal technological resources necessary to enact the program when they started it. Since then, their access to not only more computers but also to increasingly sophisticated hardware and software has grown considerably. All the teachers have at least one computer in their classrooms, all have access to a modem either in class or nearby, and all have access to an in-school technology specialist (either someone officially in that role or a knowledgeable and willing colleague). Additionally, all but two of the schools have computer labs with as many as a dozen to two dozen computers. The other two schools have a common computer area with fewer than six computers and are in the process of securing funds to create a computer lab.

Mastering the Technology

For most of our teachers, the telecommunications component was the most novel part of their technological experience. While enacting the recommended activities in the *Teacher's Guide*, teachers also had to learn new computer terminology, protocols, equipment, software, and functions. Several factors can help a teacher master these technological components of the Kids Network (figure 3). For example, some principals created a supportive infrastructure in which teachers have access to equipment, assistance, and training; developers created a software manual; and the publishers offered a toll-free computer hotline.

The rapid pace of schooling in general. and the role of the computer within the Kids Network's intense time frame in particular, put teachers in the position of having to seek quick remedies from other sources (e.g., calling the hotline or seeking help from a technology specialist). Consequently, teachers and students have limited opportunities for solving problems through experimentation. When this tendency to rely on others to solve imminent computer problems is

Figure 3.   **Factors Contributing to Mastery of Kids Network Technology Components**

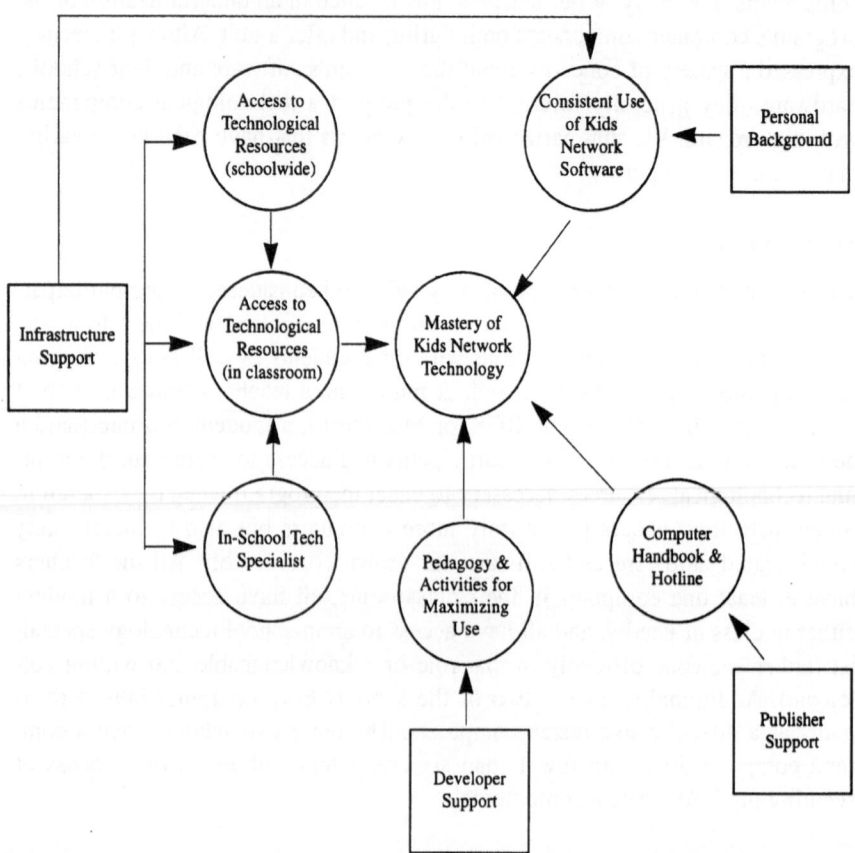

compounded with an annual 10-month absence from the Kids Network materials, the difficulty of mastering the computer components becomes even more apparent.

The Underutilization of Kids Network Technological Components

Even though the teachers in our study have infrastructures that support the development of their personal computer competencies, have access to more than the minimum technological requirements, have a Kids Network computer manual, have used the computer hotline, and have access to technology specialists, our observations indicate that the schools' technological resources are underutilized during the Kids Network. Table 14 illustrates that although teachers at five of the six schools had access to a well-equipped computer lab, only one classroom was observed using the lab for Kids Network purposes (e.g., the mapping activity).

Table 14.   **Underutilization: Computer Resources Compared to Possibilities**

Site	In-class computers	Computer lab w/12+ computers	Computer lab used or Kids Network	Primary user	Comments
1	1 w/modem	No	-	Teacher	Teacher telecommunicates before and after school; students never observed using computer
2	5	Yes	No	Science specialist, students	All telecommunications done by science specialist in science lab; mapping and graphing program done in classroom on one of five computers
3	1 w/ modem	Yes	No	Student pairs	Pairs frequently rotate on one classroom computer to telecommunicate
4a	0	Yes	No*	Student groups	Four to five students frequently rotate on one computer in computer lab right outside classroom
4b	1 w/ modem	Yes	No	Teacher demonstrations	Whole-class demonstrations on one computer with minimal direct opportunities for students
5	1 w/ modem	Yes	No	Teacher	Teacher telecommunicates before and after school
6	6	Yes	Yes	Teachers (telecommunications); students (mapping)	Computer lab observed used for mapping activity

*Only one computer loaded with modem and used for Kids Network mapping, graphing, and telecommunications.

One might expect that veteran users would have taken advantage of these labs and created systems to ensure that as many students as possible could use the telecommunications, mapping, and graphing tools as often as possible. This was not observed. For example, most students in most classes did not have a chance to use the telecommunications or graphing tools; most of the sites, however, had the technological capability of involving the entire class, working alone or in groups of two or three on their own computers. Only Site 6, however, made enough copies of the Kids Network software so that an entire class of students, working in pairs, could do the mapping activity. AN at Site 4a shares a sentiment expressed by many of our participants: "It would be nice if students could use all the hardware and software without student/teacher assistance. If they were more familiar with it, they certainly could. Unfortunately, they don't have enough opportunities to really learn it."

The students in the intern's class at Site 1 were never observed using computers for Kids Network purposes. The teacher explained that she would ask one student to periodically pick up anticipated e-mail. Toward the end of the unit, this teacher's supervising teacher connected his computer to a television monitor to lead a whole-class discussion about teammates' data. HA, at the same site, indicated that she used to take her entire class of fourth graders to the library where the one computer with a modem and the Kids Network software was located. Now she, like the science specialist at Site 2 and the teachers at Sites 5 and 6, does most of the telecommunicating before and after school. The third and fourth graders at Site 3 and the fifth graders at Site 4a regularly rotated in pairs to telecommunicate. The teacher at Site 4b typically led whole-class computer demonstrations and occasionally assigned a pair of students to telecommunicate or do the mapping program.

Although the teachers at Site 6 typically did the telecommunicating from the IIGS in the teachers' room, their entire class was observed working in pairs on the mapping program in their computer lab. This was the only case where a teacher had taken full advantage of the school's computer lab and loaded each computer with a copy of the Kids Network software. Since the final sessions of most of the participating classes were not observed, the extent to which students used the computer to analyze the network data is uncertain.

Some Technological Concerns

Even though the Kids Network was most teachers' first introduction to using telecommunications with students and provided for others their first experience using computers to gather, analyze, and share data, none of the teachers in our sample indicated that they felt intimidated by the program's technological demands. All the teachers—as well as the developers—did indicate, however, that they had one or more concerns about the technological components of the program. Three concerns raised by teachers are summarized here: inhibiting student-centered inquiry, attrition of teammates, and slow and unsophisticated software.

Inhibiting Student-Centered Inquiry. One teacher wondered whether being hooked into the telecommunications component of the Kids Network was so constraining that her students missed opportunities for thinking critically and creatively. She wondered whether her students would have more opportunities to design weather-related surveys to find out answers to their own questions if they did not have to collect the same data as all the other research teams.

TERC's Kids Network developer, Candace Julyan, agreed with this teacher's concern in an interview:

> One of the problems with telecommunications is that it requires everyone to be doing the same thing at the same time . . . you have to move pretty fast and it doesn't give you a lot of latitude to follow meandering paths. If you're not doing the telecommunications component, you can follow a lot of meandering paths.

Julyan finds the network and the fact that more than 10,000 schools are using it as two of the most important constraints of the program:

> because everyone has to do everything at the same time . . . Those two things make what would make sense to me to do in the classroom impossible, because with the network, you have to keep moving, and everybody has to be moving at the same pace. And time is a real killer with Kids Network, and I don't think time and constructivist teaching go together very well.

Narrow View of Computing. Julyan also emphasized that

> a computer as a telecomputing tool is such a tiny part of this tool's potential; it just seems trivial to me. I realize that it isn't trivial to anyone else . . . When you ask teachers what did kids learn, they'll often reply that kids learn what telecommunications is. My feeling is that's not very deep, so that's not as deep as understanding graphing by seeing how graphs can misrepresent data by changing the scale . . . If they only used the computer for telecommunications that would be a shame, because the software is really designed to help kids understand that computers aren't a toy; they're a tool.

Attrition of Teammates. The technology specialist at Site 5 indicated that his teachers are frustrated by the consistent fallout of participants by the end of each unit. From his perspective,

> One of the problems I guess that's consistent from year to year is . . . you start off with 14 members on the team . . . and you wind up with 75 percent if you're lucky, usually it's half. Where are these other teams, they paid their money, where are they? That's probably the most frustrating part of it that I hear from teachers.

HM, a teacher at the same school, shares the technology specialist's assessment of their colleagues' concerns. However, she also points out that her students were not as disappointed as she was about the high drop-out rate:

When I asked that question of the kids at the end, they didn't seem all that upset that they hadn't gotten letters. They thought it would be interesting, but there wasn't a big outcry, so it may be a personal thing, but it seems to me to be much more interesting having read that letter and then making your acid rain prediction. You feel as though you have something to go on, a small connection to it. So, only getting four letters out of the nine teams seemed to me a major flaw in the system.

Slow and Unsophisticated Software. When the Kids Network was first developed, its prepackaged, highly structured technological components were innovative and on the cutting edge of integrating telecommunications and other computer software with science inquiry. Although the software is now compatible with faster computers and modems, its original basic characteristics remain the same after some 10 years. For example, all telecommunicating is done off-line and is driven by a series of deadlines. When compared today with other telecommunications, word processing, mapping, and graphing programs, the Kids Network is relatively expensive, slow, and unsophisticated.

MD at Site 1 indicated that "up until this point, the telecommunications aspect was pretty unique . . . Now we have three telecommunications opportunities available every day for students in the class. I have a computer with a modem hooked up. I have access through the new University of New Hampshire Mentor System to the Internet, as well as to just about any similar service . . . We also have America Online through Scholastics."

The teachers tolerate the Kids Network software despite their concerns because it is a piece of a larger curriculum that the teachers do like.

Concluding Remarks About Implementing New Technologies

A number of ironies and paradoxes are inherent in technology-dependent curriculum. For instance, the high rate of technological change guarantees almost instant obsolescence. As a result, computer-literate teachers and students become quickly frustrated with "slow and unsophisticated" hardware and software. Furthermore, mastery of the computer components is not going to be easy when they are used only six weeks out of the year.

Most students, even those who never had an opportunity to telecommunicate directly during a unit, described the telecommunications component as one of the program's major appeals. One group of fifth graders was asked, "What computer changes would you make so you can be on computer more?" Their responses represent a range of ways that many of the students felt about the role of the computer while doing the Kids Network:

- Andrew didn't think the program was a computer unit, but rather about testing pH.

- Christa suggested they get a faster modem and come up with ways for all students to get on the computer.

- Ariel didn't offer a suggestion, except what can be inferred implicitly in her response, "I wasn't on computer a lot."

- Yoko also said, "I don't remember going on computer." She added that the computer doesn't help the Acid Rain unit. She emphasized that "it is better to test [experiment] and see what happens then to see if mail has come or not come."

- Nancy, who also doesn't recall being on computer, explained that she thinks that it is good to learn how to do research with and without a computer.

A couple of teachers indicated that the program could be as successful either without the students' direct participation with the telecommunications component or without the telecommunications altogether. For example, SL at Site 5 indicated that while the telecommunications component is essential to the program, "it's not essential that the kids should do it and see it." Her colleague HM agrees. She says she does not think "it's telecommunications that's exciting for them; I think it's hearing from other teams, whether it's on the computer or whether it's the U.S. mail." HM adds:

> One of the things that is really interesting is that we haven't used the computer at all . . . On the whole nobody has even looked at the computer. [My colleague] doesn't have [the computer] in her room, so obviously her kids haven't looked at it. So what's interesting is this is supposed to be a telecommunications program, and that's not how we're using it, which is kind of fascinating . . . [The telecommunications] seems to me the least important part of what the program represents.

Julyan agrees that the program can be done effectively without telecommunicating: "It would be possible to do the Kids Network and never use telecommunications at all . . . and I would still consider that a useful, productive way to use the curriculum . . . as long as they had real data that other kids had collected, and the data were not tied to a particular time frame."

◆

Student Perspectives

The everyday experiences of the students who participated in the Kids Network have been described throughout this case study. We here present students' ideas about how they compare the Kids Network to other curricula, what they like and dislike about the program, and what they would like to do next. Students were interviewed individually and in small groups during the second and third round of field visits.

How the Kids Network Compares to Other Curricula

Students compared their experience with the Kids Network to what they do elsewhere, notably in science. They described the Kids Network as unique on a number of criteria: more experimenting (and less reading and writing), telecommunicating, and studying real-life problems.

More Experimenting and Doing (Corollary: Less Reading and Writing). Luke, a sixth grader now in his second year with MD (Site 1), was considered by his teacher to be the "strongest" science student. Luke said he feels that what he does during the Kids Network is not like doing science at other times of the year "in any way possible. Because in Kids Network we do experiments and we usually talk about stuff and we type letters out, whereas in science—usually it's from the text, answer and complete sentences, write down all the vocabulary." One of Luke's classmates continued his description: " . . . and then answer about four questions at the end of the section. After you finish doing that, you just move onto the next section."

A fifth grader at Site 5 doing Acid Rain said this: "[Now] we're going outside and getting the things and then testing them instead of watching her do it . . . We're doing it." A fifth grader Site 4a agreed. He explained that the Kids Network is different from social studies, because when "we're doing social studies and stuff, we just read out of the books, but when we're doing science we get to run tests and everything."

Dana, a fifth grader in rural Vermont (Site 6), insisted that most of the time her class does not do experiments when doing science. Rather, she explained, they do research by "looking up stuff in the library, [and] doing a report on it."

Students identified "doing" and experimenting as key Kids Network activities that make it distinct from not only other subjects but their prior science experiences as well.

Telecommunications. Most students mentioned the telecommunications component as a distinguishing feature of the Kids Network. A fifth grader from Site 5, one of many students who highlighted this component, explained: "It's kind of like having a pen-pal, since you're writing to them every once in awhile. I guess we just don't do those things because we're reading or writing or something like that. It's really different."

Since pen-pal relationships depend on reading and writing, it's interesting that this student identified this feature as a distinguishing mark. Most students, including those who never had an opportunity to participate directly in any of the Kids Network telecommunications sessions, reported that this was one of the "best parts" of their Kids Network experience.

Focus on Real Problems. A number of students doing Acid Rain and Too Much Trash? mentioned the importance of solving real-life problems. For example, a fifth grader from Site 6 explained that the trash unit's focus on real problems made it distinctly different from the kinds of problems in mathematics: "With [Too Much Trash?], it's real and there's a real problem with trash, because there's so much of it. [In mathematics,] it's just a problem that someone made up and they're doing it to kind of make you smarter."

A fifth grader in HM's class (Site 5) explained why he thinks a friend cannot remember the experiments prior to the Kids Network—those experiments were "nothing of importance, just little tiny five-minute experiments."

Greg, a fifth grader from Site 6 emphasized that the Kids Network, unlike the problems they work on in mathematics, "is like really real." Hillary agreed with her classmate and added that the Kids Network presents an interesting "story . . . because we're figuring out something, not just figuring a [mathematics] problem."

Another fifth grade girl from this class explained that studying trash is so important "because we're going to have to live on the earth for the rest of our lives and we don't want to live in trash." Her classmate added, "because we have to make it nice for the next people who live here." Hillary reiterated, "I think it helps us all realize what we're doing and what we can fix."

What Students Liked

Most students said that they liked the Kids Network because it was "fun." Fun, according to the students, involves doing one of two things: experimenting or building things, and telecommunicating. In general, what students liked about the Kids Network was very similar to what they said makes the Kids Network different from other things they do.

Experimenting and Doing. Student after student reported enjoying doing the experiments. Those doing Acid Rain repeated, "I like testing the water—how much acid is in the water." And many students highlighted the making of their own rain collectors as one of their favorite activities. Students doing Too Much Trash? repeated "I liked measuring our trash." Those doing What's in Our Water? liked "doing the fertilizer experiments." And the third and fourth graders studying Weather in Action liked going outside every day to record the cloud types and take the temperature. A number of students from each site also indicated that they liked comparing the results of their experiments and measurements to their teammates' data.

Telecommunications. Most students indicated that they liked the telecommunications part of the program. Even students who never participated firsthand with the telecommunications liked this feature best. Ben, a third grader from suburban New Hampshire studying Weather in Action, said, "It's cool and it's fun to hear from different cities and different places, about what kinds of weather they're having." Similarly, a fifth grader at Site 6 said, "It's neat finding out where, like other places, how much trash they use, because they're in different places and you might think that they use more, because they're in a certain place, but sometimes they don't, and it's cool finding that out."

At least four fifth graders in AN's class (Site 4a) said that what they liked best was sending (via conventional mail) a pH testing packet and letter to a relative or friend in a foreign country and then comparing their acid rain data. This is an additional activity made up by their teacher.

What Students Did Not Like

Most often, when asked what they did not like about the Kids Network, students replied, "I don't know, it's fun"; "Nothing really bad"; and "I liked everything about it." Very few students were able to describe what they didn't like. A few mentioned feeling rushed, the slowness in telecommunicating, and having to always work in groups.

A boy studying Acid Rain in AN's class (Site 4a) said, "I hate having the teacher talk about it all the time . . . It's just like a constant thing . . . In the beginning it was really interesting, but after awhile it just gets boring, if you do it every day."

Mat, a fifth grader from the same class, said he does not like "not having a lot of time to work on things . . . I wish we could do this the whole school year." Maggie, a fifth grader from SE's class (Site 4b), echoed many others: "Well, I like least about how the computer, it goes very slowly and the line seems to always be busy. I would like it more often and more quickly then we do now."

A couple of students studying trash at Site 6 indicated that they do not like "writing the stuff down that we do in school each day." Furthermore, a couple of students insinuated they did not like working in groups. For instance, Ariel said that she would prefer to write to her own pen-pal. "Then," she added, "we can ask our own questions." Similarly, her classmate Rachel thought it would be better if students had their own materials to do the vinegar experiment.

What Students Would Choose to Do Next

Most students did not hesitate long before answering the question "If you could decide what you did next in this unit, what would you choose to do?" Their choices can be divided into one of five types of activities: experimenting, constructing, communicating, field trips, and social action.

   Experimenting. A fifth grader from suburban New York participating in Acid Rain wanted to learn firsthand about the ground and the animals that live in it. Jared explained, "I think that I would choose to, I think that I would like to learn more about the ground and the animals that lived in the ground . . . and maybe if we did it we could take samples or something of the ground and see if there was anything . . . I think that would be very interesting."

Hillary, a fifth grader studying trash in rural Vermont, wanted to do a compost experiment: "I'd want to find out how long it takes, like, an orange to compost . . . I'd like to see each day, study a compost pile, each day write what's happening to an orange or something."

Christa wanted to figure out whether paper cups might give different acidity readings than other types of cups. Eric wanted to figure out why and how the pH paper changes color. A fourth grader, having done Acid Rain the year before, said he wanted to do more testing like "testing different spots, like see if there's more acid, if you put it up high in a spot." Another student wanted to "do maybe three tests or one on the chimney and put one near the smokestack and put another one that got sun. And so we could see the difference between the smokestack and where there's not very much smoke."

Constructing. A couple of rural Vermont boys studying trash reported that they would want to construct something. For example, Greg wanted to construct a mini-landfill: "I think it would be fun, it would take forever . . . but to make a little mini tiny landfill . . . so that you can get trash from the kids and stuff and families. You could like change it into soccer fields or whatever and it would be a good thing for the school."

Shawn wanted to build his diagram of an electric plant that burned trash, including its own waste.

Field Trips. The field trip to a landfill by a class at Site 6 made a very big impression on students. Shawn is only one of many of his classmates who highlighted the field trip to the dump as what they liked most. He included other types of field trips as something he would choose to do next:

> Maybe a bit more trips that we could find out more. I mean maybe visit someone, another state's landfill, see what theirs is like. Maybe even see an incinerator. By doing that you could learn more, and just take notes during that instead of spending all your time at school and just learning from books. Which may not even be updated. So why go for a book which may not even have the actual information, when you can just go out there and find the actual number?

Social Action. The last few activities of the Kids Network units involve students' educating others about what they have learned and considering other actions they might take to reduce a problem. Even before doing this part of the unit, a number of students studying trash in rural Vermont said they wanted to educate others and offer help. For example, Hillary said, "Well, we could make something. We could make people realize it, like, hang up signs, saying 'Recycle for the Earth' and stuff like that." A classmate agreed and suggested, "Or we could do a thing where people, my aunt can't find a place to bring her plastic and stuff and we could have a thing where you bring all your recyclable stuff here. We send it to the recycling place."

Andy wanted to do a schoolwide experiment as a way to reduce the school's trash problem:

> If it was my choice what to do next, is, I would kind of make an announcement to the school and tell them about what we've been doing and tell them that they should really cut down and stuff and take a whole school measurement of all the trash, in one day of being normal and then, of the next day or so, trying to measure it all out . . . It would be really cool and then we could tell the whole school afterward how much trash there really is. They probably don't even realize that they're throwing away that much stuff.

Concluding Remarks About Student Perspectives

What students identified as unique is also what they liked about the program. This, in turn, complements the types of activities they want to do next. The key elements, according to the students, were:

- experimenting or "doing,"
- constructing,
- telecommunicating,
- less reading and writing,
- going on field trips, and
- investigating relevant topics.

In general, students' responses to the program were very positive. Most of them considered the Kids Network to be very different from other subjects. They said it involved more "doing," more experimenting, less reading and writing, and—for those doing Acid Rain, What's in Our Water?, or Too Much Trash?—learning about something "real." Most students said telecommunicating was also novel and one of their favorite parts. Some students also identified doing experiments and going on field trips as favorite Kids Network activities. Some students from all sites said they did not like the slowness of the telecommunications sessions and were frustrated by teammates who dropped out. When asked about what they would choose to do next, students had little difficulty suggesting one of four types of activities: experimenting, constructing, telecommunicating, and social action.

◆

Changes in Practice and Program

Although we visited our field sites through an academic year, we had stipulated that teachers have at least three years of experience with the Kids Network, so that we could witness more masterful models of teaching, more powerful or novel inquiries, and mutual adaptations between the school and the practice. This is not exactly what we found, although we depict below some features of progression or refinement we observed on the part of Kids Network teachers.

Initial Preparedness

The information in hand, primarily from the teachers, seems unequivocal: There was minimal initial training or preparation of any kind. This "orphaning" has important consequences. To begin with, it encouraged teachers in the Kids Network (as well as those engaged in the Voyage of the Mimi innovation) to follow the guides and manuals very closely, even if program overview and introductions called for pedagogical "adventures." This kept the comfort level high, but it drastically reduced the change level—the liberties teachers might have taken with preset formats—and, in so doing, might have lowered the potential impact on teachers and on the range and depth of experiences made available to pupils. We thus have a greater threat of unexamined misconceptions during experiments and explanations, along with a greater threat that pupils, too, will hold passively onto prior understandings of the processes and content under study. As always in such cases, the best prepared teachers were those already familiar with inquiry-centered science or mathematics and comfortable with the uncertain twists and turns it typically takes as pupils formulate and then test the problems suggested by the *Teacher's Guide*.

The absence of modeling, systematic observation, self-observation, and group discussions based on similar attempts to execute an activity that met important program goals—some of the staples of initial training—left teachers with few models to show how the program might be executed. As a result, this might have pushed them still more to the only available model, the *Teacher's Guide*. Studies of innovation mastery show clearly that initial steps are highly mimetic—that is, that users remain faithful, almost word for word, step by step, to the guides that accompany the program, then gradually develop their own repertoire. In these cases, there were fewer instances, even in later phases, of liberation from the guides' instructions for designing and executing experiments.

Why? There is a paradox here, in the contradiction between an inquiry-oriented view of science/mathematics pedagogy and a view of instructional strategies as mostly fixed, predefined, and mimetic of the models portrayed in the *Teacher's Guide*.

Theoretically, we should have found the classic S-curve of innovation mastery (figure 4), first devised by Hall and Loucks (1977) and refined by Huberman and Miles (1984). Each of the steps denotes a different phase in the mastery process and is largely self-explanatory. For the teachers in our study, it typically took two or three years to reach the phase of "initial coordination"; a few are already at the succeeding phases and are beginning to combine the Kids Network with other—sometimes new—components of their instructional repertoire. One or two are at the "refinement and extension" phase. But there are also plateaus, regressions, and sudden spurts, none of which are rendered by the shape of the curve which, at best, can only suggest the nonlinear quality of practice mastery.

In many instances, then, more experienced teachers would have a more constructivist orientation—one that puts the pupils more in the role of inquisitive scientists, often in open-ended situations, to resolve problems whose parameters have not been spelled out in advance and for which more than one solution may be possible. The role of the teacher here is to tease out pupils' underlying representations of the experiment they are working on, to confront different versions among the group, to counter-suggest or suggest modifications, and to engage pupils' schemata in more challenging and differentiated ways to further the logical and mathematical construction of pupil understandings.

Figure 4.   **The Classic S-Curve**

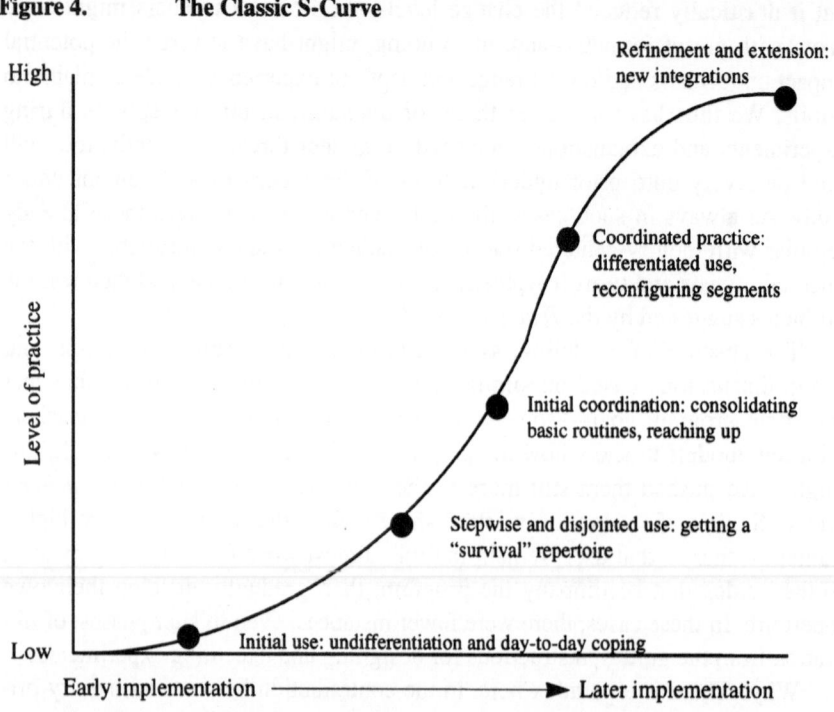

We saw little of this. Rather, we witnessed a more algorithmic mode, in which pupils follow a series of largely predetermined steps to reach correct answers. Much of this is hands-on science and mathematics; as such, it represents a new approach on the part of teachers in the sample, who worked with basal textbooks and whose forays into real-time experimentation constituted a genuine shift in pedagogy. Still, under these conditions, level of mastery becomes an ambiguous term. Moreover, to the extent that it is worked at in solitary ways, by trial and error and approximate successions, it goes unquestioned, if not unconsciously forward.

More fundamentally, it seems likely that the Kids Network will not transform the science/mathematics curricula until activities are implemented more interactively, with greater degrees of freedom for both teachers and pupils. This requires a stronger inclination to take risks, with the assistance of consultants who can help execute the cognitively transforming opportunities latent and present in the programs. Both teachers and pupils may become more sophisticated technologically, but they will have passed by the core intents of the program from a scientific and mathematical point of view.

Phases of Progression Among the Teachers[7]

During our interviews with the teachers, we asked them to characterize the phases they moved through while using the Kids Network. Many were asked to imagine the chapter titles to their Kids Network memoirs, depicting their journey with the program. What follows is a summary of their responses and a comparison of similarities and differences across the cases.

In Hall and Rutherford's (1976) work on measuring the implementation of innovations, adopters' initial concerns tend to be personal ones having to do with their own adequacy—how the innovation may affect their professional status, their sense of professional competence, their rewards. For more veteran users, these concerns change gradually to concerns about the correct execution of the innovation and the ensuing impact on students; and then to concerns about whether other innovations, or a new configuration of this program with another, can achieve the same goals more effectively. The process is often a slow one, and varies by teacher; it also has to do with the amount of assistance and support received and with a benign infrastructure (calendar, materials, regulations).

Our teachers appeared to follow this general pattern. Gradually, they seemed to move progressively from personal concerns; to matters having to do more with mastery of the program's technical aspects (e.g., telecommunications); to issues of how to use the program in ways that would have a greater impact on their students.

In almost all cases—regardless of the teacher's previous background in science, years of teaching, grade level when first implementing the program, and availability of or experience with computers—teachers in our sample characterized their feelings during the first phase of use as "overwhelmed" and "intimidated" by the telecommunications components of the program. It was a period of "intense—but largely solitary—preparation" and of "disorganization" (table 15).

Much analogous research (Fullan and Steigelbaver 1991, Huberman and Miles 1984) has documented a similar pattern for the early phases of complex implementations. At the classroom level, there are complaints of day-to-day coping, unsuccessful attempts to "make it work like it's supposed to," successive cycles of trial and error, exhaustion sometimes in getting through daily or weekly segments, and the sacrifice of other core activities (conventional science or mathematics lessons). In its own way, these authors conclude, difficulties at the outset are good harbingers. They signify that teachers are genuinely trying to come to terms with the program and to face the discrepancies between their own, congenial practice and the change-bearing features of the new program.

The difficulties seem related to the overflow of seemingly simultaneous tasks ("so much coming at me"); to unpredictability ("sometimes an experiment

---

[7]Each phase typically corresponds with a teacher's implementation of a unit. In KA's case, however, her phases refer to her journey within a single period of use.

works, sometimes it doesn't, and I did the same things"); and to a lack of under-standing as to how the program is constructed and interrelated. Practice, support, and what Fullanand Steigelbaver (1991, p. 106) call "working out one's own meaning" are key facilitators.

Clearly, the telecommunications components were the most unsettling. For HA at Site 1, it was fear of the technology that motivated her to master it; other teachers defaulted, initially, to a colleague who felt more comfortable with the technology. In most of our sample, teachers progressed through increasing levels of comfort with the technical aspects of telecommunications—and, thereby, through stages of greater technical and pedagogical efficiency.

Nonetheless, despite increasing comfort with the telecommunications and computer components, very few teachers fully profited from the software for their pupils, given the particular context of each user. In other words, even in those settings in which teachers had access to enough computers for every student, the class continued to enact the program on only one computer. This may be explained, in part, by the long period of not using the program; it is boxed and shelved for more than 10 months of the year. As a result, the annual learning curve is high with each enactment, but the depth of technological mastery remains relatively shallow over time.

Table 15 shows the first and most recent phases of progression reported by teachers since they began using the Kids Network.

"Innovation bundles" are innovations that are often introduced as single innovations, but that are actually several; Rogers (1962) demonstrates that teachers who cannot unbundle and try out parts of an innovation are less likely to be adopters. The Kids Network, for instance, is often considered an innova-tion using telecommunications in education; however, depending on the teacher, it could be considered innovative on several fronts:

- telecommunications;

- word processing;

- hands-on data collection activities;

- open-ended data analysis;

- content (e.g., acid rain, trash, water, weather, solar energy, etc.);

- small group collaboration;

- communications with other schools; and

- presentations to general public.

For most, telecommunicating was novel, but certainly not daunting. For some, the program's focus on real-world problems like acid rain, trash, and water quality was different from their prior instruction. For others, doing any type of hands-on, algorithmic activity was new. A few veteran users modified

Table 15.   **Phases of Progression: First and Most Recent Phases, Across Kids Network Cases**

Site: Teacher	No. of times Kids Network used*	Phase I	Now
1: HA	4	Ambivalence, uncertainty, and fascination with telecommunications components	Modifying a framework
1: MD	4	Absorption: "A vehicle that drove me"	Letting go and moving beyond
1: KA	1	High faithfulness to *Teacher's Guide*	Increasing student-centered activities
1: JM	3	High faithfulness to *Teacher's Guide;* dependency on colleague for technical assistance	Utilitarian (extracts the good pieces without having to use the entire program)
2: AT	2	Extreme faithfulness to *Teacher's Guide*	Anticipation
2: KW	10+	"Great expectations" of technological components	Pushed aside; not enough to do
3: KL	10+	Extreme faithfulness to *Teacher's Guide*; heavily teacher-directed	Increasing student-centered activities
4a: AN	4	Extreme faithfulness to *Teacher's Guide;* everyone on same task at the same time	Increasing student-centered activities
4b: SE	3	Discomfort with telecommunications components	Needing more latitude
5: SL	8	Extreme faithfulness to *Teacher's Guide*	Expanding; integrating other curricula
5: HM	3	Uncommitted; uncertain of program's potential	Enriched; more fun and interesting, personally
6: TA	3	Extreme faithfulness to *Teacher's Guide;* "panicky" about telecommunications components	"Adding in"; less excited
6: CM	2	Overwhelmed and confused	Confident and comfortable

*Includes time the unit was implemented during this study.

the program in ways that created other innovating experiences for themselves. Typically, the teachers worked hard to learn the content; many added resource books to the classroom library and sought out complementary field trip experiences and local resources, including speakers. As they became more comfortable with the structure, format, and technical aspects of the program, they modified the program to overcome some of their concerns and constraints. Most often, this resulted in doing fewer of the recommended activities and supplementing the program with guest speakers and field trips. The lack of equipment—such as a modem in their own classroom—was more difficult to modify on an individual basis, although teachers coped by begging, hoarding, and bringing needed equipment from home.

Only a few teachers, however, described their evolution with the program as being pedagogically significant. For example, according to KA, the teaching intern, her experience with the program was transformative. Within a six-week period, KA moved from what she called a teacher-directed "cookbook" approach to more student-directed inquiries. She evolved from having students read and follow the directions on their student activity sheets to asking students how would they would design an experiment. Her initial hesitations about the program became a stimulant. Also noticeable were an increasing number of strategies involving more student participation, particularly during discussions. She attributes this change, in part, to being involved in our study, which gave her structured opportunities to reflect on her Kids Network experience. Her pedagogical evolution also may have resulted from her supervising teacher's willingness to take care of the telecommunications and computer demands of the program.

As a science specialist at a science, mathematics, and technology magnet school, KW's journey at Site 2 is also instructive. She began with "great expectations," toned down the next year with a "twist of reality." During her third year, she literally stepped back from the program; instead of running the program in the science lab, she supported teachers in conducting it in their respective classrooms. Now she feels "pushed aside" with not enough to do, other than to continue supporting colleagues' telecommunications efforts.

MD, a veteran user at Site 1, experienced another journey, with some interesting side trips. Feeling at first absorbed and controlled by the program, he saw it as "a vehicle which drove" him. Although he initially was uncomfortable with the telecommunications component, he soon mastered it and became the local expert. By his third use, he felt he was in the driver's seat. Because of his initial discovery of faulty litmus paper used for measuring pH, he felt that students needed direct contact with scientists who were using precise measuring tools. He shifted his use to increasingly more reliance on students and professional scientists. He described his most recent phase as "moving beyond" the program and searching for ways to integrate the technology and instruction inherent in the program. MD indicated that his relatively recent understanding of and belief in integrating "outcomes-based performance" and "constructivism" had changed his orientation to the program.

JM (Site 1) also began as a colleague-dependent, *Teacher's Guide*-following user of the program. She described herself as becoming less dependent, first on her colleague for technical telecommunications advice and later on the *Teacher's Guide*. Now, after three years of use, JM is thinking about not registering for the program and instead just "pulling experiments that I enjoyed."

AN (Site 4a) described three phases she has moved through during her use of the program. In addition to the expected technical changes experienced by most of the teachers (e.g., an increasing level of comfort with the program's technological and organizational demands), AN's progression reflects a significant pedagogical change. She described her shift from having all students doing all activities at the same time to having different students doing a variety of simultaneous activities—a shift to what she describes as "putting the children in charge."

AT at Site 2 indicated that she has moved from being highly faithful to the *Teacher's Guide* during her first year to anticipating the program's demands and timing.

SE at Site 4b characterized the initial phase with the program as a period of discomfort with the telecommunications components. By the middle of her third and current year of participating in the Acid Rain unit, she wanted more latitude with the program. This is interesting because SE is the only teacher in our sample who made the program more teacher-centered and worksheet-driven than the *Teacher's Guide*. For example, she replaced a few of the hands-on activities with a variety of worksheets.

Phases of Progression: A Summary

Huberman and Miles (1984) point out that "mastery is both a process (it has several stages or phases) and an accomplishment (later phases are more 'masterful' than earlier ones)." They identify, in their participants' initial use of various innovations, three typical mastery-related difficulties:

1. feeling "overloaded,"

2. experiencing different consequences with different classes, and

3. feeling nearsighted in their inability to fully grapple with the programs' seemingly unintegrated fragments.

A number of our teachers also expressed these difficulties. However, a rough start is not necessarily fatal to implementation. In fact, Huberman and Miles judge the most successful projects as "rough starters." During later periods of implementation, users saw themselves as being more comfortable, confident, in control, and gratified.

Huberman and Miles (1984) compare implementation to a "sculpture in process, on which some parts of the body were rough-hewn, other parts already distinctive but in need of fine chisel work, and still other parts fully completed and integrated into the larger vision of the work as a whole." However, this metaphor may be misleading, for it suggests that teachers will ultimately complete the implementation process according to their larger vision. Since visions are dynamic, users of an innovation—unlike a sculptor who eventually says "I am done"—are potentially wedded to fine chisel work, a curse perhaps to some teachers and a blessing to others.

Because we cite here only the teachers' first and last phases of progression and thereby gloss over the intermediary phases, our users' progressions appear linear, as if there is a one-to-one correspondence between level of practice mastery and program experience. It makes them look as if they moved from a phase of technological incompetence to technological wizardry or from a period of complete faithfulness in the written program to high levels of spontaneity and experimentation. These interpretations are misleading; like the teachers in Huberman and Miles's study, our participants developed their practice mastery in a nonlinear way.

Changes to the Innovation

Teachers' phases of progression with the innovation also reflect ways in which they changed the innovation itself. Except for a few cases such as KA (Site 1), KL (Site 3), and AN (Site 4a), most of the ways in which the teachers modified the innovation were logistical and technical in nature. For example, two teachers began the program two weeks early. HA (Site 1) did this so that her students could get a head start on some of the concepts and technical aspects of the computer; for AT (Site 2), jump-starting the original program was her way of coping with the demands and constraints that might interfere with the program. It is interesting to note that none of the teachers talked about extending the program in the other direction—that is, past the closing of the network. For an integrated program like this, it might be expected that many student-initiated inquiries would naturally emerge from the unit, thus pushing the inquiry beyond the curriculum's time frame. This was not reported by the teachers.

Another technical change made by many of the teachers involved bypassing some of the primary suggested activities. In some cases, the teachers reordered the sequence of activities without sacrificing their telecommunications deadlines. In a number of cases, teachers provided students with more opportunities for personal or social communications across the network rather than scientific communications. AN at Site 4a is an exception to this. She added to the unit an opportunity for students to send a letter and pH testing materials to someone from another country, via "snail-mail" (conventional mail), asking them to test the pH of their rainwater.

Teachers at three sites enhanced the unit with field trips and guest speakers. For example, the students at Site 1 helped scientists monitor a lake, students at Site 6 visited a local landfill, and a few at Site 5 spent a day with a state acid rain scientist. Guest speakers ranged from acid rain experts (Site 1) and atmospheric scientists (Site 5) to a local custodian in charge of the school's trash and a local teacher who produced only three bags of trash per year (Site 6). Students described these supplemental experiences with enthusiasm.

As indicated earlier, only SE (Site 4b) modified the innovation by replacing some of the more hands-on activities with more than a dozen short-answer worksheets and readings from a traditional science textbook. A few teachers, however, described their evolution with the program as being pedagogically significant. KL, the veteran teacher at Site 3, described her shift from being highly teacher-directed when she first used the program to being more student-centered. This was evident when she supported a student's interest in creating charts that were different from and more complex than the ones provided in the *Teacher's Guide*—ones that would answer her own questions about her weather data. KA, the intern at Site 1, also modified the program during her first experience with it by supporting some spontaneous student-generated inquiries.

As a strong advocate of the program, MD at Site 1 explained that he modified two components of the program that he found to be problematic: the use of unsophisticated materials for scientific measuring and the absence of interaction with real scientists. These concerns also contributed to the development and recent publication of a lake ecology curriculum. "That's why when we had the chance to have our own expert in the class regularly, it made a lot of sense to me." He felt that these visiting scientists, as well as the ones his students began to accompany on field trips during his lake ecology curriculum, enabled students to experience more professional scientific instrumentation.

> I want kids to have authentic experiences in reading and language arts. I want them to have authentic experiences in science. Some of the experiences that they have are not authentic . . . I want them to know that when they test pool water that [the data they] distribute nationally are very inaccurate. I want them to interact with people who do it regularly, I want them to interact with people who know and can explain the difference between a calibrated pH meter and pH paper . . . that difference created a desire on my part to bridge that gap.

Overall, teachers modified the innovation to meet certain technical and practical demands, as well as to enrich the program with other adults and excursions. Interestingly, none of the teachers modified the student activity sheets, even though these sheets represent the majority of the students' time with the program. This is a crucial place where one might expect to see pedagogical modifications in the program reflecting teachers' changing conceptions of the nature of science teaching and changing levels of concern and comfort with the execution of the program. But along this plane of modification to the innovation, the teachers' journey seems particularly short. This should not be too surprising, for others have claimed that such individual or causal reforms to professionalize teaching "may not substantially change how teachers actually teach without additional efforts to enrich pedagogy" (Firestone and Bader 1992).

Finally, it is important to remember that the Kids Network program typically lasts six weeks and is not used again for another 10 months. Under these circumstances, we should expect a different practice mastery trajectory than that of the users in Huberman and Miles's study.

Changes in Teachers' Practice and Careers

The metaphor of journey does not only concern the ways in which the teachers' relationship to the program changed over time; it also charts changes in teachers' practice—and, in some cases, their careers. Miles and Huberman (1994) finds significant changes in teachers who took on innovations. He notes that users of nontrivial innovations often moved from nonteacher to teacher roles—from teacher to consultant, from consultant to coordinator—all the time building up specific skills that made it easier to go elsewhere, to move up some more in the educational hierarchy, to become better trained specialists.

MD's (Site 1) first experience with the Kids Network strongly influenced his practice and career. During that first year, he and his students discovered that the litmus paper provided by NGS was giving faulty pH readings. As a result, MD invited state scientists to come into his classroom to demonstrate more sophisticated acid rain equipment. He ended up collaborating with the state officials in the development of a curriculum about lake ecology. That curriculum addressed, in part, his concern that the Kids Network involved neither direct interaction with scientists nor precise equipment. His lake ecology program has led to the development of state and federally funded teacher's guides, video productions, and teacher training institutes.

KA, MD's intern at Site 1, credited the Kids Network with positively affecting the rest of her teaching internship. Because of the program, "I started really talking to them, what do you think, how's this going, what works for you, what doesn't? Then they really responded to that and now I do that with everything else too."

KA also attributed to the program the idea that it is acceptable to change someone else's curriculum and to use innovations as "just a starting point." JM, also at Site 1, gave less credit to the influence of the program alone than to her three colleagues: "So I think other things have sort of made me evolve . . . I couldn't really say that it was Kids Network; it was more a lot of other things that were happening."

Although KL (Site 3) said she felt that while the program is not responsible for changing her ways of teaching, it has influenced her career. Her Kids Network classes have been featured a number of times in nationally published education magazines; and she often assists NGS in presenting the Kids Network to teachers at regional, state, and national conferences on science and technology education.

The program affected HM's (Site 5) teaching primarily by making more apparent the value of bringing in outside speakers. Unlike her colleague HM, SL

(Site 5) says her curriculum and career were influenced significantly by her involvement with the program. SL views it as a model worth replicating "all across your curriculum, which is what I've done . . . I . . . replicate the sort of pattern that it creates and that to me has been the most significant thing for me . . . So, when we try to create any unit that we're doing, part of what we're thinking about is something that I got from the Kids Network."

SL is also a member of a committee of middle school science teachers and technology specialists in Vermont who have been collaborating on the design and implementation of a statewide Kids Network-like program. During the past summer, "everything we did is based on the Kids Network . . . the community letter . . . our investigations to do individually, we need to telecommunicate . . . It's more complicated than Kids Network . . . We can have six different things going on at the same time."

The topics of investigation were developed by the committee—political topics they hoped would come up before the Vermont legislature. Some were "not science-oriented," she added, even if they contained a topic with a science component to it, "or whatever we felt that the kids would be interested in. But the whole thing is sort of modeled on Kids Network."

Here, too, the theme of teachers' careers is often present in the empirical literature on innovation. Huberman and Miles (1984) and Fullan (1982) made some mileage from the notion of a "trajectory," claiming that the innovations in question brought locals to different places from where they had begun their journey. And yet it was the energy of the career aspiration, whether it was there at the outset or emerged as the project unfolded, that helped with successful implementation. Part of that energy came from what Huberman and Miles call "enhanced capacity," in which a local teacher becomes a prime computer consultant or finds that experiments that turned the initiative fully or partly over to the students are a source of stimulation and cognitive growth. This, then, becomes part of the teacher's instructional repertoire and carries over into other topics.

◆

Continuity, Institutionalization, and Spread

Continuity is not always correlated with success. Strong and well-documented projects may disappear, and indifferent ones may endure. The key lies in the protection each gets institutionally: in budget cycles, training cycles, equipment requisitions, purchases of special materials—routines indicating that it was always there. To be sure, the support of key administration is also important, but what Yin et al. (1978) call the "routinization" factor is stronger.

Our analysis of the continuity, institutionalization, and spread of the Kids Network among participating school sites seems like an incomplete story. All the characters agree. Every teacher anticipates using the program for at least the next five years. And the principals, in particular, support its spread among more faculty (table 16).

Table 16.   Institutionalization of the Kids Network—Personal Commitment, Potential to Spread, and Explanations About Future Use

Site	Degree of Commitment	Potential to Spread	Explanations About Future Use
1	High	Medium	"I see no reason why I'd be pulling out of it. I might want to just shift to a different unit some time." (HA) "I would do it again, because I like to do things like that again and see what works and make it better each time. It was fun." (KA)
2	High	High	Almost all, if not every, teacher will be using one or more of the units next year.
3	High	Low	KL plans to continue rotating units, regardless of where she ends up, following a districtwide restructuring process.
4a	High	Low	Program will continue to be adopted districtwide for fifth graders. There is no indication that districtwide adoption will include other grades.
4b	High	Low	"I wouldn't want to give it up, because there are very wonderful elements of it that I think are things that the kids will remember even if the experience isn't as full as I'd like it to be—that will transfer over into other kinds of experiences that they'll have with technology." (SE)
5	High	High	"Yes, I'd like to do a different program, though. I don't know which one. I've now done Acid Rain, Weather in Action, Too Much Trash?." (HM)
6	High	Low	"Next year we're planning to do this first thing in the year and I think that's really going to be ideal, because we can set the stage for what we're looking for from kids and the kind of attitude that we want to give to them for a year's worth of being here." (TA)

At four of the seven sites, the program's continuation depends on one or two teachers at the school. If these advocates left, so likely would the Kids Network; however, that may not be the case at Sites 4a and 4b, where the program has been adopted districtwide by fifth grade teachers. At three of the seven sites, the program has spread beyond the initial users and is expected to continue spreading to other teachers. For example, at Site 2, teachers in every grade from first through fifth are expected to be using the Kids Network next year. At Site 5, four of the seven teachers who use the Kids Network during the school year have used it for three or more years.

All the principals in our sample expressed enthusiastic support for the continuation of the Kids Network—not only for current users, but for new users as well. The principal at Site 1 gave his enthusiastic support for the program, saying that if a potentially interested teacher approached him, he would say, "Do it! Sign up for it, get started, and it'll be good for you and good for your kids." He added, "I think it offers some types of things that are just great for teachers, and it's good trying something that's new to them and a little bit different for them."

The principal from Site 4a indicated that if the two current Kids Network teachers did not want to continue the program the next year,

> they'd have to do a lot of explaining, unless something certainly went wrong, not to do it. I can't envision that right now . . . I feel that it's here to stay. I think it's in the forefront of a lot of what is happening . . . So I think that the program you're talking about is here, and probably will expand. But it does depend a great deal on staff and their involvement.

As much as the teachers and principals envisioned a long-term future with the Kids Network, a number of factors may prevent this. They involve issues like professional identity, finances, and organizational stability. One factor limiting the voluntary spread of an innovation like the Kids Network might be the degree to which teachers have a need to create their own professional identity; this might be perceived as being compromised by riding on a colleague's curricular bandwagon. KL, the third/fourth grade teacher in suburban New Hampshire, commented on this indirectly. She has helped present many Kids Network awareness workshops with NGS at state and national conferences. She has also been featured in at least two nationally distributed education magazines. Given that she has done so much outside her school for spreading the Kids Network, it was surprising that only one of her colleagues was using the program and another one had discontinued it. KL was not surprised, however, for she had never presented the program to teachers in her school or district. KL hints at the importance of a professional identity when she explains why her colleagues might not use the Kids Network: "I think each person is respected for the talent that that teacher has. They have other ways of working with children."

The spread of a nonmandated innovation through a school might also be influenced by important psychological and social issues, as well as pedagogical ones.

Financial resources can also be an obstacle to continuation. Although this was not the case at the schools in this study, it is certain that the up-front and auxiliary costs to run a Kids Network unit are prohibitive for many teachers. Even if the initial expenses are covered, the annual fee and costs of upgrading and maintaining necessary equipment may undermine lasting commitment to the program.

Organizational instability can also be an obstacle to continuation. As discussed earlier, a number of teachers and schools discontinued their participation during periods of organizational change. In some cases, a person's retirement or departure from the school led to noncontinuation of the Kids Network. In other cases, a district's restructuring seemed to make the program particularly vulnerable. When schools are better equipped with access to computers, modems, and dedicated phone lines, and as teachers' technological competencies continue to improve, programs like the Kids Network may not be as vulnerable to more tangential changes in the institution.

◆

Equity Issues

The Kids Network raises several equity-related issues, including:

1. To what extent is the program being used equitably (by schools from different SES communities)?

2. To what extent did different groups of people contribute to the development of the curriculum?

3. How are different groups of people portrayed in the materials?

4. How does the program appeal to different types of students?

These issues of access, input, portrayal, and appeal are discussed below.

Access

Our unsuccessful attempts to find Kids Network veterans from schools in poor districts in New England is one indicator that the program is distributed unevenly, favoring wealthier districts that can afford the program's costs. Dot Perecca from NGS thinks that the number of schools in New England—not including Boston—using the Kids Network is relatively low compared to other areas, because the New England schools cannot afford either the initial purchase of the program, its annual registration fee, and/or its peripheral or hidden maintenance costs (e.g., computer and modem repairs, hardware and software upgrades, etc.). Bob Tinker, TERC director, said the cost was a surprise: "We never anticipated that it would cost as much as it did . . . We genuinely thought it would be in the $50 to $75 range for the stuff."

One way in which less wealthy schools have invested in the Kids Network is for two or more teachers to share one unit and, at times, share a single registration. While many of the participants in our study shared a registration between two classes, these teachers indicated that their motivation was pedagogical or logistical rather than financial.

Another equity/access issue concerns the extent to which all teachers within a school that has invested in the Kids Network have opportunities to use the program. According to GM, the technology specialist at Site 5, his principal made the program accessible to all of his teachers because "it was an equity thing. In his mind it was equity. We had to be able to offer the stuff to all the kids, not just one particular class." Additionally, all the administrators in our sample indicated that they were willing to invest in providing the program to all interested teachers at their schools.

Equal Contributions to Development

Did students from different racial, ethnic, grade, and academic backgrounds contribute to the program's development (e.g., participate in pilot testing programs for formative purposes)? According to Candace Julyan, project director of development of the Kids Network, TERC included a range of schools and student populations in its selection of 200 pilot sites. Julyan explains,

> We wanted to have at least two sites in every state and we wanted to have a good spread of schools where there had never been a computer in the school ever before, as well as schools where kids all have computers at home. We wanted to have a good spread of suburban schools, urban schools, and rural schools. We wanted to have a good spread of ethnically diverse student populations as well as what we knew were going to be the all-white populations.

However, Julyan adds, "We didn't do SES. We let the science supervisors in each state help pick them and we trusted them." Consequently, the extent to which schools from lower SES profiles contributed to the development of the units is uncertain. Based on the difficulties we experienced in locating schools using the Kids Network from middle- to low-SES communities, we suspect that not many of such schools were part of the piloting process.

Portrayal of Equity in Materials

Another way in which issues of equity concern any innovation is in the way in which diverse types of people—along such factors as gender, race, ethnicity, class, and abilities—are portrayed in the curriculum materials. A quick glance of the images in the Kids Network Kids Handbook for the What Are We Eating? unit portrays people representing a variety of ages, ethnicities, cultures, and races. On a related note, MD (Site 1) recalled that three out of five unit scientists he worked with were female.

Some teachers attribute the opportunities for students to collaborate in small teams as well as the opportunity to telecommunicate with students from different environments and cultures as examples of how the program's design inherently respects diversity.

Appeal to Different Types of Students

A program that favors a particular group of students may not be providing different pupils with equal opportunities for success. Except for pets in the Hello!

unit, the program's topics—such as acid rain, weather, trash, water, and nutrition—are universal and do not inherently exclude any group of people. As for students with various backgrounds and abilities feeling successful with the program, a number of teachers expressed MD's (Site 1) sentiment: "Kids that may not succeed in a traditional classroom setting have an opportunity with this [the Kids Network], a big opportunity, and some of the kids who gained the most have exhibited at-risk prior to this."

Similarly, HM at Site 5 creates Kids Network teams with diverse abilities; consequently, kids with diverse needs succeed with the help of their peers. She adds that with the kids with special needs, however, "I lose them with the theoretical stuff."

According to AN at Site 4a, "My girls are just as interested as my boys" in doing the Kids Network Acid Rain unit. Similarly, another teacher at Site 1 indicated that the weather unit has the same appeal to boys and girls and meets the needs of different types of students. She believes it is the teacher, however, not the curriculum, "who provides that support for those who need it and . . . the opportunities for those who are going to take off."

MD (Site 1) says he tries to make the program accessible to all of his students:

> No matter who's doing it, I try to make them all comfortable with the process and try to take away any barriers they perceive to be too difficult or uncomfortable. I try to open the gate to make sure that they keep moving, but that's a part of my personal teaching profile that isn't really related to acid rain. It's more of a "What can I do to give everyone access?" I'm less focused on what actually happens in the package, and more focused on what I can do to make sure that the package has an impact on everybody.

Concluding Remarks About Equity Issues

The Kids Network touches upon a number of equity issues. While the schools participating in the piloting of the units varied by geography, race, and grade, the diversity in schools from different socioeconomic classes that piloted the program and contributed to its development is uncertain. All of the teachers in our sample indicated that the program appeals to both boys and girls as well as to students with a range of academic abilities. Most of the teachers indicated also that the curriculum materials such as the *Kids Handbook* respectfully represent different genders, races, and cultures. Since NGS has not maintained a database on the SES levels of the schools that have purchased the Kids Network, it is difficult to assess the extent to which poorer schools have access to innovations like this. Our unsuccessful attempt at finding poorer schools using the Kids Network may indicate that there might be many schools that are not using the program because they do not have opportunities to do so.

◆

Concluding Remarks

We have documented our analysis based on some 42 days of observations of 11 teachers implementing the Kids Network in seven New England schools. Additionally, we conducted extensive interviews with more than 20 Kids Network veteran teachers, principals, and technology and science coordinators; and reviewed and analyzed the Kids Network *Teacher's Guides,* student activity worksheets, promotional materials, and annual reports. Our analysis leads us to an image of the Kids Network not as a product, but as a journey of change.

The causal network that follows (figure 5) is "an abstracted, inferential picture organizing field study data" in a way that renders "the most important independent and dependent variables in a field study and of the relationships between them. The plot of these relationships is deterministic rather than solely correlational" (Miles and Huberman 1994, p. 132). Our causal network highlights the key determinants of the Kids Network's journey and of the journeys made by the various players it touched along the way. There is, in fact, an elaborate literature on the process of curriculum change, including very recent and large-scale empirical studies or major syntheses, much of which has been built into this study (e.g., Huberman and Miles 1984, Louis and Miles 1990, Fullan and Steigelbaver 1991, Cohn and Kottcamp 1993, Fullan 1993). The following describes the journey as framed by our causal network depicted in figure 5.

The Kids Network as a Journey

Development. In innovations akin to the Kids Network, a product typically is developed in a university or research center; is connected, in the best cases, to a theory of learning or development; and is field tested more or less thoroughly in the kinds of places it will be applied. With each corrective loop of field testing, it is readier for dissemination directly through a publisher or indirectly though an "intermediary organization" (e.g., a laboratory or university).

The principals and Kids Network teachers in our sample were actively interested in a more inquiry-based science curriculum. The developers tried to initiate and operationalize a curriculum in this direction; however, they were constrained by their publishers, who demanded a program requiring what they considered a more palatable pedagogical stretch for teachers. Consequently, the program is highly prescriptive due to the developer and publisher's "negotiated" perception that this is what teachers need and want.

Adoption. An innovation's arrival at the school means simply that it constitutes a new practice—an innovation—in the local setting, not that such practices have never been invented before, nor even that variants of the same practice are unfamiliar to a few actors in the school setting. But the derivation is not local, nor has the product been created by peers in a common network or association. It is novel to the environment into which it enters.

In many ways, in fact, innovations are what the French call "revélators"; they bring into the open opportunities, constraints, conflicts, old wounds, and

Figure 8.   **Causal Network of Kids Network From Development to Consequences**

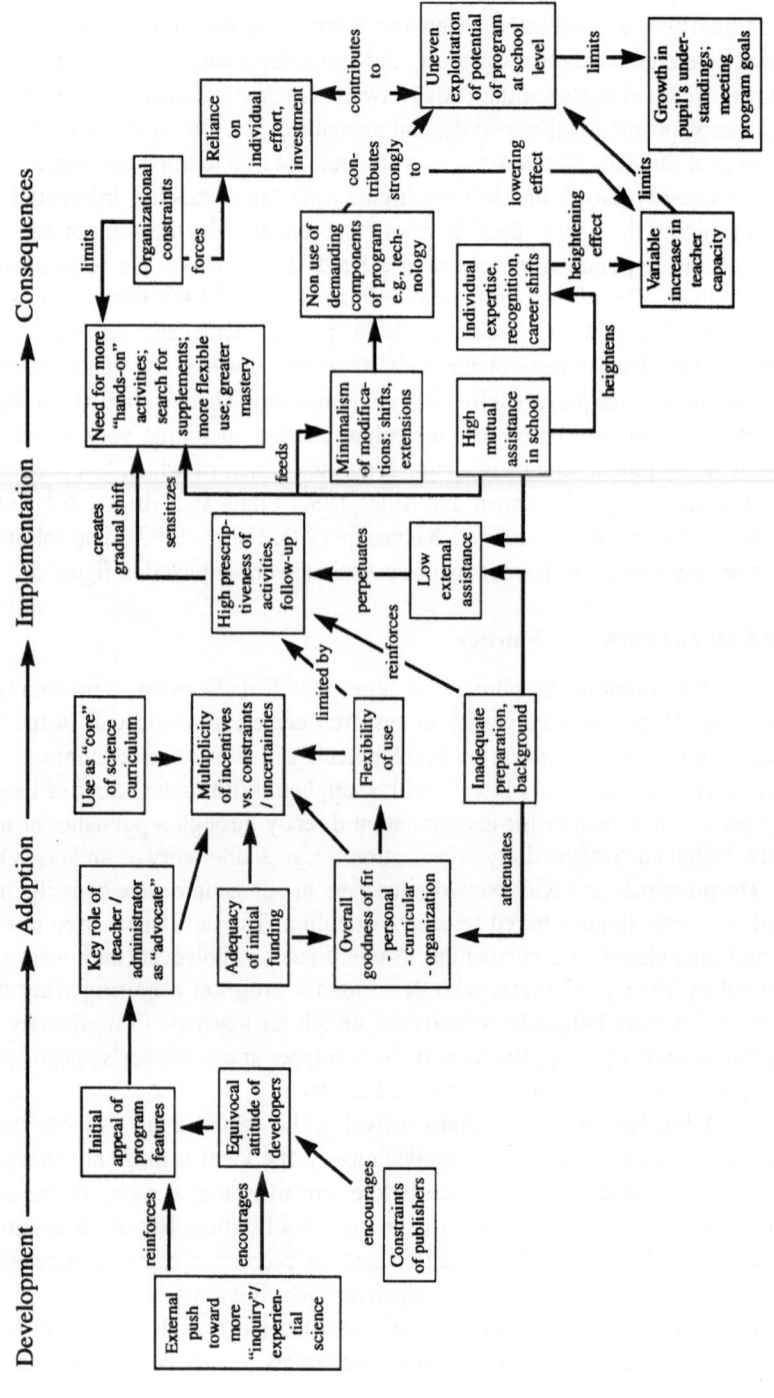

sometimes heroics that lay dormant until then. And they sometimes introduce ideas, pedagogies, or vehicles of which, up to then, few in the school had dreamed. In the Kids Network, these included telecommunications with classes in other time zones, a focus on real-life environmental and social problems, hands-on experiments nested in a series of science and nonscience experiences, and opportunities to compare data collected in a variety of milieus. These were all components of the journey of the program through the school.

The appeal of the program's features, the adequacy of initial funding, and the perception of goodness of fit between schools and the program led to the Kids Network's adoption as a significant six- to eight-week science curriculum.

Implementation. What we know about process and modal outcomes is that the succeeding phases are always somewhat adventurous and sometimes turbulent. To be sure, we have identified some of the most salient or crucial variables at each turn in the road, but they often play out differently, especially when the context of use is different. For example, "hi-tech" innovations such as those found in the Kids Network computer components do not readily nest themselves in a poor, inner-city environment where few children have a computer at home and few teachers have the requisite technical infrastructure or skills for putting the technology to use in the program. These differences in context will have different consequences on the remainder of the journey.

The process of undoing the packet and then negotiating it through personal and institutional constraints, differing conceptions of legitimate science and mathematics, various backgrounds and thresholds of risk or exploration, and available sources of support constituted the teachers' journey and determined whether the result was merely adventurous or whether it led to cognitively stretching or broadening experiences for the pupils and teachers involved. There was evidence, in fact, that the Kids Network could lead teachers to experiment with less predetermined, inquiry-centered formats; however, the *Teacher's Guide* itself provides more prescribed, often less cognitively demanding, exercises which most teachers followed closely.

But some teachers in our sample had already made parts of this more adventurous journey together or with their peers, here or elsewhere. They were not blank slates; they had experimented with teacher- or worksheet-directed hands-on learning, with new technologies, with student groups working on similar tasks, with expeditions outside the classroom, and with interdisciplinary projects. Although each teacher was differently experienced, differently eager or disquieted, and differently capable of going beyond or behind the *Teacher's Guide*, most were at least three-time veterans who have remained faithful to the original program. This then determined the conceptual or content-rich quality and depth of the pupils' journey, the one that bears the closest scrutiny.

Consequences. The journey made by the teachers, via the program, puts pupils on new courses of learning and instruction. In the Kids Network, for example, there are group-led hands-on activities, with known or partially known solutions to work out. Mathematics and language arts are embedded in various data collection and analysis exercises. There are possible expeditions outside the classroom, where students can explore their local environments (e.g., daily weather measurements on school grounds, a field trip to a landfill, a field trip with a state acid rain scientist). And there is a computer program that engages students foremost in the magic of telecommunications, and secondarily serves as a tool for graphing data. It is not simply a new program we have here, but oftentimes the potential for different ways of developing cognitive skills, reasoning processes, group problem solving, and inductive and deductive solutions from real-world exemplars.

These incentives, however, were counterbalanced by several constraints and uncertainties having to do with teachers' science education backgrounds, the technological components of the program, and an unyielding infrastructure. These factors, combined with low assistance from outside the school, led at first to minimal modifications and extensions of the program, especially of its more demanding components. Gradually, however, teachers looked for supplements and considered more flexible uses and new configurations coming from their greater mastery of the Kids Network. In some instances, teachers were recognized as pioneers by administrators, locally and statewide.

Some teachers transferred the capacities for inquiry and experimentation learned in the Kids Network to other topics and to other parts of their science curriculum. Overall, however, there was an underexploitation of the potential of the program and a corresponding growth in pupils' scientific understandings and mastery.

Even though the schools were not uniform in their capacities to take on the program, in most of the cases, the Kids Network required few organizational changes beyond dedicated phone lines and other electronic equipment. Of course, creating a technological infrastructure can have a profound organizational influence, which in turn can change classroom learning experiences. However, we should keep in mind that the Kids Network is a six-week curriculum, and thus is implemented during only a fraction of the school year.

Even if the program were a more significant part of the curriculum, organizations have a way of adulterating or thinning out significant change. There are, of course, more consequential scenarios of change, but they tend to occur in places in which the management of change has itself been methodically cultivated, studied, and refined. Here, the paths and intersections of the multiple journeys—pupils, teachers, organization, program—have actually been built into the ongoing management of the establishment (Fullan 1993 and Senge 1991), including the requisite confrontations, uncertainties and improvisations.

One of the prime roles of an innovation is to create "discrepancy," especially at the instructional levels. For example, some of the potential discrepancies in the Kids Network had to do with new science content, more hands-on activities, experimentation with groups, integrated technologies, and interdisciplinary curricula. As a welcome intruder in its new surroundings, the discrepancies of the program contributed to changes in teachers' practice. However, the potential for change was mediated by an ongoing, implicit battle for influence between the kind of change-bearing demands of the program we have outlined and the more equilibrium-preserving working arrangements by which schools manage so many moving parts.

Intrinsic Value

According to the developers, all the Kids Network units are guided by three principles:

> That students should deal with real and engaging scientific problems, problems that have an important social context. That kids can and should be scientists (students are working as scientists on real science problems). That telecommunications is an important vehicle for showing children that science is a cooperative venture in which they can participate (TERC 1987, p. 1).

For more traditional teachers, the Kids Network promises a transition from a more teacher-driven science program with textbooks and seat work to a module containing telecommunications, thematically organized connections between environmental science and social issues, some hands-on activities, and simulations of the procedures used by working scientists.

For those who identify themselves with a more student-centered, social constructivist, heuristic-based pedagogy, the program needs to be modified to realize its potential. At issue here is not whether students should focus on environmental problems, but how; the issue is not whether students should be scientists, but what that experience should look like; and the issue is not whether students should experience science as a cooperative venture, but the nature of that cooperation.

Potential

A discussion about an ongoing journey would be incomplete if we did not consider the Kids Network's potential for changing the ways in which teachers interact with children and construe concretely the fit of real-life scientific themes into the upper elementary school science and mathematics curricula. If teachers do not experience the discrepancies offered by an innovation, however, then the influence on a teacher's practice over the years is likely to be minimal. Nothing will have changed notably in the students' learning environment. If teachers experience too many discrepancies, or if they are too far-reaching, the innovation's journey will likely generate more turbulence than the teacher might be able to handle without strong assistance and changes in the school's infrastructure. In either case, the potential of the intended curriculum will not be realized. It is therefore a delicate balancing point to shake a teacher's practice, to solicit the challenge and make the risk palatable, without making the teacher tumble. It is with this caution in mind that we assess the Kids Network's potential.

We can envision the implementation of the Kids Network in ways that are more aligned with Scenario #3 in the previously set forth "Conceptions of Science Teaching" without upsetting the basic organization and structure of the original program. In this vein, teachers are facilitators of different student groups pursuing multiple but related lines of student-generated inquiries, as well as learning explicit strategies for collaboration. We imagine students being challenged to collaboratively design and rationalize formats for recording and analyzing their observations. We envision students having more opportunities for authentic communication with their teammates, and teachers making more regular assessments of their students' content and procedural knowledge and then modifying the curriculum accordingly.

The National Science Foundation and NGS have invested heavily in creating the Kids Network, and the ground is fertile. Teachers now require supplemental ideas for enacting the program with more constructivist teaching strategies without having to abandon the strengths of the original curriculum. Here is an illustration based on an activity offered in the *Teacher's Guide*.

The Grass Experiment. The grass experiment is presented during the second week of What's in Our Water? as one of three activities about pollutants. The *Teacher's Guide* lists the session's objectives:

- to review how substances can get into water,

- to introduce the idea that a substance can have both beneficial and harmful effects,

- to introduce nitrate (one potential pollutant), and

- to begin an experiment with plants and nitrate fertilizer.

The activity sheet begins with a list of the materials to be collected by students. It then describes how to prepare three cups: one containing no fertilizer, one with a low concentration of fertilizer, and one with a high concentration. For example, for the low-fertilizer solution it says, "Mix 1/8 teaspoon (about 0.5 milliliters) of liquid fertilizer with 2 teaspoons (10 milliliters) of distilled water." Students are then provided with a list of eight instructions for how to plant their seeds. They are instructed on how to label their cups (as shown in a picture), how to punch holes in the bottom of each, and how to fill them with sand and grass seed. The next set of directions tells students how to care for their grass and when to observe. On the back of the activity sheet, there are four prediction questions:

1. Will most of the seeds sprout in Cup 1?

2. Will most of the seeds sprout in Cup 2?

3. Will most of the seeds sprout in Cup 3?

4. Which cup will have the tallest grass?

These questions are followed by three charts, one for each cup, each with the same headings:

Cup 1	Observation 1 (Date: _____)	Observation 1 (Date: _____)
How many seeds sprouted? (Most? Half? Less than half?) Grass Height Grass Color		

The grass experiment is a hands-on activity in that it involves the physical manipulation of everyday objects. It involves creating a fair test, in which one variable is manipulated to determine its influence. It entails making hypotheses and observations, and recording and analyzing data. It emphasizes implicitly the importance of following directions, being precise, and being systematic. It is also algorithmic, in that it provides everything the student needs to do the activity.

Assuming everyone is faithful to the directions, the original activity has one result (the cup with the highest concentration of fertilizer will germinate the least number of seeds, and the cup with the least amount of fertilizer will germinate more seeds than the cup without fertilizer). The primary responsibility of the teacher, like the students' during the exercise, is to follow the *Teacher's Guide*, which contains an elaborated set of systematic directions. In effect, the teacher's role is to manage materials and help the students follow the directions. Except for predicting which cup will contain the most seeds, there is no explicit attempt to elicit the students' prior conceptions about fertilizers, the purpose of their experiment, or the process of fair testing. Finally, cooperative learning involves the distribution and sharing of tasks, such as deciding who is to retrieve the materials, who is to execute each direction, and who will return the materials.

An Alternative to the Grass Experiment. Now let's consider a more inquiry-centered alternative to the grass experiment. The alternative's objective would be the same as the original activity—to learn about the effects of various amounts of fertilizer (nitrate) on plant growth, and more abstractly, that pollutants can be too much of a "good thing."

The approach, however, would challenge students to design and conduct their own experiments to learn about the effects of various amounts of fertilizer on plants. Students can work in small groups to design a fair test and then propose their procedures to a peer-review board (the rest of the class) whose job is to help groups improve on their original designs. Thus, the students would be responsible for creating a chart, with headings they themselves conceived, that will enable them to record and compare their data so they can determine the effects of various amounts of fertilizer. After the experiments are completed, students would present their data and findings to their classmates for review and determine the extent to which their conclusions conflict or corroborate with those of other groups. At that time, the teacher would highlight points of discrepancy and determine with the students how they might proceed, if at all, in resolving those discrepancies.

Like the original grass experiment, this alternative is also hands-on; however, it is heuristic in nature, in that it encourages students to discover some key underlying principle for themselves. While it provides students with a problem or challenge, it does not furnish procedures for how to solve it. The materials, directions on how to use them, and an organization for recording observations are conceived and developed by the students, under the teacher's direction. Consequently, the activity consists of many problems that students must solve using their own strategies: which materials and how to organize them, how to control variables, when to observe and how, how to record observations in formats that can help analysis, how to present their findings to their peers—to name just a few. This alternative not only allows for the application of many methods to the same problem, but many solutions as well. Finally, the alternative also enables students to build on one another's experiences and findings and thus socially construct an understanding about the effects of different amounts of fertilizer on plant growth and an appreciation for a variety of experimental, recording, and analyzing methods.

During this activity, a teacher's primary responsibility would be as an orchestrator and fellow investigator, searching for answers to unanticipated questions and helping students by evaluating the integrity of their processes, approaches, and findings. Inherent in this alternative are many opportunities for students to reveal their prior conceptions about the content and process of their investigation. Finally, with respect to cooperative learning, this alternative requires groups collectively to design and execute the conceptual and mechanical aspects of their experiment. This type of instructional sequence presents teachers with new challenges, such as eliciting students' ideas, supervising a variety of simultaneous activities, and providing work for those who complete activities before others. Trying to manage eight groups simultaneously on three different tasks is daunting. This alternative, therefore, would need to include effective strategies for how to approach these new pedagogical challenges as opportunities for learning.

Planned Obsolescence and Meaningful Reform

We have centered on the life cycle of an innovation in science and mathematics education. It might or might not resist the next round of seductive curricular innovations. This constant barrage of new materials creates an atmosphere of planned obsolescence. Consequently, the next "solution" makes for a relatively short lifespan for any innovation and short-circuits the upgrading of existing curricula.

Bob Tinker from TERC reminds us that "no one intervention is going to cause [fundamental] change" and that change is more likely when supported from a multiple of directions. Tinker notes,

> You perhaps lower your sights as a developer, thinking that curriculum itself is going to make the difference. What you hope is that the curriculum contributes to the change . . . You have to approach all sides; you have to break into the ring from the professional development, the assessment, the curriculum, the organizational points of view. All of those have to be working for there to be the change. There's no curriculum, nothing that we could ever write, put in a box, and send off to schools, that is going to cause the schools to automatically change.

The most vulnerable innovations are like the one we have studied—the Kids Network. As technology develops, they will become technologically obsolete, but not necessarily learning-poor. It is the depth, quality, and transformation of students' learning experience that form our barometer.

References

Berman, P., M. McLauglin, et al. 1975-77. *Federal programs supporting educational change.* 8 vols. Santa Monica, CA: Rand Corporation.

Borko, H., and R. Putnam. 1995. Expanding a teacher's knowledge base: A cognitive psychological perspective on professional development. In *Professional development in education: New paradigms and practices*, ed. T. Guskey and M. Huberman. New York: Teachers College Press.

Brickhouse, N.W. 1990. Teachers' beliefs about the nature of science and their relationship to classroom practice. *Journal of Teacher Education* 41 (3): 53-62.

Brophy, J., and J. Alleman. 1991. Activities as instructional tools: A framework for analysis and evaluation. *Educational Researcher* 20(4): 9-23.

Brooks, J. Q., and M. G. Brooks. 1993. *In search of understanding: The case for constructivist classrooms.* Alexandria, VA: Association for Supervision and Curriculum Development.

Cohen, D. K. 1990. A revolution in one classroom: The case of Mrs. Oublier. *Educational Evaluation and Policy Analysis* 12(3): 327-45.

Cohn, M. M., and R. B. Kottcamp. 1993. *Teachers: The Missing Voice in Education.* Albany, NY: State University of New York Press.

Cowley, S. 1994. National Geographic memo to Dot Perecca, 2 October.

Cuban, L. 1986. *Teachers and machines: The classroom use of technology.* New York: Teachers College Press.

Driver, R., and B. Bell. 1986. Students' thinking and the learning of science: A constructivist view. *School Science Review* (March): 443-56.

Driver, R., E. Guesne, and A. Tiberghien. 1985. *Children's ideas in science.* Buckingham: Open University Press.

Driver, R., and V. Oldham. 1986. A constructivist approach to curriculum development in science. *Studies in Science Education* 13: 105-22.

Duffy, T. M., and D. H. Jonassen, eds. 1992. Constructivism and the technology of instruction: A conversation. Hillsdale, NJ: Lawrence Erlbaum Associates.

Fine, C.S., and L.B. Friedman. 1991. Evaluation report of Kids Network in IOWA: 1990-91, North Central Regional Educational Laboratory.

Firestone, W. A. Why "professionalizing" teaching is not enough. *Educational Leadership* 50(6): 6-11.

Firestone, W. A., and D. M. Bader. 1992. Why "professionalizing" teaching is not enough. *Educational Leadership* 50(6): 6-11.

Fullan, M. 1982. *The meaning of educational change.* New York: Teachers College Press.

————. 1993. *Change forces: Probing the depths of educational reform.* Bristol, PA: Falmer Press.

Fullan, M., and S. Steigelbaver. 1991. *The new meaning of educational change.* New York: Teachers College Press.

Gardner, H. 1991. *The unschooled mind: How children think and how schools should teach.* New York: Basic Books.

Guba, E. G., ed. 1990. *The paradigm dialog.* Newbury Park, CA: Sage Publications, Inc.

Hall, G., and S. Loucks. 1977. A developmental model for determining whether the treatment is actually implemented. *American Educational Research Journal* 14(3): 263-76.

Hall, G., S. Loucks, W. Rutherford, and B. Newlove. 1975. Levels of use of the innovation: A framework for analyzing innovation adoption. *Journal of Teacher Education* 26(1): 52-56.

Hall, G., and W. Rutherford. 1976. Concerns of teachers about implementing team teaching. *Educational Leadership* 34(3): 227-33.

Havelock, R. 1969. *Planning for innovation through dissemination and utilization of knowledge.* Ann Arbor: University of Michigan, Center for Research on Utilization of Scientific Knowledge.

Heaton, R., and M. Lampert. 1993. Learning to hear voices. In *Teaching for understanding*, ed. D. Cohen, M. McLaughlin, and J. Talbert, 43-83. San Francisco: Jossey-Bass.

Hesse-Biber, S., T. S. Kinder, P. R. Dupuis, A. Dupuis, and E. Tornabene. 1993. HyperRESEARCH [computer software]. Randolph, MA: ResearchWare.

Huberman, M., and M. Miles. 1984. *Innovation up close: How school improvement works.* New York: Plenum.

Julyan, C. L. 1989. NGS Kids network investigate: January 1989 field test. Cambridge, MA: Technical Education Research Centers.

Julyan, C., and S. Wiske. 1994. *Learning along electronic paths: Journey with the NGS kids network.* Cambridge, MA: Technology Education Resource Center.

Krajcik, J. S., P. C. Blumenfeld, R. W. Marx, and E. Soloway. 1994. A collaborative model for helping middle grade science teachers learn project-based instruction. *The Elementary School* Journal 94(5): 483-97.

Loucks, S. F., and G. E. Hall. 1977. Assessing and facilitating the implementation of innovations: A new approach. *Educational Technology* (February): 17-21.

Louis, K. S., and M. B. Miles. 1990. *Improving the urban high school: What works and why.* New York: Teachers College Press.

Mander, J. 1991. In the absence of the sacred. *Whole Earth Review* Winter: 5-16.

Miles, M., and M. Huberman. 1994. *Qualitative data analysis: A sourcebook of new methods.* 2nd ed. Thousand Oaks, CA: Sage.

Nussbaum, J., and S. Novick. 1982. Alternative frameworks, conceptual conflict and accommodation: Toward a principled teaching strategy. *Instructional Science* 2: 183-200.

Osborne, R. J., and P. Freyberg. 1985. *Learning in science, the implication of children's science.* Auckland, New Zealand: Heinemann.

Pintrich, P. R., R. W. Marx, and R. A. Boyle. 1993. Beyond cold conceptual change: The role of motivational beliefs and classroom contextual factors in the process of conceptual change. *Review of Educational Research* 63(2): 167-99.

Posner, G. J., K. A. Strike, P. W. Hewson, and W. A. Gertzog. 1982. Accommodation of a scientific conception: Toward a theory of conceptual change. *Science Education* 66(2): 221-27.

Ragin, C. 1987. *The comparative method: Moving beyond qualitative and quantitative strategies.* Berkeley, CA: University of California Press.

Rogers, E. 1962. *Diffusion of Innovations.* New York: Macmillan.

Senge, P. 1990. *The fifth discipline.* New York: Doubleday.

Stake, R. E., and J. A. Easley. 1978. *Case studies in science education.* Urbana, IL: Center for Research and Curriculum Evaluation.

Stepien, W., and S. Gallagher. 1993. Problem-based learning: As authentic as it gets. *Educational Leadership* 7 (April).

Strike, K. A., and G. J. Posner. 1985. A conceptual change view of learning and understanding. In *Cognitive structure and conceptual change*, ed. L. H. T. West and A. L. Pines, 211-31. Orlando, FL: Academic Press.

Technology Education Resource Center (TERC). 1987. *National Geographic Society Kids Network Project: Annual report October 1, 1986-September 30, 1987.* Cambridge, MA.

————. 1988. *National Geographic Society Kids Network Project: Year 2 annual report: October 1, 1987-September 30, 1988.* Cambridge, MA.

————. 1989. *National Geographic Society Kids Network Project: Year 3 annual report: October 1, 1988-September 30, 1989.* Cambridge, MA.

————. 1990. *National Geographic Society Kids Network Project: Year 4 annual report: October 1, 1989-September 30, 1990.* Cambridge, MA.

Winner, L. 1991. Artifact/ideas and political culture. *Whole Earth Review* (Winter): 18-24.

Worth, K. 1994. Memo: Comments on paradigm time warps. Personal communication, 15 December.

Yin, R., S. K. Quick, P.M. Bateman, and E. L. Marks. 1978. *The routinization of innovations.* Santa Monica, CA: Rand Corporation.

Appendix A: Case Study Team

Michael Huberman, formerly a professor of education at the University of Geneva (1972- 1993) and now professor emeritus, is presently director of research at the Swiss Federal Institute of Professional Education, and since 1991, visiting professor of education at Harvard University. He also was a senior researcher at the National Center for Improving Science Education where he led this case study. His areas of interest are qualitative research, research use, longitudinal studies of teaching, and educational innovation. He is a co-author, with Matthew Miles, of *Qualitative data analysis: A sourcebook of new methods*, and also recently published *The Lives of Teachers*.

Jimmy Karlan is an assistant professor of environmental education at Antioch New England Graduate School. He recently received his doctorate from Harvard University and was the principal field researcher on the Kids Network case study while a research associate at the National Center for Improving Science Education. His research interests lie particularly in aspects of environmental education, including students' ecological concepts, curriculum development, professional development, and design of informal learning exhibits and experiences.

Sally Middlebrooks is director of education at the Association of Science-Technology Centers. She recently received her doctorate from Harvard University, after having been a research associate at the National Center for Improving Science Education and the Bank Street College of Education in New York. She has worked in the area of childrens' learning, notably in science, in non-formal settings such as playgrounds, parks and in their own homes. Her dissertation on this topic is soon to be published by Teachers College Press.

Appendix B: Participants' Roles and Profiles

Site	Participant	Participant Role and Profile
1 Urban NH	MD	Combined fifth/sixth grade teacher; enthusiastic science teacher; developed and published ecology curriculum modeled after some Kids Network components as well as resolved some constraints; discovered faulty litmus paper provided by publishers; supervised student teaching intern who was responsible for implementation of What's in Our Water?
	KA	First-time Kids Network user; student teaching intern responsible for execution of Kids Network; actual classroom practice not observed by supervising teacher
	HA	Fourth grade teacher; provides consultation within and outside of school district on issues of literacy and portfolios; completed doctoral program in education at Boston University during observation period; dissatisfied with implementation of weather unit this year
	JM	Fifth grade teacher; decided not to do program this year (district office foul-up); gets assistance from MD on computer components; uses MD's ecology curriculum; feels there are not enough hands-on activities in the water unit
	MB (Principal) when ready	Supportive of teachers and knowledgeable about Kids Network; would like more teachers to use Kids Network
2 Urban NY	AT	Third grade teacher; first-time user of innovation, but her use judged exemplary by principal and science specialist; enthusiastic about the innovation, choosing to do two units this year; characterized herself as new to technology, but has taken time to teach herself; works well with science special - ist, whether exchanging ideas or receiving assistance; mathematics so much the core of her teaching that students talk about themselves as mathematicians
	KW	Science specialist; introduced the innovation to the school in 1990; taught units herself for 3 years, then turned over to classroom teachers in 1993; currently acts as go-between, facilitating scheduling of the units and the telecommunication activities that follow as modem is in her science lab; said she often photocopies letters received from other sites to look at herself; characterizes herself as strongest in physical sciences; chosen to attend NASA teacher workshops this summer

Appendix B: Participants' Roles and Profiles
(continued)

Site	Participant	Participant Role and Profile
	GC (Principal)	Although she said she relies on advice of her science specialist (KW) in such matters, she was conversant both with what the innovation offers and the varying abilities of her teachers to use it; as a magnet school vying for pupils, she said this innovation fits well with the school's technology focus; moreover, it distinguishes them as offering something special
3 Suburban NY	KL	Initial user; has used four different units; used as many as three different units in one year with third graders; presents workshops for NGS on personal experiences with Kids Network
	JC	Third grade teacher and technology specialist; designed and installed school-based LAN; in charge of computer hardware and software purchases; maintains school's computer lab; registration for Kids Network fouled up by district office
	SR	Discontinued Kids Network after one experience with Acid Rain unit with third graders; felt unit lacked hands-on components
	AH (Principal)	New principal; uninvolved with innovation
4 Suburban NY	AN (School 4a)	Fifth grade self-contained classroom; enthusiastic about program; collaborated with P to share one Kids Network registration between their two classes; facilitated most joint class meetings; very comfortable with computer components, including e-mail; supports P's use as well as SE's at another school in district
	P (School 4a)	Fifth grade self-contained classroom; second year doing Kids Network; collaborated with AN with whom her class shared one team name
	SE (School 4b)	Only user in school; uses Kids Network with all three fifth grade science classes
	SF (Tech Specialist/ Principal)	As district technology specialist, played a critical role in initial adoption and implementation; as principal of another school in district, continues to provide technical and curriculum support for teachers at both schools
	MS (Principal)	Uninvolved in implementation, but supports Kids Network by securing budgets for paying annual fees; however, plays an insignificant role during its implementation

Appendix B: Participants' Roles and Profiles
(continued)

Site	Participant	Participant Role and Profile
5 Rural VT	SL	Fifth grade teacher; received national award for science teaching; rewarded computers for her classroom; more experienced with Kids Network than HM, with whom she collaborated and shared Kids Network registration; very comfortable with computers and readily sought support from technology specialist
	HM	Fifth grade teacher; formal science background; collaborated with SL; comfortable with computers and readily sought support from technology specialist
	KB	Job-shares a third grade class with another teacher; has primary responsibility for implementing Kids Network while other teacher assists, but is beginning to transition out
	GM (Tech Specialist)	In charge of school's computer lab which is used by classes throughout school day; developed and installed school-based LAN; recently spent about $5,000 for developing school's software library; extremely knowledgeable about technolo-gical/electronic needs and operation of Kids Network; assists teachers in use of software and telecommunications components
	Principal	First year as acting principal, was previously elementary teacher in same school; provides peripheral support for Kids Network teachers; wishes more teachers wanted to take it on; reports that school board is supportive and generous
6 Rural VT	TA	Fifth/sixth grade teacher; fourth-time user of Kids Network; team teachers with CM in open classroom environment and collaborates with CM on all aspects of implementation of Kids Network; more experienced with Kids Network and computer components
	CM	Fifth/sixth grade teacher; second-time user of Kids Network; team teaches with TA in open classroom environment and collaborates with TA on all aspects of implementation of Kids Network; faithful to *Teacher's Guide*
	DF (Principal)	Has general understanding of Kids Network's basic com-ponents; has asked school board to invest in one laptop per teacher; reports school board very supportive financially in purchasing Kids Network equipment and materials; provides support on demand for teachers on implementation; also sup-ports teachers as a chaperone on a trash unit field trip to a neighboring landfill

Appendix C: Analysis Products Summary

Memos

More than 62 memos have been written. Memos have emerged from interviews, observations, document summaries, contact summaries, and colleagues' memos. Memo topics include:

- the challenge of applying constructivist teaching strategies to postpositivist curricula,

- children's conceptions of Kids Network unit content,

- distinguishing algorithmic and heuristic approaches to the teaching/learning of science,

- the meaning of hands-on science and mathematics,

- the high speed at which the innovation is executed,

- teachers' efforts at eliciting and understanding student conceptions of science and scientific processes,

- teachers' antithetical practices,

- methodological suggestions and concerns,

- Kids Network teacher training workshop,

- the paradoxes of systemic and personal reform,

- technology education: tool versus design,

- dissemination strategies, and

- tentative themes within and across cases.

Document Summaries

As of June 1994, document summaries had been written for the following materials:

Borko, Hilda, and Ralph Putnam. 1995. Expanding a teacher's knowledge base: A cognitive psychological perspective on professional development. In *Professional development in education: New paradigms and practices*, ed. Guskey and M. Huberman. New York: Teachers College Press.

Brickhouse, Nancy W. 1990. Teachers' beliefs about the nature of science and their relationship to classroom practice. *Journal of Teacher Education* 41(3): 53-62.

Brophy, Jere, and Janet Alleman. 1991. Activities as instructional tools: A framework for analysis and evaluation. *Educational Researcher* 20(4): 9-23.

Cohen, David K. 1990. A revolution in one classroom: The case of Mrs. Oublier. *Educational Evaluation and Policy Analysis* 12(3): 327-45.

Fine, Carol S., and L. B. Friedman. (1991). Evaluation report of Kids Network in IOWA: 1990-91, North Central Regional Educational Laboratory.

Firestone, William A. Why "professionalizing" teaching is not enough. *Educational Leadership* 50(6): 6-11.

Julyan, Candace L. 1989. NGS Kids network investigate: January 1989 field test. Cambridge, MA: Technical Education Research Centers.

Loucks, Susan F., and Gene E. Hall. 1977. Assessing and facilitating the implementation of innovations: A new approach. *Educational Technology* (February): 17-21.

Mander, Jerry. 1991. In the absence of the sacred. *Whole Earth Review* Winter: 5-16.

Stepien, William and Shelagh Gallagher. 1993. Problem-based learning: As authentic as it gets. *Educational Leadership* 7 (April).

Winner, Langdon. 1991. Artifact/ideas and political culture. *Whole Earth Review* (Winter): 18-24

Kids Network Units

The following Kids Network student and teacher guides (produced by Technology Education Resource Center and published by National Geographic's Educational Media Division) were analyzed:

- Hello!
- Acid Rain
- Weather in Action
- What's in Our Water?
- Too Much Trash?
- What Are We Eating?
- Solar Energy

Appendix D: Thematic Codes, Examples

ALGORITHMIC VS. HEURISTIC	Examples and definitions distinguishing between algorith mic and heuristic approaches to teaching and learning
COGNITIVE SCHEMATA OF TEACHER	Domains of knowledge (cognitive schemata) teachers draw on for planning and instruction: pedagogical, students, sub ject matter, pedagogical content, other content, curriculum, educational aims
CONSTRUCTIVIST CONCERNS	Constructivist approaches expressed and demonstrated: theory and practice
COOPERATIVE LEARNING	The ways in which cooperative learning is presented in cur riculum and experienced by teachers and students
DISTINCT FROM OTHER CURRICULA	Features of curriculum that make it distinct from other cur riculum
DIVERSITY	Diversity issues raised in the development and/or implementation of the curriculum; appeal of program to a diversity of students
INTERPERSONAL CONSIDERATIONS	The possible effects of interpersonal relations and dynamics
INTERDISCIPLINARY ISSUES	Interdisciplinary issues raised in the development and implementation of the curriculum
KIDS AS SCIENTISTS	References regarding the value of kids doing what scientists do
KN CONSTRAINT	Constraints of Kids Network expressed by developers, publishers, users, researchers
MONEY NEEDED	References to the amount of money needed to operate curriculum
OBSOLETE INNOVATIONS	The nature of innovations to create their own obsolescence
ORPHANING THE TEACHER	Ways in which the teacher has been "orphaned" in the adoption and implementation phases

Chapter 4

Science, Technology, and Story:
Implementing the Voyage of the Mimi

Sally H. Middlebrooks
Michael Huberman
James W. Karlan

Harvard University,
National Center for Improving
Science Education

Contents

4

Science, Technology, and Story:
Implementing the Voyage of the Mimi

Introduction

For several years, educators in member countries of the Organisation for Economic Co-operation and Development have aimed to stimulate and improve work in science, mathematics, and technology. In the American context, an area of special interest has been the middle school curriculum (grades 5-8). Promising projects have been designed, field tested, and carried out—unfortunately, however, usually leaving few tracks for us to study closely in their wake. Although some of these projects are still in operation, the memory of their adoption and the keys to explain the success of some and the demise of others are missing.

In the United States, these innovative projects reflected a quest for a redefined science content, one less factually packed and that tried to replicate basic scientific structure through exercises and experiments such as scientists themselves might conduct: expeditions, transformations of materials, predictions of effects from the manipulation of variables. In mathematics, programs were scouted out that emphasized more of a problem-solving or hands-on approach— an application of mathematical knowledge and skills to real-life dilemmas whose meaning could be probed at any point in the experimentation process.

It was commonly held that these approaches should be applied by teachers with requisite subject matter knowledge specialized for teaching, including the choice of appropriate strategies and representations, the anticipation of children's insights, and the interpretation of students' conceptualizations. Finally, there was a particular interest in the use of new learning technologies—from videos to computers—whose use makes the environment more authentic, expands interaction and collaboration with others through networks, promotes laboratory-like investigations, and emulates the tools experts use to produce artifacts. Such technology also allows students to manipulate, construct, and revise their own representations and artifacts easily and in several media, including text, graphics, video, and audio.

Given its objectives and program, it was logical that implementation of the *Voyage of the Mimi*—a multimedia curriculum package combining a television series, software, and print materials to present an integrated set of concepts in mathematics, science, social sciences, and language arts—be identified as one of the eight U.S. case studies selected for closer study. The focus of the study was threefold:

- Closely describe the enactment of a program that—potentially at least—had students actively and thoughtfully engaged with important issues of science and mathematics. What did such an enactment look like when observed closely? How could such a program be carried out in garden-variety schools, as opposed to hothouses of experimentation, since that is where future executions would take place?

- Find out how such a program had been discovered, adopted, implemented and stabilized. What was the real story behind Mimi, in multiple—and therefore different—settings? How had the staff been prepared and assisted on an ongoing basis? What kinds of organizational shifts were required? How difficult was the process of mastery by which teachers hopefully got to the richer parts of the program, and how long did that take? How much was the original program modified in the course of its use? While many of these questions have been addressed in studies of the "innovation implementation" process, an area of specialization of the authors, there was evidence that work in science education had underestimated the complexities and difficulties of implementing inquiry-centered programs.

- Determine how the promising parts of this enactment of Mimi could be replicated elsewhere. While this could not be fully answered by the study, detailed descriptions of the ecology of local use and the renditions of the process by which a set of schools implemented the program would provide other school districts with the requisite information for deciding whether the match between their context and the six settings investigated in this study warranted their adoption of Mimi.

◆

Methodology

A case study is precisely that: the study of a case, in much the same sense as a lawyer or forensic scientist is said to study a case. A case can be a person, a group, a school, or other organization. It can be an event, a type or instance, a process. In this study, we are looking at particular cases of "innovation," but looking at them in different settings. In effect, cases are studied in their natural surroundings to better understand their histories, their contours, their dynamics, their core events and actors, and—in the best instances—the reasons they take the shape they do.

When several cases (teachers, schools) are studied, as here, each is given somewhat less time and attention, but the core facets of each are studied careful-ly and over time, and the comparisons bring to light insights that would have been overlooked given a single exemplar. When, for example, one set of teachers works routinely from their Overview Guide and a second set turns over the choice and design of an expedition to groups of students while peppering them with questions and apparently incongruous findings, we see significant differences in pedagogy, learning opportunities, and—most of all—the gamut of experiences and understandings one curriculum makes available.

A case study is best conducted as a process of gradual understanding of the core processes and determinants at play. This is achieved through multiple questioning, combined with a process of intelligently observing different arenas of activity. Informed by the research questions, researchers progressively become more locally knowledgeable. These observations can be highly formal, accompanied by instruments or formal notetaking; or they can be more informal. In most cases, as here, observations are mixed with other types of data-gathering devices, especially interviews—typically prepared in advance, but shifting as some questions take on more importance and some explanations begin to come clear.

In the study that follows, the research team looked closely at six exemplars of the *Voyage of the Mimi* during the school year 1993-94. The case study method allowed for a fairly systematic construction of the project's history; a description of the physical, social, and cultural environments of each school site; and, above all, observation of the project's life under natural conditions, month to month. In other words, this methodology let researchers become comfortable with the particulars of local sites, while becoming familiar figures themselves. Collectively, researchers and informants could reconstruct the local genesis of Mimi; tell the story of its implementation through the eyes of teachers, administrators, students, and other involved parties; document the constraints and supports along the way to daily practice, including many factors still present in the environment; and witness its daily enactment within the mathematics and science curriculum. The visits by the research team gave local participants the opportunity to describe their perceptions of Mimi from the inside out, and to tell their personal history of involvement with its materials and activities. (See p. 9 and appendix A for information about the researchers.)

The team spent much of its time observing how Mimi was acted out: how the teachers' guides were actually used; how teachers modified or otherwise experimented with the curriculum; and, above all, what pupils and teachers actually made of program activities designed to deepen understandings of scientific experimentation and mathematical principles applied to everyday, and some exotic, settings. To understand Mimi was to grasp its local surroundings; understand the strategies of, capacities of, and constraints on its staff; observe the nature, variety, and depth of its practices in different contexts of use; and then compare the six research sites on meaningful dimensions to tease out important commonalities and differences among them.

Sample Selection

Initial Design and Criteria for Sites. Aware that six sites are not representative of the hundreds of sites that constitute the *Voyage of the Mimi* population, our initial criterion was to look for diversity of use and of local conditions.

A sample matrix was developed, divided by the following criteria: location of the school (urban, suburban, rural); socioeconomic status and racial mix of student population; and variations in local practice. In addition, four other parameters were set:

- elementary or middle schools, grades 3-8;

- use of Mimi for three or more years;

- self-contained classrooms or teaching within a subject area, for example, science;

- connections or collaborations with other users within the school or district.

Site Selection Process. Site selection was initially complicated by the difficulty of acquiring lists of Mimi users. Finally, a staff person with Mimi's current publisher, Wings for Learning, randomly chose 50 sites (54 sites were eventually named), focusing on Eastern Seaboard cases. Further basic data on these sites (e.g., student population, names of teachers using the curriculum, etc.) were collected by telephone.

There were relatively few problems in locating suburban sites to include in the study. In fact, pilot visits made to 10 classrooms were in schools that fit this category; furthermore, the users on the list provided by the Mimi publisher were predominantly suburban. It was more challenging to locate urban and rural sites. Several strategies were used to fill the urban cells of the matrix, for example, telephone contact. The impression gained from this effort was that few schools in urban areas use Mimi.

Although follow-up conversations by telephone were helpful, visits to potential sites were key to making the final selection. These visits provided opportunities to observe teachers teaching a Mimi lesson; see a school's computer lab; interview teachers, principals, and specialists; and, in some instances, listen to children describe their experiences. These visits also revealed those who seemed comfortable with as well as interested in the project, and those who seemed best able to coordinate access to key players at the site.

From a total of 11 exploratory site visits, 6 were selected for the study. It is noteworthy that all sites finally selected—after a wider search—were in the Northeast. This was done both for convenience and because, after careful examination of alternative sites, it was determined that the standardized format of the program meant that the variety captured within the Northeast was as wide as would be possible anywhere else. It is, nonetheless, a limit on the sampling universe.

The six sites are located in five states in the Northeast: two are in rural areas, two in suburban, and two in urban settings; three are elementary schools, and three are middle schools. Sites 1, 2, and 5 teach the first *Voyage of the Mimi*; Site 3 teaches the second voyage. Both voyages are taught during the school year at Site 4; and at Site 6, the first voyage is taught one year, and the second, the next. Table 1 summarizes key characteristics of the six sites selected.

Table 1.   **Main Characteristics of Study Sites**

Site	Name	Type, Location	Student Population*	SES**	Mimi Use Observed
1	Greenfield Elementary	Rural, NH	250 students, white	Low to high	3rd grade, Mimi 1
2	Central Middle School	Urban, NY	800 students, black	Low	6th grade, Mimi 1
3	Lakeview Elementary	Rural, ME	460 students, white	Low to high	6th grade, Mimi 2
4	Union Middle School	Urban, MA	425 students, black, Asian	Low	6th grade, Mimi 1 & 2
5	Sherwood Elementary	Suburban, MA	440 students, white	High	5th grade, Mimi 1
6	Riverside Middle School	Suburban, VT	220 students, white	Low to high	5th/6th grade, Mimi 1 & 2

  * Student populations of fewer than 5 percent are not shown.
 ** Socioeconomic status (SES) is derived from information provided by teachers and principals.

Instrumentation and Data Collection

Several instruments for observation and data collection were developed and pilot tested. The final set of instruments included the following:

- interview schedules for developers, school-level and district-level administrators, teachers, pupils, and assistance providers near to or, in some cases, far from the site;

- observation schedules and checklists for classroom activity;

- guidelines for nonparticipant observation and participant observation (work in the classroom);

- formats for document analysis and work samples, both from the site and beyond;

- formats for analytic memos ($n = 75$) and summaries of site visits ($n = 20$).

During the period September 1993 to June 1994, the six schools were visited at least three times, one to three days at a time, for a total of 45 days on site. All visits were planned in cooperation with the sites. The first day of a site visit was used for orientation, with the researcher conferring with each participating teacher on such matters as confidentiality and goals of the study, and introductions to the school principal and key support staff, etc. Moreover, at this time, background information was confirmed.

Visits to sites were conducted throughout the school year; for example, Site 2 was visited in September, November, March, and April; Site 3 was visited in October, November, and March. Site visits typically began with the start of the school day and ended with school closing. The researcher was therefore able to observe not only lessons and activities related directly to the *Voyage of the Mimi*, but to compare them with lessons in mathematics, social studies, and reading as well.

Sites varied regarding the number of teachers using the innovation and the number of support staff. For example, at Site 1, an urban middle school using the *Voyage of the Mimi* to integrate mathematics and science on the sixth grade level, both science and mathematics teachers were observed and interviewed; in addition to the school principal, the school science specialist, the district science coordinator, the district grants writer, and a sample of students.

In general, formal and informal interviews with teachers were conducted during each site visit. In order to fit into teachers' schedules, interviews ranged in length from one half-hour to three hours. Between interviews, the researcher reviewed her notes and transcripts, if available; this process allowed for follow-up questions to be asked when the interview resumed, whether on the next day or during the next site visit.

When appropriate, one to two interviews were conducted with principals and/or other administrators and science and technology specialists at each site. These interviews were typically conducted during school hours and were between one half-hour and one hour in length. Interviews were conducted with samples of students at each site; these interviews ranged from 15 to 30 minutes in length; all took place during school hours.

Two to six observations of Mimi-related activities, one half-hour to two hours in length, were made for each participating teacher. The researcher's handwritten notes were often supplemented by an audio recording, all transcribed verbatim.

In summary, an estimated 100-plus hours of formal interviews with 17 teachers, 66 students, and 11 principals and other support staff were carried out, and an estimated 135 hours of observations (45 visit-days at three hours each) were conducted. In addition, the researcher joined classes from three sites for their all-day field trips: an ecology camp, a whale watch, and a regional Mimi Festival. See Appendix B for other data collection activities.

Research Questions

Numerous research questions were developed for this study. These were derived from previous work on Mimi; studies of the innovation process; and our desire to capture the history, implementation, and patterns underlying the day-to-day enactment of Mimi, notably in relation to the opportunities for learning made available to students. Our research questions covered the following categories:

- origins,
- background and history,
- the process of local assessment and adoption,
- developers'/publishers' perspectives,
- organizational fit,
- innovation characteristics,
- principals—role and perceptions,
- school district personnel—role and perceptions,
- assistance,
- early implementation,
- later implementation: teacher/classroom variables,
- pupils—role and perceptions, and
- organizational outcomes.

   Origins. Our questions under this category involved the historical context of program development; the arrangements between funding sources and developers; the dominant paradigm of mathematics, science, and technology; and the initial characteristics of the program. We also wanted to look closely at the development of the innovations and at how developers contacted potential participants.
   Background and History. Here, we wanted to examine the circumstances by which the innovation was adopted at the site, including the chronicle of local adoption—sequences, key events, determinants—as perceived by the main actors and the prime advocates of the innovation. We also looked into the local context at the time of adoption: the social, economic, and political frames; school and neighborhood characteristics; and the innovation history of the school .

The Process of Local Assessment and Adoption. Under this category, we sought to determine the motivations and incentives to adopt, along with the initial perceptions of the innovation (perceived merits, promises; doubts; limitations) and the expectations invested in it. To this were added questions about the orientations of teachers using the practice, their professional profile, and their stance toward science/mathematics teaching—e.g., more constructivist, more prescriptive, or structured.

Developers'/Publishers' Perspectives. We were interested in knowing whether developers and publishers had different views on the content, components, format, and teaching strategies of Mimi materials; how conflicting views were resolved; and how they influenced the materials.

Institutional Issues: Organizational Fit and Innovation Characteristics. We wanted to know more about the fit of the innovation to the organization: its nesting in the organizational structure and working arrangements of the school, e.g., with the curriculum in science and mathematics, the sequence of preceding and following curriculum units, and school/district policy. A series of important questions related to the innovation characteristics included its philosophy of mathematics and science, its view and practice of technology, its multicultural sensitivity, its vision of science expertise and inquiry, its epistemology, its complexity and difficulty, its portrayal of learning and motivation, and its components and their links.

Roles, Perceptions, and Assistance. We wanted to examine the roles played by the principal and district office personnel, especially their knowledge of the innovation and their behavior in times of strain; their technical and social support; their candid assessment of Mimi; and their special role in the crucial assistance network required to sustain such innovations in moments of preparation, uncertainty, and outright turbulence.

Implementation. Several questions pertained to the actual implementation of the innovation, as it interacted with the local infrastructure. For example, did the project influence core aspects of teachers' work, bringing about changes in repertoire; changes of knowledge base in mathematics/science; and/or transposition of the project's content into instructional strategies, pupils' understandings and perceptions.

Pupils—Role and Perceptions. We wanted to know what the innovation looked like on a daily basis from the pupils' point of view; e.g., the typical activities, the materials in use, the tasks undertaken. We questioned also the level of task engagement or involvement, task progression, and assessments made by the students themselves. To the extent possible, we also tried to assess the short-term, observable effects of the innovation on pupils.

Organizational Outcomes. Finally, we asked about the extent to which the innovation was settled or stabilized in the school. We also asked in which ways the school organization and yearly routines were likely to affect the innovation's structure, operations, and substance.

Analytic Procedures

Data analysis in qualitative case study research is largely iterative—i.e., it proceeds as the data are collected and reorients the future waves of data gathering and interim analysis. Interim analyses were driven by the original research procedures (translation of provisional research questions into codes, creation of provisional data collection instruments, multiple field visits, production of audiotaped or handwritten interviews and observations, examination of key documents and photographs). The procedures were dictated mainly by the structure of the site visit, which allowed for day-long contact (multiple observations and multiple interviewing of the same and different informants) and for multiple-day visits. Between visits, researchers could review their tapes and notes, identify meaningful trends, devise provisional hypotheses, and prepare follow-up questioning and observations for the next day, etc.

We worked with 17 functional coding categories, regrouping 32 thematic codes and 173 general codes. A sample of thematic codes is included in appendix C.

The thematic material, keyed to the original and revised research questions, was then subjected to a question-by-question analysis using templates devised by Miles and Huberman (1994). In particular, checklist matrices, event listings, and time- and role-order matrices were used to gather similarly plotted data on one chart. The same was done with conceptually clustered matrices and causal chains. Comparative analysis then allowed cases to be clustered, significant similarities and differences established across cases, and—from there—systematic sources of variation to be identified. This analysis provided us with the explanatory material we sought to supplement the narrative, more contextualized, material. This analysis enabled us to account for the enactments of Mimi across and within six settings—in particular for its history, evolution, instructional activity over time, and the reasons for significant similarities and differences among sites. Appendix B provides detail on some analytic procedures.

To strengthen the validity of our findings, and to learn more about Mimi in other contexts, we talked to teachers and district curriculum directors in six states outside of our sample. From among these, we briefly visited two districts where the Mimi curriculum was not only being used, but was said to be flourishing.[1] Table 2 summarizes the key characteristics of these two sites.

Research Team

The team was comprised of three researchers, all based at both Harvard University and the National Center for Improving Science Education. Michael Huberman, a specialist in the areas of innovation, learning theory, and qualitative methodologies,

---

[1]Peter Marston, who plays Captain Granville in the Mimi video serial, assisted us in selecting these replication sites; his recommendations were based on his travels to "Mimi fests" over the past several years.

Table 2.   Main Characteristics of Secondary Study Sites

Site	Type, Location	Student Population	Mimi Use	Years of Use	Inter- viewees	Constraints Identified
7	Rural/ suburban, NE	Families said to have fled the "problems of the city"; white	5th grade, Mimi 1: Show videos within 6- to 8-week period preceded and followed by a kit-based science package	9: Started by teacher of the gift- ed in 1986; currently part of dis- trictwide science	Principal; 2 teachers from 1 school	Lack of materials
8	Urban/ suburban, MI	District divid- ed between stable immi- grant popula- tion and slightly larger white popula- tion	4th & 5th grades, Mimi 1: Show all videos within short time period as part of language arts; show selected expe- ditions when related to science kits	'9: Started by teacher of the gift- ed and others in 1986; changed districtwide 3 years ago from SCIS to Mimi plus another kit-driven curriculum package	Curriculum director; testing advi- sor; 9 teach- ers from 9 schools	Statewide science test, time, pull- outs, full curriculum, access to computers, training

directed the study. Research associates Jimmy Karlan and Sally Middlebrooks were responsible for the science and mathematics components of the study, both from developmental and school-related perspectives. Karlan was also instrumen- tal in much of the software development (and debugging) for the study and was especially involved in the epistemological and pedagogical issues raised by the study. Middlebrooks added significantly to the phenomenological dimension of the study, i.e., the perceptions and meanings derived from teachers and students. (See Appendix A for additional information about the research team.)

Possible Limitations of the Study

The research team members all shared an orientation toward social-construc- tivist approaches to teaching mathematics and science—i.e., student-centered approaches emphasizing the construction of knowledge by students through the use of open-ended experimentation, class discussion, and independent work. Aware of this bias, members sought deliberately to be as descriptive as possible in their observations and to ensure complete accuracy in the conduct and transcription of interviews.

Other possible study limitations warrant mention:

- the restriction of the sample to the Northeast;

- the use of a framework concentrating on the process of implementating change;

- as in all multiple case studies, noncontinuous presence at the field sites;

- restriction of the sample to three years of experience;

- disproportionate number of more experienced teachers in the sample;

- reliance on self-reported data for earlier events for which no documents were available; and

- lack of interviews with principals who were felt not to be knowledgeable (e.g., recently hired, too remote from the project).

The strengths of the study arise indirectly from these limitations, however—that is, a presence at multiple field sites; a focus on the process and nature of specific enactments of the Mimi curriculum; and the attention paid to the perceptions of key actors, the teachers and students. Further the study's nonevaluative perspective provides a unique and instructive framework for understanding how innovative projects move to and look at the core level of application—the classroom—and what learning opportunities are actually realized.

◆

The Program

The *Voyage of the Mimi* combines a video series, software, and print materials designed to present an integrated set of concepts in mathematics, science, social sciences, and language arts. There are two separate packages, the *Voyage of the Mimi* and the *Second Voyage of the Mimi*. The first voyage takes students on a study of whales off the coast of New England, where students are meant to apply ideas of proportional reasoning, triangulation, and navigation to solve the problems that arise; they also contend with on-board issues of diversity and equity among the crew. In the second voyage, students accompany an archeological expedition in the Yucatan Peninsula of Mexico, working with the Mayan number system, the Mayan calendar, the relationship between the earth and sun, and discussing various social issues which arise during the trip.

Notably, both the *Voyage of the Mimi* and the *Second Voyage of the Mimi* are designed as multimedia packages that supplement, but do not replace, standard curricula in the upper elementary grades. The mix of television, print, and computer materials are understood as offering "hooks" into science, mathematics, and other subject areas. Believing all children to be naturally curious, Mimi's developers sought to provide "an activity base that promotes hands-on exploration and independent investigation" (Martin et al. 1988, p. 174). The developers were, however, less "sanguine" about teachers' capacity for learning nonscripted curricula; thus, they developed specific instructional materials that might fit and be "adaptable to different classrooms" with—as Mimi originator Samuel Gibbon notes[2] —"a variety of teaching styles." Consequently, teachers in neighboring classrooms teach Mimi differently, as we shall see.

Mimi's developers intended to show, graphically and realistically, what conduct of a scientific enterprise looks like. Gibbon's recalls:

> We were comfortable if boundaries around science and scientists got blurry . . . We really wanted kids to imagine a kind of scientific activity that included scientific observation, messing around in the data, looking for patterns . . . We intended to suggest that you could be curious about anything . . . [That] what distinguished scientific curiosity was suspension of belief, questioning of data, challenging of authoritative statements, continuing to keep an open mind about things.

It is helpful, before looking at the actual dynamics of the program, to examine its background. The following sections describe Mimi's development process (funding sources, constraints, formative evaluation, and training) and components.

From Development to Publication

The development, funding, and publication of Mimi is a story of how curriculum developers, of necessity, masterfully stitched together public and private funding sources to cover the period of development. In 1981, the Bank Street College of Education received $2.64 million from the U.S. Department of Education for development costs. Three years later, the *Voyage of the Mimi* was published by Holt, Rinehart and Winston. In 1986, a new $4.5 million grant from the U.S. Department of Education and the National Science Foundation was awarded to develop the *Second Voyage of the Mimi*, which was published in 1989 by Wings for Learning.

Developing each piece of the Mimi package involved a large number of people. At the core, however, was a small group of writers and researchers nested in a college of education known for its emphasis on child development and social relevancy. And, at the center of this core, both conceptually and organizationally, was Samuel Gibbon, Jr., Mimi's originator and executive director.

Teachers also played a part in Mimi's development. In the design of software materials, for example, teachers participated as curriculum advisors, field testers, and codevelopers. As Char and Hawkins (1987, p. 215) report: "[T]eachers

---

[2]Quotes from Samuel Gibbon, when not otherwise referenced, are drawn from our interviews with him.

became agents of the software experience and provided numerous insights, suggestions, and innovative uses for software."

Mimi was not developed unfettered, however. Mimi entered classrooms, according to Gibbon, "at a time of angst about the United States' measuring up;" he continues:

> We got from the publishers very strong pressure to put paper-and-pencil tests in all the material. Of course, their other materials did that. We saw this as perverting the Mimi materials. All that the episodes and expeditions programs offered would be contravened by tests that said, "These are the important and, therefore, the only things you ought to teach to . . . " Conventional testing does not fit with Mimi.[3]

Formative Evaluation

Although students were able to manage the Mimi software, evaluators found that they did not necessarily understand the underlying phenomena (e.g., temperature; see Char et al. 1983, p. 22). Comprehension of material, then, was an important issue and the result of students' difficulty in detecting what they might be doing right or wrong as they used the software materials and teachers' difficulty in detecting student problems (Char et al. 1983, p. 58). Extended use of the software by all students depended on teachers—both experienced and inexperienced—seeing Mimi's software programs as both content rich and "a vehicle to introduce various topics in math, science, and social studies" (Char et al. 1983, pp. 52-53).

At the same time, during this field testing stage, researchers learned firsthand about school conditions that dramatically limited full use of Mimi's multimedia package:

> [W]e learned a good deal about the limited computer, media, and teacher resources in most schools, [and were forewarned about] the limited amount of time usually devoted to science and mathematics instruction, the considerable range in elementary school teachers' training and expertise regarding mathematics, science, and computers, and the importance of good teacher training workshops and support (Char and Hawkins 1987, p. 217).

An important outcome of field testing software prototypes with teachers and students was the creation of "preparatory" software materials which, subsequently, led to more complex simulations. For example, to accompany the module *Rescue Mission*, three software games were eventually developed—each with a particular mathematical or navigational focus.

---

[3]This lack of assessment formats in the Mimi materials may contribute to the low level of use of Mimi's computer components and, specifically, of their underuse among girls. This point is expanded on later in this report.

Training

The Mathematics Science and Technology Teacher Education (MASTTE) project was the first and only attempt by Mimi's developers to devise a teacher training component for Mimi users, new and experienced. Funded by the National Science Foundation, the three-year project (1984-86) was conceived as an in-service training model illustrating ways for teachers to use the first voyage to integrate technology into their classrooms. It also offered 35 hours of training on the computer software programs developed for the first voyage. Various types of support were offered to participants as follow-up the next year, notably telecommunicated or print support and on-site observation and training.

Martin et al. (1988) summarize what researchers found during the MASTTE training. Noting that Mimi lessons are adaptable to a range of situations, they point to perceived changes in teacher thinking and practice as a result of group activities with the computer software programs. They also describe how, after MASTTE training, children in classes where Mimi was being used began to offer spontaneous comments, ask questions, and generally take greater overall initiative in the scope and flow of discussions (Martin et al. 1988, p. 182).

Outside of Bank Street, a course was offered to Mimi users for four summers during the late 1980s by Lesley College in Cambridge, Massachusetts. Groups of teachers from across the United States lived for a week aboard a research sailing vessel. They viewed episodes and expeditions, tried some of the computer modules, and took turns being "sailors." Three teachers from two sites in this study participated; they describe their experiences as unusual opportunities for learning and exchange. PW at Site 5, for example, recalls:

> We sailed out of Boston, and went up to Gloucester, and then we sailed across to Provincetown, and back into the Boston Harbor. We spent a week learning how to sail a ship, learning how to use the computer program for the *Voyage of the Mimi* and different activities to do for each of the expeditions and episodes. It was fantastic; it was unbelievable. I've never had a course like that.

Over the years, teachers at Sites 3, 5, and 6 have attended workshops related directly or indirectly to Mimi. These include National Audubon Society courses in marine biology and both formal and informal exchanges among Mimi users. Two of the six sites regularly participate in regional Mimi fests which draw schoolchildren to a local maritime museum. These events typically include a morning of various activities related to whales, whaling, and navigation led by museum educators followed by students and teachers joining Peter Marston (Captain Granville in the Mimi series) for singing and a question-and-answer period.

Instructional Strategies

Essentially, the instructional strategy of Mimi consists first in getting the children hooked on the story (trying to solve the mystery of an ancient Maya site, for example); then, buoyed by student interest, to have teachers follow up with more conventional approaches to teaching science, supported by enrichment activities, vocabulary lists, and preview and follow-up questions provided in the *Overview Guide*. Gibbon explains: "If we could get children curious . . . [then teachers can] lead them to want to know more . . . What teachers did with it might be much more rigorous and sequentially organized . . . might look much more like conventional science."

Gibbon describes the material in Mimi's *Overview Guide* as "conventional"; he gives two reasons for a rather scripted approach: "In fact, classroom materials—the print materials—are quite conventional . . . [We were] dealing with a publisher who wanted to get things into schools, and teachers who had a certain way of proceeding, and [we were] already bold in asking teachers to show videos and use software."

In the developers' eyes, videos and software were assumed a stretch for most teachers, whereas the print materials were assumed a more likely fit for teachers with more conventional instructional styles.

Main Components

Both Mimi voyages feature common themes (e.g, exploration of aspects of the undersea world and concern for endangered animals and ecosystems) and common main components. These components are:

- a video series made up of episodes and expeditions;

- an *Overview Guide* for teachers;

- a *Student Book* featuring an illustrated version of the video series and suggestions for activities and projects;

- microcomputer software learning modules; and

- navigational charts, maps, and posters.

Mimi's multimedia package builds on findings about children's proclivity to become engaged with a story and to retain information presented in a story format; thus, the video series purposefully "combines a serialized scientific adventure story with a series of 'Expeditions'" or documentaries that show real people studying "interesting pieces of the real world" (Gibbon 1985, 1989). Similarly, a narrative format was chosen for the software materials whenever possible. As for the *Student Book*, it is a retelling of the episodes and expeditions in print, plus suggested activities for children to do on their own.

Technological Components—Television Series. At the core of Mimi is the television series[4]—a serialized adventure story having "all the elements of

---

[4]The series originally was broadcast but since has been made available in video and laser disc forms.

Hollywood melodrama" (Gibbon 1989). The episodes for each series have a large cast of characters, diverse in terms of age, gender, race, and ethnicity. Also, in each series there is one character with a physical disability. "The intended message must be plain; science and math are open to all" (Gibbon 1989, p. ix).

Accompanying each episode is an "expedition" or documentary; there are 13 of these in the first voyage and 12 in the second. These expeditions are intended to show children "asking questions of adults in a curious, spontaneous way" and to model "how the teacher can use the drama to provide starting points and problems for further inquiry and discussion" (Martin et al. 1988, p. 176).

Students viewing the episodes and expeditions see many types of technology. For example, in the first voyage, students may be surprised to see a computer on board (it's used to record whale sightings). Also in this series, one character, Ramon, tags a whale with a monitoring device while two other characters, Rachel and C.T., use an XBT (expendable bathythermograph) to collect data on temperature at various ocean depths. The expeditions in the first voyage present an array of technology for mapping the ocean floor, collecting marine organisms, tracking whale migration, recording the "songs" of whales, overcoming human hearing loss, collecting and plotting weather data, studying the human body under different environmental conditions, purifying seawater, and so on.

Technological Components—The Learning Modules. An important component of the multimedia Mimi package is its microcomputer software. These computer modules involve students in gamelike formats that are intended to replicate what scientists do: for example, inputting data, displaying graphs, and using a computer-simulated model to understand the complex relationships within an ecosystem. They were developed to complete other, "off-line" activities (e.g., mapping, orienteering, exploring shadows, discussing food chains) so as not to appear as a separate package of self-contained games but rather an extension of the videos which additionally offered hands-on experiences. To encourage group interaction, the modules are designed for more than one student to use. Each module includes both teacher and student guides.

The learning modules for the first voyage are:

- *Maps and Navigation: Pirate's Gold, Lost at Sea, Hurricane!, and Rescue Mission.* Four computer simulations build on navigational aspects presented in the television series—finding whales and sailing a boat. These introduce students to longitude and latitude; methods of locating objects on a map and chart; and relationships among speed, time, and distance. Designed as a progression of activities requiring increasingly more complex concepts and skills, *Rescue Mission* is the culminating program. Students work as a team to free a whale caught in a fishing net somewhere at sea. Given only its location, they must determine compass heading, speed, and time needed.

- *Whales and Their Environment.* Students explore important physical properties in real time. They use sensors attached to the computer to measure light, sound, and temperature. Measurements are displayed on the monitor as graphs.

- *Ecosystems: Island Survivors.* Following a game format, students develop concepts about simple relationships in ecosystems, including food chains. Graphs display increases and declines of plant and animal species. Students strategize in teams to maintain a balanced ecosystem.

- *Introduction to Computing* (no longer sold). Students learn critical programming concepts through a series of games using the LOGO programming language.

The learning modules for the second voyage are:

- *Maya Math: Glyph Trek Maya Calculator, Maya Calendar.* Students act as "mathematics archeologists" to explore the Mayan base 20 number system and calendar. They convert Mayan calendar dates into our own system and vice versa.

- *Sun Lab.* Students explore relationships of the earth and sun through a simulation/animation computer program. Accompanying print materials involve students in measuring shadows, building a sundial, and investigating maps and globes.

- *Scuba Science.* With the ecosystem of the coral reef as background, students use sensors to explore relationships between pressure and water depth; and experiment with light, temperature, and sound.

Print Materials. The *Student Book* follows the video series: each episode and expedition is written in book format with accompanying illustrations and photographs. Among the six field sites in this study, only Site 5 supplied books to all students (a total of four classes, or 90 children). Some students at Site 5 reported using their books as aids in answering homework questions; others said they enjoyed the pictures and being able to read ahead.

The *Overview Guide* is for teachers.[5] There is one for each voyage. Each comes in a looseleaf notebook. This format is convenient, since it allows a teacher to take out the pages with information, questions, and activities related to specific episodes and expeditions. Additionally, the format seems to encourage teachers to make additions to the notebook (inserting photocopied pages on how to build miniature lighthouses, for example). Teachers share these added pages with each other.

Mimi's teachers' guides provide a mix of information and activities for each episode and expedition (see table 3). Key features of the guides are discussed below.

---

[5]We use the terms *Overview Guide* and "teachers' guide" interchangeably in this report.

Table 3.   *Overview Guide* **components for a sample Mimi episode.**

Title of Episode
Content Objectives
Vocabulary
Program Components

Activity Materials
   Video Series
      Review of Previous Episode
      Summary of This Episode
    Background for Teacher
    Preview Questions
    Follow-up Questions

Navigation Chart [called "Map" in second voyage]
Activities [in *Student Book*]
Enrichment Activity
Extending Concepts [called "Related Topics" in second voyage]

Pre- and Post-Questions. The teachers' guides provide teachers with a mix of discussion questions to choose from: some relate to specific science concepts, some encourage consideration of the values underlying scientific research, and some focus on the social dynamics among Mimi's crew. For example, following the first episode in the first voyage (called "Setting Sail"), six of the nine questions focus on science and scientists and three relate to whale research. In the final episode ("Rolling Home"), none of the 10 questions suggested is science-related (e.g., "How do you think individuals' feelings about themselves and others had changed since they had been on the island?").

Typically, the pre- and post-questions suggested for expeditions pertain to science-related content and issues. For the first expedition in the first voyage ("Planet Ocean"), questions focus on oceans—percentage of the earth covered, human influences, etc. For that same expedition, follow-up questions ask about oceans in relation to human life, research, and adaptive features of whales. The final question of the nine asks students if they would find marine biology an interesting profession.

Enrichment and Hands-On Activities. The enrichment activities suggested in the guides range from crossword puzzles to ideas for field trips. The hands-on activities suggested in the *Overview Guide* and *Student Book* include making a simple solar still using a small plastic bag filled partially with saltwater, for example.

Differences Between the Guides for the First and Second Voyages. The teachers' guide to the second voyage includes the same introduction as the first, plus an additional two pages of text. The focus of the second voyage (archeology) is explicitly linked to the focus of the first (whale research). Both voyages, according to the text, are "scientific," and both encourage an interdisciplinary approach to the material. The guide for the second voyage—but not that for the first—follows the introduction with detailed descriptions of Mimi's components. Pre- and post-questions, for example, are said to be designed "to encourage lively and in-depth class discussion." Some of the questions, the text continues, are about the story while others "call for speculation, hypothesis-making, and reasoning." Later, the guide reiterates the importance of "discussion," as follows: "A major part of your students' learning is likely to take place in these discussions as they try out ideas; hear the ideas of others; and examine and evaluate their own reasoning that is being formed, validated and tempered by the ideas and reasoning of others."

Although the *Overview Guide* for the second voyage is essentially the same as for the first, it does feature several additions. For instance, mixed into the text are shaded boxes of various sizes. Some give a "behind-the-scenes" look at the making of the Mimi; others contain classroom management tips. The box on page 2, for example, refers to a character named Pepper, who joins the Mimi crew in the voyage's first episode. An expert scuba diver, she has had one leg amputated because she has had cancer. This box says, in part: "If you watch your students as they watch this episode, you can be aware of their reactions to Pepper and the manner in which her disability is presented and dealt with . . . Creating an atmosphere in which students feel free to express their feelings is important. Getting whatever discomfort there is out in the open will help dispel it all the sooner."

The *Overview Guide* for the second voyage has more material for both the episodes and expeditions than the first. In addition, it includes 17 pages "About the Maya," followed by 20 pages "About Archeology." These are followed by a set of blackline masters (recipes, crossword puzzles); an eight-page glossary; 17 pages of references by topic (the sea, rain forests, venomous creatures, etc.); and, at the end, a two-page list of resources (films, videos, maps)—a total of 202 pages. "Learning," both guides seem to argue, is primarily through discussion, which is itself stimulated by questions provided in the guides and asked by the teacher. "Doing science" is primarily accomplished by following experimental formats, which are also provided in the guides.

Assessment Component. There is no formal assessment component in the Mimi materials. The developers report their strong resistance to applying what they judged as flawed elementary-level science tests to the Mimi material, as noted earlier.[6]

---

[6]This lack of assessment components in the Mimi materials may be one reason for the low level of use of Mimi's computer components and, specifically, of their underuse among girls. This point is expanded on later in this report.

Most teachers, however, are required to give a grade for science at several intervals during the year. In the absence of either suggested assessment strategies or forms to follow in Mimi, and with no alternative formats developed within their schools and districts, the teachers in our study report having to rely on traditional methods of assessment. Typically, student participation is the crucial variable, along with discussions and successful completion of assignments. (Although students sometimes work in groups, assignments—and, consequently, grades—are individualized.)

For example, students at Site 5 explain the assessment methods used by their fifth grade teacher, PW:

> First Student: "Plus" is really, really good; and "checks" are average; and the "check minus" is pretty bad. "Incomplete" is if you are really bad.
>
> Researcher: So how come you got a plus?
>
> First Student: I explained more, and it was longer.
>
> Second Student: I like to make it things that people would really say, not like all this totally scientific stuff. Like . . . I always say in the beginning, like, "Hi, this is Arthur Spencer [a Mimi character] bringing you your local evening news . . . " You can do it in any form that you want to. She [PW] just wants a journal, and she just wants to know what happened; that's basically what she wants. I don't think she really cares if you give stories in your own words and like a talk show or news . . . She just wants it to be like you know what you're talking about.

A teacher at Site 2, EC, explains how she evaluates whether all her sixth graders understand a particular mathematics concept:

> I sort of look in their faces and, you know, it's like [thinking], oh, God, that's what I looked like sitting there [as a student], "I have no idea what you're talking about, lady." So what I do is, I try to find three or four different ways to explain the same thing. I tell them that as soon as they understand, don't listen to me anymore. And so far, knock on wood, it's working somewhat.

◆

Conceptions of Science

In this section, we explore conceptions of science as understood and enacted by developers, teachers, and students. We begin with the developers' intentions to change the definition of science and ways to teach science in upper elementary and middle school classrooms. We describe Mimi developers' notion of the "ideal teacher" and the constraints they see as limiting this vision. Second, we listen to teachers' views of science and the scientific enterprise as it is exemplified in Mimi; we then give examples of how teachers use Mimi materials to guide their instructional activities. Third, we turn briefly to the students to find out how they view science and the scientific enterprise based on watching Mimi episodes and expeditions.

At least three tensions permeate this discussion:

- There is a tension between science as a specialized body of knowledge composed by others versus science as something "anyone can do" and, according to this view, something that everyone is doing everyday.

- There is a tension between a particular way of doing science—sometimes referred to as the scientific process—and science as "messing about."

- There is an uncertainty about how much teachers need to attend to mastery of science content and how much importance they need to give to fostering student curiosity, reasoning, and motivation.

Science From the Developers' Perspective

At the heart of Mimi is an adventure story, serialized and televised. The series is seen, on the one hand, as appealing to students; and, on the other, as presenting few problems for teachers. Gibbon (1989, p. vii) writes that "The instructional function of the video series is to provoke and sustain the interest of students by bringing into the classroom some inherently interesting pieces of the real world and by showing some attractive and interesting people engaged in studying them."

The video series succeeds in portraying science as a human enterprise that is exciting, often fun, and always purposeful. Both the story and the documentaries (that is, the episodes and expeditions, respectively) show that the motivation for doing science—that is, for being a scientist—comes both from wanting to find answers and from a desire to benefit others. In addition, both make the case that the scientific enterprise includes men and women of diverse ages and backgrounds. Finally, the video series shows that the scientific enterprise often involves more than one person and that collaboration, although not easy, is beneficial.

So "hooking" students is the developers' main aim; teachers, they assume, will follow. The assumption is that children are naturally curious and want to know about the world around them. The developers report, however, that they did not sufficiently take into account teachers' beliefs, styles, and constraints, that they failed to find ways to overcome some teachers' inability, unwillingness, and/or lack of support in allowing knowledge to be "constructed," rather than "transmitted" and "managed." It turns out that although learning may be "natural," teaching and its surrounding conditions are not. Gibbon explains, "We wanted teachers to welcome opportunities to say 'I don't know.' [We] underestimated how hard it is for teaches to feel out of control of the intellectual sequence: the 'What are we going to learn next?' question."

Mimi's formative evaluation team made a similar point as early as 1983 after observing 13 teachers in seven schools who piloted Mimi episodes and computer programs (Char et al. 1983). Although their findings focus on the factors that affect the use of Mimi's software components, those factors are pertinent to Mimi as a whole:

- The availability of hardware (we would add equipment and materials) significantly influences students' learning opportunities.

- The role teachers choose to take (demonstrator, manager, guide) shapes students' experiences.

- Teachers who are more knowledgeable about the scientific concepts inherent in the program are better able to affect students' experiences through relevant preparation, probing questions, and appropriate help.

- The program itself, as well as the teacher, strongly affects learning opportunities for students.

At the time of their evaluation, the Mimi evaluators concluded that "generalist teachers," as opposed to teachers with strong science backgrounds, would likely be the primary users of Mimi materials. They pointed out that this would have important consequences for the level of prescriptiveness or direction in Mimi's *Overview Guide* for teachers. After defining the teaching styles among the small number of classrooms they observed as either teacher-centered or child-centered, the evaluators then dropped this bomb: computer activity is inherently challenging to a teacher-centered style (Char et al. 1983, p. 76). Hence, they argued that the main challenge facing the Mimi developers was how to make their materials as "comfortable" to as many teachers as possible, whether they are teacher-centered or child-centered.

Today, Gibbon frames the decisionmaking process around the development of Mimi's multimedia package in terms of the most appropriate use of federal funds—reaching the largest number of teachers, offering the broadest range of styles. Gibbon maintains, "It's a tough call to what extent it is appropriate to use federal funds [to say] 'Teach this way or don't use the stuff.' Our choice was to say: 'Here's some useful material we think could be valuable to kids in a wide variety of classrooms and with a variety of teaching styles; use them in any combination—you can teach conventionally, more adventurously.'"

Teaching "adventurously," however, would entail teachers' taking liberties with most of the accompanying instructions or suggestions; and this, in turn, presupposes a more "constructivist" or student-centered frame and probably a stronger science background. It also involves changes in the instructional formats that many teachers use to sequence and orchestrate their instruction. For many teachers these demands are hard to meet, as Char et al. (1983) found regarding the software component and we learned more recently from teachers. For example, EC, a teacher at Site 2, refers to using Mimi for teaching sixth grade mathematics: "I can go out of sequence and out of things like that for certain things, but without certain concepts that came before, it's hard for me to just jump right into it."

Similarly, enacting the cooperative learning format, as opposed to a conventional group format, was often a dramatic change for the students and, thereby, for instructional management. KN, at Site 2, appreciates that matching high- and low-achieving pupils, as is done in Mimi, is beneficial, but notes that the students "have never been trained for that." This makes group work difficult, if not impossible, while sustaining class management and scheduling limitations: "You can bang your head— especially in the middle school, especially if you've seen them only 40 minutes a day—trying to get them to understand that concept and trying for them to realize that that bottom person, if that bottom person doesn't do it, then they all sink."

In contrast, RS, a colleague at the same site, takes a different stance, seeing science as a process that demands cooperative learning—no matter what the struggle for students or their teacher:

> As far as the kids are concerned, I think they understand what I try to get them to understand: that science is a process, it's not a body of knowledge . . . It's the process that's more important . . . In order to do this, they have to work in groups, teams, cooperative groups . . . There has to be cooperation and there has to be self-discipline among the individuals so that there isn't bickering and fighting and to ultimately work towards a goal. I think that's the main thing I try to get across, and I think they have a sense of it.

RS notes that having children construct their own knowledge entails an understanding that "goes way beyond the test," even though it will "take them longer to get it." He concludes: "You have to allow them time not to look so good in the beginning." Although RS refers to students here, he may agree with us that teachers and a new program also need time and should be allowed "not to look so good" in the beginning.

Science From the Teachers' Perspective

In this section, we organize excerpts from our interviews with teachers into how Mimi fits with their views of science and science teaching. Teachers say Mimi shows science as (1) part of "everyday" experience and science as in everything around us; (2) a series of steps that must be followed; and (3) an enterprise that is sometimes adventurous, sometimes boring, but built on a passion to know and find out.

Mimi Shows "Science" as It Is—Part of the Flow of Everyday Experience. A teacher at Site 3 describes Mimi: "It's the way the world really is . . . Everything combines to one thing. That's what Mimi is. When you sail a boat, does the Captain say, 'Gee, I'm going to do my math now, I'm going to do my science.' You know, they have to sail the boat and they're doing all those things. So, that's what Mimi prepares kids for—to do stuff like life is." For this teacher, "Mimi is more real world" than school where teachers and students say, "Right now, I'm doing science, I'm doing math; now I'm doing social studies." "Mimi," he sums up, is more "like life."

Mimi Has a "Gentle Way" of Showing the Scientific Process. A Site 1 third grade teacher states:

> There are specific steps that you have to follow in science. Mimi is a nice, gentle way of showing this to the kids without saying to them, "You must, you must do that." I think at this age that's scary for them. So, I think this is presented in a very positive light, and yes, as the series goes on, there are certain things you have to do. It's like when they go aground. All of a sudden the machines don't work, and something is wrong, and what's wrong? Well, we don't know, but we have to find out. and there are specific ways that they go about finding out.

Mimi Shows the Romance and Passion in Scientific Work. A teacher at Site 6 refers to the last expedition in the first voyage, which introduces students to the real "Captain Granville." The actor's main occupation is being a physicist at MIT.

> It's really wonderful to see a real scientist playing with questions, and then taking them, having the ability or facility to take them really interesting places. When that happens—that he is all of a sudden a scientist doing research—that's really exciting to me. It reinforces to me why I want kids to do research. Just hoping that they find something that they really like, that really makes them ask lots of questions. Get excited about asking lots of questions, go home and talk about that with their parents.

Another teacher at Site 6 contrasts how science and scientists are differently portrayed in episodes and expeditions: in the former, it's adventure and romance; in the latter, it's real life—and it looks boring—but not, this teacher surmises, for scientists:

> Probably in the episodes, the work of scientists is pretty romanticized . . . you don't see the hard work and the years of collecting and all of that kind of thing, "Oh, next year we'll get it, won't this be wonderful?" . . . I think that the expeditions, on the other hand, really present a pretty good picture of what it's like to do research . . . [She alludes to an expedition in which a female scientist is inside, sitting, seemingly doing nothing very exciting.] "Oh, it's just so boring," [the kids say]. And it is boring, but she just loves that. So it's exciting to her.

Science From the Students' Perspective

In this section, we give examples of students' perspectives of science and scientists based on excerpts from our interviews.

Changes in Ways Students Think About Scientists. Our analysis shows that students' thinking about scientists changes because of watching Mimi; an example from Site 6 follows:

> Student 1: When I thought of science, it's more lab testing.
>
> Student 2: Yes. Taking samples.

Researcher: Right, so now [after watching Mimi] what do you think of scientists?

Student 1: I think they can do anything: detecting, they can detect a lot of things, and they're not always in white coats in a room, they're out exploring how to—

Student 2: Yes, you can't find about everything sitting somewhere—you have to go out.

The Nature of the Scientific Enterprise. Students' descriptions of the scientific enterprise range from remarking on where scientists go to do their work, the characteristics of that work, and how that work differs from students' own class work. A sampling of their comments follows:

- Scientists find out something—even if it takes a long time and isn't easy—but it looks like fun: "They're studying. They're studying and worrying about stuff. They're looking at it, cleaning it off, picking up stuff. If it's broken, they try to put it back together to see what it was, just like it was before."

- Compared to what students do in school, scientists do the real thing: "Scientists are actually out there and getting the real thing. They get to really see what the whales are doing, they're not reading about it."

- Science is useful; it matters: "They know when they're trying to find out stuff from the past and stuff that could be something that we could use. It could help us like—if we knew more about our ancestors, we could figure out how they made their stuff and we could maybe try to recreate some of that stuff and maybe more people would understand why they did it."

Students, unlike their teachers, refer to the mismatch between scientists as presented in Mimi and their everyday experiences in classrooms. Students say they are unlike scientists because they stay inside, read about what others have researched, are engaged—except for short periods of time—with unimportant matters, and don't struggle to find out and learn. Their experience, in one student's words, is "not the real thing."

◆

Settings

This section documents in greater detail the context for the implementation of Mimi in our six study sites. All site names are psudonyms. Appendix D provides profiles of participants at each site.

Site 1—Greenfield Elementary School, Rural New Hampshire

Greenfield is an elementary school for grades 1-5 situated in a rural New Hampshire community which is economically dependent, to a large extent, on tourism. The driver of the school bus, for example, also owns a local motel. Bordered by woods and fields, the school is at one end of a small town; at the other end is a feed store, grocery store, gas station, and "bed and breakfast." Lining the road that links school and crossroads are single-family houses, each with its own woodpile.

As part of a regional district that encompasses seven towns, Greenfield is one of four elementary schools feeding into one middle school and one high school. The school was built in the late 1980s to meet a growth in student population. Adding one or more classes each year, Site 1 has grown until now it has 250 students; 11 teachers; 4 full-time specialists (in art, music, physical education, and reading); 3 part-time specialists (in learning disabilities and speech therapy); and its administrative staff. Science and technology specialists are at the district level and are strong supporters of Mimi. There are 20 students in each of the two third grades. HP, one of two teachers participating in the Mimi project, has been assigned three children who are identified as having special needs.

The principal at Greenfield describes himself as facilitator and evaluator. Although he came on board after the *Voyage of the Mimi* became part of the third grade curriculum, he describes himself as "cheerleader": a supporter of both Mimi and the two third grade teachers in his school (KC and HP) who teach it. He sees Mimi as building on children's natural interest in the ocean and as offering teachers a "whole package" from which they can "branch out" into reading, writing, measuring, and counting. He further attributes Mimi's potential at his school to the two third grade teachers' enthusiasm, hard work, and personal interest in teaching science. They, in turn, characterize the principal as someone who encourages, rewards, and praises.

The majority of teachers at Greenfield have taught between 10 and 20 years. It is noteworthy that staff development in recent years has focused on cooperative learning and teacher effectiveness training, and less on science and mathematics. The student population at Greenfield is uniformly white, but represents many ethnic groups; the range of socioeconomic classes is from poor to upper middle class. The 12 third graders interviewed for this study were enthusiastic about watching the Mimi television series.

The two third grade classes are mixed for most Mimi-related activities. There are practical and pedagogical reasons for this. HP's classroom is larger than KC's; its size and HP's possession of a television set and laser disc player are practical reasons why KC's students go to watch Mimi episodes and expeditions across the hall. These viewings are typically followed by a group discussion led by KC, and next by an activity that mixes students from both classes into smaller groups or pairs of students working together on an activity. KC and HP say Mimi's crew provides a model for cooperative behaviors and group work.

Choosing to do Mimi together, however, necessitates frequent exchange between KC and HP. For HP, who sees herself as less familiar with Mimi and lacking KC's sailing experience, KC has played a mentoring role. HP says, "I am very fortunate to have KC because she has all this incredible knowledge and she's done this a lot longer than I have, so she's much more familiar with the activities. I can run in and say, 'Okay, how do I do this? What do I do with this? Where do I go with it?'"

HP and KC make little use of the school yard and nearby environment for instructional purposes. They instead take three field trips, all connected to Mimi, during the year. Teachers, parents, and students work to raise additional money for trips to a marine environment, the weather station at the top of Mt. Washington, and a whale watch. Thinking about the whale watch, KC reflects, "It's much easier not to." She considers field trips as "very hands-on" for children and in sharp contrast to the passivity of watching Mimi episodes on television.

Each third grade has two computers; none of the Mimi software programs is in use. Instead, children use class computers on their own for drill and practice in spelling and computation. In addition, with assistance from the district's technology specialist, a small group of students in HP's class are creating books with pictures and sound. In KC's view, computers assist and complement—but do not replace—her instruction.

As detailed in the next section of this report, Mimi's arrival in the district in 1984 stems directly from one teacher's enthusiasm for it. Currently, third grade teachers use four kits, not all Mimi-related, one for each of the science units they are expected to teach at some point during the year. Kits come with posters, books, films, and ideas for hands-on activities. Kits are reviewed yearly by a committee composed of teachers and the district's science specialist. He explains: "We have a science committee . . . adding new units and taking out some that weren't very successful, and so it's a very dynamic curriculum in the sense that it's ongoing and changing." The tapes for Mimi are included in the kit on oceanography. KC and HP, however, have their own set of Mimi episodes and expeditions on laser disc and choose to show them, in sequence, over the course of the school year. As for the other three science kits (the units are the solar system, plants, and rocky shore), KC and HP choose to teach them in a sequence that matches with Mimi's story line. For example, in late spring, children watch Mimi's crew explore and find food in a marine environment. They visit a similar environment on their field trip and have a cookout as well.

In summary, teachers at Site 1 appear to be faithful and efficient Mimi users. They show episodes and expeditions in sequence; they ask questions and do all the hands-on activities suggested in Mimi's *Overview Guide*. They find ways to do two things at once: for example, make Mimi's vocabulary words double as a dictionary drill. They see in Mimi a way to bring up issues related to group work and interpersonal relations. Choosing to teach Mimi not as a six-week unit as others do, but across the school year, KC and HP look for connections between Mimi and the other science units.

Site 2—Central Middle School, Urban New York

Central is a middle school covering grades 6, 7, and 8; it is in a large, urban school district. Like many inner-city schools, it strikes the first-time visitor as somewhat grim. It has seen many shifts in the racial and ethnic makeup of its student population, but little in its primarily all-white teaching staff. The school currently draws its 800-plus students, almost all of whom are black, from two neighborhoods, one poor and one working class—"the projects" and, the other, one- and two-story residences.

Until five years ago, there was general agreement in the district that Site 2 was "at the bottom of the barrel." The district's grants writer had put it this way: "When teachers come to work in this district, that school is often last on the list of where they want to go. It has a reputation of being a poor school, with children with multiple problems, much of which comes from poverty."

Her assessment predated the injection of large doses of federal funding. Five years later, success is uncertain. The ambitious goals of integrating science, mathematics, and technology and the change of science teaching to include more hands-on and cooperative activities continue to be out of reach at Site 2. In the end, the traditional separation of subject areas and faculties, the 40-minute period, mandated curricula, and standardized tests still appear to predominate.

What does emerge, however, is that at least for a small group of teachers this was a privileged time—perhaps the only time in their careers—when they were deeply engaged in issues of teaching and learning across subject areas. In addition, for three teachers (RS, KL, and MX, all male) at Site 2, this was an opportunity to leave the classroom and assume responsibilities as "staff developers/instructional leaders" (their term): "After 20 years of being in a classroom," says KL, "it was a good opportunity to move, to grow." And finally, this was an opportunity for the teachers to develop expertise in the use of new technologies and produce a body of material the school and district were eager to spotlight.

Whether changes at Central are perceived as large or small, fleeting or enduring, a small group of teachers has thrived, and the principal now says proudly of them: "We're envied all over the district . . . my teachers are known all over the country." Moreover, for the past five years, sixth graders have enjoyed watching Mimi episodes, participated in more experiments, and become familiar with computers as tools.

The principal of Central Middle School grew up not far from the school. Perhaps to demonstrate the difficulties of his job, he shows visitors a sample of the knives he has taken from students over the years. He talks of his failure to have welcomed with open arms the changes the federal grant might bring. He says:

I didn't want to lose my best teachers in science and math out of the room . . . so I was resistant. And I am happy to say I was wrong, because what we got out of this—in terms of what the kids learned, in terms of the technology that was brought into the school, in terms of the publicity and public relations that the school got—was enormous, far beyond my ability to grasp at the time.

Central's principal is expected to retire soon. Those involved with the recent changes in teaching science fear that support for teacher-leaders and for changes made in science may dwindle under new leadership.

Six of the seven mathematics and science teachers involved in some way with teaching the first *Voyage of the Mimi* at Site 2 are white and so do not mirror the racial and ethnic backgrounds of the school's current student body. Five of the seven members of this group are experienced teachers, having taught for 15 to 25 years; the other two have taught for three to five years. None, however, began as science or mathematics teachers, but rather as social studies and art teachers. Thus, science and mathematics teaching needed strengthening. RS, a science teacher-leader and former science teacher, explains: "The school needs bodies. My first day . . . someone said, "You want to teach here?" I said, "Yes," and they just grabbed me. I was a living, breathing, body."

Turnover, illness, and maternity leave have each contributed to staff changes in general at Central, and loss of those trained to use Mimi in particular. The district grants writer sees these as "administrative problems" and "the realities of school change in an urban district . . . in a school that does not get the cream of the teachers."

Recent immigrants from the Caribbean make up the majority of Central's student population. Although these students speak English, many come from rural backgrounds, and their prior schooling is judged as inadequate. Seventy-nine percent of the students are eligible for the free lunch program. The students at Central are tracked. In the sixth grade, there are eight groupings, labeled from 6.1 to 6.8, with the lowest group more often the most recent immigrants. Several teachers describe their students as "more low functioning" with each passing year; they attribute the school's ineffectiveness with these students to family and social environment. EC notes: "Some of these kids . . . the mother is working nights. So when they come home from school, they're either making dinner or taking care of the other kids."

Site 3—Lakeview Elementary School, Rural Maine

Lakeview is an elementary school, grades K-6, located in rural Maine. The majority of the adults in this community of 5,000 people work elsewhere. Local industries include boat building, technology, and tourism. Site 3 was built in the late 1950s and enlarged 30 years later to accommodate a growing student population. The school's architecture, carpeted hallways, and color scheme give it the appearance of a modern office building. It has a library, art and music rooms, and a small gym that doubles as the school lunch room and local community assembly hall. There is a well-stocked computer lab with a full-time specialist.

Lakeview has 25 teachers; the majority are experienced. In addition to the teaching faculty, there are 12 specialists, some of whom are full time and some part time. There are no external curriculum specialists to provide assistance to teachers; also, statewide workshops and conferences, once regularly attended by teachers, are no longer held. Consequently, these days, the three sixth grade teachers participating in the study (MG, RJ, and EH) rely on one another and form a tight group. They typically arrive an hour before school starts, and, over coffee in RJ's room, they talk and plan.

The student population numbers 450 pupils and is white. Family incomes range from poor to prosperous; approximately 20 percent of the students participates in the free lunch program. Class size at Site 3 is small: 17 to 19 students in general classes with fewer students in each of the three special education classes.

Although students attending Lakeview live within an hour's drive of the Atlantic Ocean, they are said to "know very little about ocean life." Instead, they are familiar with the lakes, fields, and woods that surround their single-family homes.

Science curriculum at Lakeview is a local effort with no outside pressures. There is, for example, no state-mandated curriculum. A standardized science test is given in the sixth grade, but because students' scores are consistently above the national average, this test is not perceived as a force for changing science content. Furthermore, because Site 3 graduates go on to enroll in middle schools outside the community, it would be difficult for Site 3 teachers to teach with any specific scope and sequence in mind.

One of the four wooden dories built by previous sixth grade classes is displayed in the school's foyer. On the walls outside the three sixth grade classrooms, teachers have displayed photos of that fall's week-long experiences at a nearby camp facility. The photos show students canoeing, painting personal glyphs on t-shirts, and participating in a nature walk to learn examples of medicinal and edible plants indigenous to their area. Photographs also show students and teachers conducting an archaeological dig at the camp's former dump site. (Among the finds were parts of a metal bed frame, a leather shoe, chinaware showing the camp's logo, and a small, blue medicine bottle).

There are three sixth grade teachers at Site 3. MG and EH are first generation Mimi users, whereas RJ began teaching the second voyage for the first time during the year of our study. Although alike in their strong advocacy for Mimi, MG and EH are strikingly different from each other in several key ways: their knowledge and use of technology, classroom set up, teaching styles, and the location of Mimi within their teaching. MG is an experienced teacher of 18 years (all at Site 3). His classroom was used originally as the school's science lab. The space is half the size of other classrooms and looks too small for 17 sixth graders, three cages housing guinea pigs, a large aquarium, and four computers. Colorful posters, mostly of animals, cover every inch of wall space. In addition, perhaps because MG was one of the two original Mimi users, also on display are all the graphic materials that come with Mimi for both the first and second voyages. For MG, Mimi models a way of integrating subject areas and skills. Mimi is at the core of his teaching: "It's just the way I think now." He continues: "Mimi is the way the world really is, as opposed to what schools do."

EH teaches only two more students than MG, but his classroom seems spacious in comparison. The adjectives "neat" and "orderly" come to mind. There are posters on the walls, but only a few are directly related to Mimi. Students sit at desks in clusters of twos and threes; all face the blackboard. The small space behind a tall shelf provides a private space for students. There is one computer.

EH has taught for over 20 years, the majority of those at Lakeview. Unlike MG who puts Mimi at the center of all he teaches, EH sees Mimi as only one part of the sixth grade curriculum, and a dispensable one at that. He says: "We chose to have it a year-long program . . . but is it right to have the children engaged for an entire year . . . ? That's why we've, I branch off and do the dissections, lighthouses, do other kinds of things, to get kids involved in other kinds of sciences . . . Maybe that's why I do it, because I'm not real comfortable with doing Mimi all year long." One consequence of this attitude is that EH looks for activities outside Mimi's *Overview Guide* that involve students in hands-on activities that grow out of their interests. He calls this way of teaching "teacher-driven."

Site 4—Union Middle School, Urban Massachusetts

Union is a middle school for sixth, seventh, and eighth grades; it is located in an urban neighborhood several blocks from a busy intersection. Tall trees and multifamily houses share the same street. One house is boarded up: "A drug haven; closed by the police," is the explanation.

Evidence lingers that the building was first used as a vocational school. For example, in the upstairs kitchen where teachers hurriedly microwave their lunches, adolescent girls once were taught to cook, sew, and iron. The most recent changes at Site 4 followed the coming of a new principal in 1993. His changes included a longer school day (teachers and students arrive an hour earlier for a period of "enhancement"), new rules (no talking in the hallway during class changes), and new office space for himself.

Students at Union number 425 and represent great diversity. For example, the 20 sixth graders in the study listed Cantonese, Spanish, English, French Creole, German, Portuguese, Cape Verdian, Japanese, Vietnamese, and Swahili as languages either they or their family members speak.

FJ, a veteran teacher of over 25 years, is the single remaining Mimi teacher at Union. She reports that when she started teaching Mimi nine years ago, she had little knowledge about the content areas of either voyage. She found Mimi's teachers' guides "exceedingly helpful," especially as she felt less comfortable teaching science than mathematics. She recalls those early years of teaching Mimi as one of close collaboration with peers. FJ and two other sixth grade teachers attended a summer Mimi institute at Bank Street College of Education. They returned to school eager to find ways to extend Mimi across disciplines. She recalls: "We all agreed that interdisciplinary education was the only way to reach middle school kids. What we found in Mimi was a program that was already interdisciplinary."

Until the change of principals and the departure of colleagues, cross-disciplinary teaching evolved. It included telecommunicating with a school in Hawaii to report whale watch experiences and annual field trips to a local aquarium. Each year, teachers developed curricula focusing on a different theme, including the history of whaling and endangered animals.

The year of our study was FJ's second year teaching an advanced group of sixth graders. Although students go to art, computer, and French classes, they spend the majority of each day with FJ. The hope of these students is to pass the citywide examinations and thus become eligible the next year for elite public schools. To help make that possible, FJ devotes a sizable portion of the fall term to review, drill, and practice test-taking. Although compared to previous years, less time is devoted to Mimi—for example, the only field trip was to a local Mimi fest—FJ still showed all the episodes and expeditions for both voyages; she also introduced more of Mimi's computer progams than any other teacher in the study.

FJ's classroom is a storehouse for her collection of materials. Shelves of paperbacks for students to borrow line the back wall of the classroom. The blackboard up front is barely visible behind stacks and boxes of books, workbooks, and papers. A large cabinet on one side of the room is always open, filled too full to be closed. From these seemingly disorganized shelves and piles, FJ seems magically to pluck a video, a computer disk, a book. She says, "I'm to the point where I know where I'm going."

Over a two-day period during one of our visits, she showed two Mimi episodes, led a discussion about food chains (the concept is presented under the "Background for the Teacher" section in the *Overview Guide*), and instructed students to work in pairs to complete a written assignment taken directly from the Mimi booklet "Ecosystems." She also introduced the whole class to a computer simulation game, Island Survivors.

TJ is a teacher who once shone, who felt stimulated, supported, and appreciated by colleagues and administrators. Next fall, having passed the citywide exams, these children will be in another school; their teacher will remain. How long she will continue with Mimi is another matter.

Site 5—Sherwood Elementary School, Suburban Massachusetts

Sherwood is an elementary school for grades K-5. It is located in a small, affluent suburb in southeastern Massachusetts with a stable population of approximately 4,000 people. Like many of the homes in this community, Site 5 sits back from the main road, secluded within a woodland setting. Parental involvement is strong at Sherwood: "One parent works," says the principal, "and the other devotes her time to their child." Parents have no other public school choice—Sherwood is the only school in town.

The interior space of Sherwood shows that there has been a visible educational shift over the school's history, specifically, the open quality of one wing of the building. None of the classrooms here has doors, but instead open out onto a balcony that overlooks a large library. During the year of the study, the four fifth grades had moved into this area of the building, three along one open corridor, the fourth on the corridor opposite.

The most recent emphasis at Site 5, according to the school's new principal, is toward instituting a common language among teachers. To meet this goal, an educational consulting firm was recently hired. Charts displayed in the teachers' lounge show that managing student behavior was a prominent topic of interest at their fall workshop. Following the workshop, parents received a booklet of expectations—among them, that children would have 50 minutes of homework nightly.

Slightly fewer than 450 students attend Site 5. Five children are black and are bused in each day as part of a voluntary integration program. Except for a small number of children who are Asian, all other students at Sherwood are white. Twenty students is the average fifth grade class size.

The teaching faculty at Site 5 is supported by librarians; art, music and computer specialists; and numerous parent volunteers. Three of the four fifth grade teachers (PW, HG, and DF) have taught for 15 or more years; the fourth (NM) was beginning her third year of teaching during the study.

Mimi is the science curriculum for both fifth and sixth grades and is taught over the course of the school year. The first voyage is taught to fifth graders; the second, to sixth graders. (The sixth grade classes had been relocated into a middle school out of the community and therefore were not included in this study.) Teachers' experience in teaching Mimi and the extent to which it is embraced as a total curriculum vary among the four fifth grade teachers. For example, PW began teaching the first voyage in 1988 as a way to replace what she saw as a worn-out science text "that made students groan." On the other hand, because she does not think Mimi offers enough hands-on activities, she supplements it with AIMS program materials and encourages her colleagues to do likewise.[7]

HG has taught the first voyage for seven years, but is a reluctant user; he views the former science curriculum as "pure science" and Mimi as yet another add-on to an already full agenda. DF has taught for 20 years. The year before the study, she was a sixth grade teacher and consequently taught the second voyage. During our study, because of her change to fifth grade, she was teaching the first voyage for the first time as well as trying to keep up with PW next door. The fourth teacher who participated in our study is NM; the year of the study was her

---

[7]Activities to Integrate Mathematics and Science (AIMS) are K-9 materials produced through an NSF grant.

third year of teaching and her third using Mimi. Unlike PW and HG who each attended a week-long institute for Mimi users sponsored by Lesley College in 1988 and 1990, respectively, neither NM nor DF have had any formal training with any of the Mimi components. Consequently, they rely on their own resourcefulness and PW's leadership.

Students' use of Mimi's computer programs is minimal at Site 5. Among the factors contributing to this are (1) one computer is shared among three classes (the computer is on a cart and rolled from classroom to classroom), and (2) two of the four teachers did not know of Mimi's computer software and had not received any training in its use. Whereas PW may demonstrate a computer program to her whole class, NM and DF choose specific students from their classes to watch and learn so as to teach their fellow classmates. HG, following a different strategy, allows students to teach themselves when they have finished their other work: "These kids sit down and intuitively figure out what's supposed to be done . . . It's amazing. I could probably do it myself if I had time enough after school."

The principal at Sherwood looks upon Mimi favorably, primarily for what he sees as its technological features. He seemed uninformed, however, about the minimal day-to-day use of Mimi's computer programs within each of the four fifth grade classrooms. He clarified that Mimi was not "mandated" for the fifth grade. Each subject area is scheduled for review in an orderly and systematic way; a review of the science curriculum is not scheduled for several more years.

Site 6—Riverside Middle School, Suburban Vermont

Site 6, called Riverside in our study because of the town's location along a major river, is a middle school for fifth and sixth graders. The three-story building sits on open land at the outskirts of a town in lower Vermont, with a population of 5,000. The decision to corral fifth and sixth graders into one school is a recent one made by the town's school committee. Relevant to this study, this policy helped create a logistical nightmare for teachers who must juggle schedules and share a limited supply of Mimi materials, equipment, and needed resources. The student population at Site 6 numbers 200; it reflects the community in terms of race (white) and range of socioeconomic backgrounds (from poor to affluent).

At Site 6, Mimi is the science curriculum for both fifth and sixth grades and is taught over the entire school year. Both voyages are taught in an annual rotation. During the year of our study, classes combining fifth and sixth graders watched the first voyage; the sixth graders had seen the second voyage the year before.

The principal at Site 6 is a strong and articulate advocate for Mimi. A graduate of Bank Street College of Education, he secured the first voyage in the mid-1980s for the fifth grade teachers at the elementary school where he was then principal. He describes Mimi's appeal and how, through sustained "conversations," he encourages teachers to expand upon its possibilities:

The beauty of the *Voyage of the Mimi* is that it's not a curriculum in and of itself . . . It's not something already packaged and you just hand out the books, you look at the tapes, and you do it . . . It's a context that a teacher can then develop for really exciting and really engaging curriculum with kids . . . In my experience, even those teachers who have expanded it and extended the opportunities would benefit from continuing to have conversations around Mimi. Fortunately, they've been doing that with each other, but I am not as big a part of that now as I would like to be.

RB, a teacher at Site 6, presents a different view: teachers are *not* talking among themselves, and she wishes they were. Furthermore, the principal plays a diminished support role.

RB: I think his role has changed completely since we've come here. We never have an opportunity to talk. It's a terrible loss. But I don't think his posture towards Mimi has changed, in that he certainly supports what we want to do and we all go on whale watches.

Researcher: So in the past, when you would talk with him, what were those conversations like?

RB: I'm particularly interested in questions, so I might bring a circumstance and then we'd talk about what other kinds of questions I could ask; [or] what we want kids to be thinking about . . . [or] how do we incorporate things like journal writing and all these other pieces that can enrich how we're teaching.

Mimi fits with RB's ideas of what she terms "emergent" and "integrative" curriculum. She characterizes Mimi as open and flexible for her as a teacher and, consequently, for her students: "I don't have to follow the guide and use the whole program . . . there's flexibility for me as a teacher to think about how kids might be involved with it in different kinds of ways."

Some of these "different ways" are reading children's books which appeal to the age group and allow for further study of the marine environment (e.g., *Island of the Blue Dolphin*). She provides materials for both group and individual art and design projects (designing glyphs on clay tiles, building a model of a Mayan temple with blocks, making books for younger students, etc.). Most of RB's Mimi-related science teaching is discussion-based or having students write reports. Some students use books from among her extensive personal collection.

Regarding the use of Mimi's computer learning modules, the principal and teachers at Site 6 are in agreement that little attention is given to them, but they give different reasons for this limited use. The principal attributes the minimal use to teachers' lack of training with computers generally, and with this software specifically; JK says lack of hardware is key. RB describes the programs as "old" and lacking in visual excitement; she also voices a philosophical difference as motivating her unwillingness: "They remind me of those old-fashioned, programmed readers, where you fill in the blanks . . . My other personal thing with a lot of the computer stuff is that I really think that it's important for kids to talk and reflect on what they're doing. You can't do that with a computer . . . I don't use the computer stuff. I let the kids use it."

◆

The Adoption Process

Thus far in this chapter, we have set the context for the case study: presenting our methodology, describing Mimi's components and their development, and the conceptions of science that Mimi portrays from the perspectives of the developers, teachers and students. Then, we briefly described the schools and classrooms visited.

We now chart the innovation process as a set of intersecting journeys, beginning with the process by which Mimi was adopted locally. We then move on to teachers' readiness and willingness to implement Mimi's components. From there, we chart early and later implementation. Within these aspects of implementation, we study the roles, on the one hand, of teachers' concerns, risk-taking, and experimentation; and, on the other hand, the more institutional constraints and limitations they faced. We look at the presence of support and assistance, then—briefly—at the later phases of use and the likelihood of continuity. Through this, our ultimate concern is with what teachers and pupils were doing with Mimi during the process: the type, interest, and fertility of the enactments for scientific understanding and conception mastery.

The Arrival of Mimi at Local Sites

Key Actors, Main Events. At the sites participating in this study, Mimi was purchased as a result of only one or two people advocating its use as a replacement for, or an enhancement to, the science curriculum then in use. Thus, a single person led—or, in the case of administrators, directed—Mimi's purchase and use by others. Table 4 summarizes sites' adoption process in terms of key actors and critical events.

As the table shows, event chains vary in length from one to four links. For example, the sites with the longest and shortest links (Sites 2 and 4, respectively) are both urban and part of large systems. A major difference between these sites is the presence or absence of a mandated science curriculum. For example, teachers at Site 2 experience curriculum choice as a domino effect: sitting at the end of the line, they feel pressured to implement local and state curriculum mandates. In contrast, the experienced teacher at Site 4 feels no outside pressure from the district level—she doesn't recall ever seeing a district curriculum guide for teaching science.

The table also makes clear that except for Site 2 where Mimi was the choice of the district's grants writer, Mimi's adoption was initiated at a grassroots level, either by a principal or by teachers. In each instance where Mimi's adoption was initiated by one or more teachers (Sites 1, 4, and 5), principals' support was

Table 4.   **Key Actors and Events Leading to Mimi's Adoption**

Site	Actors and Events
1	Teacher ⟶ District ⟶ Teachers
	Teacher pilots Mimi; her enthusiasm fuels local adoption.
2	District ⟶ Principal ⟶ Teachers ⟶ Teachers
	District writes grant and persuades principal that integrating science, mathematics, and technology will bring his school a success not known before; principal reluctantly releases best teachers from classroom responsibilities as mentors to those remaining in classroom.
3	Principal ⟶ Teachers
	Principal returns from conference having purchased Mimi; he provides money for curriculum development; first generation teachers are enthusiastic, second generation users receive with mixed feelings.
4	Developers ⟶ Teachers
	Teachers attend training with Mimi developers; excited, they use Mimi as way to integrate subject matter areas.
5	(Principal) Teachers ⟶ Teachers
	Principal returns from conference with reports of Mimi; two teachers raise funds for training and materials; response of next generation of users is mixed.
6	Principal ⟶ Teachers
	Principal introduces to teachers; provides counsel and funding.

important, both in the short and long term. Principals encouraged and supported teachers' efforts and provided the funding needed to get the basics: the Mimi package, a VCR, a television monitor, and one or more computers. Specifically, the principal at Site 3 channeled money into curriculum development, paying teachers for their work during the summer; the principal at Site 2 made scheduling changes that allowed teachers to meet on a weekly basis; and the principal at Site 6 counseled Mimi's new users.

What principals did not provide in the way of training and staff development, teachers provided for themselves. Teacher initiatives took several forms: graduate course work in technology and education (Sites 2 and 3); participation in the Lesley College institute aboard a teaching/research vessel (Sites 1 and 5); attendance at local and national workshops and conferences in science and social studies (Sites 2, 3, 5, and 6); and participation in MASTTE, the training program offered by Mimi's developers through Bank Street College of Education (Site 4). Furthermore, when principals and school systems did not provide computers for classroom use, teachers at Sites 3 and 4 secured computers through grants and by bringing in their own from home.

Stimulants to Adoption. Across the six sites, Mimi is championed for its intrinsic qualities, its format, and its high appeal to students in grades 3-6 (see table 5). In addition, the teachers and administrators we interviewed found that Mimi matched teachers' styles and seemed to be a good fit with the required science curriculum. Finally, they called Mimi flexible and a "springboard" to other subject areas and activities. All of these qualities provided stimulus for Mimi's adoption at the sites.

The following excerpts from interviews with teachers at Site 6 expand upon the information presented in table 5, providing a richer view of Mimi's initial appeal:

*   **Mimi's intrinsic qualities.** "I get the feeling that there was a lot of thought, effort, planning, and rethinking in it, before this chunk of curriculum was presented to me. To me, that feels comforting. I feel like there's some dignity and respect within the presentation of it. It's based on experience and not just someone's wish that people would teach this."

*   **Mimi's high appeal.** "I always learn things, no matter how many times I see the episodes and the expeditions . . . I know that I can't see everything when I look at it once. I mean, I go to museums and look at the same painting a lot of times."

*   **Mimi's close fit.** "It felt like somebody was doing what I had been trying to do . . . getting people really involved, showing people who care about something . . . [I felt] that somebody was catching on to what it is that I felt was important in education."

Uncertainties and Constraints. The local adoption process included uncertainties and perceived or real constraints (see table 6). These can be divided into three main categories, as follows: (1) the teacher feels unready and/or unable to implement the program, (2) there are perceived inadequacies in Mimi, and (3) teachers feel limiting pressures or that they are being asked to do too much.

Compared to studies of innovations consistent with the magnitude and demands of Mimi, the number of uncertainties is considerable. This is noteworthy—a factor to keep track of as we progress through the chronicle.

Mimi's Arrival: Three Portraits

The Teacher as Advocate. Greenfield Elementary School (Site 1) is one of four elementary schools in a school district where Mimi has been integral

Table 5.   Incentives to Adopt Mimi

Intrinsic Qualities of Program	
Scientific, serious, high quality	"Mimi presents science as a common, everyday occurrence."
Culturally rich	"Shows women as scientists." "Field trips are a natural extension." "The second voyage provides opportunities to compare cultures."
Relevance: real-world, credible	"Kids did not like what we had before: we were studying light waves, sound waves, doing cells, talking about oceanography . . . but it never made any connections. There wasn't a story line to what we were doing . . . Mimi really makes the connection to the kids' lives."

Format of Program	
Interdisciplinary, integrative, thematic	"It's a perfect place for me to start and jump off from." "I just didn't know what integration was until I used Mimi."
Television component appealing	"Kids like to watch television."
Computer components appealing	"Students love 'Island Survivors.' I have kids who will stay after school to play it."
Good suggestions for activities	"Having it perfectly laid out [in the guide] made it real easy . . . they even have preprepared questions to ask the kids." "Activities are kind of spelled out for you in the teachers' guide. The kind of materials that they want you to get are very easy to acquire, and the activities are very easy to do."
Good background material, guides	"The guides give you a basic idea, and you just add on. In fact, to encourage that, they have that Mimi newsletter that shows what other schools do."
Opportunities for cooperative learning	"Supports discussion around interpersonal relations."

High Appeal to Students	
Highly motivating for students	"Just enough scientific information to interest the kids, but not boggle them." "It says to kids: 'You, too, can be a scientist.'" "Kids like mystery, the ocean, whales, and a good adventure story."

Table 5.   Incentives to Adopt Mimi (continued)

Good Fit for Teachers, Curriculum, and Structure	
Good fit with curriculum	"Fits with our unit on the ocean." "Can extend content into using computer as a tool for collecting and sorting data."
Science curriculum needed replacing, updating, enhancement	"Science specialist wanted more hands-on activities, less lecturing." "Students used to groan when I told them to open the science textbook."
Good personal fit	"Mimi accommodates a variety of teaching styles." "As soon as we saw the whales, we were hooked." "I can see the episodes repeatedly and still learn something new." "Matches my interests in environmental issues." "If I had to study air pressure, for example, without having Mimi as a vehicle, I don't know if I would be so interested; maybe, maybe not."
Good structural fit	"It can be adapted."
Modifiable, Pliable, Extendable	
Modifiable, pliable, open-ended	"You can make it what you wish." "It's flexible." "This is really open for us." "It accommodates a variety of teaching styles." "That's why we have ownership because we put our own stuff in, but we've also used their guidelines."
Springboard for other activities	"I can make choices based on my own and students' interests."

to the third grade science curriculum since 1985. The site's interest in Mimi stems from a single teacher.

AA, an experienced teacher, returned from a regional science conference in 1985 wanting to convince her superintendent to let her teach Mimi. At the conference, she had viewed parts of Mimi episodes and was eager to show them to her class of third graders. She didn't purchase the materials, taping Mimi shows on PBS instead. According to AA, she soon had children "singing the Mimi song" in the hallways. "It was so successful," she continues, "that everybody wanted to do it . . . My enthusiasm was contagious." The next year, money was included in the science budget to purchase one copy of Mimi materials (the district has never bought the *Student Book*). This copy was used throughout the district, one school after another, as part of an eight-week science unit on oceanography. Also that year, AA presented workshops to her colleagues on how to teach Mimi. In retrospect, AA maintains that 10 years ago, there was nothing compa-

Table 6.   **Uncertainties and Constraints in Adopting Mimi**

Teacher Unready, Unwilling, Unable to Implement in Present Form	
Hard work; Mimi "looked overwhelming"	"At first, I was timid." "I had to learn the content and think about how to teach it." "You need courage to leap in."
Depends on teacher's background, skill, initiative, pedagogy , training	"Few teachers have adequate science background." "Some resistant teachers have a 'drill-and-practice' mentality." "Asked to do something new, some teachers are afraid of failing." "Teachers must be skillful in leading discussions." "Teachers leave; replacements need to be trained." "It takes a 'special breed.'"
Insufficient time, materials, equipment	"It's often impossible to find time when the whole class can watch together."
Insufficient time to experiment with, to execute	"We were asked to come up with a curriculum too fast." "There's not enough time to do all the software programs."
Little to no assistance or networking	"No means to talk to Mimi users outside the district." "I was unaware that he [another teacher] had a computer I was supposed to use, too."
Untrained in technology	"Only *some* people want to use computers and be trained to do so."
Perceived Inadequacies in Program, Its Background Material, Information, and Suggestions	
Inadequate suggestions for hands-on activities	"The skills covered are not listed for each activity."
Inadequate science background provided, inappropriate science activities suggested	"The technology used by the Mimi scientists is out-of-date." "Students leave without having acquired a fundamental background in science."
Inadequate math background provided, inappropriate math activities suggested	"We must water down the math to meet a third grade level." "The math in the Mimi does not fit with what students have been taught."
Needs supplements, more activities	"I think the AIMS program brings in heavier science- and math-specific skills that you want the kids to work on. AIMS actually lists the skills at the top [of each worksheet]."

Table 6.   Uncertainties and Constraints for Adopting Mimi (continued)

Uneven quality of software	"It's on the order of 'fill-in-the-blank.'" "It's very slow and the graphics aren't very graphic." "It's not as good because it's more like drill."
No training from publishers	"There is no training, and that's an essential because teachers are very hesitant to do something other than what's in print."
Pressure to Execute	
"Locked in" to a curriculum	"It's a commitment to a single curriculum."
Pressured to try it out	"Mimi is another item on a long list of what I'm supposed to teach."
Low cognitive functioning of students, resistance	"Students don't know the difference between a thermometer and a ruler." "Some students say: 'Just give me a book and the questions.'"
Computers—not enough or not up to date	"Lack of resources to promote the technology." "None of the Mimi software works on my computer . . . I've got this cheapo thing."

rable to Mimi: "It filled a hole, so people made time for it." She offers the following reasons for Mimi's success with students and teachers: (1) Mimi is an adventure story that involves the children right away; (2) there are lots of spin-off possibilities (e.g., field trips); and (3) Mimi is open-ended—teachers have lots to choose from.

As "second generation users" (i.e., after the original introduction of the program), KC began teaching Mimi four years ago, and HP, one year later. Although both teachers agree with AA's positive assessment of Mimi, they responded to their status as second generation Mimi users in ways slightly different to each other and to teachers in their district. Less experienced with the material and uncomfortable teaching science, HP was grateful for KC's taking the lead in class discussions before and after their combined classes watch episodes and expeditions. Moreover, when on her own, help is close by. As for KC, she credits both her long experience as a sailor and her ability to do well those things she finds most difficult, including teaching science. In contrast to their colleagues at other schools, KC and HP teach Mimi throughout the year. They have been given their own set of Mimi tapes. The VCR, a monitor, and—most recently—a laser disc player remain in HP's classroom at the end of the hall. Interestingly, no one asks to borrow them.

The Principal as Advocate. The *Voyage of the Mimi* also has a long history at Lakeview Elementary School (Site 3). Mimi's arrival at Lakeview differs, however, in two respects from its arrival at Greenfield. First, Site 3's principal was Mimi's first champion; second, the teachers who were told to use it then continue to do so with this modification—they teach the second voyage; the first is now taught in the fifth grade.

In the spring of 1985, the principal at Site 3 returned from a conference where, having seen parts of Mimi, he bought it on the spot. (Note that the events at Site 5 follow a similar pattern.) Back at school, the principal asked his two sixth grade teachers (MG and EH) to revamp their science program based on the Mimi and paid them for three weeks in the summer to do so. Although they were reluctant at first, MG and EH soon changed their minds: "You can't watch episode 4 and not be in love with whales," EH recalls.

Over that first summer, EH and MG developed year-long lesson plans. The next year they team-taught and revised their plans. In addition, they planned a week-long outdoor experience modeled around the Mimi episodes in which the crew is shipwrecked on an uninhabited island. During the second year, and for three years thereafter, the sixth grade classes built a wooden dory. The project involved adult volunteers from the community, helped to raise money for class field trips, and was featured on local television.

Major changes occurred when Mimi's new publishers gave EH and MG a complementary copy of the *Second Voyage of the Mimi*. The next year, 1989, EH and MG tried teaching first one voyage, then the other. This hook-up of voyages did not work out to their satisfaction. So, with the backing of parents, the new principal agreed that the sixth grade teachers would teach the second voyage, the fifth grade teachers, the first. Along with the tapes and *Student Book*, the sixth grade teachers supplied the fifth grade teachers with their thick notebook of worksheets, one for each episode and expedition. The fifth grade teachers, as the second generation, did not welcome the change. Having developed a science curriculum of their own, they saw Mimi as an intruder and as rather a slap in the face—what was wrong with what they were doing? As a compromise, the newest fifth grade teacher (RJ) volunteered to teach the first voyage to each of the three fifth grade classes in 12-week units, 40 minutes each day.

The District Administrator as Advocate. Mimi's infusion into the science curriculum at Central Middle School (Site 2) deserves special attention as an example of adoption from the top down: district office to principal to teachers. Having a "vision" is the way the district grants writer speaks of herself. She wrote the grant that funded Mimi; she convinced the principal at Site 2 to participate and then kept up the pressure when he wanted to back out, wary of the consequences of taking his best teachers out of their classrooms. For these reasons, she sees her role as being an agent of change: "Sometimes you need that outside person to give a little push. Because when you're in a school, the day-to-day life of the school and the emergencies take over."

Also at the district office was AD, the district's science coordinator. AD saw Mimi as a means to increase hands-on activities and opportunities for small group work in middle school science classes. She argued successfully that as long as the state curriculum was covered, the sequence could follow Mimi's story line. The grants writer explains: "Follow the scope of the science curriculum for grade 6, but not the sequence . . . adjust the sequence to match the Mimi. And then . . . adjust the math curriculum to match the science we have selected for the Mimi."

Bypassing Mimi's *Overview Guide*, AD and a small group of teachers worked over the summer writing lesson plans to fit the mandated curriculum for sixth grade—physical science. (Taking a less radical approach, the sixth grade mathematics curriculum was reshuffled, as opposed to rewritten.) AD worked long hours revising these plans in order to submit them to the funding agency by the fall. During the next three years, she was at Site 2 one day each week.

Project Director KL recalls: "Mimi gave us something we could get into right away; there's a lot of material in it." However, there were two main problems in using the first voyage. As noted, Mimi's material—with its focus on whales and whale research—did not fit the mandated curriculum. Making Mimi fit was not easy. "It was a struggle," comments KL.

The second issue of fit pertains to technology. The modernization of Site 2's computer lab was with equipment incompatible with Mimi's software modules. An effort similar to rewriting Mimi's curriculum guide was undertaken in order to create and/or revise its software programs. New to the technology and new to programming, four teachers (RS, KL, MX, and KN) began to teach themselves about programming and to write curriculum. "We didn't realize how much time we'd end up spending . . . to make sure that it's as good as it could be. It's almost like writing the books, and the editor comes back for the writing, makes his changes, back and forth—it could take years until it comes out right," notes KL.

Summary of Factors Related to Mimi's Adoption

Five interrelated factors appear to make Mimi's adoption more palatable. These factors are cost, a school or district's decision that Mimi will be the core of its science curriculum for a particular grade, the absence of assessment in general, the absence of a standardized science test in particular, the state of the science curriculum that Mimi is replacing, and the career trajectory Mimi offers.

Cost. The cost of the basic Mimi package is modest. It was pointed out to us more than once that one Mimi package costs less than textbooks for two to three classes of children; further, while textbooks are not usually shared, tapes, a VCR, and a monitor can be. Moreover, activities suggested in Mimi's *Overview Guide* and *Student Book* require minimal expenditures, and most teachers are reimbursed; the bigger expenses are, of course, field trips.

Core of School's Science Program. A related point is that by making Mimi the core of the science program, teachers teach it throughout the year. Showing an episode one week, followed by an expedition the next week, adds up to 26 weeks—the usual length of a school year.

Assessment. Generally speaking, at each of the sites in this study, teachers' assessments of student understanding is based on students' responses to questions asked by the teacher either in classroom discourse or in the form of written assignments. At only one of the six sites in this study (Site 3) is a standardized science test given to students during or immediately following the Mimi experience. Students are said to do consistently well on the test, but to what extent Mimi does or does not contribute to this success was acknowledged as difficult to gauge.

Science Before Mimi. Mimi replaced science textbooks published in the late 1970s and early 1980s. Scientific advancements outpaced district budgets and the purchase of up-to-date materials. In addition, available textbooks were judged as failing to promote hands-on activities, address issues of diversity, or promote group work. Moreover, textbooks seemed distant from the growing emphasis on technology. Finally, they lacked appeal to students. The most graphic description of how bad things were before Mimi comes from PW at Site 5: "We knew the old science program wasn't working. Kids hated science. Kids didn't want to do science. They hated the book. They would groan as soon as we took it out."

◆

Implementation Factors

Many uses were made of Mimi but, given the possibilities of the program, perhaps fewer than might have been. In this section, we review some of the background factors that might have accounted for the ways teachers worked with the program. We begin by discussing teachers' background in science and mathematics and follow with their beliefs about science and science teaching. We conclude with a discussion of teachers' roles and responsibilities, and changing practices.

Teachers' Background in Science and Mathematics

Ideally, Mimi relies on teachers having a reasonably strong academic background, combined with experience in hands-on mathematics and science. This leads to more varied and cognitively stretching opportunities for student learning and, at the same time, assumes for mastery of the content matter associated with grade-level requirements. Some of these skills, of course, may be acquired in the process of working with the program. On the other hand, the more conventional or "algorithmic" formats and design of the Mimi program as built into the teachers' guides may require less academic training in science and less experience with more inquiry-oriented practices.

Only two teachers—RS at Site 2 and RB at Site 6—said they had strong academic backgrounds in science. Also, only women volunteered that they did not have as strong a background in science as they wished, attributing that lack to having grown up at a time when girls and women were not supported in taking science courses or in pursuing science careers.

Site 2 presents an interesting case. On the one hand, three administrators judged that teachers had succeeded in developing their science program, with these teachers (all male) having previously taught art and social studies. Only one of those teachers is now certified to teach science. He argues that only with stronger science backgrounds will teachers have enough "deep knowledge" to respond adequately to children's questions as they arise in the Mimi or in projects taking similar approaches.

Teachers at Site 1 speak of having felt "uncomfortable" teaching science and having not followed training or course work as much as learning from observing a colleague and reading on their own. Theirs was a kind of "by-my-own-bootstraps" method. The female teachers at Sites 3 and 4 all said that they take more time planning than they usually do when it comes to preparing science lessons. The female teachers at Site 5 did more; namely, they added activities from AIMS.

In general, although Mimi teachers may report poor academic backgrounds in science and mathematics, they don't see themselves as handicapped . Rather, they feel capable of overcoming perceived limitations by (1) teaching themselves science content through books, (2) making careful plans on how to teach the content, (3) devising multiple teaching approaches, (4) relying on Mimi materials, and (5) relying on fellow teachers who are more experienced. The quotes below recount the experiences of three women, each an experienced teacher:

> Going through school as a junior high and high school student, it was always a tough subject for me . . . I didn't have any special training in science and math, but because it was a more difficult area . . . once again, I'm a female, and science and math weren't all that important for females in the 1960s really. I mean, we were coming off Sputnik and so forth still, and we still had a male-dominated math and science industry, so it wasn't a big deal if it wasn't a strength of mine. Now I consider it a real important thing for both boys and girls (KC, Site 1).

> My background in science was not that great. I had high school chemistry and biology, I had college field biology, but I never took a chemistry or a biology course in college. No physics, none of that; they didn't teach girls physics in those days. So I've had to rely on learning it before I teach it. The biology I had had, you know, classification of whales and maps, was easy to understand. I knew how the orders were broken down, that was easy . . . I probably spend more time planning science lessons than I do all the other subjects combined; just so I know what I'm doing (FJ, Site 4).

> As a kid, [when the subject was] science, I said: "I don't do that, I don't know how to do that, I don't understand it, and therefore I can't do it." It's taken me years to get over that. But, still, I sit tight. But now, as an adult, I can say, 'Okay, if I get books and I can use resources and I can understand what I'm

doing, then that's fine'; I'm much more comfortable . . . to have something in front of me and go to and say, 'Okay, I can do this activity and it's not going to be a disaster and it's really going to work and I'm going to be able to work this through with the kids' (HP, Site 1).

Teacher's Beliefs

Although the teachers observed in this study are apparently similar in the ways they enact the Mimi, there are significant differences in their thinking about key aspects of their teaching. For example, RB contrasts her approach with that of her colleague, JK, next door: "My interests are probably different from his, and my thinking about kids' experiences to a certain degree is different from JK's; so what I do is different."

In the discussion that follows, we turn to what RB calls thinking "different-ly"—what we might call beliefs—about learners and learning, subject matter, and roles and responsibilities as teachers. Teachers have different perspectives on these subjects, and their constructs influence their understanding and conduct of Mimi. For some:

- Learning scientific processes and content is (or is not) acquired and learned one step at a time.

- Students have (or do not have) the same capacities for learning this material.

- There is (or is not) one "best" way to teach Mimi.

- Learning this material can (or cannot) be reliably assessed.

- Mimi is (or is not) "teacher driven."

- Science is (or is not) a fixed body of knowledge.

- Teachers need (or do not need) deep understanding of subject matter to do justice to the several components of Mimi.

Learning. The learning process can be construed holistically or incre-mentally toward the eventual understanding of a problem. A teacher at Site 2 explains: "Students build, they continually build on what they did before. So, even if they didn't quite understand it when they did it, somewhere later on, it makes sense to them—even the things they didn't understand."

In other words, the adoption and implementation of Mimi depends on the "instructional significance teachers perceive in [their own and] a child's emo-tions, interests, capability of choice, and social interactions" (Bussis, Chittenden and Amarel 1976, p. 4). For some Mimi teachers, age is the key factor; for others, scores on standardized tests signify capacity. JK maintains an essentially biologi-cal model of child development and adjusts his teaching accordingly: "This is like their last, best chance to be different than they are . . . I want them to learn how to think about more than just themselves. They are capable of that at this age."

Similarly, KN teaches science to five classes of sixth graders in an inner-city middle school where students are tracked according to their test scores. He views himself as an energetic and dedicated teacher. KN maintains that students' capacities for learning have worsened each year:

> This is a much lower functioning sixth grade than I have seen. It's almost like you can lead a horse to water, but can you make it drink? We give them all the tools, but depending on their level of development and abilities, will they make the connection or not? Remember, you're dealing with kids who, in 6.1 have about an 85 to 90 percent reading score; [but] the 6.5 class, most of them are in the 20 and 30 percentile. There is a big, big difference between what they grasp and what they don't.

Believing moreover, that in the final analysis, students—not teachers—are in control of what they learn, KN teaches each student the same way: "I give all the classes the same thing, and what they internalize is what they internalize and get."

HG, teaching 25 fifth graders in an affluent suburban elementary school, believes that each child needs to feel successful. Accordingly, he dispenses high grades for assignments (other than tests): "I want the child to succeed, and when they do their projects, their activities, I make sure each kid can find some way to score high."

Subject Matter. Teachers have their own views about the conceptualization of school mathematics and science. They also have strong views about the instructional process that is the most appropriate to mastering the requisite skills, contents, and investigative processes. As teachers experiment with Mimi, they decide whether its programs represent the kinds of mathematics and science they feel are appropriately taught in this manner, at this level, or within the curriculum for which they are responsible. They then make choices that affect how they use Mimi, how close to the *Overview Guide* they remain, and whether the components of process or content are reconcilable in the curriculum they teach.

For example, HG at Site 5 says he's learned—with difficulty—how to cross subject areas: "There are some teachers here who are able to do the whole thing and mush it all together and accomplish the goals—but not do it step by step. Then there are people like me who feel a lot of safety in following a set map. Mimi's given me a chance to cross the curriculum safely . . . Next year we're going to do it more and more." He describes the science in Mimi as "watered down" by "soft" activities: "Much of the science is not really hard science. We're not talking physics and biology . . . The activities are somewhat soft, but pass for science . . . We made signal flags. We had to do research, paper had to be cut, colors had to be filled in. But that's not pure science."

In contrast, MG at Site 3 finds Mimi compatible to his view of science and science teaching because it portrays the world as it really is: "Mimi is the way the world really is, as opposed to what schools are." He believes the separation of subjects in the curriculum interferes rather than aids students with their science learning.

Integral to the instructional process are methods of assessment. How, for example—to use KN's term—do teachers know what their students internalize? To examine this question, we look next at what teachers say should be assessed and how. Previously, we noted that HG at Site 5 assesses students' projects and activities differently from more traditional tests. KN (Site 2) uses multiple-choice tests to grade what his 150 students grasp; JK (Site 6) believes in his ability to measure student learning by talk and by observing students' physical responses. JK explains: "It's conversation. The looks on their faces. I spend a fair amount of watching kids watching the Mimi and seeing when they laugh, seeing when they're puzzled, or when their eyes are glued or when they're playing around."

Teachers' Roles and Responsibilities. Different, even conflicting, visions of teachers' roles and responsibilities exist within our sample. RS at Site 6, for example, says that her way of working is to have students work independently on research topics they choose themselves: "I'm most interested in kids not knowing everything, but being able to get that information, access that information, by going to the library, finding information in the books, or asking questions they need to ask in order to get what they want to know—and that's not easy to do."

At Site 5, uniformity is emphasized: students must organize their Mimi notebooks and sequence their written reports according to their teacher's directions. Group work, too, is consistently the same for each group. HG's sense of Mimi's prescriptiveness provides a possible explanation: "Basically, it's a descriptive program: this is what you do. You can change it around and modify it, but it pretty well follows procedure already established."

In contrast, EH at Site 3 believes Mimi is "teacher-driven"—by which he means that teachers, not the curriculum package, make the major difference. He describes the role he plays: "More than the others, I want to have kids involved with what we're doing and that means hands-on . . . I have a sense maybe that I'm the one who comes up with ways and means and activities and makes sure that we do them."

Change. Teachers believe that they are in the process of changing their minds and their practices. Teaching itself is a powerful influence: trial and error, others' input, and student responses each contribute to instigating changes in teacher practices. PW notes: "Someone else comes up with a new idea, a new way of trying out something or presenting it, and we try it and see if it works. So it seems to be somewhat of a fluid program, with us, too, constantly changing."

Presenting himself as a more traditional teacher before Mimi's arrival, HG describes how the program influenced him to experiment with a more interdisciplinary way of teaching:

It's made me a lot less rigid. Now I see that, it's quite safe to move across the curriculum, because my whole training was: first you do math, you do spelling, you do science. These things are pretty separate. But now I can see where it is very possible to go right across the curriculum and do math, science, spelling—do the whole thing all mushed up and jumbled together, in one exercise.

Some teachers do not change how they teach Mimi, according to HP at Site 1. At other schools in her district, teachers have consistently taught Mimi in the same way: "I know three teachers in one school who have been doing Mimi since day one, and they've been doing it the same way as day one." Unlike these teachers, HP teaches Mimi differently now than when she started. She believes several factors contribute to the change process for her: team teaching with a more experienced Mimi user, enthusiasm for the topic (she collects books on whales and marine life both for herself and her students), and easy access to hardware (she has a laser disc player in her classroom). Moreover, she feels she's less "directive": "Rather than saying, 'Okay, this is what you've learned today,' this year I've felt more like a guide."

Other Factors

A host of factors account for the relative ease and mastery of teachers who take on new practices. As we have seen, many are background factors; they may even predict to some extent how use early on or later will turn out. Others have more to do with the participants themselves, as they take on—unevenly, gradually, sometimes anxiously—the components of the program and react to the discouraging, surprising, or exciting incidents as the class runs through a new set of discussions, experiments, and assessments. Still other factors are more pertinent to institutional parameters; a final set has to do with the infrastructure for assistance and support (this last is discussed in a later section).

The core issue still remains, however, that of the journey of the pupil. Thus, as teachers feel lower levels of concern, derive help, and show greater skill and versatility in the use of the Mimi program, do their pupils confront more challenging, cognitively stretching tasks; do better at them; seek out more of them; devise critical experiments; and bring in stronger evidence for their claims? Furthermore, as Mimi promotes, do students develop a better understanding of other cultures and human diversity? Do students increase their awareness of equity issues?

In drawing our sample, we stipulated that participants must have at least three years of experience with Mimi so that we could witness more stabilized experience with the program. This is not exactly what we found, however: years of experience is *not* tied neatly to optimal use of Mimi's components.

The information in hand, primarily from the teachers, seems unequivocal: there was little initial training or preparation of any kind to teach Mimi. This "orphaning" has important consequences. First, it encouraged teachers to follow the guides and manuals closely, even if the program overview and introductions called for pedagogical "adventures." This conservative approach kept the comfort level high, but it reduced drastically the change level—the liberties teachers might have taken with pre-set formats. In so doing, it might have lowered the potential impact on teachers and on the range and depth of experiences made available to pupils. We then have a greater threat of simplistic or unexamined misconceptions on the teachers' part during experiments and explanations—along with a greater threat that pupils, too, will hold passively on to prior understandings of the processes and content under study. An example follows.

HG finished Mimi weeks before his colleagues. He explained to us how he directed students in making a water still, an activity suggested in the *Student Book*. Recalling the connection to the Mimi story, he reports:

> HG: One of the activities that the survivors on the island had to do was to figure out how to make a water still so that they could evaporate water and have water, so we made one. Each individual in the class made a water still with nothing more than a plastic baggy, filled with water, taped to a window. We could see how the dew would evaporate to the top of the plastic and then a bead would form, and small, sort of like little cloud driblets would form.
>
> Researcher: Did you use saltwater or freshwater in the plastic bags?
>
> HG: Freshwater. You could use saltwater.
>
> Researcher: What did the Mimi crew use? [The problem the Mimi crew faces is how to produce a supply of freshwater for drinking when all they have is saltwater.]
>
> HG: That's a good idea for next year, because then it would leave some of the salt behind. Then the kids would see. I didn't think about that. You could show the kids how the water rises, but the heavy salt particles are left behind. You can see the salt on the bottom. I didn't think of that.

Often, the best prepared were those already familiar with inquiry-centered science or mathematics, and those comfortable with the uncertain twists and turns this approach typically takes as pupils formulate, then test out the problems suggested by teachers' guides or added by the teacher. The absence of modeling, systematic observation, self-observation, and group reflection based on attempts to execute a similar activity left teachers with few concrete models to follow in imagining how the program might be executed in various ways. One consequence is an increased reliance on the only available model—the teachers' guides. Studies of innovation mastery (e.g., Fullan 1991) clearly show that, even in the best cases, initial steps are highly mimetic—i.e., users remain faithful, step by step, to the guides that accompany the program, then gradually develop their own repertoire. In these cases, there were fewer instances, even in later phases, of latitude taken from instructions contained in the guides for designing and executing experiments, expeditions, and simulations.

Theoretically, though, we should have found the classic S-curve of innovation mastery, first devised by Hall and Loucks (1977) and refined by Huberman and Miles (1984). See figure 4 in this volume's case study of Kids Network. The figure is a simplification, of course; it underrepresents the nonlinear quality of practice mastery: the plateaus, regressions, sudden spurts, long moments of latency, etc. Generally, however, each of the steps indicated on the figure denotes a different phase in the mastery process. Hall and Loucks (1977) found that it took a good three to four years before teachers began to get beyond the routine phase.

It typically took one of our Mimi teachers two to three years to reach a phase of initial coordination, but a few are already at the succeeding phases, and are beginning to combine Mimi with other—sometimes new—components of their instructional repertoire. One or two are at the refinement-extension phase. A few are adopting a more systematically inquisitive orientation, one which puts the pupils in the role of scientists, often in open-ended situations, to resolve problems whose parameters have not been spelled out in advance and for which more than one solution may be possible. Virtually all teachers remain more in an algorithmic mode—as in the earlier phases indicated in the figure—in which pupils follow a series of largely predetermined steps to the correct answers. Missing are hands-on science and mathematics—actual hypothesis-testing, derivation, inferential marshaling of evidence—these are alternatives to the approach typified, in general, by teachers in the sample.

◆

Teachers' Progressions Through the Program

The Journeys of Three Teachers

In this section, we follow the journeys of three experienced Mimi teachers. The most experienced user, JK from Site 6, welcomed Mimi's arrival as validating his thinking about science teaching. In contrast, HP from Site 1 was less sure of her capacity to teach science, aware that her strengths were in social studies and language arts. The third teacher, PW, is an experienced teacher from Site 5. She championed Mimi from the start as a replacement for an obsolete and "boring" science curriculum built around textbooks.

JK's pattern is one of initial excitement—but with no formal training—followed by limited time to prepare his materials. His teaching time is shortened by children being pulled out of class and by having to work with limited resources. His repertoire includes few hands-on activities and little use of Mimi software modules. However, as JK states, his greater familiarity with the Mimi story convinces him of its power to motivate students. For example, he sees himself as allowing students to follow personal interests both in class discussions and in their choice of research topics. An example proves the point:

Researcher: What are you working on?

Student: I'm putting my notes into categories . . . like what manatees look like, what they eat—that kind of stuff.

Researcher: Why did you pick manatees for your report?

Student: I've always had an interest in them; they're gentle, not fierce, and they don't scare me.

HP is experienced, but relatively new to Mimi in a school district which adopted Mimi for its third grade five years before her arrival. Hesitant at first, she speaks of a slow—but gradual—growth in science content knowledge through her own efforts: "If I get books and I can use resources and I can understand what I'm doing, then that's fine, I'm much more comfortable." Co-teaching Mimi from the beginning with a female colleague, HP reports that she "sat back" during the first year. Feeling more comfortable with the material in her third year, she then began active co-teaching. Currently, she sees herself as less directive. She explains: "I guess this last year, I've felt more like a guide, kind of working them through it, rather than be a directive teacher, more of trying to get them to realize the information they have, more on their own than my standing up and saying, 'Okay, this is what you've learned today.'"

Following our prompt to think of her teaching with Mimi as chapters in a book, the third teacher, PW, provides us with the following progression:

- Chapter 1—Awareness (of the program)

- Chapter 2—Writing a Grant to Purchase the *Voyage of the Mimi*

- Chapter 3—Training (Lesley College course, "Voyage at Sea")

- Chapter 4—Jumping Into the Program

- Chapter 5—Designing an "Adventurous" Experience (taking an overnight trip to Cape Cod and a whale watch)

- Chapter 6—Making an Interim Evaluation (and finding that the program needs more "meat"; in this case, students are instructed to write journals following each episode)

- Chapter 8—Extending Mimi Still More Through Curriculum Materials (e.g., through AIMS and by looking for speakers from the community)

- Chapter 9: Moving Into an Assistance Mode (e.g., conducting an AIMS workshop for colleagues)

- Chapter 10—Extending Mimi *Still More* by Including "Project Jason" (more recent uses of technology in oceanography)

- Chapter 11—Choosing an Overarching Theme (e.g., the water cycle and showing how this theme runs through science, social studies, and language arts curricula)

- Chapter 12—Focusing on Lighthouses (as an important feature of local history)

- Chapter 13—Systematic Comparisons With Other Teachers

- Chapter 14—Considering a New Focus for Next Year

PW presents herself as a teacher who continually looks for improvement by changing the content of Mimi's basic program. She is the protagonist: she has championed Mimi all along as the core of the school's science curriculum, while still introducing changes and additions; instructional issues are not a core problem for her.

Phases of Progression

Hall and Rutherford (1976) describe how adopters' initial concerns in implementing innovations tend to be personal ones, having to do with their own personal or professional adequacy—how the innovation may affect their status, their sense of professional competence, their rewards. These concerns change gradually to concerns about the correct execution of the innovation and the ensuing impact on students, and then to concerns about whether other innovations or a new configuration can achieve the same goals more effectively. This process varies somewhat by teacher; for example, the degree and quality of assistance and support received can change the profile, as can a benign infrastructure (favorable class assignments, good materials, supple regulations).

We asked all the teachers to characterize, if possible, the successive phases they went through while using Mimi. Below is a summary of their responses and a comparison of similarities and differences across cases.

Our teachers appear to follow Hall's general pattern. They seem to move progressively from personal concerns to mastery of the material, then to matters having to do more with the program's technical aspects (e.g., computer modules), then on to means for extending or deepening the program. Following is a composite of the experiences described by two fifth grade teachers at Site 7:

> **Beginning:** Told that the *Voyage of the Mimi* was a new part of the fifth grade curriculum—handed the teachers' guide. Showed the episodes and expeditions and used the follow-up activities. Three weeks spent on entire unit. Few computers available. After the first year, the district encouraged the fifth grade teachers to get together over the summer to work on extensions and supplements.

> **Middle:** I used ideas from our "summer workshop"; I implemented more technology and began to integrate Mimi in all areas of the curriculum. Took my family on a whale watch out of Boston.

> **Nine Years Later:** Mimi is the basis for our science curriculum and is integrated into social studies, mathematics, fine arts, and literature. I team teach with the other fifth grade teacher. Fifth graders spend nine weeks or more on Mimi. There is more group work than before, and I use some of the AIMS materials. We hold a Mimi fest, and the custodian, a former merchant seaman, gets involved. We each do presentations across the state.

In almost all cases—regardless of the individual teacher's previous background in science, years of teaching, grade level, availability of or experience with technology—the teachers in our sample characterize their first phase of use as feeling both excited and "overwhelmed"; the latter feeling is attributed to having to learn unfamiliar content and how to teach it. RJ at Site 3 comments:

> Well, in fact it does take preparation, it does. It's not stuff that you automatically know; it's not like, I mean, I already know how to write, how to do long division and fractions . . . I don't have to learn these skills in addition to figuring out how I'm going to teach them. But when I do Mimi stuff, I have to learn the information as well as teach it, because I don't naturally know about the Maya civilization . . . I didn't know that much about whales.

Much analogous research (Fullan 1991, and Huberman and Miles 1984) has documented a similar pattern for the early phases of complex implementations. At the classroom level, these authors observe, there are complaints about day-to-day coping, unsuccessful attempts to "make it work like it's supposed to," successive cycles of trial and error, exhaustion sometimes in getting through daily or weekly segments, and the sacrifice of other core activities (conventional mathematics or science lessons). In their own way, these authors conclude, difficulties at the outset are good harbingers. They signify that teachers are genuinely trying to come to terms with the program and to face the discrepancies between their own, congenial practice, and the change-bearing features of the new program.

The difficulties seem related to the overflow of seemingly simultaneous tasks ("so much coming at me"); to unpredictability ("sometimes an experiment works; sometimes it doesn't, and I did the same things"); and to a lack of understanding as to how the program is constructed and interrelated. Practice, support, and what Fullan (1991, p. 106) calls "working out one's own meaning" are key facilitators.

Regarding Mimi's multimedia package, clearly, the software modules are the most difficult aspect. Except for teachers at Site 2, who developed their own software by necessity, only two teachers in our study—one at Site 3 and one at Site 4—became knowledgeable and regular users of the software. Innovation "bundles" (Rogers 1962) are innovations that cannot be taken apart; they are harder to adopt and implement. Mimi, however, *can* be unbundled. Teachers can simply choose, for example, to leave out the computer modules but include the hands-on activities and extend through field trips.

For most, following a serialized television program was novel, but certainly not daunting. For others, doing any type of hands-on, inquiry-oriented activity was new. A few veteran users modified the program in ways that created other experiences for themselves and their students. Typically, the teachers worked hard to learn the content; many added resource books to the classroom library and sought out complementary field trip experiences and local resources, including speakers and museum kits. As they became more comfortable and flexible with the structure, format, and technical aspects of the program, these teachers modified it to overcome their concerns or constraints. For example, before showing each expedition, PW at Site 5 selects five to eight vocabulary words from those listed in Mimi's teachers' guides and writes them on the board. Students write a summary of the expedition using these words. In this way, she ensures that the science in Mimi is not lost in the story: "The kids are not aware of the science in the videos unless you talk with them about it ahead of time . . . So we talk about the fact that scientists do research, and this is the vocabulary they use." In constrast, lack of equipment—TV monitors and computers—was more difficult to modify on an individual basis. Teachers coped by begging, hoarding, and bringing needed equipment from home.

Changes to the Innovation

Teachers' phases of progression with the innovation also reflect ways in which they changed the innovation itself. The major change was simply bypassing the computer modules. Other modifications to Mimi materials include add-ons, enhancements, and substitutions. It is important to remember here that at all sites in our sample, Mimi is used as the core of the science curriculum. It does not necessarily follow, however, that teachers are teaching science more often. A brief review of Mimi in relation to science follows.

At Site 2, state science curriculum requires that additional topics be covered during the year. Teachers finish Mimi by the spring of each school year to allow time to cover other science units not seen as connected with Mimi. At Sites 3 and 5, teachers regularly supplement Mimi with hands-on activities they find in other curriculum packages. Some teachers do this more than others, according to EH at Site 3: "I feel more the need than others to have kids involved . . . and that means hands-on . . . and more than what we bought with Mimi ."

At Site 5, teachers use activities from AIMS as their main source for more hands-on work, children working in small groups on the same assignment. Additionally, PW shows the series "Project Jason" in order to present more up-to-date technology in the field of oceanographic exploration.

Teachers at Site 3 developed their own worksheets for each video (these are used as homework assignments). One teacher at Site 5 follows a similar routine, but without worksheets; and the science teacher at Site 2 assigns homework from questions listed in a science textbook.

At all sites except two, field trips are used to enhance Mimi themes. These are usually all-day trips to museums, lighthouses, and aquariums. Three of the sites take classes on whale watches and one attends a regional Mimi fest. Moreover, Site 3 has developed large projects to enhance its Mimi program, including construction of wooden dories and a week-long camping program.

Compared to the other sites, teachers at Site 2 have made the most direct and substantive changes to Mimi materials. Instead of focusing on whales and whale research, they look for the physical science in the series: energy and sound, for example. Instead of using Mimi's software modules, they have created their own. In addition, with an extensive media library at their disposal, teachers at Site 2 frequently substitute films and videos for those Mimi expeditions judged as "talky."

Finally, one major modification of the Mimi package at Sites 1, 2, 4, and 6 stems from their decision not to purchase the *Student Book* for every student. Without copies of the *Student Book*, students at these sites do not have opportunities to review and study material already seen or to do the activities suggested after each episode and expedition.

Experienced teachers consulted at the replication sites and those in the sample reported decreasing dependence on the questions and activities suggested in Mimi's teachers' guides. Contrary to this report, we observed experienced teachers in our sample referring to the guide before and after showing an episode or expedition. Over time, teachers become more familiar with the material, yet many continue to follow the scripted format provided. Overall, what is striking is the high degree of faithfulness with which most teachers at most of the sites show videos in sequence, use the vocabulary lists given for each episode and expedition, ask the suggested questions, and follow the activities suggested in Mimi's teachers' guides—even after three to four years of program use.

Assistance

Some Mimi teachers reacted differentially to a low level of initial support. One said that she would never choose to be "trained" in a formulaic approach, but would attend a plethora of workshops. Others profited from the ability to use their own discretion in choosing Mimi units on which to focus. Still others sought out experience-based training events ("On Board the Mimi"), or college course work and in-service workshops (e.g., technology use and managing cooperative groups); almost all found a "buddy" or a team and multiplied ongoing conversations, experimentations, diagnoses, and modest attempts to experiment with the program. For some Mimi participants, in fact, without that network of mutual assistance, the program would have collapsed in most places in which it was introduced. This is especially true for the technological components: at several sites, one teacher became the local specialist—resource, trainer, and cheerleader for the others in using Mimi's software components.

Following are two matrices related to the enabling effects of assistance (see tables 7 and 8). The first presents internal sources; the second, external sources.

Internal sources of assistance across the six sites, as table 7 shows, come in different forms and amounts, but their enabling effects look very much the same. Initially, for example, financial assistance enables the purchase of Mimi materials; and, in some cases, provides minimal group preparation time to get started. Moreover, there is some attempt by principals for greater flexibility in schedules. Later on, reimbursements for materials and field trips become important for further implementation to take place. While financial assistance is a formal procedure handled at the district and school levels by grants writers, curriculum specialists, and principals, teacher assistance is more likely to be informal and reciprocal. A Site 3 teacher explains: "The first year we worked as a team to generate ideas and activities that would be most appropriate for each of the episodes and expeditions. So it's a real collaborative approach." A second teacher from that site adds, "Now, we often have our ideas at 7:30 in the morning and 4:00 in the afternoon when we're all sitting around talking: 'Did you try this? Let's try this.' And we finally hit a match for the activities that we want to do."

Table 7.   Location, Type, and Enabling Effects of Assistance: Internal Sources

Source	Assistance Type	Enabling Effects
District-level administration	**Financial** 1. Providing seed money (Site 2)	1. Allows purchase of Mimi materials: "The district came up with a (sizable federal) grant . . . They really had the vision." (Site 2)
	2. Updating with additional materials and equipment (Site 1)	2. Allows for greater ease of use
	3. Underwriting staff development (Sites 1,2)	3. Allows collective preparation to launch Mimi: "We had a couple of summers where we looked at the Mimi and took it apart, episode by episode, to fit what we wanted to cover over the year." (Site 2)
	Administrative Flexibility 4. Allowing changes in sequence of mathematics and science content (Site 2)	4. Allows Mimi to dictate sequence of mathmatics and science curriculum: "We sort of got like the Pope's dispensation." (Site 2)
	Stretching Instructional Practice 5. Providing change agent (Site 2)	5. Legitimates change: "Sometimes, you need that outside person to give a little push (beyond) the day-to-day life of the school." (Site 2)
Principal	**Financial** 1. Initiates purchase of innovation (Sites 3, 6)	1. Initiates implementation
	2. Underwriting staff development (Sites 3, 5)	2. Further strengthens launching of Mimi and, later, a second generation of users
	3. Reimbursing money spent on materials (Sites 1, 2, 3, 4, 6)	3. Supports more hands-on activities

Table 7.   Location, Type, and Enabling Effects of Assistance: Internal Sources (continued)

Source	Assistance Type	Enabling Effects
	4. Covering expenses for field trips (Sites 1, 3, 5, 6)	4. Extends Mimi beyond the classroom
	Administrative Flexibility 5. Allowing for flexibility in schedule and tasks	5. Facilitates teachers' active collaboration (joint activities)
	6. "Championship"—boosting the innovation and its users (Sites 1, 2, 5, 6)	6. Legitimates Mimi to outside world and broadcasts success of staff
Colleagues	**Leadership** 1. Teaching others	1. Helps colleagues master aspects of innovation: "In my opinion, she has become a leader of the whole thing here. She's helped to keep other people up to speed." (Principal, Site 5)
	Collaborations 2. Peer coaching	2. Encourages initial mastery; lowers discomfort: "When we first started this, the other teacher and I peer coached each other on every single lesson for two years." (Site 3)
	3. "Touching base"	3. Aid and advice nearby: "I can run in and say, 'Okay, how do I do this? What do I do with this? Where do I go with it?'" (Site 2)
	4. Sharing	4. Ongoing mutual information, help: "Every Thursday, we meet for about 45 minutes to talk about anything we'd like to share, upcoming events, field trips, who's scheduling what." (Site 5)
Local staff development personnel	1. Modeling implementation (Sites 1, 2) "The science coordinator is available as a total resource if we needed him." (Site 2)	1. Allows users to observe various instructional strategies (Sites 1, 2)
	2. Providing materials and equipment (Site 1)	2. Bolsters hands-on activities
	3. Providing mentoring relationship to future trainer (Site 2	3. Initiates into a larger community: "The district science coordinator introduced me to a lot of books . . . we had a lot of conversations . . . and she introduced me to people noted in the field." (Site 2)

Table 8.   **Location, Type, and Enabling Effects of Assistance: External Sources**

Source	Assistance Type	Enabling Effects
Developers	1. Allowing new users to learn from Mimi software producers (Site 2)	1. Provides training and establishes mentoring relationship. (Site 2)
	2. Allowing new users to experience the innovation (Site 4)	2. Evokes enthusiasm and provides concrete instructional activities; helps cement bonds among users. (Sites 4)
Publishers	1. Providing materials for Mimi's second voyage. "In appreciation for what we had done with Mimi 1, they had sent us all that stuff." (Site 3)	1. Allows teachers to pilot new series (Site 3)
	2. Replacing software. "They never charged us for the upgrade." (Site 3) "I just send it away and they replace it; they're really good about that." (Site 5)	2. Allows continued use of Mimi software programs. (Site 3)
	3. Staying in touch. "Every once in awhile, they'll have someone call from California to ask how it's going." (Site 3)	3. Makes user feel supported.
External trainers	1. Providing introductory immersion. (Sites 1, 5) "People were looking for new ways of doing things . . . it was a wonderful group of educators who were anxious to do things better." (Site 1)	1. Evokes enthusiasm, provides concrete instructional activities, models teachers' reciprocal learning. (Sites 1, 5)
	2. Providing relevant science content (Sites 3, 5, 6); training in use of Mimi software. (Site 3) "The state had all kinds of conferences and workshops for science and math." (Site 3)	2. Deepens and/or widens teachers' knowledge base. (Sites 3, 5, 6) "It was a wealth of information that we shared." (Site 3)
	3. Providing theory and practice in classroom management, cooperative learning, etc. (Sites 1, 5)	3. Facilitates trials with group (Sites 1, 5)

Table 8.   **Location, Type, and Enabling Effects of Assistance:**
External Sources (continued)

Source	Assistance Type	Enabling Effects
Other	1. Funding sources.	1. Multiplying efforts: "Mimi came in under a federal grant. It was well over $1 million for teachers, hardware, all the technology. There were after-school components, summer components . . . There was even a parent component." (Site 2)
	2. Networking.	2. Increases repertoire: "We just shared our information with other districts and they with us. That's how I built up all these activities." (Site 5)

Figure 1 shows the sources of assistance at Site 1. In the outer layer (called the "District) are the two districtwide coordinators—one for science and the other for technology. Arrows are directed at the recipients of their assistance: general assistance to the school, and direct assistance (hardware in this case) to one teacher. The thickness of the arrows denotes the relative interaction among the players. In the inner circle are the two third grade Mimi teachers who receive strong support from their principal.

Figure 2 displays the assistance map at Site 2, where federal funding provided seed money for efforts to integrate science, mathematics, and technology. The players involved included the grants writer (she is credited with having the "vision"); the district science coordinator (her goal was to increase hands-on activities, and she visited the school weekly); the principal (who was reluctant initially, but who made time for teachers to meet each week); a sixth grade science teacher, RS (after experimenting with Mimi in his classroom, he became a lead teacher, advising others); the coordinator of the computer lab, KL (formerly the computer lab specialist, he became project director); a new science teacher, KN (hired to work with a mathematics partner, EC). Again, arrows show direction of assistance and their thickness, relative amounts.

Besides mapping the internal sources of assistance at Site 2, figure 3 shows an external source of assistance as well: a member of the team who developed Mimi's computer software programs and who tutored RS and KN. Evaluation of her assistance is mixed: while RS says she opened his eyes to the unique possibilities of software development, KN disagrees. He makes two points: her help was too early—he wasn't ready for it—and too late—she taught them on Macs, but their new lab was equipped with IBMs.

Table 8 shows external sources of assistance, beginning with Mimi's developers. Three teachers at two sites report direct contact with Mimi developers, in two ways: as a participant in the summer institute at Bank Street College (FJ at

Figure 1.   **Sources of Assistance at Site 1**

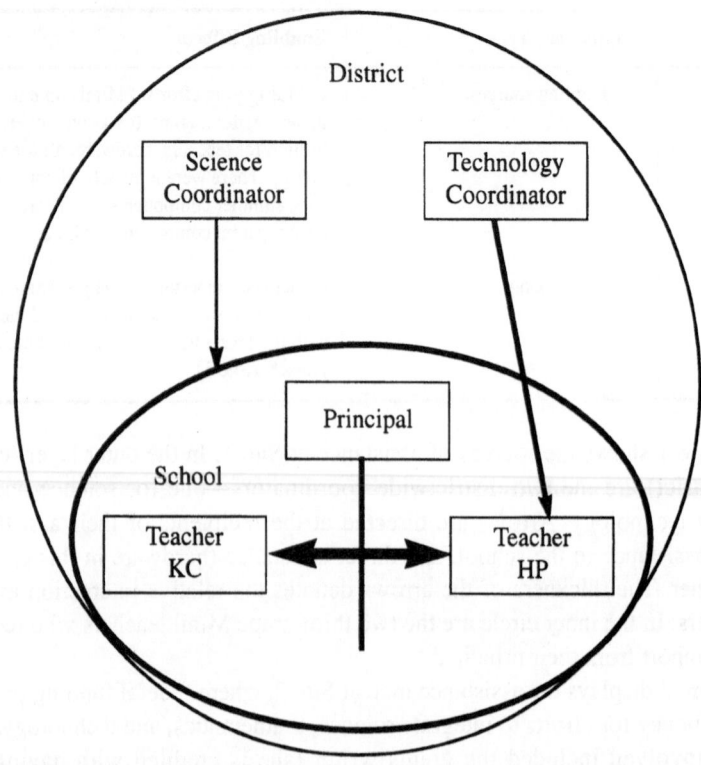

Site 4) and working with Mimi software developers (RS and KN at Site 2). Although short-lived, both FJ and RS report that their contact with members of the development team not only provided them with opportunities to learn new skills, but converted them into Mimi advocates.

Although several sites report contact with Mimi's publishers, this does not tend to be continuous. The experience of MG at Site 3 is thus unusual. His contact with the publishers has included receiving updates of software and even the tapes for the second voyage—at no cost (MG attributes this to the publishers' wanting to continue to highlight innovative uses of Mimi in their sales materials). RJ, at the same site as MG, is new to Mimi; she points out that receiving a newsletter from the publisher (called the *Crow's Nest*) is not as useful as receiving training would be. She says: "As far as actual people-assistance [from the publisher], at this point, it isn't happening." PW at Site 5 reports good luck in receiving replacements for software from the Mimi publishers; however, she is the only teacher among four at her site who understands the procedure for doing this. Teachers at Sites 7 and 8 report that Mimi's current publisher, Wings for Learning, supplied them with all the computer disks they needed for their Mimi fests.

Figure 2.   Sources of Assistance at Site 2

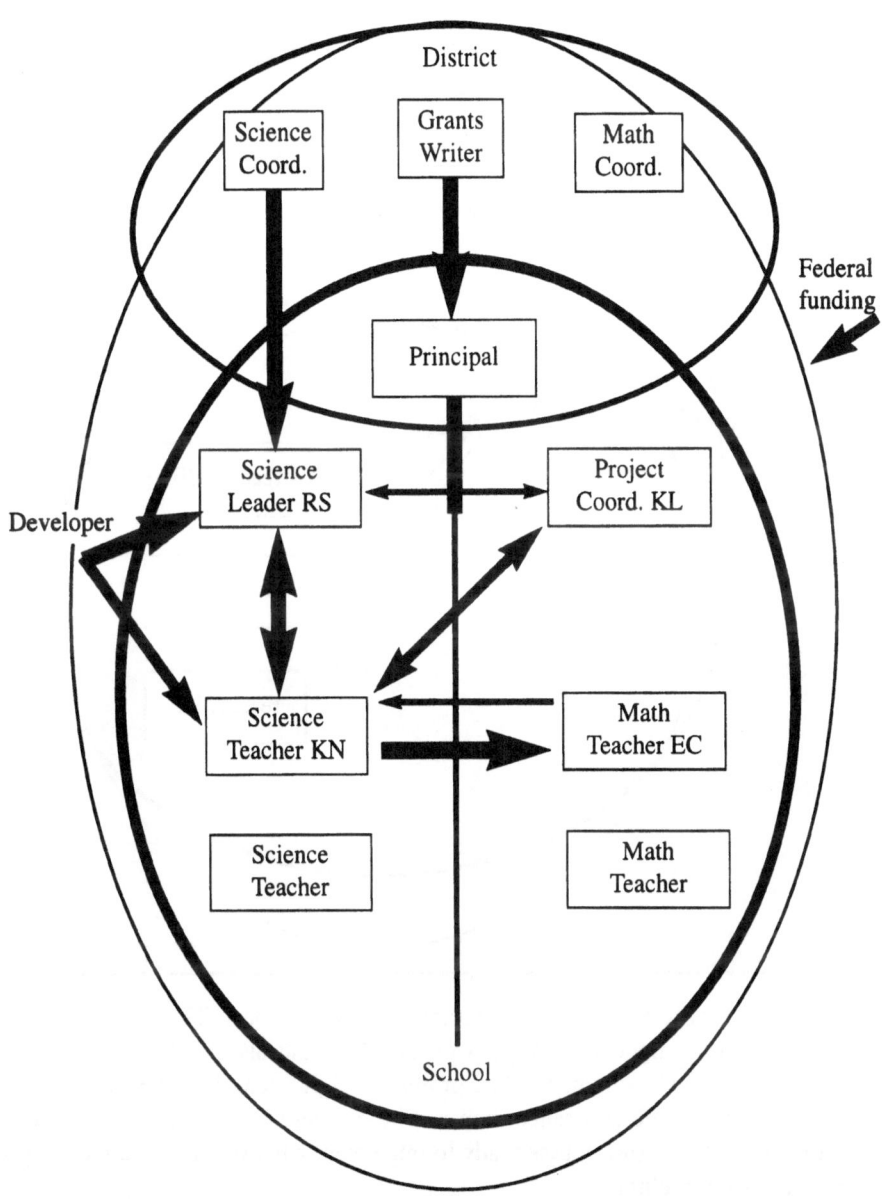

Regarding external training, by far the most enthusiastic reports come from three teachers who participated in week-long summer courses offered by Lesley College in Cambridge, Massachusetts. The teachers are KC at Site 1, and PW and HG from Site 5. PW describes her experiences during the summer course: "There were about 15 other teachers throughout the country . . . It was new to all

Figure 3.   **Sources of Assistance at Site 5**

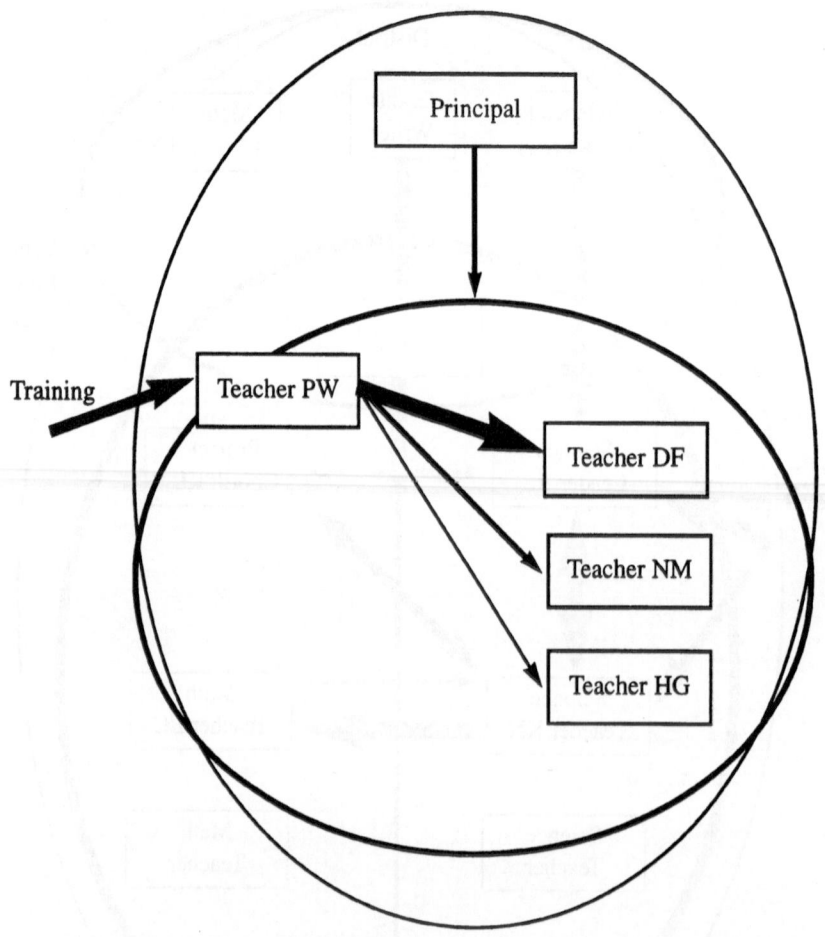

of us. They taught us the entire program while on the boat. The boat was the classroom. Every night we saw parts of the Mimi and discussed it, and in the daytime it was a classroom. We had professors right there that lived with us." Figure 3 displays the assistance map at Site 5 where fifth grade teacher PW returned from the Lesley course ready to implement what she saw as a radically new science curriculum.

These experiences helped jump-start the new science program at Site 5 that first year. PW has continued to provide guidance and materials to teachers new to the program ever since: "It was fantastic . . . I've never had a course like that . . . That's how we started it. It was a huge jump, because Mimi is not a very set program: it's very open-ended, and you can add and subtract and you can focus on what you want to focus on. It was really a very different program for us. So we just jumped in and we did it."

This training is no longer offered; currently, except for a new initiative begun in 1994 by Peter Marston called Mimi Connections, very little is available that is directly related to Mimi. Teachers report looking for opportunities indirectly related to Mimi; these include lectures and workshops in marine science and the Mayan culture.

◆

Enactments

This section presents six activity sequences observed in Mimi classrooms. In general, the sequences represent ways in which teachers were observed to mix and juggle what is available to them in the Mimi package with other options, taking into account classroom management issues and instructional goals. Some teachers are faithful to the guides, others add on where they see gaps, and some integrate Mimi with outside materials and learning experiences. An abbreviated classroom observation is included for each sequence.

Enactment A: Faithful Use

Enactment A presents the "faithful" user—the teacher who follows closely Mimi's *Overview Guide*. Typically, presentation of an episode or expedition is preceded by preview questions and followed by selected follow-up questions. A teacher-directed activity may be next; at Site 5, the activity may be from AIMS materials; at Site 1, from the *Overview Guide*. At Site 3, a worksheet devised by the teachers is assigned as homework following Mimi episodes.

The two third grade classes, 40 pupils, gather on the rug in HP's room at Site 1; the adults (teachers HP and KC, a student teacher, and the paraprofessional who works with children who have special needs) sit or stand around the edge of the group. The children ignore the large TV in one corner and face KC as she asks questions suggested in Mimi's *Overview Guide* for expedition 5 related to animal adaptation. When the TV is turned on, the children hum to the theme music that accompanies shots of the boat under sail. The children stop as soon as the Mimi character Rachel explains that she will be visiting the Marine Biological Laboratory at Woods Hole, Massachusetts, to learn how marine animals are collected and used for research purposes.

After the show—and more humming—KC continues with more questions from Mimi's teachers' guide. She then goes over several vocabulary words, including "chemoreceptor," and suggests that, at lunch, students try holding their noses while eating their sandwiches (her addition). Probably because a horseshoe crab is featured in the film, one boy interjects, "I know why the horseshoe crab got its name," adding that he has the shell at home. KC does not suggest he bring his specimen to school.

Toward the end of the period, KC asks the group for their "thoughts and feelings" regarding the removal of animals from their habitats for the purpose of research and experimentation. Unlike other questions she has asked, this one is

Figure 4.   Faithful Use

1. Teacher poses several questions to class ⟶ 2. Class watches a Mimi episode
3. Teacher poses several follow-up questions. ⟶ 4. Students work alone or in pairs on a predesignated assignment.

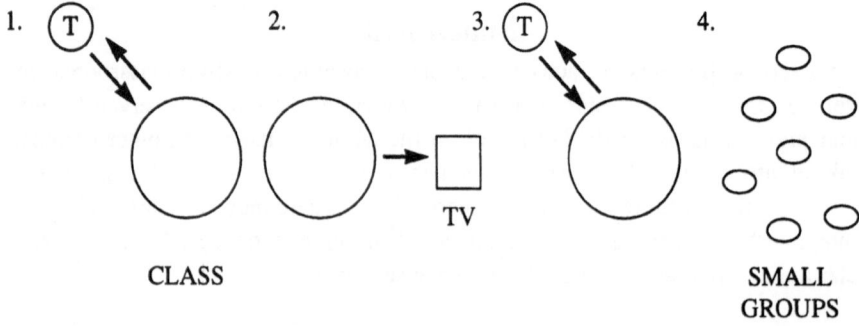

not among those suggested in the teachers' guide. Children's responses are diverse, and many seem eager to voice an opinion:

- "I wouldn't want to experiment on an animal."

- "I wouldn't want to be that animal experimented on."

- "If don't do it, don't find out things that are helpful to us."

- "Even if the animal is returned to its habitat, I don't think it will find its family again."

KC then wraps up: We get information from experimenting on animals, and that helps us, and we shouldn't take more than is necessary. The two classes separate. The students remaining in HP's class soon are using dictionaries to look up the short list of words they have been given and to write out their definitions. They work in pairs with friends.

Later, in conversation, HP speaks of the activity as a "double lesson": with a simple activity, children can learn scientific vocabulary as well as how to use a dictionary.

In this observation, KC asks the questions, staying close to those suggested in Mimi's teachers' guide. In this way, the teacher keeps the pace of the discussion moving and chooses what is "important." Students respond with short answers, so a list of topics and concepts gets covered. Students' idea that marine animals have adapted "on purpose" to different conditions, for example, goes unchallenged either by the teacher or by other students. In addition, by concentrating the pre- and post-commentary on a single expedition, reference to previous expeditions or episodes are discouraged; links between this expedition on animal adaptation and students' recent work (drawing a marine mammal, concentrating on shape and parts) are not evoked.

Enactment B: Wizardry

Enactment B is an example of teacher as demonstrator. In this instance, the teacher combines lecture, demonstration, and technology to direct content and sequence.

At Site 2, the science teacher, KN, intersperses his questions with a whole-class demonstration. In the observation that follows, objects and displays using diverse technologies are masterfully intertwined with talk. One characteristic, however, is an increase in students' inappropriate behaviors, so the teacher often interrupts with threats and reprimands.

As usual, KN is in someone else's classroom (over the course of each week, he uses eight different classrooms). This morning, for 40 minutes, he teaches sixth graders in the other sixth-grade science teacher's room. Hanging on her walls, he points out, are his Mimi materials. KN has brought two rolling carts. A large TV monitor sits on one; an overhead projector sits on the other. The objects KN will use to demonstrate the science topic he plans to cover for several weeks are also there.

As 33 students settle into their desks, the TV is turned on. A student is overheard to say: "That's not the *Voyage of the Mimi*." The student is correct. They watch an episode almost every other week, but today KN will teach about sound using an overhead projector, a tape recorder, a record player, and a laser disc player. KN is well-organized; he has a plan in which he will explain through demonstration, by using one item after another, that sound is vibration. Students will be asked to brainstorm what they know about the topic, to answer KN's questions, to close their eyes and listen, and to watch their teacher's demonstrations.

KN starts by walking around the room to collect last night's homework

---

Figure 5.   Wizardry

1. Teacher demonstrates

1.

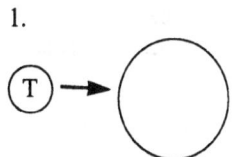

assignment. The noise level is high. Now and throughout the period, he stops repeatedly to reprimand students, individually and as a class. KN gives the new homework assignment and reminds the students about grades:

> Now your homework tonight is just like the other homework, except it's going to be a little more strict. On pages 106 and 107 [in a science textbook], there are three blue questions. If you do not write the questions, you only get half-credit for them. Now please, all those little halves add up to a lot of missing wholes . . . When I have to do your report card, your grade will be lower, your mark will be lower; remember, I am looking for things like class participation, homework, things like that.

> Okay, we've started the topic sound . . . Shh. [KN attempts both to get quiet and to get their attention to the topic.] And the first thing I want to know is what comes to your brain when you hear the word "sound"?

Students raise their hands. Sometimes KN calls on those with raised hands, sometimes on others. As students give answers, he writes each short response on a transparency that is then projected on the board. The list students generate includes noise, motorcycle, cars, and fire truck.

A student from class 6.1 momentarily stops the flow by his entrance and request for a chair. (We are told later that because so many parents push for their children to be in 6.1, the "top" group, that it is larger than other classes.) After "waiting until it is quiet," KN calls on more students whose ideas include a bell ringing, a moving car, and an explosion. KN then says to the students in 6.2: "What you have done, whether you know it or not [students become quiet] is that you have described different examples of where you hear sound, but none of you has described what is sound. And we will be discussing it over next few weeks: What IS sound?"

He next instructs students to be "absolutely quiet" and explains what he means: "No movement, no shuffling of papers, no anything, no whacking pens; what I want you to do [pause], I want you to listen to the sounds you hear, okay, for 30 seconds. As soon as you are quiet, we'll start . . . on your mark, as soon as you stop . . . on your mark, get set . . ." The room is quiet.

Students, mostly boys, call out answers. After a firm exchange with a pupil, KN returns to his questioning. "What else did you hear?" he asks.

> Students: People breathing . . . Hum of lights . . . I heard giggling . . .

> KN: Have many of you, lying in bed, alone, the lights out, total total silence. [He gives a warning to a male student.] What are some of the sounds that you hear?

> Student: Your heart beating.

> KN: Heart beating. If you are ever in the house alone, go in a dark closet or somewhere, you can hear your heart beat.

Students make other suggestions ("people talking," " a hum in your ears . . . "). KN then plays an audiotape of 10 sounds; these are interspersed with some laughter and looks of recognition. KN asks the class to identify the next sound (a toilet flushing). After a fragment from an opera, KN says, "I always thought it was Kelly singing in the shower." Children laugh, having understood the reference to a classmate.

KN says, "Now, the reason [he pauses until there is quiet, then proceeds, but must stop again], the reason we did it—you are breaking my heart [he pauses again]. Shh. The reason we do this is for you to realize there are all different kinds of sounds depending on what you do." Saying this, KN picks up differently sized tuning forks from the cart. The children appear to be watching closely. He then demonstrates what one student correctly calls "vibrating." KN says, "He used this word, 'vibrating,' but very hard to see. I'm going to show you. We have this string, and all I did was glue this ping pong ball on it. I'm going to whack the tuning fork and put it next to it. I'm not going to move it or anything like that."

KN strikes the tuning fork on the table and then holds it next to a ping pong ball attached to a string hanging from a stick. At first, the ball bounces slowly, then faster as it comes into contact with the vibrating tuning fork. The tuning fork, however, does not appear to be moving. The students go from silence to a general "oooooh," and then laugh as the ball bounces. The teacher demonstrates twice more, then adds a twist, claiming that students are not yet understanding why the ping pong ball bounces. He then does a similar demonstration with a cup of water. When KN holds the cup of water just slightly over one student's head; those nearby put their arms over their heads as if to prevent getting wet. There is excitement. From the other side of the room, a boy pleads: "Come over here!" In a voice louder than he has used before, KN demands quiet; otherwise, he warns, he will not proceed with the finale. Soon, he summarizes:

> Thank you. This is a vibration, and the most important thing, you can put this down [meaning, "Write this in your notebook"]. Any time there is a sound, something is vibrating. You cannot have a sound if nothing is moving. So, if you hear something, somewhere, something is moving. And we had some examples of it. For example, let's take a look at the tuning fork. Want it in English or Spanish, Luce? We'll do Spanish first, then English.

The children turn toward the TV to watch a short sequence on how moving air molecules create sound with the narration in Spanish. Some children laugh, and one says, "Is it in Chinese?"

> KN: Now we'll do in English. [Segment is reshown with English narration.] So question is, when I hit the tuning fork, what is hitting your ear?
>
> Students: Sound waves.
>
> Teacher: No, air molecules. Since we live on air, here are some examples.

At this point, KN plays segments from *Windows on Science*, an encyclopedia on CD ROM. One short sequence shows a drum being beaten as pieces of paper attached to it flutter. KN follows by playing a record. He holds paper folded into the shape of a funnel, a straight pin at the end, to the spinning record; and music plays. "What's going on?" he asks the class, then stops to send a boy to the office. He then continues:

> KN: What is making sound? Something on the record . . . What you have on the record is this: you have a groove . . . the groove gets smaller and smaller. Know what is inside that groove making the sounds?
>
> Student: Words.
>
> KN: Ready for the answer? Tiny little bumps. You can't see them without a microscope. The tiny little bumps bump the needle and—what's the big word?—vibrate. If there are no vibrations, then there is no . . . [He waits as some students say, "Noise" then "Sound"]. Sound. If no sound, there is no music on this planet. We'll find out more about that later when we find out how sound travels.

The bell rings. After a short sequence and a reminder about homework, students are out of their seats walking toward the door. KN puts things back on the carts to change classes, too.

KN's use of the 40-minute period is carefully orchestrated. However, as the above observation shows, he spends a lot of instructional time on management issues that take the form of warnings, reprimands, and, finally, dismissal. It's as if KN's mastery of technologies is partially designed to hold the students' attention, rather than to develop the underlying constructs. Having amazed them with his wizardry, it is difficult to ask students to respond thoughtfully—and only with words.

A classroom demonstration like the one above is followed another day by a classroom lab and another by time in the computer lab. (KN organizes each lab experiment in terms of its focus, what equipment two to four children will use, who will work with whom, and ways data will be collected and recorded). Activity in each of these three settings is fast-paced and orchestrated. Connections between these experiences and Mimi are not made explicit, nor are students asked to discuss what links they see. Presumably, having so little time within a period for teaching, KN sees this choice as necessary; he may assume, moreover, that "telling" through demonstrating helps to ensure that students are learning and, by seeing lots of examples, are making connections.

Enactment C: Demonstrations and Computer Activity

This enactment is a variation of teacher as demonstrator. In this instance, however, classroom talk is explicitly aligned with Mimi's story line; moreover, it is an example of a teacher introducing one of Mimi's computer modules.

Figure 6.   **Demonstrations and Computer Activity**

1. Teacher asks for definitions of concept ——▶ 2. Teacher selects 2-3 students to work
along as she demonstrates software module ——▶ 3. Teacher returns to concept and asks for
refinement

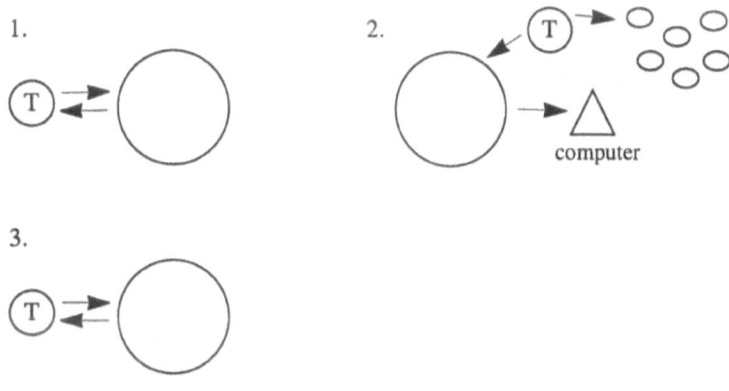

FJ at Site 4 has successfully solved a problem of connecting her computer to a working TV monitor and is ready to proceed. Over a two-day period, she shows two Mimi episodes, leads a discussion about food chains (the concept is presented in the *Overview Guide* under the section "Background for the Teacher"), and instructs students to work in pairs to complete a written assignment taken directly from the Mimi booklet "Ecosystems." She also introduces the whole class to the computer simulation game "Island Survivors", segments of which follow next.

The lights are turned off and students move closer to the TV screen and to friends to watch episode 11, "The Feast." Shipwrecked on a deserted island, the Mimi crew gathers food for a celebratory meal. When algae is gathered, students express a range of responses—from recognition, to disgust, to knowledge: "Seaweed . . . Ugh . . . That's in ice cream." When a rabbit is shown, two boys pretend to be choking something and chopping it. When the rabbit is shown skinned and on a spit, student response is again varied: "I saw blood on it . . . Any meat on that thing? . . . I know why Rachel didn't want to eat the rabbit. She must have had a pet when she was a little kid and killed it."

Mimi's theme song signals that the episode is over. The words "Island Survivors" appear next on the television monitor and are greeted with "Yeahs" from students. But before FJ explains how to play, she pauses to ask students about the concept "food chain." In this way, she links the episode they have just watched to a computer game that is new to most of them; she says:

FJ: While you were watching the episode, I was watching you. I heard someone use the term "food chain." I know that's something you've talked about a little—what can you tell me about the food chain?

Student: It's the process of bigger animals eating the smaller ones.

FJ: All right, someone add to that. [She calls on a girl who has her hand up and then other students offer their additions.]

Student: Smaller animals eaten by bigger and the larger eat the second . . .

Student: Predators eat prey.

Student: It's the cycle of food.

FJ: What other cycles have we studied in nature—in science?

Student: Carbon.

FJ: The carbon dioxide cycle, yeah.

Student: Oxygen.

Student: Water.

FJ: But this isn't a "cycle," this is a "chain." Think of the word chain—what would you add to the definition?

Student: Like, pretend like each piece of the chain is a species of animals and all connected in a way.

At this point, FJ chooses three boys—previous users—to come forward and demonstrate "Island Survivors", a computer simulation that builds on simple relationships in ecosystems, including food chains. Graphs display increases and declines of plant and animal species as the "survivors" meet their needs for food, shelter, and warmth. As they play, the remainder of the class watches on another, larger, monitor set up by FJ for this purpose. She says, "They will play, and we all will watch so everybody knows how." Students shift their seating to have a better view of the screen. FJ starts with key commands and follows with the main ideas.

FJ: Okay, wait; don't go any further because I need to explain this. Wait. Now, the main menu. The directions. Like all Mimi games, use Return to . . . The premise of this game is that three people are stranded on a island, and there are certain things that have to be done—just like the episode that we looked at. What was the first thing they did?

Student: Looking for food.

FJ: No, first thing.

Student: Made shelter.

FJ: So that's one of the choices on the menu. So the next major thing they did?

Student: Made water.

FJ: And the third thing?

Student: Food.

FJ: So shelter came first, then came water, then came—

Student: Food.

The three students at the keyboard begin by selecting four plant and animal species that will live on their island. Immediately, there is a discussion among class members about what turkeys eat; the display they are given is difficult to decipher, and some students suggest it tells them that turkeys eat acorns, while others say flowers.

FJ steps in, telling them they can press "H" for help. "H" is pressed, but information about how to play the game is displayed, not information about what turkeys eat. This sets a precedent and checking information is abandoned thereafter.

The choices to make from the menu are offered by students, and four animals are chosen. At this point, FJ reminds them that "everything must be supported; supported meaning that everything has something to eat." Acorns and turkeys are now connected. Many want to choose the bear, but before they do, FJ brings up the notion of a "pond food web."[8] When a tall plant is displayed, FJ suggests they wait and find more information. She does this by asking one player to press the question mark to learn about the food value of the cattail. The students look doubtful; she says they have probably seen it growing near the highway.

The sense of excitement creates its own momentum, and FJ adds to it by warning: "You've got to survive—like Oregon Trail—if you don't pick the right thing, you won't survive." As plants are collected and fish caught, the vertical bar graph on the right of the screen moves up and down. "He must be eating a lot," offers one student. FJ reads aloud from the screen that one of the players is ill with a cold and cannot work all the next month. This is greeted with disbelief and then a sense of concern:

Student: But he didn't do that much.

Student: How did he get a cold in the summer?

Student: Collect firewood to make him healthy again.

---

[8]That the darkened circle in the larger oval on the screen is a freshwater pond is probably news to most students. Also potentially confusing is the emphasis in the game on freshwater plants and animals since they may be relying on the Mimi episode which showed Mimi crew members collecting seaweed and lobsters in a marine environment. It is not surprising then that students think the creature displayed on the screen is a lobster, when it should be a crayfish.

As the next month proceeds, different features of the game become apparent. For example, there is both a pond and a mountain on this island and the dark line connecting them represents a stream. In addition, the "square" is understood to be a kind of container. Play continues. Although only one boy among three at the computer is manipulating the commands, most of the class appears involved.

FJ: Shh. Now listen, what more can you tell me about "food chain" now?

Student: What's done to one affects the others.

FJ: That can happen—that what's done to one species affects other species. Okay. More?

Student: That if you keep eating one animal or plant, it may go extinct.

FJ: Okay.

Student: Food chain is passage of energy through the organism.

FJ: Okay, another comment.

Student: Everything eats something else.

FJ explains that, following the usual way of working, class members may use the software program when they have completed their other work.

In this observation, one of Mimi's computer modules is demonstrated to the whole class both by teacher and students. In other words, although only three children demonstrate how to play Island Survivors, members of the class quickly offer suggestions and strategies, and the teacher acts as guide and prompter. At one point, for instance, she tries to slow down play in order to gather more information before preceding; however, the momentum of the game leaves her suggestion behind—at least for now. By allowing students to be at the controls, FJ gets procedures across and has a front row seat; there she can see what may typically get players in trouble, what they tend to ignore, and what engages them. She can also build on their experiences as she does when she returns to the concept of "food chain" following their use of the computer simulation.

Enactment D: Modified Use—Extensions

This enactment is an example of modified use and is a variation on "teacher as faithful user." Mimi's episodes continue as the core of the sequence, but longer projects involve students over several weeks and take off in a variety of directions. Two examples are an ecology camp developed by teachers at Site 3 and marine mammal reports as organized by PF at Site 5. The marine mammal reports involve students in varied activities, skills, and subjects; these include writing, speaking, drawing, modeling, mathematics, science, and geography. The week-long ecology camp for the three sixth grade classes at Site 3 takes place in the early fall at a nearby camping facility on a large lake. Says one Site 3 student, "It's like a camp that we go to for a week and learn about the environment and working together and how the food chain and other things work together."

Having already seen the first voyage the year before, these sixth graders begin the year at the camp with a Mimi-like experience themselves—"marooned" in an unfamiliar place. Activities are mixed: walks to learn about medicinal and edible plants, canoeing, painting personal glyphs on t-shirts, and sing-alongs. There are also opportunities to participate in an archeological dig (at the camp's former dump) and to play games that foster collaboration (the activity referred to below as "Wellness"). Students record their experiences in daily journals. One student writes "I liked "Wellness" the best because the instructor made the people who lost feel like winners. He made us work as a team and made us all have confidence . . . Some people were scared at the beginning, but in the end, people were almost crying because they felt so good about themselves."

Figure 7.   Modifed Use — Extensions

1. -2. Mimi episodes and expeditions shown ——► 3. "Projects" follow with students working individually, in pairs, and in small groups ——► 4. Students present their work.

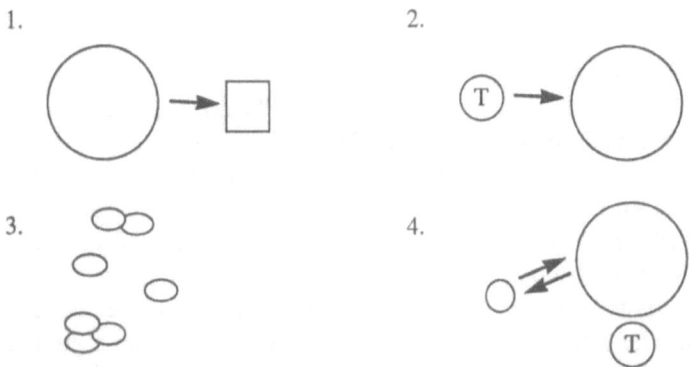

During our visit in late October, we observed key aspects of the project that PW calls "marine mammal reports": namely, students revising their written reports, students making both a diagram and a large paper cut-out of their animal (what PW calls "pillows"), and students' oral presentations. Each student draws the name of an animal out of a bag to research, draw, write, and speak about. The observation that follows begins with PW returning drafts of students' written reports (each was checked by another student, then by PW).

At 9:50 a.m., students put their homework away, and PF reviews how to do oral reports as students sit with notes and marine mammal models made from large sheets of black and gray paper.

PF: What am I looking for?

Student: Eye contact.

PF: Eyes on noses.

Student: Talk slowly.

PF: We need some interesting facts . . . Try to avoid ums . . . don't laugh.

PF goes to a large world map hanging in front of the room to show a boy where his whale lives; another boy joins them. Their reports begin. The protocol is to stand before the class, talk for several minutes, show the paper model made the day before, and take three questions from the class. Despite eight interruptions (four by adults and four by students from other classes), the young speakers gave clear, coherent, and smooth recitals (some children later report that they had followed their teacher's advice and practiced the night before in front of a family member, the mirror, or a toy animal). Class members appear to listen attentively to each report, but their clapping declines as time goes on. The girl reporting on the narwhal is asked more than three questions, but several students are not asked any. The size of the animal and its color are two points much emphasized overall, and about half the class uses the world map to show their whale's primary location. Little is said about whales being endangered. All speakers seem interested in the marine mammal they report on; one girl caresses her paper model as she talks about it.

As the culminating activity of this project, PW asks each student to write a summary of what he or she has learned. Her purpose, as she explains to us later, is to elicit and then assess each individual's experience. Later that day, PF instructs students to write her a letter about their experiences so far. She tells them to work alone and not share with a neighbor. A special sheet of paper, lined with whale flukes at top, is handed out for this purpose.

This activity at Site 5 and the camping experience developed by teachers at Site 3 illustrate ways in which teachers extend Mimi. They do so not to modify the basic material, but rather to build upon it. Mimi's teachers' guides suggest "enrichment activities." For example, an activity suggested for episode 5 in the second voyage is "Garbage Pail Archeology." These examples illustrate what teachers see as an important feature of Mimi: its flexibility as a vehicle and starting point for a wide range of activities.

Enactment E: Integrated Use

This enactment is an example of the ways teachers integrate Mimi material across the curriculum, with language arts being the subject area of greatest intersection. One reason for Mimi's permeability is its varied content. Looking at the first voyage, for example, some teachers take up the social-cultural history of the whaling industry, while others use the personal interactions among Mimi's crew members as a starting point for discussing classroom behavior. This illustrates

Figure 8.   Integrated Use

In class discussions, teacher brings in Mimi or allows students to do so, although the subject is neither science nor a Mimi lesson.

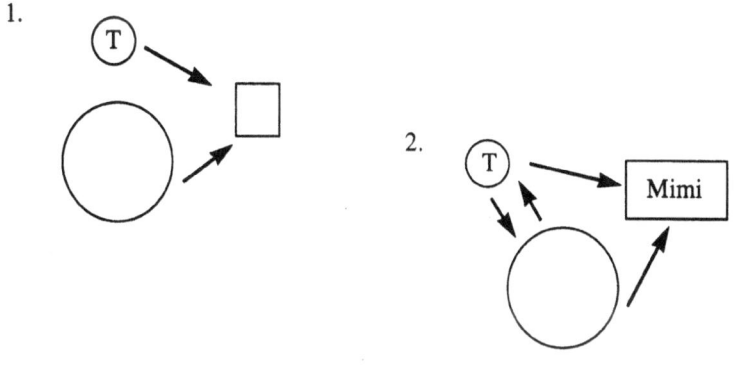

what teachers see as another important feature of Mimi: the integrativity of its story into several subject areas.

At Site 3, EH says: "We chose to have it a year-long program, but is it right to have the children engaged for an entire year . . . ? That's why we've, I branch off and do the dissections, lighthouses, do other kinds of things, to get kids involved in other kinds of sciences . . . Maybe that's why I do it, because I'm not real comfortable with doing Mimi all year long." One consequence is that EH looks for activities outside Mimi's teachers' guides that involve students in activities growing out of their own interests. He calls this way of teaching "teacher-driven." EH's definition holds for class discussions as well. For example, in March, he leads the class in reading and discussing the book *The 21 Balloons*, by Pene du Bois. The story is of a balloonist blown off course to an island that does not appear on any maps. There, he must negotiate and adapt to a place and people whose lives appear similar, yet, at the same time, different from what he has left behind. To negotiate understanding of this group and its ways, he acts in part as a detective, in part as a researcher. As EH teaches it, finding connections to the Second *Voyage of the Mimi* is encouraged. Interestingly then, although EH may not put Mimi at the center of his work, he allows for and encourages students to make connections across subject areas (in this instance, literature, social studies, and science); the students, given this latitude, often use Mimi as their connecting link, as the following observation illustrates.

> EH: Look at the coat of arms . . . The diamond-shaped emblem. [He reads.] . . . and tropical setting . . . symbolic of the islands . . .
>
> Student: It's like a glyph!

EH: Exactly! Tell me how that represents a glyph.

Student: A glyph represents a place or a person like what they like to do, and this represents the island; what they do.

EH: Even though we know the Maya gave this name to a group of symbols, it can be in any culture, can be in any group. What's an example?

Student: Our flag.

EH: Exactly. Our flag is an excellent indication or representation of that.

Student: Like the Mimi. They had an emblem on the jeep.

EH: What was that?

Student: Had a canoe, water . . . like gangs have their own.

EH: Gangs do that. Representative of what a group does. What was that written on the door of the jeep?

Student: Latin.

The discussion continues for another 10 minutes. It includes talk about Maine's new license plate and how to order one, and a student saying that the members of some gangs wear certain kinds of boots. EH follows that different Mayan groups had different glyphs; he then sums up with a statement that it "all depends on which group you belong to."

Enactment F: Complements

This enactment shows the various ways teachers extend, enrich, and complement Mimi's materials using outside resources. Visiting museums and going on a whale watch, for example, are modeled in Mimi's expeditions. Teachers go further: for example, on two- and three-day field trips that include whale watching and visits to museums and lighthouses (Sites 4 and 5) and exploring a nearby seashore (Site 1).

> To go on a whale watch with a bunch of nine-year-olds—it was a leap, it was a real risk that we took, to try to get this field trip off the ground. It's much easier not to. It's much easier not to leave here at 6:30 in the morning, with 40 third graders who are more than likely to get seasick . . . But I say the experience of hearing the blows and seeing the whales far outweighs getting up at whatever time you need to get up.

Students from Sites 3 and 4 also attend Mimi fests. Started by Peter Marston, the actor who plays Captain Granville—and who is, in real life, the owner of the Mimi—these events are often held at maritime museums and are attended by hundreds of schoolchildren. More recently, Marston has visited fests in Nebraska, Michigan, and Texas.

Figure 9.   Complements

Teacher chooses enrichment activities from outside the Mimi materials, including the following: field trips to maritime museums, natural history museums, and the seashore; a camping experience; inviting outside speakers to the classroom; attending Mimi fests.

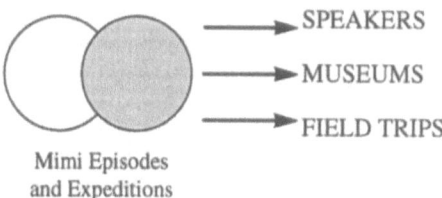

Mimi Episodes
and Expeditions

(Based on drawing by FJ, Site 4)

Enactments as Described by Students

We here give examples of students' descriptions of their experiences. Student accounts and evaluations of their Mimi experiences offer another perspective, differing from their teachers' goals and observations. To a large extent, we found that students' observations were more like our own.

To a great extent, "doing science" is experienced somewhat secondhand by students. They watch Mimi characters observe whales or, in the second voyage, listen to others' theories about Mayan culture. In the expeditions, students see the labs and workplaces of scientists and hear about others' work. Infrequently, they replicate what they see the Mimi characters doing—but in simpler versions.

From the students' viewpoint, however, this can be a rewarding process. Notes one Site 6 student: "Making our report is just like we're scientists, because you have to find out all you can and do the best as you can by it." But some students, like these at Site 3, feel that they are operating at one level removed from experiments they have observed:

Researcher: The research makes you a scientist? Doing research?

Student 1: It doesn't make us a scientist.

Student 2: Scientists have computer disks and everything; they have all the information on it . . . We're getting notes from them . . . they're taking notes from the Maya and then they're saying that on the video . . . they're getting information from the Maya stuff and then we're learning about that.

Nor is there the sense of a "real problem" to solve. Following the directions in the Mimi materials, including the simulations, appears to be so simple that it seems not to encourage questioning or making decisions—students basically watch. For example, two teachers were observed making a solar still, a life-saving activity shown in a Mimi episode and then suggested in a much simplified version in the *Student Book*. One teacher faithfully implemented the activity suggested in the Mimi materials, and the other led a class discussion about how to make a device to produce freshwater. Although the first teacher seems to be doing hands-on science, the students are essentially following a recipe. And, although the second teacher evokes the actual functions of such an experiment, he does so verbally, with no firsthand activity on his or the students' part to illustrate how complex the conduct of actual experiments can be.

However scripted, there is actual hands-on experimentation for students in some classes, unlike what most of them have done in their previous work in science. All the same, one could call them "vicarious" experiments." Students themselves are aware that there are differences between what they do in class and what they see scientists doing in Mimi episodes and expeditions. A Site 6 student remarks, "It's not the same because we can't go to the same places." A student at Site 1 notes that " We don't study as hard as scientists, because if they're not done studying something, they stay over night at the lab and they keep on studying it until they're done studying it."   In the following, a sixth grader at Site 6 contrasts two ways of experimenting: scientists in Mimi do experiments; students in classrooms watch and discuss them:

> Researcher: So, have you been a scientist while doing the *Voyage of the Mimi*?
>
> Student: Sort of, in a way.
>
> Researcher: What do you mean?
>
> Student: Cause like, after watching it awhile, we have to experiment on what's going to happen next and why they did this and why they didn't do this, and think about why they didn't do it, like what would have happened if they did do it. We kind of have to do experiments by watching it.

Several students express interest in learning for themselves. One student describes the Mimi character C.T. as modeling for him what being a scientist really means. Although only a kid like themselves, C.T. is a learner, and he prefers to *do* what scientists do as opposed to being *told* what scientists do: "He knows the background information and he's learning about what he, what the, it seems like he wants to know what the scientist knows, he wants to get in and do what they're doing, not just watch them, not to tell him all about it all the time," explains a student at Site 3.

In a memo, we tried to render a conversation with students from Site 5. The teacher asks the questions, selects the activities, and decides on how children will work. Hence, although students often work in pairs or groups, the whole class works on the same project at the same time. At the end of the interview with the students, we wondered aloud about the possibility of their exploring the wooded area that is part of a nature reserve next to the school. The students, instead, suggested exploring a nearby pond and measuring water temperature at different depths—just as the Mimi crew does, but in the ocean. They had several ideas about how to do this, and speculated that the water would get cooler, the lower the depth. Also, they talked about how to do it—in groups. In addition, the various groups would do different things based on preference. One of the students then said: "I mean, why leave all this stuff to Einstein and like Edison? I mean, why can't we be creative and try to invent something, too?"

Enactments: Three Interpretations

For some teachers, teaching with Mimi as the heart of the science curriculum is essentially a watered down version of teaching science. HG, at Site 5, explains: "Of course Mimi isn't pure science. There are so many other things that are brought in, there's sociology, human relationships; there are so many different things brought in, it's not pure science." He also notes, "Basically the episodes are a very sugar-coated version. Sort of like a seductive way of pulling a child in to learning about science."

Looking at the instructional and learning sequences, however, suggests two other possible interpretations, as follows:

- Mimi teachers have a more "teacher-centered" (demonstrative, explanatory, predetermined) conception of the science curriculum and are generally translating it into this particular program.

- Some teachers are more "learner-centered" or "experiential" in their conception, but are still mainly following Mimi's teachers' guides for a variety of reasons—including comfort with a guided accompaniment; lack of adequate science background; self-consciousness of taking liberties with the guide; students' familiarization with a more "scripted" format; fitting Mimi more readily into the mathematics and science curricula; lack of mastery of and ease with approaches that are, in many ways, more complex, demanding, and uncertain in the ways they play out in class, etc.

There are consequences in the classroom—ones observed in detail in the observations above. Note, that typical activity sequences are, in the main, heavily teacher- and script-centered. Such were not the intentions of Mimi's developers. Their promotional or introductory materials, the orientation of Bank Street College of Education to science learning, and the open invitation to be adventurous in teaching all call for activities that "teach for understanding." Furthermore, Mimi's advocates maintain that the teachers' guide—when contrasted to typical basals—present possibilities rather than scripts.

However, in the guides and in the television series, there are no models for teachers to employ that organize classroom discourse around powerful ideas and the relationships among these ideas; there is also no modeling of how to manage opportunities to "process information and construct meaning through classroom discourse and learning activities . . . including opportunities for students to conduct inquiry, solve problems, make decisions, or engage in other higher order applications of the content" (Brophy and Alleman 1991, p. 12). This, we argue, is best done through a more equal balance of teacher- and student-managed activities, group and individual work with known and unknown procedures and outcomes, replicating more closely the actual conditions of experimentations conducted by scientists, mixing the different Mimi formats, and, more generally, "acting like Einstein."

◆

Use of Technology in Mimi

"'Technology,' you always think of the up-and-coming stuff, not stuff that's been around for awhile . . . That's the best word in education right now . . . it's the new stuff . . . Technology is not cheap . . . it's going to be a world of haves and have-nots—unfortunately"—RJ, Site 3

Mimi is a multimedia package. At its heart is a series of television programs; print materials (*Overview Guide*, *Student Book*, navigational chart or map, and posters); and three or four computer software modules.

In this section, we explore the possible connections among teachers' access, use, and orientation to technology across the six sites. We begin by looking at teachers' access to the technology needed to work with Mimi's multimedia components—specifically, the computer modules. We then examine the core of Mimi—its story (called episodes) and, to a lesser degree, its documentaries (called expeditions)—to examine teachers' views and the use of television as a teaching tool. Throughout, we present students' comments and reported experiences with Mimi's technological components as well as those of teachers and administrators.

It is important to note from the start that, to a large extent, each teacher's definition of technology corresponds to his or her experience. We learned that those teachers who were enthusiastic learners vis-a-vis computers were enthusiastic users as well. Enthusiastic users found a way to obtain additional hardware—either through grants or by bringing in what was needed from home. Overall, however, training—or its lack—proved important in every case. We also found that for second generation users, Mimi's computer components may be an unused part of the package. Furthermore, not knowing that computer modules are part of the Mimi package is a common experience for second generation teachers. HP at Site 1 remarks: "I guess KC had them, but I didn't know she had them. So, for a long time, I never got into the computer programs at all."

Finally, training in the use of Mimi's computer modules is absent or—at best—comes in the form of a brief introductory workshop. At the four sites where the modules were to some degree being used, assistance was typically provided to newcomers and nonusers by a teacher more experienced with the Mimi materials and with computers. For example, one teacher demonstrated a module to another teacher's class. Other strategies involved students teaching one another, with or without any introduction or further monitoring by a teacher.

Mimi's Computer Modules

Knowing the extent of teachers' access to computer hardware and their training with Mimi's computer modules is useful in understanding the degree and ways in which teachers use Mimi or Mimi-related materials. Also important is the distinction between those teachers who brought Mimi into a district or school (the first generation users) and those who inherited Mimi as integral to the science curriculum (second generation users). Table 9 displays these factors, teacher by teacher, across the six sites. We follow this table with a discussion and profiles of four female teachers. We conclude with interpretations before turning to a second type of technology prominent in Mimi—television.

Three main points emerge from table 9. First, there is a limited number of computers in classrooms across all sites. There are, at most, two computers per class; at Site 5, three teachers share one computer on a rotating basis. Moreover, in order to have computers (and TV monitors) available, teachers at Sites 3 and 4 have brought in their own. While four of the six sites have well-stocked computer labs, each with its own specialist, only at Site 2 do students work systematically with Mimi-related materials. (As may be recalled, computer hardware at Site 2 was incompatible with Mimi, and a small group of teachers wrote programs; thus, students at Site 2 are not using the software programs that come with the Mimi package.) In summary, in none of these computer labs are Mimi's learning modules in use.

Second, except for MG at Site 3 and FJ at Site 4, teachers other than those at Site 2 received little to no training in the use of Mimi's learning modules, relying instead on assistance from colleagues—or simply not attempting to use Mimi's software at all. RS, a former classroom teacher at Site 2 who is now a lead teacher, explains below the kinds of technology use likely to be exhibited in a school's teaching faculty:

> You have the same problem with all types of technology: people who will not use an overhead projector, will not write on a piece of acetate sheet—only write on a chalkboard. And just refuse to use it. Don't see any value in it. I'm not saying it's good or bad. I'm saying this happens all the time with technology . . . They don't want to touch it. They don't want to turn a knob or anything. With Mimi, we had a small group who were really into it and very productive. And we had people, another layer underneath, who can use the material that someone else makes, and it can be pretty good at that. And then we had a handful at the bottom who just did not want to know about it, just didn't want to.

RS identifies the "mistake" contributing to teachers' resistance to technology. He argues that the problem of their resistance to computers can be addressed by putting computers first into the hands of teachers, not students:

> I just feel that the mistake was made many, many years ago, and it's still being made today. It is the rush to get children working with computers . . . The first set of computers that came in should have gone on teachers' desks. They should

Table 9.   **Access to Technology and Training and the Use of Mimi's Multimedia Package**

Site	Teacher	Generation	Access	Training With Mimi	Degree and Types of Use
1	KC	2nd	2 computers	Minimal	None: Management concerns; has used in past
	HP	2nd	2 computers; laser disc player and TV in her room	None	None: Unaware of availability
2*	KN	2nd	Computer lab	High	High: Students complete Mimi-related assignments (work-sheets) designed by school staff
	EC	1st	Computer lab	High	High: Students complete Mimi-related assignments (work sheets) designed by school staff
3	MG	1st	3 computers; shares software; shares TV	High	High: Demonstrates; students required to complete worksheets developed by NH and FG; students work individually and in pairs on Mimi learning modules
	EH	1st	1 computer; shares software; shares TV	None	None: His choice
	RJ	2nd	1 computer; shares software; TV is her own	None	Low: Depends on MG to demonstrate programs to her class

Table 9.   Access to Technology and Training and the Use of Mimi's Multimedia Package (continued)

Site	Teacher	Generation	Access	Training With Mimi	Degree and Types of Use
4	FJ	1st	2 computers (her own software); TV & monitor installed in room	High	High: Demonstrates; students work alone or in twos and threes as reward for completing other assignments
5	PW	1st	Shares computer and software; shares TV	Minimal	Some: Demonstrates; students use on own during breaks
	HG	2nd	Shares computer and software; shares TV	Minimal	Students teach themselves; students use in between required assignments
	DF	2nd	Shares computer and software; shares TV	None	None: Unaware of availability
	NM	2nd	Shares computer and software; shares TV	None	None: Unaware of availability
6	RB	1st	2 computers; shares software; shares TV	None	Low: Students allowed to use on their own
	JK	1st	2 computers; shares software; shares TV	None	Low: Students allowed to use on their own
	MP	2nd	2 computers; shares software; shares TV	None	None: Management concerns

* This school does not use Mimi's computer modules, but instead made its own to make up for a lack of compatible hardware.

have gone home with teachers even, so that teachers could get used to them, get to be comfortable with them. And then after they were comfortable, they would have seen that computers can do things for them. And they would have naturally gravitated towards writing their lessons, writing their tests, maybe using a spreadsheet to calculate something, using a database to compile information, even if it's just to do their phone book at first. But they would have seen the connection between the tool and the ultimate use. They don't see that, because we wanted to get the kids to have them. So we rush in. We give it to the children. The kids absorb it much faster than the teachers do, so now the kids are ahead of the teachers. This is threatening. This is uncomfortable in the least. And that I feel is a major, major problem.

Looking more closely at use in terms of assessment, we find that, not surprisingly, the fewer the computers, the more limited is students' use of Mimi's computer modules. However, use also relates to issues of assessment. For example, in general, students are not required to complete Mimi or Mimi-related computer modules as part of their routine assignments. MG at Site 3 is an exception; he assigns computer work and accompanies it with a worksheet. As he explains, time on computers for sixth graders should no longer be looked upon as time to "play" or something to do only if you wish to do so: "In the lower grades, when they go to the computer, it's to go play or it's free time—it's not yet to instruct. Now, when you get them up to sixth grade and you're saying, 'Look, you've got a different software, you've got to fill out a worksheet for it; you are graded on what you're doing on that software.' It's like it's a whole different idea."

We will return to assessment in our section on diversity and equity. Specifically, we there address how the difference between "assignment" and "option" appears to affect strongly the extent to which girls use Mimi's computer modules less often than boys.

Profiles of Four Teachers

Profiles of four sixth grade teachers, all female, follow. These profiles provide a closer look at similarities and differences across sites related to access, use, and training and assistance.

Reluctant User. At Site 2, Central Middle School, sixth grade classes visit the computer lab twice weekly for 40-minute periods—once with their mathematics teacher (EC) and once with their science teacher (KN). EC says she "hates computers" and is thankful to KN for his help. Still, learning to teach with them has been "a lot of work" during the past three years. Says EC:

Well, the way things are going, if you don't know how to use a computer, you're dead. . . . I didn't know how to use computers at all so it was, 'okay I want to learn something, too;' It's a new way of teaching, and let's face it, everything is computers today, it really is. So I figured it's a good opportunity for me, it's a good opportunity for the kids . . . It's a lot of work. It's a tremendous amount of work on the teacher's part.

Students work in pairs in Site 2's computer lab. A good deal of the 40 minutes is used up in organizing students and then giving instructions about the day's activity. It was observed that for some students, much effort is made to "get it right" so as not to fail, be yelled at, or have to start again. Hence, at least for these students, using the computer is rarely a time for asking questions, making choices, or experimenting. Similarly, collaboration is more an issue of keeping your partner on task—and both of you out of trouble.

Ready User. FJ, at Site 4, was already familiar with computers when she was introduced to Mimi's software as part of a summer training institute at Bank Street College of Education in the mid-1980s. Unlike EC, she presents many of Mimi's learning modules in her own classroom. Conversant with each program, she introduces a new one when it seems most appropriate. She usually does so by choosing three of her sixth graders to sit at the keyboard as the remainder of the class sits as close as they can to the television monitor. As the students work, she talks them through the program. She allows students to play before school and during class time if they have completed their other work. While some students choose to do so, others do not. Completing a Mimi-specific software program is not required. Using the computer is popular in FJ's classroom for both boys and girls:

> Student 1: We work in teams, like three players, on some games.
>
> Researcher: When does this happen?
>
> Student 1: When things are less, she lets us use it; some people use it—whoever is finished . . . It happens almost every day. The computer is plugged in already, all she has to do is load the disk.
>
> Student 2: We've been playing games about it, like "Island Survivors" and this other game, it's just like steering the ship, you've got to move the degrees and how far front you want to go to find whales and you're supposed to take pictures in order to save them and match them.

Needy User. Lakeview Elementary (Site 3) has a large computer lab, but Mimi is not incorporated into its curriculum. Among the three sixth grade teachers at Lakeview, RJ is the newest user. She must rely on classroom computers she describes as "ancient." With no training either in computer technology or in Mimi's package, RJ introduces programs sparingly and only when another teacher, MG from next door, is available to demonstrate the program to her class. Two students in RJ's class recall doing the program Whale Rescue, but it is clear that they think they drew little from it:

Student 1: It was fun, but I don't think it was detailed enough.

Student 2: All we had to do was like find where these whales were.

Researcher: Oh yes, that one.

Student 2: That was Whale Search.

Student 1: Yes, Whale Searching, that's the only one I ever did.

Although the school's computer specialist told us he would be happy to incorporate Mimi modules into his lesson plans, both RJ and MG, waging a campaign for placement of more computers in classrooms, choose not to do so.

Professed Nonuser. At Riverside Middle School in suburban Vermont (Site 6), RB is an experienced teacher of over 20 years and an experienced Mimi user. She chooses not to incorporate Mimi's computer components into her teaching but does allow those students who are interested (mainly boys, she says) to play on their own when their other lessons are completed. She gives two main reasons for her choice. The first relates to the quality of the programs; because they were developed in the 1980s, she assumes that they are unexciting to children. The second is her view that the programs are more like "traditional" forms of teaching—workbook-like—and thus not compatible with her pedagogy: "The games are 10 years old, and I don't want to spend time on 10-year-old things . . . I think students won't find it very exciting, because it's not what they're used to seeing in video," she explains. RB adds, "It reminds me of those old-fashioned programmed readers—where you fill in the blanks . . . To make it worthwhile, you would have to have conversations with students about what they are doing and that's not how I want to spend my time. That's a personal choice. Someone else may find it wonderfully compelling and that would be fine."

Interpretations

Typically, teachers at all sites except one (Site 2) separate Mimi into two parts: the episodes and everything else. For these teachers, doing Mimi's computer modules is perceived as an option, dependent in part on availability of hardware and software, comfort with computers in general and specific training with the Mimi materials in particular, and beliefs about computers as instructional tools.

Teachers who choose to use Mimi's computer modules do not use them all—in most instances, they only use one or two. Their usual mode of introducing a module to students is by demonstration; students are then left on their own. At Site 6, for example, modules are not introduced, but are simply "available" to students. In the conversation that follows, a male student at Site 6 shows a keen understanding of "Island Survivors," illustrating how opportunities to play repeatedly (in his case, over a two-year period) allow, in fact, for an understanding of the underlying concepts:

Researcher: Have you done "Island Survivors" by yourself yet?

Student: Yes. A lot of times. I'm going to start a new game. You've got to select different plants and animals.

Researcher: Four plants and animals, four of each? Four types of plants and four animals?

Student: Yes, you can select anything here as long as it works in the food chain.

Researcher: Tell me what you're selecting and why.

Student: I'm going to select the blueberries because you can eat them and so can animals. Select the deer, because the deer are good for me and they can eat the blueberries. I'm going to select the okra, whatever that is, because the deer can eat them, and people can eat them and the bear can eat them. I'm going to select the bear because this will work.

This student's experience is rare. In most cases, students do not have opportunities to return to a particular program. Similarly, teachers do not return to a program after it is introduced. However, as Cuban (1986) points out, to use a technology—in this case, computers—requires a teacher to evaluate it, become familiar with it, and consider how to use it—each of which takes time. Additionally, Cuban raises the issue of teachers' beliefs. He argues that for some teachers computers interfere with the relationship between teacher and student. Although teachers in this study voice strong opinions about the need for students to become familiar with computers, one, RB at Site 6, expresses an uneasiness with computers in her classroom: "I think that it's important for kids to talk and reflect on what they are doing. You can't do that with a computer."

In Mimi's episodes and expeditions, students see scientists using computers as aids to navigation and as tools to make graphs, record data, and make maps. The purposes of Mimi's computer modules, according to the developers of the teachers' guide to the first voyage, are (1) to allow students to explore the natural world further; (2) to experience how scientists use simulation, data analysis, etc., in their work; and (3) to become familiar with ways technology is "reshaping our understanding of the world and the ways we represent it . . . " (Gibbon 1985, p. iv). One student at Site 4 drew our attention to a difference between experiencing the gamelike qualities of Mimi's programs and experiencing what's "real"; she says:

We play games and see if we can survive—Island Survival . . . It'd be funner if it would like to teach us how to really survive, so if we do go on a trip and we need to survive, we might . . . A computer can help us a little, but it can't tell us what's real because they don't tell us every plant we can eat. It can't tell us that: what's going tobe there, right? . . . It's okay, but it's just a fun game . . . it's not reality.

Technology: Television as a Teaching Tool

In this part, we look across sites at teachers' orientation to and use of Mimi's episodes and expeditions as examples of television as a teaching tool.

A Paradox. Watching TV in school is perceived paradoxically: on the one hand, teachers say it is a "passive" activity; on the other, they say Mimi's story motivates, captivates, and, by so doing, allows them to quickly engage students in content matter. KC, at Site 1, explains, "Kids are engaged in noninteractive activity enough hours of their day already . . . But on the other hand, if you use that as springboard for interactive activity, then we're using it to its full potential."

The Power of the Visual Image. Presented as a continuous narrative, the episodes hold students' attention—even when spaced out over a year, with one episode shown about every two weeks. This is the case across the six sites. KC speaks for other teachers in describing the power and the "reality" of the visual image:

> I think that the episodes are a core component for sure. If they weren't there, it wouldn't be Mimi. It wouldn't be the same for me to read the episodes from the *Student Book*. Because no matter how I tried, I could never put across all the facial expressions and all the subtleties that are included in the episodes. There's just no way that I could mimic that. And seeing people survive on an island is different than hearing about them surviving on an island. Seeing C.T. climb the tree is different than hearing that he climbed the tree.

An Instructional Tool New to Teachers but not to Students. Teaching with technology is new to many teachers across sites, including teaching with television. Although teachers have shown a film or video, a serialized story is new. "It was very new for me—in format. I'd used a video as a teaching tool, but not a continuous one, if you will. Not one that had many episodes and so forth. So that was new," notes KC from Site 1.

Watching television, however, is not new for students. Both teachers and students agree that children approach Mimi as experienced TV viewers. EH at Site 3 remarks, "They like videos. It's a very video-oriented group of kids that are coming through here." A student at this site adds, "Kids are supposed to watch TV. And kids remember everything on tapes."

Repeatedly, teachers and students spoke of Mimi as appealing: it is fun, exciting, interesting, and—for all these reasons—engaging: "Kids are hung on the story line. When we watch the episode 7, they want to see episode 8. They really want to see episode 8 and that's good, it captures their imagination, it allows them to think through perhaps what's going to happen next," says EH. Site 3 students concur:

Researcher: Which is better, reading the book (*Student Book*) or watching on TV?

Student 1: Watching TV. Because the books, you have to read. It's work. With television, like movies, you just sit back, relax, watch.

A student at Site 6 points out this difference: "It is watching TV. But when you're at home, you don't really learn anything; when you're watching this, you do."

Access, Scheduling, and Maintenance of Equipment. Teachers must have access to television monitors and VCRs in order to show Mimi's episodes and expeditions. As noted earlier, RJ at Site 3 brought equipment from home to make scheduling easier; three teachers at Site 5 share one television; KN at Site 2 negotiated the needed hardware on a rolling cart through crowded hallways; and teachers at Site 6 dealt with scheduling but had to carry the needed equipment up and down stairs: "I have to share; another teacher wanted it this morning, but he didn't sign up, I had signed up. That's a pain in the neck and it's also hard to move, it's up and down stairs, but it's worth it," comments RB at Site 6.

In contrast, HP at Site 1 uses the latest technology to show Mimi's episodes and expeditions—a laser disc player. Obtained through a grant for purposes of mathematics and reading instruction, the equipment is available to all, but is never used. (HP is at the end of a hall, the equipment would have to be scheduled, it would have to be unplugged, and, perhaps most critical, no one else has been trained in its use.) "It's just delightful to have Mimi on disc rather than on videotape . . . with the laser disc, it's a piece of cake to do, you just click it in and there it is. With video it's a nightmare trying to fast forward until you find the right spot," explains HP.

There are two main points to emphasize. The first relates to ease of access; the second, to the power of the moving image. In this case, it is the former that allows the latter.

Teachers need a television set and VCR (or laser disc player) to show Mimi's episodes and expeditions. This hardware is only slightly more accessible to teachers than computers. At Sites 1, 2, 5, and 6, teachers share equipment. At Site 3, teachers supply their own; and at Site 4, FJ uses a large monitor—supplied by the Whittle Corporation—that is attached to the wall. KN, the science teacher at Site 2, uses an array of other technologies (overhead projector, CD-ROM, laser disc player, audiotape cassette), but he is unusual.

Appeal and Power. Except for one student who told us that she would rather read than watch TV or a video because then she could imagine the characters and the story, students, teachers, and administrators agree that the video medium is familiar, accessible, exciting, and entertaining. Additionally, most teachers show the episodes, uninterrupted by questions, a fact students appreciate: "It teaches us geography and about animals, but it's funner because you're watching a movie. You're not sitting there bored, [the teacher asking] 'What country is this?' and stuff like that," points out a Site 4 student.

Moreover, for students whose academic knowledge is limited by geography and circumstance, visual media bring a measure of equality. Books are words and some photographs; but film shows you, as this student from the Site 4 inner-city middle school explains: "It shows all the words . . . It shows everything. Kids like shows. [They like to see] what they're doing and how. Like if they say something like, 'A navigator's map' or the name of that thing they were using to test the water, kids probably want to see what it looks like."

Finally, as these two examples show, the story seems real and the characters are engaging:

Example 1—Site 5

Student 1: It's not like you're just sitting and watching—you really get into it. You're close to it.

Student 2: Yes, with the whales; I get a kind of a funny feeling about it. It seems kind of special even though I know it's just a film.

Student 1: Yes, it looks like you're right there.

Example 2—Site 3

Student 1: I think C.T.'s like us because if I were on that boat with those people, I think I'd be like him and want to find out as much as I could.

Student 2: Yes.

Student 1: He just said in the last episode we saw. "This is turning out to be quite an adventure." He just went to be his grandfather's mate to take people go out fishing, but now he's into this whole mystery.

◆

Equity/Diversity

A curriculum package such as the *Voyage of the Mimi* necessarily emphasizes some things and downplays others. Curriculum is inherently the end result of a series of choices about inclusion. This informs the questions addressed in this section, to which we have already attended from other perspectives. Several questions stand out above the others: Do teachers find Mimi compatible to teaching in diverse settings and with diverse student populations? How were these factors taken into account in the conception and production of the program? How are issues of politics and ethics handled? And, finally, on a slightly different plane, how much mathematics is contained in a program meant to include these components as well?

Context and Response of Developers to Science and Gender

The introduction to Mimi's teachers' guides provides a concise insight into the developers' thinking about the equity issues related to science and mathematics education and to technology. This project, writes Mimi's executive director, Sam

Gibbon (1985), is a response "to the general concern about the state of science and math education" in the United States as exemplified in the increasingly "male and white" domination of both fields. The *Voyage of the Mimi*, Gibbon continues, will address the problem of exclusion by providing a multimedia package that engages all children "in finding out how the world works." At its "heart" will be a "scientific adventure story" which presents "ideas in a rich human context . . . [and is] particularly appropriate for modeling roles and attitudes" (Gibbon 1985). "It's difficult for kids," Gibbon maintains, "if all they see is an edifice with no way for them to enter . . . it's a chilling experience."

In one of our interviews with him, Gibbon recalls that the integration of technology and science was a primary goal for the funders, whereas issues around gender were secondary. He relates making a decisive shift in this balance: "We elevated gender to equal status." In addition to showing women scientists, Gibbon talks of a conscious effort to present people who work "on the fringe of science: the technicians" and "people who wear jeans; who mess about."

A second source of information related to gender issues is found in Char et al. (1983). Focusing on teachers' use of the software in 13 settings, the authors report that student use varies according to class size and computer availability (Char et al. 1983, p. 49) and that no matter what the constraints, boys use computers more than girls (Char et al. 1983, p. 51).

This report is consistent with our observations. Except at Site 2 where students used Mimi-related programs in a computer lab two to three periods weekly, student use of computers at the other five sites was confined to "free time" (time available when students completed other assignments) and indoor recess. At these times, teachers told us, mostly boys, not girls, are involved—despite deliberate efforts on their part to show that technology is gender-free: "When I'm demonstrating a program, I call up girls to use the computer. So I try to spell out the fact that it's not just men—it's men and women who use it. But it's true, it's mostly boys who want the technology part of Mimi," comments PW at Site 5.

In contrast, access to and use of computers plays out differently in MG's sixth grade class at Site 3. In MG's classroom, all students must complete worksheets on selected Mimi modules. MG posts a schedule, and students sign up. Because completion of these assignments is part of the assessment process, MG is obliged to ensure time for students to use one of the three available computers. With each student provided with time on the computer, access is generalized.

Issues of Diversity as Perceived by Students and Teachers

Our data capture clearly Mimi's success in appealing to students, regardless of their gender, race, and ethnicity; geographic location; grade level; or school achievement. For example, HP at Site 1 speaks for the majority of teachers interviewed when she praises Mimi for offering not only "a wonderful science unit," but an opportunity to cover the "differences" that are part of real life:

I think that's one of the reasons I like it so much. It not only covers a wonderful science unit, but it also covers a lot of different social things in life: how people get along. The fact that there's a deaf person on the boat and how that person is dealt with by the others. A lot of social things: the divorce of Rachel's parents. And so I think Mimi covers a multitude of things for us as classroom teachers, and we try and plug into those differences.

In table 10, we provide examples of differences shown in Mimi, as perceived by teachers and students. Following the table, we describe diversity as it plays out in Mimi's episodes (the fictionalized adventure story) and then in its expeditions (the documentaries that take viewers behind the scenes to visit scientists on the job, in their places of work). We discuss two examples of what we call Mimi's "silences," and then address Mimi's compatibility with diverse settings.

Mimi's Episodes. As the developers promise in the introduction to Mimi's teachers' guides, the episodes model attitudes and behaviors that provide opportunities for teachers to enter into meaningful discussions about real-life issues. Below, two girls—the first, a third grader in a rural school; the second, a sixth grader in an urban school—confirm, as Gibbon puts it, that science is a "human activity . . . and subject to considerations of human values":

Example 1—Site 1

Student: Mimi is interesting. It's fun. It's not rude: it doesn't talk about other people.

Researcher: Tell me more about that.

Student: In some cartoons, I don't really like how they talk about themselves and how they trick each other.

Researcher: So the people in Mimi don't trick each other?

Student: They don't set traps for each other.

Example 2—Site 2

Student: A lot of people don't think it's nice to do that to the whales [tag them] because they think it might hurt them. I don't think it would hurt them, but it might hurt a pregnant whale because they have the baby inside of them and it could hurt the baby. But if it was male, it won't hurt as much as it would to the female.

The story format of Mimi's episodes allows character development, such that students can change their minds about them. RB, who teaches fifth and sixth graders at Site 6, explains:

One of the things that works is focusing on the characters to see how they change or how your perception of them changes . . . For those students who had seen the first voyage and now are seeing the second voyage this year, the first

Table 10.   **Diversity in Mimi as Perceived by Teachers and Students**

Age
They celebrate each other's successes. The adults can learn from the young, and the young can learn from the adults. (FJ, Site 4)

Researcher:	Who is your favorite character?
Student:	C.T.
Researcher:	How come?
Student:	Because he's a kid. He's doing his best to help other people. (Student, Site 1)

Gender, Race, Ethnicity
"They cover everything."
"It's very politically correct . . . They show women in leadership roles; they have different ethnic groups represented, different races represented." (KC, Site 1)
"It didn't surprise me, but it's just that when you think of scientists, you really think of men more than women. But it was good to see that women were aboard and not just men; that she [the character Ann] could take the control and not just men." (Student, Site 2)

Geography and Race
"A black person these kids come across is more likely to be from a city than from a rural area like this one. So Arthur's being from the city and feeling comfortable and C.T. being from the country . . . it's not blatant [the way it is presented], but there all the time." (EH, Site 3)

Student:	I liked Arthur best. Like at the beginning, Captain Granville says to him "What's that?" And he goes, "A suitcase." And it's a big boom box.
Researcher:	I think he called it his briefcase. It came in handy later, right?
Student:	Yes. Because he took it apart and hooked up the radar. (Student, Site 3)

Physical Disability
"I hear kids talking about Sally Ruth [character in the first voyage who is deaf]. It's like Sally Ruth is in their class. She's a real person to them." (principal, Site 6)

thing that some students said was, "Well, the Captain wasn't as mean as he was at the beginning of the first voyage." And then someone else said, "He wasn't mean at the beginning of the first voyage, he was concerned about his boat, and he didn't know that those people were going." That's a wonderful way to have conversations with kids about their behavior and my behavior and other kids' behavior.

Mimi's Expeditions. Mimi's expeditions are also successful in deemphasizing the boundaries between scientific disciplines and showing the variety of people called "scientists." Asked who the scientists are in the second voyage, a student at Site 3 describes the work of the archeologist as systematic and purposeful: "They're studying and worrying about stuff. They're looking at it, cleaning it off, picking up stuff. If it's broken, they try to put it back together to see what it was—just like it was before."

Moreover, by focusing one expedition in each voyage on the character with a physical disability, the expeditions model how discussions can take place frankly and respectfully. RB at Site 6 explains: "It helps kids and adults to be able to talk about difficult things and to have an opportunity to do that . . . It opens kids up, it doesn't shut them down unless you don't deal with it." The response of students at Site 3 is an example of the openness, stimulated in part by their watching specific expeditions, participating in classroom discussions, and hearing a former telephone linesman describe how he was electrocuted and suffered severe burns to his arms and hands. When we talked with students later about the visiting speaker, their responses were very much in the vein of Mimi's objectives:

> Weird. That's real, buddy. We have two arms and he doesn't. He has an arm, but it's artificial with a hook, so he picks up his stuff. And he showed us like all the artificial hands that he's used—like the thing that you pull the cord back and it closes and the one where you press a button and it opens back up. It's just pretty neat. And what he uses to unbutton his shirts and stuff like that . . . It's hard. It looks hard.

The expeditions do not succeed, however, in showing "real" scientists to be diverse in terms of race and ethnicity. The episodes are fictional; the expeditions, in contrast, mirror society as it is. Hence, although the expeditions for both voyages begin with women as their focus, except for one (a medicine woman), the featured scientists are European-Americans. This incongruence between fiction and fact is not made explicit in the instructional materials that accompany the expeditions.

Silences. There are other silences in Mimi as well. In the first voyage, for example, discussion about whales as an endangered species is encouraged in the teachers' guide; but missing in the second voyage, which relates to archeology and ancient Mayan civilization, is discussion of present-day Mayan culture. Gibbon responds:

> There are lots of political issues we could have introduced but didn't. But it's not a question of ruling it out. This was a science and math series and we wanted to get as much of the surround to support that. And had we been doing a truly integrated curriculum, we probably would have felt obliged to do that. We were not eager to invite controversy of that sort, but I don't think we shut our eyes to it. It's a matter of how you use the time—every line of dialogue is jealously guarded.

Mathematics is mentioned four times in the introduction to the teachers' guides: it is said to be in a "sorry state" and increasingly male and white; it is presented as a means to understand "technical matters" and as integral to doing science. For teachers in this study, however, Mimi is a science program; they don't see mathematics as obligatory. When asked about their use of activities presenting mathematics concepts, as suggested in the teachers' guide, they say their students aren't ready (Sites 1 and 2) and point to their own mathematics program that follows a basal textbook (all sites).

Mimi's Compatibility Within Diverse Settings

"Most teachers are not curriculum-producers, and they shouldn't be. They need materials to work with . . . What's unusual about Mimi is that it's quite comprehensive . . . It's this powerful story that a teacher doesn't otherwise have access to in the classroom."—S. Gibbon

Teachers concur: Mimi's adventure story does what they cannot do as easily (and perhaps as well): provide information in a visual, narrative form and create a rich context for their teaching. Moreover, we heard repeatedly from teachers that Mimi is "flexible"; it meets the needs and interests of diverse student populations.

At each site, we were told of students who blossomed because of this unique match between Mimi's structure and teachers' practice. Two testimonies elaborate this point:

> Third grader Clyde is very bright, but can't get much done on paper. And yet if you get into a discussion with him, he can probably tell you more about whales than any child here. He really got into the whale unit and just loved it. So, I put him into a discussion group with somebody who might be academically higher but does not have the motivation to know about whales, and he's able to inform those other people (HP, Site 1).

> Several years ago, one of the teachers doing the second voyage turned her classroom into a rainforest. There was a student in her class who had some pretty severe learning and self-esteem problems. But we discovered in this project that he also had a real interest in and an ability to create music and so he did the music for the rainforest. It was a really wonderful way for him to become connected to and invested in this study . . . He was being able to reveal an aspect of himself and get lots of attention and praise for it from students and adults alike. There was room for that in this study (principal, Site 6).

Summary

Mimi, like all curriculum packages, raises some issues and is silent about others; models some possibilities and not others. By Mimi's crew successfully modeling diversity, Mimi enlarges the circle around who does science. Apparently, there is room for almost everyone on this "ark," including people with physical disabilities and girls and women. Physical disability is directly and frankly addressed in both the episodes and the expeditions; gender equity is implicit in the episodes only. Regarding the latter, students recognize that a boy and his grandfather are the lead characters in the episodes, but that women and girls are experts and leaders. Mimi is less open in other areas. For example, not made explicit is diversity among the crew in terms of race, class, and culture. As a consequence, although the story line of the episodes includes conflict, it centers on personality differences.

Observations at all sites show that teachers build on what Mimi models explicitly and less so in areas where there are silences. Teachers feel the stage is prepared for discussing disability, gender, and age; but not for taking up race, class, or culture. Mimi's voice and Mimi's silences are theirs. This holds for models of classroom practice as well. Whereas the episodes portray science as an adventure story, transference to classrooms is weakly supported in the teachers' guides. The model presented there is less than adventurous when compared to imagining oneself as an actual member of Mimi's crew, really involved in an important research enterprise that has the feel of a good mystery.

◆

Causal Net

Figure 8 in the Kids Network chapter of this volume summarizes how that innovation moved from development to adoption to implementation and to consequences. Our study found these elements to be very similar in the Mimi story. In narrative form, the Mimi story reads as follows:

- The schools in the sample are actively interested in a more inquiry-based science curriculum.

- The developers try to initiate and operationalize program features in this direction but are constrained by their publishers' demands and a text censorship on their own part. Still, they create a serialized adventure story with strong appeal both to children and adults. Refusing the form of pencil-and-paper assessment as recommended by the publisher, they offer no substitute.

- The guidelines to the program itself are highly prescriptive due to the developer-publisher's perception that this is what teachers need and want.

- The appeal of the program features, the adequacy of initial funding, and the perception of goodness of fit between school characteristics and the program lead to Mimi's adoption as the core of the science curriculum. Teachers especially like the flexibility of use incorporated in the program.

- Mimi's first generation of users enjoy time for experimentation. Often with another teacher, lessons are tried out and modified, and a notebook of possible activities grows thicker. Interactions among teachers in this early stage—and later—are personal and reciprocal with ideas, materials, and equipment freely exchanged.

- These incentives are counterbalanced by constraints and uncertainties, many having to do with the limited science backgrounds of the teachers, the novel technological components of the program, and an infrastructure that remains unyielding. These factors, combined with low assistance from outside the school, lead at first to minimal modifications and extensions of the program, especially of its more demanding components. Gradually, however, teachers look for supplements, more flexible use, and new configurations coming from their greater mastery of Mimi. Teachers are, in fact, recognized as "pioneers" by administrators (on both local and statewide levels), who themselves bask in the glow of their innovative staff.

• Some teachers transfer the capacities for inquiry and experimentation learned in Mimi to other topics. But overall there is an underexploitation of the potential of the program and apparently a modest corresponding growth in pupils' scientific understandings and mastery.

◆

Continuity, Institutionalization, and Spread

Continuity

Mimi seems to have staying power. Having taught Mimi for a decade, Mimi teachers at one of the focus sites speak for the majority of teachers at other sites when they say: "We've been teaching Mimi almost ten years now and it's not outdated! . . . There's lots of meat; you can chew and chew and chew."

Although Mimi appears well-entrenched at all sites except one (Site 4), a closer look reveals that Mimi's continuity and "institutionalization" depend on a set of contingent factors. In this section, we explore these factors. We ask: How is the institution changed because of Mimi's presence? And finally, What is the likely pattern of future use? Table 11 displays sites relative to their degree of commitment to Mimi and other pertinent factors.

When discussing continuity and institutionalization, the recurring theme of flexibility is salient. Reflected throughout this report is the evaluation of Mimi by teachers, administrators, and developers as highly adaptable, whether to diverse grade levels, teaching styles, contexts, or science curricula. This contributes to Mimi's longevity within a school or system.

Adaptability, however, is not always adequate when factors external to Mimi come into play. The most dramatic is the shift of FJ's situation at Site 4; once part of a thriving team of Mimi collaborators, she was gradually the sole user, without the support of colleagues or the new principal. When she transferred to another school in 1995, she brought Mimi with her, finding herself again the only user. This time, however, she reports some administrative support.

The list of more administrative and financial factors facilitating the continuity of Mimi within the school context is a long one, including scheduling decisions; staff deployment; training and development opportunities; assessment choices; and access to resources in the form of computer hardware and software, a well-stocked library, and a science specialist. Funding and staff stability are two further requisites.

Table 11.   Continuity of Mimi—Degree of Commitment, Factors Influencing Future Use

Site	Degree of Commitment to Mimi and Explanations about Future Use	Key Factors Affecting Continuity
1	**High** Administrative support: "Some administrators encourage their teachers to be all they can be . . . other administrators really enjoy the status quo." (KC)	Lack of networking apparatus for Mimi teachers
2	**Mixed** Resistance: "If given their choice, some teachers would never walk into the computer lab, never." (KN)	No new funding; staff mobility high; new principal support unknown; teacher resistance
3	**Mixed** Mixed response: "All teachers aren't innovative; they say, 'Give me that textbook, let me do my thing.'" (EH) Competing programs: "I could teach some other science program and the students would do just fine." (EH)	Resources low; principal minimally supports; greater appeal of other science programs
4	**High** Commitment: "I'm teaching against the odds." (FJ)	Administrative support low
5	**Mixed** Too much to do: "The more things that are added, the more things are suddenly dropped. It's almost an evolutionary process." (HG) Parental power: "They would come in and say, 'How come you're not doing Mimi?'" (DF)	Technology support low; competing agenda
6	**High** Keep on growing: "I feel like the more I use Mimi, the better I use it . . . I really don't get tired of it." (RB)	Collegiality low

Funding has many guises. At Site 1, for example, HP received the "gift" of a laser disc player from the district technology specialist; profits from consultations with her colleague, KC; and benefits from extra adults in her classroom (a paraprofessional works with children with learning disabilities and a student intern from a local college assists). At Site 2, competition to be the first in line for "gifts," for limited resources, or for grants is called "politics." Notes a Site 2 teacher: "School districts are very, very political. The people who are in power control the money and control the allocations for teacher resources—it's very, very dependent on politics."

Additionally, grants themselves, and the promises made to receive them, may impede implementation. RS, a teacher at Site 2, notes: "Our number one constraint was the fact that the grant mandated that we do certain things—by its definition. And number two, we promised we would do certain things."

Teacher turnover, like funding, contributes to Mimi's levels of continuity and institutionalization. The causes of teacher turnover are numerous. During the time period encompassed by this study, critical turnovers occurred at four of the six sites. For example, at Site 2, in addition to loss of staff because of illness, death, and reassignment to the district office, there was an expectation that the principal would retire; his replacement's support for Mimi is uncertain.

Institutional Change/Reform

Change at the institutional level appears negligible in spite of Mimi's recognition as a success. At Site 2, a middle school where science is taught in 40-minute periods, schedules are rearranged to accommodate a weekly meeting of four teachers and to place students in the computer lab for two periods each week. At Site 3, students and teachers experience an intensive week of study off-campus. Overall, however, it's Mimi that is adapted to the institution.

Change in science curriculum and science teaching is viewed as more dramatic. In general, teachers and administrators consider the Mimi material up to date and, more importantly, more appealing to the students than the preceding curriculum. Additionally, they attribute the increase in hands-on activities to the Mimi material—and to other commercial components they have added. They feel encouraged to cross subject area boundaries more freely. Greater collegiality is also a measurable effect.

Another effect is the absence of science textbooks at Sites 1, 3, 5, and 6 since Mimi's arrival—with two exceptions. At Site 2, the sixth grade science teacher continues to use selected parts of the science textbook for homework assignments. At Site 4, with a small budget for new materials at her disposal, FJ supplements teaching Mimi with material from a commercial textbook series based on themes (e.g., patterns). At Sites 1, 7, and 8, Mimi is part of a larger science curriculum based on kits—a major difference being that at Site 1, kits are developed by teachers, while at Sites 7 and 8, the kits are commercially produced and complement textbooks.

Spread

At every site except one, Mimi is integral to the science program at the district level (the exception is Site 4 where there is no enforced districtwide science curriculum). Moreover, with the adoption of the second voyage, students experience both voyages at some point in their schooling. For example, at Sites 1, 2, 3, and 5, students have opportunities to see the first voyage in an early grade and the second voyage one to five years later (students at Site 1 see the first voyage as third graders, the second as eighth graders). At Site 6, the sequence of voyages varies. There, the two voyages are shown on a rotational basis to mixed fifth-sixth grade classes. Consequently, fifth graders may see the second voyage before seeing the first. Teachers at Site 8 show all the episodes from the first voyage twice, both in the fourth and fifth grades; they show expeditions selectively, following a sequence given to them by the district.

At all sites except Site 4, teachers at particular grade levels are expected to teach the first or second voyage or both. The term "mandate," however, is not used either by teachers or by administrators to describe Mimi's position in the science curriculum. For many teachers, mandated or not, they say they would choose to teach Mimi anyway:

> Researcher: Is it mandated?

> I think so; but I don't know that, to be perfectly honest, because I would do it anyway. So it doesn't make any difference to me (KC, Site 1).

Finally, when asked why Mimi teachers say little concerning the conditions that limit and constrain their teaching, RJ at Site 3 says: "Mimi is not where I have problems." To which her colleague MG adds: "There are so many problems, I have to stay optimistic. Mimi helps me think I make a difference."

◆

Mimi as Journey

There is an elaborate literature on the process of change, including very recent and large-scale empirical studies or major syntheses, much of which has been built into this study (e.g., Huberman and Miles 1984, Louis and Miles 1990, Fullan 1991, Cohn and Kottcamp 1992, and Fullan 1993). In innovations akin to the *Voyage of the Mimi,* a product—typically developed in a university or research center—is connected, in the best cases, to theories of learning and development; and is field tested, more or less thoroughly, in the kinds of places in which it will be applied, usually with input from teachers. With each corrective loop, it is readied for dissemination directly through a publisher or indirectly though an intermediary organization such as a laboratory, center, or university. The succeeding phases are always somewhat adventurous and undetermined.

In the case of the *Voyage of the Mimi* innovation, we have to be sure, identified some of the most salient or crucial factors at each turn in the bend, but they often play out differently, especially when the context of use is different. These differences in context will not necessarily determine the remainder of the journey, but they will call forth salient skills, demands, tensions, and ambitions that were not—up to then—as prominent in another school. In many ways, in fact, "innovations" are what the French call "revelators"; they bring out into the open opportunities, constraints, conflicts, old wounds, and—sometimes—heroics or collective adventures. They may introduce ideas, pedagogies or vehicles of which, previously, few in the school had dreamed, except at a distance: video sequences tied both to scientific experimentation and expeditions and to larger questions of diversity and ethics; and experiments off the school site in which pupils are meant to pose the key questions, debate the likely solutions, collect the requisite data, and interpret scientific meaning and local significance. These are all components of the sometimes turbulent, always eventful journey of the Mimi program through the school.

Teachers' Journeys

Mimi is a packet of novelty, both substantively and pedagogically, and has the potential of changing the ways in which teachers interact with children and construe concretely the fit of scientific themes into the school science and mathematics curricula. But did it? The process of undoing the packet, then negotiating it through teachers' constraints, conceptions of legitimate science and mathematics, background and sense of risk or exploration, as well as through available sources of support, is what constitutes the teachers' journey. These factors determine whether the experience was merely adventurous or whether it led to cognitively stretching experiences for the pupils and teachers involved. There is evidence, in fact, that the program invites teachers to experiment with less predetermined, inquiry-centered formats, but that the *Overview Guide* itself allows for more prescribed, often less cognitively demanding exercises which most teachers follow closely.[9] This then determines the quality and depth of the pupils' journey—the one that warrants the closest scrutiny. For many teachers, such new programs may then represent little more than a succession of packaged curricula during their career. As they reach out to embrace a new mode of instruction and learning, they reach out with their old professional selves, including the mostly intact ideas and practices they have carried along with them through the years.

Teachers in our sample gained mastery of Mimi by pulling themselves up by their own bootstraps—often in the company of colleagues, always with the blessing of their principal. Because of Mimi and, to a greater extent, because of

---

[9]Some Mimi teachers interviewed in our secondary sites do not agree. They see the guide as open, so much so that teachers used to more traditional materials are put off by it. For some Mimi teachers, at least in their view, the guide presents "possibilities" rather than a strict "how-to-do-it." However, this view came from a small set of informants whereas our finding is based on multiple interviews and observations at our primary sites.

the support of their school or district, they tried new initiatives (e.g., small group work), new technologies (videos and computers), made expeditions outside the classroom (a whale watch), and developed interdisciplinary projects (turning the classroom into a rainforest). When support was missing (e.g., training with Mimi's software and access to hardware), they were less likely to experiment. The same can be said for experimenting with alternative ways of teaching. All remained enthusiastic about Mimi. All saw themselves as special—teachers willing to put in the time necessary to learn new material, eager to spread Mimi throughout the school year and watchful for additional materials and activities.

Pupils' Journeys

The journey made by the teachers, via the program, puts students on what are often different courses of learning and instruction from those experienced previously. In Mimi, for example, there are group-led hands-on activities, with known or partially known solutions to work out. There are mathematics problems embedded in the story of fictionalized characters. There are possible expeditions outside the classroom, where students can explore natural environments and view collections of living and nonliving things. There are computer modules that engage students in exploring mathematical and ecological concepts. Mimi does not simply offer a new program, but oftentimes new ways of developing cognitive skills, reasoning processes, skills in group work, and inductive and deductive solutions for real-world problems. Moreover, there is the challenge of translating this material and these experiences into the conventional mathematics, science, and social science curricula.

Schools' Journeys

Schools are neither uniform nor infinitely flexible in their capacities to take on new instructional programs. They have structural and working arrangements, decision trees, informal loops of power and influence, "gatekeepers" who can kill or accelerate new initiatives, finite resources, variable forms of leadership and delegations, curriculum and testing regulations, powerful external constituencies, changing mixes of pupils, variations in size, pupil performances, teacher backgrounds, etc. In other words, programs such as Mimi enter a more or less accommodating arena at a more or less favorable moment. As they are put into practice, they are forcibly modified, both in order to fit the surrounding context and to accommodate local constraints and priorities ("We force-fit Mimi to fit our needs," says one teacher at Site 2). Several teachers worried, for example, about coverage of science and mathematics content, about insertion of Mimi into the science or mathematics curriculum, about upcoming citywide tests ("the bottom line is, nobody cares about anything else," notes a Site 2 teacher, who stopped teaching Mimi in order to prepare students for tests). Here, the organization exerts its influence on the program.

At the same time, innovations like Mimi are usually welcome intruders in their new surroundings, and one of their prime roles is actually to create discrepancy, especially at the instructional levels. For example, some of the potential "discrepancies" in Mimi had to do with new science content, more hands-on activities, experimentation with cooperative learning, integrated technologies, interdisciplinary curricula, and materials showing a more prominent role for women and other minorities. Some innovations are even potential institutional change-bearers—for example, at the level of the mathematics or science curriculum or when some of the instructional formats venture into other subject matters.

Without such discrepancies, very little will change; with too much, the journey of the innovation will generate more turbulence than the institution can handle. This results in an ongoing, often implicit, battle for influence among the kinds of change-bearing demands of the programs we have outlined, and the more equilibrium-preserving working arrangements by which schools manage so many moving parts. What will prevail? When the innovation is limited to within-classroom activity only, host schools will not be pressured to modify the conditions of learning, instruction, scheduling, and instructional management that, in fact, deliver significant learning outcomes. Rather, the program's journey through the institution results in watered-down or "domesticated" innovations, as opposed to change-bearing ones. The process of local appropriation is thus a dialectical one: often painful, uneven, uncertain, even periodically unmanageable; yet only successful when reciprocal accommodations are made—ones that allow the program to alter the learning process in significant ways. Otherwise, on returning to a site several years later, we may find few or no traces of influence of the original innovation—on the teachers, the pupils, or the organization (Yin et al. 1978). There are, of course, more positive scenarios of change, but they tend to occur in places in which the management of change has itself been methodically cultivated, studied, and refined. Here, the paths and intersections of the multiple journeys—pupils, teachers, organization, program—have been built into the ongoing management of the establishment (Fullan 1993 and Senge 1990).

Let us stay a moment, however, with conventional schools and this problem of local fit. As change-bearing devices for improving instruction and learning, innovations fit locally in a variety of ways. Where exactly does Mimi go in the science curriculum? Or is it part of the social science program? Or the English program? At each of the six sites, Mimi was the principal piece of the science curriculum: it provided continuity for different content areas; it was the "vehicle" driving the science curriculum. But there remained the problem of transition from a more teacher-driven science program with textbooks and seatwork to elaborate instructions for watching Mimi episodes and expeditions, using computers, and doing hands-on activities. Under these conditions, was the video series simply a replacement for the science textbook? When teachers, faithful to

Mimi's *Overview Guide*, asked scripted pre- and post-questions, handed out worksheets, and assigned vocabulary words to be looked up for homework, this seemed to be the case.

◆

Summary of Principal Findings

- Mimi replaced a science curriculum that was textbook-based, lifeless, and outdated. Mimi, by contrast, is perceived as offering exciting material that catches and holds students' interest and provides teachers with many possibilities, both for teaching science and for integrating science with other subjects. Either way, its adoption within our sample was either administratively driven (Sites 2, 3, and 6) or introduced by teachers (Sites 1, 4, and 5).

- Teachers with limited science backgrounds were insecure but willing to "bootstrap" methods to teach Mimi effectively. That students found it exciting, and that teachers themselves liked the material, sustained their momentum in moments of risk and awkwardness.

- Mimi users can be divided into first and second generations. Successful users were in either group, but crucial to their success was the time accorded or demanded (1) to become familiar with the program, (2) to fit it to their needs, and (3) to experiment with it before being judged on the results. During the period of experimentation, more successful teachers worked with a peer: they shared materials, took turns teaching, talked during spare moments, and made changes to the program format. Exchanges were both about practical matters and collegial support for venturesome teaching.

- Mimi users gradually supplemented the program with other commercial materials; they felt comfortable in doing so because they saw Mimi as flexible. Their prime dilemma—and opportunity—was making choices about what supplemental materials to include, omit, or modify. Teachers supplemented the Mimi materials either because they perceived gaps in the curriculum or because there were dimensions they wanted to extend. Gaps included more hands-on activities or experiments and substitutions for expeditions that failed to hold students' attention. Teachers extended Mimi by taking field trips to natural environments, lighthouses, maritime museums, and weather stations; attending a Mimi fest; developing an ecology camp; building wooden dories; showing films; and choosing specific children's literature. In addition, some teachers supplemented the enrichment activities suggested in Mimi's teachers' guides. For example, teachers at Site 5 chose materials from the commercially available curriculum package, Activities that Integrate Math and Science (AIMS); and the science teacher at Site 2 presented a lab every other week based on material he had developed.

- Few Mimi users integrated the computer modules. Lack of training, lack of access to computers, little time, and the belief that computers were poor substitutes for teacher-student interaction were the main reasons given. Those few teachers in our sample who did use the software typically demonstrated a program to the whole class, and then allowed students to work in twos and threes on their own when other class work was complete, thereby undermining the potential for the whole class of the well-designed simulation component of the program. At two sites, students used the school computer lab. At one of these, all students worked on Mimi-related material at the same time; at the other, a small group of students used the computing Mimi material while the majority of the class worked on something else.

- There were indications that Mimi changed the way students thought about scientists—who they are and what they do. The episodes, more than the expeditions, showed science to be open to all, regardless of age, race, ethnicity, gender, or physical ability or disability. Students saw that scientists work in varied settings and locations. The work of scientists looked exciting—a combination of an adventure story and mystery but with real problems to be solved. At the same time, scientists appeared to study and work hard, often over long periods of time, students claimed.

- Although students typically were enthusiastic about watching the episodes and the expeditions, they noted differences in what they do in school and what scientists do in Mimi. "Real scientists," as one Site 6 student says, "are actually out there and doing the real thing. They get to really see what the whales are doing. They're not reading about it." In contrast, students characterized their work in school as researching the research of scientists. Their work was reading and doing assignments, including worksheets: "When you are doing your work, you are not really figuring it out, you're just thinking about finishing it," explains one student at Site 6. For the most part, students read, write, and talk: "We watch either an episode or expedition and then we talk about what we saw and what we think would happen next. If it was an expedition, we would talk about what we learned, about air pressure or something," notes a Site 6 student. They then move on, often rapidly. Adds a student at Site 2, "The teacher has to go on to another subject at a certain time . . . so we have to move on, a lot faster."

- Mimi did not substantially change the way teachers taught science—science instruction continued to be short, mostly talk, and teacher-directed. For example, to move the discussion along, preferred answers were short and specific, leaving little time for reflection, the airing of different perspectiveness, or expansion of the topic. Additionally, hands-on activities—although more numerous than previously—followed scripted material. Students at Site 1 recall doing four experiments during the year, each taken from the Mimi materials. Their teacher reports many more "vicarious" experiments as well: "Even if I may not be comfortable having my kids do experiments, we can analyze the experiments that are done in the episodes . . . So it gives me a vicarious experiment rather than an actual hands-on." Hands-on activities were welcomed by students. At these times they manipulated some materials, ran through experiments, and interacted with other students on procedures and results. Even though all the students in the class followed the same directions for the same experiment, there was excitement and suspense.

To a large extent, the choice of task, procedure, and evaluation rested with the teacher. Missing were students' self-initiated questions and activities. Doing science became largely talking about what they saw scientists doing. We observed few opportunities for students to translate their ideas or hunches into experiments, imagine the pertinent procedures to test them, experience the consequences, and make appropriate explanations or revisions. In the day-to-day life of the classroom, students had few direct experiences with material and phenomena and little opportunity to act as one another's resource for experimentation and learning.

- Even operating in this mode, the administration tended quickly to proclaim Mimi users as "successful." The gradient of change, however, was low; and day-to-day experience in the science classroom was not fundamentally changed. Still, the program brought many teachers beyond the basal text stage into multimedia environments, greater student involvement, and more experiential or vicarious work with scientific constructions and content.

  Some Mimi users were characterized as "stars"; principals and, in some instances, district administrators pointed to them as examples of where a school or district was headed. Local success was followed by recognition on a broader landscape. Mimi teachers attended and made presentations at state and national conferences. There were several consequences. Stars moved out of the classroom, taking consultative or administrative positions (Sites 1 and 2). In addition, the program was not assessed closely. Only a handful of teachers had been able to take Mimi beyond the prescribed guidelines into new territories for themselves and their students—and this in spite of a faulty or ill-fitting infrastructure. However, at the school building level and within the district, issues around scheduling, curriculum choices, assessment mechanisms, funding, staffing, tracking, and so on, went unchallenged.

- Although the number of years Mimi was in place varied from five years at Site 2 to eight years and more at Sites 1, 3, 4, and 6, the term "mandated" was avoided by administrators and teachers alike when describing Mimi's place in the curriculum. At each site, the combination of Mimi's longevity and staffing changes created a second generation of users whose enthusiasm appeared to vary according to what had worked for them in the past and to the extent they—like the first generation of Mimi users—were given time to experiment and to change the program to fit themselves (what teachers often referred to as their "teaching style"). For many of the second generation who were handed a program already judged as successful, Mimi was not seen as highly flexible.

◆

Acknowledgments

We are grateful to David Kennedy, Jerry Pine, and Karen Worth for keen advice on this work; the product is much the better for their input. We also appreciated the give and take among our colleagues in the OECD project. Special gratitude goes to Senta Raizen and Ted Britton of the National Center for Improving Science Education in Washington, D.C., and thanks to our production team at The NETWORK, Inc., in Andover, Massachusetts: Mary Poulin, Mary Stenson, Sandra Thibodeau, and Colleen Simonds.

We are especially grateful to the teachers, children, and administrators who participated in this study. At each site, teachers often met with us before and after school hours and organized their school day for our benefit. We learned a great deal from each of them and are deeply in their debt. In this report, we try to present their stories respectfully and fairly, but we realize they may not always agree with our interpretations. We are also appreciative of the willingness of Mimi's developers to talk with us as we tried to understand the context and challenges of the curriculum-making process. And we would like to thank Peter Marston for inviting us to his first Mimi Connections institute where we met Mimi educators from across the United States.

Finally, a main intent of this report is to stimulate further thought about the day-to-day life of the classroom, giving special attention to teachers who make changes in their conceptions and practices of science teaching and learning. While this focus may appear to be a narrow one, it is our best opportunity to follow the changes in national and local policy and, at the same time, to represent the classroom-eye view of those directly concerned.

References

Berman, P., et al. 1975-77. *Federal programs supporting educational change.* 8 vols. Santa Monica, CA: Rand Corporation.

Brophy, J., and J. Alleman. 1991. Activities as instructional tools: A framework for analysis and evaluation. *Educational Researcher* 20(4): 9-23.

Bussis, A. M., E. A. Chittenden, and M. Amarel. 1976. *Beyond surface curriculum.* Boulder: Westview Press.

Char, C., J. Hawkins, J. Wootten, K. Sheingold, and T. Roberts. 1983. *Voyage of the Mimi: Classroom case studies of software, video, and print materials.* Phase 1. New York: Bank Street College of Education.

Char, C., and J. Hawkins. 1987. Charting the course: Involving teachers in the formative research and design of the *Voyage of the Mimi.* In *Mirrors of minds: Patterns of experience in educational computing,* ed. R. D. Pea and K. Sheingold, 211-22. Paper from the Center for Children and Technology, Bank Street College. Norwood, NJ: Ablex Publishing Corporation.

Cohn, M. and R. Kottcamp. 1993. *Teachers: The Missing Voice in Education.* Albany, NY: State University of New York Press.

Crandall, D., et al. 1983. *People, politics and practices: Examining the chain of school improvement.* 10 vols. Andover, MA: The Network.

Cuban, L. 1986. *Teachers and machines: The classroom use of technology.* New York: Teachers College Press.

Fullan, M. 1991. *The new meaning of educational change.* New York: Teachers College Press.

———. 1993. *Change forces: Probing the depths of educational reform.* Bristol, PA: Falmer Press.

Gibbon, S. 1985. *The Voyage of the Mimi: Overview Guide.* Scotts Valley, California: Wings for Learning, Inc.

———. 1989. *The second Voyage of the Mimi: Overview Guide.* Scotts Valley, California: Wings for Learning, Inc.

Hall, G., and W. Rutherford. 1976. Concerns of teachers about implementing team teaching. *Educational Leadership* 34(3): 227-33.

Hall, G., and S. Loucks. 1977. A developmental model for determining whether the treatment is actually implemented. *American Educational Research Journal* 14(3): 263-76.

Hall, G., S. Loucks, W. Rutherford, and B. Newlove. 1975. Levels of use of the innovation: A framework for analyzing innovation adoption. *Journal of Teacher Education* 26(1): 52-56.

Havelock, R. 1969. *Planning for innovation through dissemination and utilization of knowledge.* Ann Arbor: CRUSK, University of Michigan.

Heaton, R., and M. Lampert. 1993. Learning to hear voices. In *Teaching for understanding*, ed. D. Cohen, M. McLaughlin, and J. Talbert, 43-83. San Francisco: Jossey-Bass.

Huberman, M., and M. Miles. 1984. *Innovation up close: How school improvement works*. New York: Plenum.

King, G., R. Keiohane, and Verba. 1994. *Designing social inquiry: Scientific inference in qualitative research*. Princeton: Princeton University Press.

Louis, K.S., and M. Miles. 1990. *Improving the urban high school: What works and why*. New York: Teachers College Press.

Louis, K.S., and S. Rosenblum. 1981. *Linking R&D with schools: A program and its implications for dissemination and school improvement policy*. Washington, DC: U.S. Office of Education.

Martin, L., et al. 1985. *Interim report, mathematics, science and technology teacher project, Phase I: Evaluation of training*. New York: Bank Street College of Education.

Martin, L., et al. 1985. *Progress report, Bank Street College of Education mathematics, science and technology teacher project*. New York: Bank Street College of Education.

Martin, L., J. Hawkins, S. Gibbon, and R. McCarthy. 1988. Integrating information technologies into instruction: The *Voyage of the Mimi. 1988 AETS Yearbook: Information Technology and Science Education*, 173-86. Columbus, OH: ERIC Clearinghouse for Science: Mathematics and Environmental Education, The Ohio State University.

Miles, M., and M. Huberman. 1994. *Qualitative data analysis: A sourcebook of new methods*. 2nd ed. Thousand Oaks, CA: Sage.

Pintrich, P. R., R. W. Marx., and R. A. Boyle. 1993. Beyond cold conceptual change: The role of motivational beliefs and classroom contextual factors in the process of conceptual change. *Review of Educational Research* 63(2): 167-99.

Ragin, C. 1987. *The comparative method: Moving beyond qualitative and quantitative strategies*. Berkeley, CA: University of California Press.

Rogers, E. 1962. *Diffusion of Innovations*. New York: Macmillan.

Senge, P. 1990. *The fifth discipline*. New York: Doubleday.

Yin, R., et al. 1978. *The routinization of innovations*. Santa Monica, CA: Rand Corporation.

Appendix A: Case Study Team

Michael Huberman, formerly a professor of education at the University of Geneva (1972- 1993) and now professor emeritus, is presently director of research at the Swiss Federal Institute of Professional Education, and since 1991, visiting professor of education at Harvard University. He also was a senior researcher at the National Center for Improving Science Education where he led this case study. His areas of interest are qualitative research, research use, longitudinal studies of teaching, and educational innovation. He is a co-author, with Matthew Miles, of *Qualitative data analysis: A sourcebook of new methods*, and also recently published *The Lives of Teachers*.

Sally Middlebrooks is director of education at the Association of Science-Technology Centers. She recently received her doctorate from Harvard University, after having been a research associate at National Center for Improving Science Education. She was the principal field researcher on the Voyage of the Mimi case study. Middlebrooks previously was a research associate at the Bank Street College of Education in New York. She has worked in the area of childrens' learning, notably in science, in non-formal settings such as playgrounds, parks and in their own homes. Her dissertation on this topic is soon to be published by Teachers College Press.

Jimmy Karlan is an assistant professor of environmental education at Antioch New England Graduate School. He recently received his doctorate from Harvard University and participated in the Voyage of the Mimi case study while a research associate at the National Center for Improving Science Education. His research interests lie particularly in aspects of environmental education, including students' ecological concepts, curriculm development, professional development, and design of informal learning exhibits and experiences.

Appendix B: Data Collection and Analysis

Data Collection

In addition to data collection described in the chapter, we also:

- interviewed six members of the team that developed *The Voyage of the Mimi*; in this group were the Executive Director, 3 members of the research staff, the editor and the coordinator of classroom materials;

- analyzed innovation-related documents and products created by or used by teachers and students during implementation; these included worksheets, vocabulary lists and children's writings. In addition, commercially produced materials used by some teachers to supplement Mimi materials were collected.

Document summaries

- major reports produced by developers

    Char, C., et al. (1983). "Voyage of the *Mimi*": Classroom Case Studies of Software, Video, and Print Materials. Project in Science and Mathematics Education. Bank Street College of Education: New York.

    Martin, L., et al., (1985). Interim Report, Mathematics, Science and Technology Teacher Project, Phase I: Evaluation of Training. Bank Street College of Education: New York.

    Martin, L., et al., (1985). Progress Report, Bank Street College of Education, Mathematics, Science and Technology Teacher Project. Bank Street College of Education: New York.

- teacher and student guides produced by developers and publishers

    The Voyage of the *Mimi*: Overview Guide

    The Voyage of the *Mimi*: The Book (student guide)

    Ecosystems with Island Survivors

    The Second Voyage of the *Mimi*: Overview Guide

    The Second Voyage of the *Mimi*: The Book (student guide)

- articles pertinent to study

    Anderson, R. D., et al., (1992). Issues of curriculum reform in science, mathematics and higher order thinking across the disciplines. Boulder: University of Colorado.

Char, C. & Hawkins, J. (1987). "Charting the course: Involving teachers in the formative research and design of The Voyage of the *Mimi.*" In Mirrors of minds: Patterns of experience in educational computing. Pea, R. D. & Sheingold, K. (Eds.). Ablex: Norwood, NJ.

Char, C., et al., (1983). Voyage of the *Mimi*: Classroom case studies of software, video, and print materials. Bank Street college of Education: New York.

Martin, L., et al. (1988). Integrating information technologies into instruction: The Voyage of the Mimi. ERIC: Columbus, Ohio.

Webb, N.L. (1992). Mathematics accessible through technology: The Voyage of the Mimi as an interdisciplinary and technologically-based program. The National Center for the Improvement of Science Education: Washington, D.C.

Analytic Procedures

More specific analytic devices included the following: coding, contact summaries, "memos," document summaries, meetings, and interim reports.

a. coding of transcripts and field notes, including the identification of new codes, the discarding of inapplicable codes and, above all, coding for recurrent themes, which were then discussed at team meetings. Data were coded using HyperResearch, a qualitative analysis software program.

b. contact summaries, written after each field visit. The contact summary corresponds to a set of field notes, highlights potentially significant themes and other ideas the researcher found interesting or illuminating during a visit. The summaries also include a set of questions that need to be addressed during the next field visit, as well as a list of collected documents.

c. regular production of analytic "memos," elaborating a theme or idea that struck the researcher during the interview or field visit. Memos were also written about readings and off-site observations, e.g., of training events.

d. document summaries, written about innovation-specific materials (e.g., teachers' and student guides, workbooks, etc.), as well as literature relating to science, mathematics, and technology education.

e. meetings at regular intervals, with analytic work explicitly on the agenda.

f. interim reports. Several such reports were prepared, mainly for overall project coordination and as information to other research teams. The preparation of the reports forced out provisional findings, more thematic or super ordinate codes and reviews of memos for recurrent themes, promising relationships, clusters and explanations, to be explored subsequently.

Appendix C: Thematic Codes

ALGORITHMIC VS HEURISTICS	examples & definitions distinguishing between algorithmic and heuristic approaches to teaching
COGNITIVE SCHEMATA OF T	domains of knowledge (cognitive schemata) teachers draw on for planning and instruction: pedagogical, students, subject matter, pedagogical content, other content, curriculum, educational aims
CONSTRUCTIVIST CONCERNS	constructivist approaches expressed and demonstrated: theory and practice
COOPERATIVE LEARNING	the ways in which cooperative learning is presented in curriculum and experienced by teachers and students
DISTINCT FROM OTHER CURR	features of curriculum which make it distinct from other curriculum
DIVERSITY	diversity issues raised in the development and/or implementation of the curriculum; appeal of program to a diversity of students
INTERPERSONAL CONSIDERATIONS	the possible effects of interpersonal relations and dynamics
INTERDISCIPLINARY ISSUES	interdisciplinary issues raised in the development and implementation of the curriculum
KIDS AS SCIENTISTS	references regarding the value of kids doing what scientists do
MM CONSTRAINT	constraints of MM expressed by developers, publishers, users, researchers
MM NARRATIVE APPEAL	appealing characteristics of the role of story in Mimi
MM STRETCHES LANGUAGE	ways in which Mimi stretches students' language
MONEY NEEDED	references to the amount of money needed to operate curriculum
OBSOLETE INNOVATIONS	the nature of innovations to create their own obsolescence
ORPHANING THE TEACHER	ways in which the teacher has been orphaned in the adoption and implementation phases

Appendix D: Roles and Profiles of Participants by Site

Site 1	Participants	Roles and Profiles of Participants Greenfield Elementary School
Rural NH, grade 3, Mimi 1	CF	Principal; cheerleader for work KC and HP are doing. Praised innovation as high-quality package having focus (the ocean) that children find highly interesting; saw innovation as integrative—giving teachers an opportunity to teach a "double lesson," evaluated technology use in school as low; made link between majority of teachers being women and their discomfort in teaching science.
	DM	District's science specialist, responsible for assisting teachers, supplying some materials, and coordinating science kits that are shared among the schools. Said that district supplied science kits as replacements for science textbooks leading overall to an increase in comfort level and teachers using more hands-on activities. Evaluated KC and HP as more committed to innovation than teachers in other schools and their level of teaching of the innovation as superior.
	KC	Third grade teacher; required to teach five units of science of which innovation is one. She sees direct links to units on plants and rocky shore and finds links in units on solar system and ice and snow. Said hers was a "bootstrap" approach to teaching science; that is, essentially self-taught. Has strong personal experience with sailing. Attended the intensive workshop offered by Lesley College; emphasis on the social aspects and issues in innovation.
	HP	Third grade teacher; arrived on scene after KC and innovation were in progress; overwhelmed by material at first, sees herself as slowly taking on more responsibility for joint-teaching and bringing her strength in social studies and language arts to the program; classroom includes children who have been evaluated as having learning disabilities.

Site 2	Participants	Roles and Profiles of Participants Central Middle School
Urban NY, grade 6, Mimi 1	IM	District grants writer; secured NSF funding to develop integration of science, mathematics, and technology on middle school level. Chose Mimi as optimal medium. Recently secured private funding to increase telecommunications capacities among schools in district resulting in the continuation and expansion of earlier project, in which Mimi continues to be key element.
	AD	District science coordinator; primary force in production of curriculum material for this science-driven initiative. Provided weekly on-site assistance during initial stages; spoke of need for more hands-on science as first step to more constructivist teaching.
	CS	Principal; recalled that he tried to back out of project early on, but now sees it as putting his school on the map.
	RS	Former sixth grade science teacher who became science specialist for the project; taught himself computer programming skills and is familiar with latest educational technology. Credited Mimi with provoking major pedagogical changes in his teaching; spoke of need for schools to reach out to private funding sources if are to stay current with technology.
	KL	Former computer specialist who became project director.
	KN	Science teacher for past years. Not in at beginning of project but quickly became person others turned to for writing new software programs to replace and extend those provided by Mimi. Without a room of his own, he daily pushes carts loaded both with materials needed for his demonstrations and supplementary audiovisual equipment. Class discussions are typically in a question-and-answer format that include teacher bringing in childrens' everyday experiences.
	EC	Mathematics teacher for past years who, feeling pressure that students—especially those in the highest track—do well on the citywide mathematics test, suspends any attempt to integrate Mimi into her lessons. Handles computer lab duties competently but without enthusiasm—would prefer "old-fashioned" drill.

Site 3	Participants	Roles and Profiles of Participants Lakeview Elementary School
Rural ME, grade 6, Mimi 2	EC	Principal and superintendent; unfamiliar with innovation but has insights to offer on earlier innovations (e.g., Man, a Course of Study). Stance toward innovation seemed to shift toward end of school year to more positive and as key element within the curriculum.
	BK	Computer specialist; taught a third sixth grade early on when innovation was new. Now as computer specialist, does not use any of the innovation's software [he said teachers should ask him to use it, MG and RJ said computer lab cannot be well-connected to class work and computers should be in classrooms].
	NJ	Sixth grade teacher; has taught parts one and two of innovation. Developed boat building program, week-long ecology camp, then ran out of steam; innovation is at center of his teaching. He chooses lessons in math to fit with the series as it unfolds (e.g., symmetry). With a master's in educational technology, he is the most competent and frequent user of the innovation's software programs.
	EH	Sixth grade teacher. With NJ, started teaching part 1 of innovation reluctantly but then caught on; has taught parts one and two of innovation. Sees other teachers as jealous of their success. Sees need for innovation to be "teacher-driven"; sees it as lacking in hands-on activities and supplements. He moves in and out of innovation and other subject material, stimulating students to look for connections between ancient Mayan government and the rest of the curriculum.
	RJ	Sixth grade teacher; new to innovation two years in a row: last year when other fifth grade teachers were reluctant to teach part 1, she taught part 1 to each of the three fifth grades within 12-week periods; this year, as a sixth grade teacher teaching part 2, she had to learn new material and then decide ways to present it; she brings ideas from course work for master's degree into her teaching; the only computer in her classroom is not compatible with innovation.
	CD	Fifth grade teacher, teaching part 1 this year. As sixth grade teacher last year, taught part 2 with NJ & EH.

Site 4	Participants	Roles and Profiles of Participants / Union Middle School
Urban MA, grade 6, Mimi 1 & 2	FJ	Sixth grade teacher of "advanced" class; early user of both parts of the innovation that included contact with developers; early use brought small group of teachers together and included overnight field trips and whale watching. With change of school administration and loss of comrades, she is now alone in teaching innovation. Her class of "advanced" sixth graders is, except for several subjects taught elsewhere, self-contained and allows her flexibility; she considers herself strong in both science and mathematics content, better able to teach mathematics than science. Her classroom is stocked with treasure trove of materials which she draws upon; she is well-versed and competent, and comfortable in use of innovation software, parts one and two.
	PV	Science specialist; no longer uses Mimi materials, but important member of small group that brought Mimi into the school. No longer at site and has new position with Board of Education.
Site 5	Participants	Roles and Profiles of Participants / Sherwood Elementary School
Suburban MA, grade 5, Mimi 1	DN	Principal; details of innovation largely unfamiliar to him; he hoped innovation would make technology stronger; would not call teaching of innovation "mandated."
	PW	Fifth grade teacher; earliest user of innovation and its chief booster; has added numerous activities from AIMS to make up for what she sees as a lack of hands-on activities in the innovation. Secured local funding to attend the intensive at-sea course offered several years ago by Lesley College. Other teachers come to her for advice both regarding innovation and AIMS. Has access to one computer one day each week; modest use of innovation software.

	HG	Fifth grade teacher. Also one of the early users and also attended intensive at-sea course offered by Lesley College; expressed longing for a slower pace to teaching as day is overloaded with too many demands; prefers "discussions" as means of engagement, in which he does not take a leading role, often to relate personal stories. Little if any use of innovation software.
	DF	Fifth grade teacher; first year teaching part 1 of innovation although had taught part 2 the previous year as a sixth grade teacher. Guided by PW, uses activities from AIMS. Little if any use of innovation software.
	NM	Fifth grade teacher; relatively new to teaching; teaching innovation for the second time. Draws theme of endangered species from innovation, modifying activities she has learned in workshops. Little if any use of innovation software.
Site 6	Participants	Roles and Profiles of Participants Riverside Middle School
Suburban VT, grades 5 & 6, Mimi 1	AL	Principal; articulate spokesperson for what innovation offers. Sees himself as supporting teachers interested in changing their practice as well as making field trips possible.
	RB	Fifth/sixth grade teacher; 22+ years of teaching. Early user of the innovation, now alternating between parts one and two. Most of her Mimi-related science teaching is discussion-based or having students do research using books. Little collaboration with other Mimi users.
	JK	Fifth/sixth grade teacher; 20+ years of teaching. Also an early user of both innovations. Most of the Mimi-related science experiences are discussion-based or students doing research through books. Little networking with other teachers in school or district.
	MP	Fifth/sixth grade teacher; new to teaching and to innovation.

Chapter 5

ChemCom's Evolution:
Development, Spread,
and Adaptation

Mary Budd Rowe
Julie E. Montgomery
Michael J. Midling
Thomas M. Keating

Stanford University

Contents

ChemCom's Evolution:
Development, Spread, and Adaptation

Introduction

Who Should Read This

For people interested in how curricular innovations find their way to the people who could benefit from them and why some curricula survive and keep growing and adapting while others whither and die, this case study of a survivor chemistry curriculum has many lessons to teach. There are lessons for policymakers who want to protect and nurture investments and for administrators seeking to make the best use of in-service monies in a changing curricular landscape. For developers, there are useful lessons to be learned about dissemination, about relations with publishers, and about adapting to a variety of state curricular frameworks and testing initiatives as well as to competitor programs.

For university educators, the case study poses questions. The interdisciplinary nature of *Chemistry in the Community* (ChemCom), with its focus on applications of knowledge, reflects a growing trend in curriculum that raises questions about what appropriate education of science teachers in the preservice years should be. Can the natural and social science departments of colleges and universities provide the necessary education to prospective teachers? As yet, only a small number of them are trying to develop programs that are interdisciplinary and action-oriented and where the prevailing ethos supports this kind of experimentation.

For scientists, there are lessons to be learned about how to collaborate with teachers in the context of a professional society to support wider access to quality science programs. For industry scientists and executives, the supportive atmosphere and activities provided by the industrial sector of the chemistry community should prove instructive. For teachers trying to bring about change in their own school districts, there are some suggestions. For everyone, a few messages are the same: It takes much more time than you think; persistence is vital; teaming up with others is essential; and a continuing source of money is an absolute necessity.

Methodology

We developed our ChemCom story using information from a variety of sources. In the first place, we read all the course-related textual material we could find: both editions of the textbook, the Teacher's Guide, advertising material, descriptions of workshop content, the course materials for trainers, published papers about ChemCom. We found a teacher-initiated electronic bulletin board focused on ChemCom and archived its contents periodically. We also reviewed *Chemistry at the K-12 Level* (ACS 1993b), an analysis done by an American Chemical Society (ACS) task force. It analyzes the chemistry content in the Project 2061 benchmarks document and the Scope and Sequence Content Core document of the National Science Teachers Association (NSTA) to determine how these standards "map" on to ChemCom (AAAS 1993, NSTA 1993). We also studied a one-semester college chemistry course developed by ACS using the expertise of its academic members as well as a middle school science program that can be considered another ChemCom outgrowth. We also received *Chemunity News*, an ACS chemistry education newsletter which contains ChemCom-related articles and communications by and for teachers.

Second, we identified four schools that we could visit periodically to see ChemCom in action. We made one-time visits to three others. We particularly wanted to interview regular classroom teachers who also were ChemCom trainers, and to observe their teaching. This turned out to be a particularly ellusive goal since four of the people we identified were not teaching but were instead doing training and serving as resource specialists for their districts. Our sample was derived of convenience—i.e., to stay within budget, our sites for observation were limited to four schools in the Northern California Bay and Peninsula region.[1] They served somewhat different constituencies with respect to the ethnic composition of their student bodies. In addition, taking advantage of another study being conducted in Florida by our senior author, it was possible to do a small case study within this larger case study. The smaller study focused on a rural/suburban district with six high schools that was in the process of piloting, adopting, and adapting ChemCom.

When we observed classes, we generally looked for the chemistry and mathematics content presented by the teachers. We tried to visit laboratory as well as class discussion sessions. All of our school sites were visited by some team member on two to five occasions over a two-year period. Periodic phone interviews kept us apprised of changes in offerings, enrollments, and other relevant background variables—e.g., the change in incoming student mathematics backgrounds that influenced what teachers did. In the Florida case, we talked to administrators as well as to the science supervisor and students. In Florida, we were in touch for three years so that we saw the whole adoption process take

---

[1]The research team was based at Stanford University; see appendix A for biographical information.

place from awareness, to piloting by two teachers, to district approval for use in all high schools, to diffusion and training of all six chemistry teachers in the district.

Third, we interviewed Sylvia Ware, director of the ACS Education Division. Throughout the study, she has supplied us with information, names, and reactions to our writings. At times, we also talked to other people connected with either the writing of workshop materials or the evaluation of workshops. In addition, we talked with two people at Kendall/Hunt, the publisher of ChemCom, to verify descriptions we had from other sources of what the company does in marketing and disseminating the program. We obtained approximate sales information from Ware since the publisher, not surprisingly, treats that information as confidential. What we were trying to find out is not just market penetration but whether patterns of adoption were related to participation in ChemCom workshops.

Fourth, a major additional source of information for us came from talking to people at state, regional, and national NSTA meetings. We attended NSTA sessions presented by ChemCom teachers as well as presentations at other professional meetings. This gave us an opportunity to get some picture of how the views that ChemCom teachers gave when making conference presentations comported with observations and interviews at the school sites we were studying in California and Florida. We listened to or interviewed more than 80 individuals who made conference presentations. They came from different parts of the country. Some had ChemCom training; others did not. The average length of time they had been teaching ChemCom was a little over two years. As a matter of fact, it was difficult to find people who had been teaching the course for more than three years, since the ChemCom textbook only came out in 1988, and then was marketed without regard to state or district adoption schedules. Thanks to a state science supervisor, we found a teacher in the state of Washington and another in Oregon who were part of the original writing team and were still doing ChemCom with enthusiasm. Both were ChemCom trainers. We also attended a portion of two ChemCom training programs where we talked with staff and participants.

Our interviews focused largely on curricular and instructional aspects of ChemCom:

- How and why schools adopt ChemCom—How teachers first heard about it; the number of sections taught, the grade levels and background of students taking the course.

- How teachers use the curriculum to teach chemistry—Which ChemCom modules do they choose to teach? What supplemental things do they add, if any? We also looked for indicators of whether ChemCom ideas influence any other chemistry courses they teach as well as influences on ChemCom instruction from those courses. This information answered to our interest in adaptation over time.

- What teachers think about the ChemCom approach—the textbook, laboratory activities, comparison with other chemistry courses they teach regarding perceived demands for quantitative and reading skills, the applications orientation, the focus on decisionmaking and small group work.

- What students think about ChemCom—We interviewed students during site visits. Some of them responded to a questionnaire that our team constructed. In addition, two teachers, one in California and one in Florida, constructed their own questionnaires to collect student impressions. They shared these with us. At the Florida site, one of the teachers added some questions to her final exam that asked students for comments and feelings about the ChemCom course. She did this for each of the two years she taught the program.

We wrote up our interviews and, in many instances, sent these writeups to interviewees for their feedback. We made changes when appropriate based on this feedback.

◆

Development of ChemCom

Chemistry in the Community is a high school chemistry course developed by the American Chemical Society for the college-bound, nontechnical student and for the bright student not planning to attend college. It began as a series of free-standing special-topic modules that could be inserted into existing chemistry classes and evolved eventually into a full course with a textbook. The new program broke with tradition in its goals, its organization and choice of content, and its recommended approach to instruction.

In 1977, Dr. Anna Harrison, then president of ACS, hosted a conference on teaching chemistry to the general public. The conference proceedings included a recommendation that ACS develop a high school chemistry course that would appeal to a broader range of students than did the traditional chemistry courses. Harrison felt that the goal should be to attract a wider variety of students. This might be done, she thought, by choosing topics with more social relevance and interest to students. ACS would sponsor development of curricular materials if it could get support.

ACS is a large professional society made up primarily of industrial and university chemists as well as a small but vigorous group of chemistry educators. The role that the main body of the society played (and continues to play) in ensuring the well-being and growth of the program over time is an important part of the ChemCom story. In 1979, ACS hired Sylvia Ware, a former teacher and industrial chemist, to serve as the manager of a newly established staff Office of High School Chemistry. The Harrison conference paper inspired Ware to approach the National Science Foundation (NSF) about possible funding of a new approach to high school chemistry. After receiving a preliminary proposal from Ware, NSF encouraged ACS to submit a full proposal. This proposal was written by W. T. Lippincott, then of the University of Arizona, who was the editor of the *Journal of Chemical Education*, which is published by the ACS Division of Chemical Education. It requested funding for development of "chemistry for citizenship" curricular materials. Lippincott, a respected chemist, would be the principal investigator for the proposed project and ensure its intellectual quality. Harrison would chair the project's steering committee, reporting directly to the ACS Board of Directors. Ware would serve as project manager.

The proposal called for preparing chemistry modules rooted in cultural/ community contexts. Teachers doing traditional chemistry could select modules to insert in their courses. The proposed materials would be oriented around major societal issues involving chemistry. Water quality issues, for example, provide a context within which chemical information appears in the curriculum. The issues surrounding water dictate what chemical content will be given prominence. Chemistry content would be introduced on a "need-to-know" basis—i.e., relevant to the particular issue to be explored. As students move from one unit to another—i.e., from one class of issues to another—new chemical information becomes relevant. As a result of this method for choosing chemical content, some cherished features of chemistry disappear or are downgraded in terms of the amount of attention they receive. This project's senior investigator heard the pained cry of a young teacher paging through the book, "But where are the orbitals? Who ever heard of a chemistry course without orbitals? "

Figure 1 shows our encapsulation of the design philosophy. Instead of banking chemical knowledge against a future, rather undefined, need—as is done in a standard chemistry course—content would be introduced *as needed* to pursue an identified issue, e.g., water pollution. Obviously, there must be a judicious choice of issues to ensure some reasonable spectrum of exposure to chemical topics.

ChemCom was to be much more than just another applied chemistry approach; instead, it would focus on major environmental and health issues that had some footing in chemistry and good potential for student action in their communities. The issue-driven vision that would guide development was broader and more interdisciplinary in concept than was typical in either standard or applied chemistry courses at the time.

Upon receipt of an NSF award, ACS formed three separate writing teams, each consisting of a university chemistry educator and a group of four talented high school chemistry teachers. During the summer of 1982, each team developed a different topic. The approaches taken by the first development teams were shaped to some extent by the text *Science and Society* produced by the Association of Science Education and a project called Interdisciplinary Approaches to Chemistry (IAC) done at the University of Maryland, College Park. Both these initiatives incorporated science, technology, and society themes in chemistry programs. IAC was especially influential in the early stages of ChemCom development since some of these initial writers had worked on it. The work of Alex

Figure 1.   **ChemCom Design Philosophy**

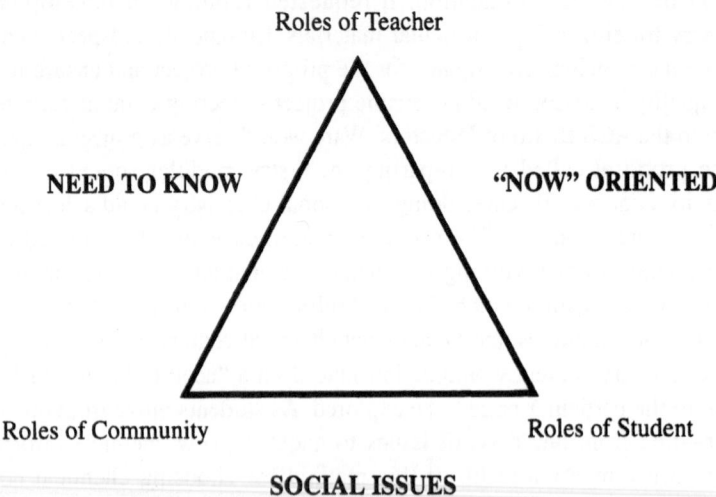

Roles of Teacher

NEED TO KNOW   **"NOW" ORIENTED**

Roles of Community   Roles of Student

**SOCIAL ISSUES
AND CHEMISTRY**

Johnstone and Norman Reid in Glasgow, Scotland, was also influential. They had produced interactive teaching and role-playing units as part of science instruction.

During the school year following the initial summer of writing, the three teams, working largely independently, tried out their modules at several school sites and collected feedback. At the end of the year, after examining the feedback, the development team elected to use the water chemistry module developed by the team at the University of Maryland headed by Dr. Henry Heikkinen as the template or model for all successive modules. Five additional writing teams took on the preparation of new modules and their testing in the following school year. Many of the high school teachers on the writing teams had previously been involved in curriculum development projects such as IAC and/or ChemStudy, a chemistry reform curriculum developed in the 1960s. They were very experienced teachers with a sound background in chemistry. Many were ACS members or were involved with other professional organizations. Some of the college chemists who headed each team had also previously taught in high school. Within ACS, they were identified as "chemical educators," having good contacts with their local high school communities.

Up to this point, the intent was to develop a set of eight modules that could be infused, inserted into, or appended to existing chemistry courses. The choice and arrangement of chemistry content in the modules was driven by the issues. The curricular approach in each was interdisciplinary. There were implicit—if not explicit—social action implications for students that could emerge from study of any of the modules. Some modules provided a context for genuine community—based problem solving—e.g., students could deal with real local dilemmas connected with pollution of one kind or another.

Compared to traditional chemistry, the strategy for choosing and introducing content of the modules was unusual. The choice of issue would dictate the choice of content. Moreover, the recommended teaching strategies were far from common practice in chemistry. Instruction suggested for the modules emphasized teamwork, collaboration on problems and tasks, discussions of alternative solutions, and student decisionmaking. ChemCom modules required a major shift in the organization and conduct of instruction as well as the breadth of content to be treated. But since they were only modules, one or two of which might be taken up in a year of traditional chemistry, they did not entail a major alteration in the content and practice of chemical education. But that was about to change. At the end of the second year of development, the planning team for ChemCom decided to knit the modules together into a full course with a textbook.

Switching to a Full Course

The possibility that the set of modules could be put together to form a full chemistry course emerged when a "synthesis" committee mapped the concepts in each module. This analysis showed where there were redundancies that could be eliminated. It also suggested how the topics might best be sequenced if they were transformed from modular to textbook and full course status. Two questions arose: Could the developers move directly from module A to B or did they need to insert additional supportive material? When and how could various essential concepts—e.g., ionic and covalent bonding—be introduced given the design imperatives, namely, the introduction of concepts on a "need-to-know" basis?

The steering committee chaired by Harrison and the editorial group headed by Lippincott chose to go the course and textbook route. The modules would be the base for a chemistry course that would attract able students who might ordinarily avoid chemistry. (It is interesting to note that in the second edition of the ChemCom text, the goal is broadened to include all students, not just a special subset of the able science "avoiders.")

Taking the eight disparate units and forging them into a cohesive whole, as mapped out by the synthesis committee, became the task of an expanded writing team, working at the University of Maryland. The team was headed by Dr. Henry Heikkinen, who then became the chief editor for the project. Over the course of a year, five or six high school teachers, some still in the classroom, others working on a graduate degree in chemistry education, worked under Heikkinen's leadership to develop the field test version of ChemCom—both the student and teacher materials.

Laboratory activities, experiments, and investigations are placed directly in the ChemCom text. There is no separate laboratory manual. This arrangement departs from the more usual chemistry text and separate laboratory manual arrangement. The goal of the laboratory activities, as described in the introduction to the *Teacher's Guide,* is to motivate student interest in chemistry and its applications, to introduce and develop key chemical concepts, and to provide practice with laboratory skills.

ChemCom supplies a very comprehensive teacher's guide. It gives lab and teaching tips, expected results for laboratory activities, safety guidelines, answers to questions, supplemental reading resources, advice on organizing instruction, and tests for each unit. The *Teacher's Guide* repeatedly emphasizes the importance of active student participation in decisionmaking and evaluating data and arguments.

In appendix B, we provide the preface from the second edition of ChemCom (ACS 1993a, pp. xi-xii). It briefly describes the history and characteristics of ChemCom.

Getting and Keeping a Publisher

The decision of the steering committee to transform ChemCom from a set of free-standing modules meant to be infused into existing courses to a formal textbook that would be the basis for a new chemistry course oriented around issues, meant a shift away from an infusion/insertion strategy to a diffusion/replacement change strategy. Moreover, it meant ACS had to find a publisher willing to take on responsibility for a text and a whole course rather than just a collection of modules. Financial risk for the publisher is greater when a wholly new course is involved. In the case of ChemCom, the course was a substantial departure from the standard content and instructional practices in chemistry. In addition, the course was meant to attract a broader cadre of students. That meant a new market would have to be reached.

Once the eight modules were organized into a draft textbook, ACS carried out a large-scale, year-long field test funded by NSF. The field test was critically important to the development of the first commercial version of ChemCom. It allowed the developers to determine what worked or didn't work well in the classroom. It allowed for refinement of the laboratory activities and redesign of the decisionmaking problems. It also permitted the ChemCom team to determine the extent to which there was a need for teacher workshops to support ChemCom's implementation. The field test included both "teacher-supported" and "teacher-unsupported" sites. At the seven supported sites, a select group of college chemists—under the leadership of chemistry professor Dr. Dwaine Eubanks, then at Oklahoma State University—each conducted summer workshops to prepare the field test teachers for the new program. Each of these seven field test directors also met with their local teachers once a month throughout the school year. This served two purposes: it allowed for feedback on the units as they were taught, and it allowed the field test directors to provide continuing support for the next unit to be taught. Teachers at the four unsupported sites received no special training, being given the text material only.

The field test served an additional important function in that the sites developed a nucleus of teachers supportive of ChemCom across the country. In fact, field test teachers were responsible for many of the initial sales of ChemCom.

After the first field test, Lippincott and Heikkinen, together with Ware and a professional science writer, Mary Castellion, produced the final version of the first commercial edition of ChemCom. Lippincott and Ware then met with a number of commercial publishers, all of whom were invited to submit a proposal to ACS for the right to publish ChemCom.

It is important to realize that publisher-developer relations have been problematic for innovative, government-sponsored curricular projects for a quarter of a century. Publishers are leery of innovative projects for a number of reasons that are ultimately driven by bottom-line considerations. Developers work with teachers and students. They have content expertise as well as results of field tests upon which to base curricular and instructional decisions. Publishers have marketing issues, relatively small operating margins, and costs that developers do not appreciate. The cost/benefit ratios for publishers are often questionable, in part because the cost of preparing and marketing a really innovative product is high. Further, if the publisher has a field sales force, sales commissions can be a problem. The sale force would rather sell a standard program familiar to teachers because they have to spend less time per school selling it—and that means more commissions in a shorter time. New programs, with their increased demands on marketing and—in the case of ChemCom—the need to create a new niche, put commissions at risk.

Lippincott and Ware invited some 15 or so publishers to a meeting in New York to learn more about ChemCom and the kind of agreement ACS was willing to consider with a publisher. Most projects lose editorial control on signing with a publisher. ACS had no intention of losing either this control or the copyright to the work. Eleven companies expressed an interest in ChemCom. ACS selected Kendall/Hunt, a small publishing company with no large national sales force, as a result of this search.

ChemCom Finances

Royalties from sales go to ACS, which to date pumps them back into ChemCom activities. The marketing budget of the publisher includes support for some level of information dissemination. Kendall/Hunt and ACS collaborate on awareness and short course training ventures. If the publisher wants to sponsor an awareness session in some part of the country, ACS identifies a trained and experienced ChemCom teacher to run the session. The Kendall/Hunt-ACS association has lasted through two editions and is now moving into the preparation of a third. Each edition is revised after five years based in part on information collected from teachers in the intervals between editions.

Table 1. Summary of ChemCom Funding

Year	Grant & Source	Activities: Development, Evaluation, Dissemination	Product
1981-1982	$193,000 NSF	• Development of 3 modules • Field testing in 10 pilot schools & revision of modules	3 modules
1983	continuation of original	• Revision of first 3 modules; development of 5 new ones • Field testing & revision of all 8 modules • ACS meetings & publications used to spread ChemCom message	8 modules, stand-alone/ supplemental
1984	$470,000 NSF	• Develop a course from the 8 modules • Field testing in 7 supported sites & 4 unsupported • 1 week of workshop training for field testers & monthly meetings during year (4-day meeting over winter break)	Draft course textbook
1988-1991	$551,000 NSF	• Evaluation of summer workshops with feedback used to improve subsequent workshops • Master teacher strategy: four 10-day residential workshops for 20-25 teachers each over a three-year perio; • ChemCom newsletter	Kendall/Hunt publishes first edition of ChemCom
1990-present	Other publishers and ACS/ Russia	• Development of ChemCom in Japanese, Russian, Chinese, & Spanish • ACS team and 20 high school teachers train 140 Russian teachers. • Workshops in Mexico by ACS	Some Chem-Com transla-tions com-pleted, others begun
1991-1993	ACS funds & royalties	• Compilation of five separate ACS newsletters into one with 4-6 page section for ChemCom • Evaluation of summer workshops • Teacher training workshops continue (increasingly funded by ChemCom royalties) • Free newsletter 5 times per year, circulation over 20,000	Chemunity News news-letter
1992	NSF supported	• Development of FACETS modules for grades 6-8 • Field testing & revision of FACETS modules • Funds are available for field testing of FACETS but not yet for teacher training	FACETS to be published in 1995
1993	ACS funds & royalties	• Feedback from first edition used to revise & publish second edition • Evaluation of summer workshops • Teacher training workshops continue	Second edition of ChemCom published
1994	$250,000 from ACS funds &	• Revision for third edition • Develop multimedia version or supplement • Evaluation of summer workshops • 10 5-day summer workshops	*Chemistry in Context* college course published

As Table 1 shows, funding came in the beginning from NSF and later from ACS itself, which included a chemical education initiative in its general fundraising campaign. (Thanks in part to its large industry-based membership, ACS is a relatively well-funded organization.) ACS has chosen to return royalties to the ChemCom activities. In addition, the publisher's advertising budget and the increasing willingness of school districts to pay for training have further enlarged the project's dissemination/diffusion activities. The policy decision ACS made to continue to plow royalties and capital campaign funds back into ChemCom activities is a key factor in the continuing spread and survival of the program. We discuss these activities in more detail in a later section.

◆

A Course to Attract More Students[2]

Our intent in this section is to convey the critical features of ChemCom: What makes it a noteworthy innovation? We begin by constructing a vignette of how ChemCom transpired in one classroom. The story is a fictitious scenario, but it draws out the innovative features of ChemCom and incorporates implementation issues consistent with the field work in this study. Before presenting the vignette, however, we should note that ChemCom is being used in quite varied settings, as illustrated by the following descriptions of two sites we visited.

- A suburban school district in our study is upper-middle class; the median price of homes is about $400,000. The school's physical plant is as close to the image of a "country club" as one could expect from a public school in today's California. The campus has neatly kept lawns and attractive buildings. Unlike inner-city schools, the buildings are free from vandalism and graffiti. Class size is generally small. The two classes we observed had attendance of about 20 students. There is a relaxed atmosphere at the school; there are no outward signs of security forces or security problems, and guest registration is a simple procedure that involves a quick and friendly exchange and a signature. Our first impressions were that the students and teachers have a cordial and easy attitude in their interchanges both inside and outside the classroom.

- In stark contrast to the suburban school described above, a large inner-city school in our study is located in the heart of a predominantly Hispanic district of San Francisco. All of the doors to the building are heavy metal and locked. We were admitted into the front door by a big security guard. When we went to register as visitors at the principal's office, the secretary/receptionist joked that we didn't really want to do that, or we'd have to be searched. Instead, she called the teacher, and he sent a student down to meet us. Walking

---

[2]Much of this section is excerpted, with minor modifications, from Lynch and Britton (1993, pp. 59-65).

up the stairs, we could see that despite a new paint job, the graffiti on the windows was still visible. Class enrollment for ChemCom is 41, exceeding the capacity of the old lab room. Except for a few modern conveniences such as an overhead projector, the room could easily pass for a pre-World War II relic. Ceiling titles were missing; the stools looked as if they would fall apart at any moment—in fact, one did break while we were there.

A Classroom Vignette

The following fictitious story describes what Mr. Benson and his students experienced during their first ChemCom unit.

During the first day of class, Mr. Benson distributes ChemCom textbooks and has the students read the opening section, excerpts of which follow (the full article is a page and a half).

Water Emergency in Riverwood: Severe Water Rationing in Effect. Water engineers and chemists from the County Sanitation Commission and the Environmental Protection Agency (EPA) will search for the cause of a fish kill discovered yesterday . . . Mayor Edward Cisko, citing possible health hazards, today announced the shutdown of the Riverwood public water pumping station and cancellation of the "Fall Fish-In" that was to begin Friday . . . Councilman Henry McLatchen described the decision as a highly emotional and unnecessary reaction. He cited the great financial loss that town motel and restaurant owners will suffer from the fish-in cancellation as well as the potential loss of future tourism revenue due to adverse publicity (ACS 1988a, p.4).

After spurring the class to speculate on why the fish died and to debate the merits of Riverwood's response, Mr. Benson explains that the article is the storyline tying together the first ChemCom unit: a progression of fictitious newspaper stories about Riverwood's crisis. Students will investigate possible causes of the Riverwood situation, learn the chemistry needed both to understand the issues and to collect data required to discuss them, and make decisions about water quality problems. The first homework assignment, to be continued throughout the unit, is to scour newspapers and magazines for actual stories on water resource issues (water pollution, water supply, water use). The bell rings.

Students' exiting chatter tells Mr. Benson their curiosity is piqued. But he's uneasy, even a little insecure. After all, today's class was a far cry from his standard introduction to the traditional chemistry course: a wow-'em chemical demonstration show and a lecture delineating chemistry and the scientific method.

During the next weeks, the territory most familiar and comfortable for Mr. Benson is helping students learn the traditional chemistry concepts and laboratory procedures needed to delve into water resource issues:

Traditional Chemistry Topics

common metric units
physical properties of water
types of mixtures and solutions and their properties

molecular view of water (atoms, molecules, compounds, bonds, properties)
protons, neutrons, electrons, ions
solubility of solids, gases; solution concentration
characteristics of acids, bases; pH
common acids and bases: names, formulas, and uses
common ions and ionic compounds: names, symbols, charges
molecular explanation of dissolving solids, gases

Traditional Laboratory Procedures

using a graduated cylinder to measure liquids
filtering liquids with filter paper, funnel, and ring stand
qualitative analysis of aqueous ions: Cl^-, Ca^{+2}, Fe^{+3}, SO_4^{-2}
graphing data and interpreting graphical data
testing solubility of solutes in polar and nonpolar solvents

But the unit also takes students through a wealth of applied chemistry topics never broached by most traditional chemistry texts. In fact, most of this information is new to Mr. Benson:

Applied Chemistry Topics

types and amounts of water usage in geographical areas of the United States
the water cycle: natural purification of water
demand and supply of dissolved oxygen
sources and effects of heavy metal contamination: Pb, Hg, Cd
hard water and water softening
municipal water purification
chlorination of water

Laboratory Investigations for Applied Topics

testing the purity of foul water
keeping a detailed diary of home water use
classifying solutions using Tyndall effects
testing ways of softening water

Moreover, some of the learning activities in ChemCom really put Mr. Benson and his students into uncharted waters. There are the typical yet necessary questions and exercises, called Your Turn, which help individual students check their understanding. But other types of activities are rather novel. Many students relish the little puzzlers, ChemQuandrys. For example, why does it take 450 liters of water to put a single egg on your plate, or 120 liters to produce a 1.3 liter can of juice? In one of the You Decide activities, Mr. Benson has groups of four students pore over the Riverwood articles, separately listing reported facts and questions prompted by them. Subsequently, students advance possible causes for the fish kill and decide whether sufficient information is reported to definitively substantiate or refute each one.

Mr. Benson often groups students for laboratory procedures, but grouping students effectively for You Decide activities is more challenging. To foster energetic, productive group dynamics, he must consider students' general abilities, verbal skills, ability to work cooperatively, etc. Further, Mr. Benson is more used to being an authority than a facilitator.

The unit's culminating activity, Putting It All Together, further exercises students' group activity skills. The activity's topic for this unit is "Fish Kill in Riverwood—Who Pays?" The entire class prepares for and stages a mock meeting of the Riverwood town council. Mr. Benson has groups of students assume the roles of council members, power company officials, scientists, engineers, chamber of commerce officers, and officers of Riverwood's taxpayer association. The ChemCom text suggests numerous specific facts and issues that each group should consider before the meeting. On meeting day, each interest group has two minutes for presentation and one minute for rebuttal. Following the meeting, each group either prepares an editorial letter to Riverwood's newspaper or prepares the group spokesperson for a simulated television interview.

Mr. Benson feels a sense of accomplishment because the students, most of whom wouldn't have taken his regular chemistry class, are enjoying and learning chemistry—and more. ChemCom's integrated treatment of science, technology, and society isn't just a hook to engage students (which it does). It also teaches concepts worthwhile for every person to know—e.g., the capabilities and limits of science and technology; the interactions between science, technology, and society; the pressing scientific/societal issues facing our communities, the nation, and the world; and the possibility of individuals making a difference through their collective actions. In fact, Mr. Benson is already wondering how he can infuse some of ChemCom's features into his regular chemistry classes.

ChemCom's Innovative Characteristics

What makes ChemCom an innovation? ChemCom's goal is to help students become scientifically literate citizens rather than only provide a base of knowledge for studying chemistry in college. Hence, the ChemCom curriculum treats scientific, technological, and societal topics in an integrated fashion. Further, many ChemCom activities require small student groups to make decisions on issues having scientific, technological, and societal factors.

ChemCom designers argue that for students to be scientifically literate citizens, they must understand technological and scientific aspects of chemistry in concert with traditional science concepts. Such citizens need to understand the nature of the scientific enterprise and the interactions that exist among science, technology, and society. The following four (out of 32) major ChemCom concepts illustrate this principle (ACS 1988a, p.xxix).

- Expert agreement upon underlying scientific and technological facts related to a societal or technological issue does not necessarily imply that experts will agree on a particular "fix." Social, political, economic, and ethical values influence the opinions and advice of experts.

- All technological benefits are associated with some level of risks/costs/burdens.

- Individual actions that may seem insignificant considered alone can have major societal and ecological impact when multiplied by similar actions of many individuals.

- Our present state of knowledge about any given societal/technological issue is likely to contain imprecision, inaccuracy, and uncertainty. Society must act upon the best available information with the understanding that additional information may call for subsequent reevaluation of an issue and/or previous solution.

ChemCom units use realistic community issues (e.g., Riverwood's fish kill) to interweave scientific, technological, and societal topics. While discipline-based science topics comprise all or most of a traditional chemistry course, ChemCom also includes related technological and societal topics.

Although less course time is available for traditional science content, ChemCom still addresses most traditional chemistry topics, generally by developing each topic in less depth. ChemCom's design principle is that content is introduced on a need-to-know basis—i.e., chemistry concepts are developed to the extent students need them, along with technological and societal concepts, to understand the units' community issues.

Some notable differences exist, however, between the science topics covered in ChemCom versus traditional courses. ChemCom covers much more biochemistry, organic, and nuclear chemistry than do traditional courses. But it does not teach atomic and molecular orbitals, kinetics, equilibrium, or energy in reactions; and only briefly treats molecular structure.

Finally, decisionmaking activities are integrated into every unit in Chem-Com. Through such activities, ChemCom authors believe students develop the reasoning skills needed to function in a world driven by science and technology. Also, they believe that placing the decisionmaking exercises in a setting of cooperative learning leads students to "own" the content, not struggle with mastering it: By having to teach and present concepts to each other, they "buy into" their own learning. Examples of student-centered, cooperative learning activities include surveys, interviews, simulations, and debates. The developers stress that the teaching approach for ChemCom differs from traditional chemistry courses. In ChemCom, teachers are not central to the classroom activities. Rather, they guide students, who work in small groups in cooperative learning activities. Table 2 summarizes key differences between traditional chemistry and ChemCom.

Table 2.   **Comparison of Traditional Chemistry With ChemCom**

Traditional Chemistry	ChemCom
Preparation	Popularization
Generating knowledge	Applying knowledge
Discipline focus	Societal issue focus
Science on lab bench	Science in the community
Model building	Decisionmaking
Mastery of content	Ownership of content
Individual problem solving	Small group work

Source: Ware (1990).

Elements of ChemCom

The Units. ChemCom is delivered in eight units, listed in table 3. Each is based on societal issues. Eight modules are more than can be taught in one year. (Five was the average number of modules completed by the teachers we interviewed.) The first four units need to be taught in sequence, but the final four can probably be taught in any order. Some teachers elect to drop the fourth unit, go straight to unit five, and then select from the remaining three.

The Topics. ChemCom covers the science concepts listed in table 4. Throughout the course, when a concept or exercise builds on previously learned information, the specific reference is given. This reinforces the original lesson and helps teach the new one. Table 4 indicates where the concepts are introduced, and where they are elaborated and applied. The ChemCom text includes a glossary.

The Labs. The curriculum is about 50 percent laboratory-based, and laboratories are included in the ChemCom text. A typical unit includes five laboratory activities. These are integrated into each lesson to emphasize their relevance to the particular social issue or problem. The laboratory is inquiry-based, meaning that the students are not given the answers, but must find them through analysis and experimentation. According to one ChemCom teacher (Berry 1988), the laboratory setting allows students to learn by doing. It is the appropriate setting to learn the scientific method as well as manipulative skills, data collection, and data analysis.

Laboratory supplies for ChemCom labs are no more expensive than those used in traditional chemistry labs. They are usually obtained in hardware and grocery stores rather than chemistry supply companies.

Table 3   ChemCom Units With Illustrative Sample Topics

ChemCom Unit	Sample Topics
Supplying Our Water Needs	Water quality, supply, and demand
Conserving Chemical Resources	Properties, sources, and uses of chemical resources
Petroleum: To Build To Burn?	Source, uses, and alternatives to petroleum
Understanding Food	Food and nutrition, metabolism, world hunger
Nuclear Chemistry in Our World	Radioactivity, pros and cons of nuclear power
Chemistry, Air, and Climate	Properties of gases, threats to the atmosphere
Health: Your Risks and Choices	Chemistry in human metabolism, threat of drugs
The Chemical Industry: Promise and Challenge	Industrial processes, electrochemistry

The ChemCom *Teacher's Guide* explains how to microsize laboratory activities, an increasingly common practice in traditional chemistry (Gross 1989). Microsizing (scaling down in amounts of sample, reagent, and equipment for a protocol) is a common laboratory modification which has a number of benefits: less time needed to perform laboratory exercises, less expensive, smaller amounts of reagents and samples needed, less storage space needed, and less waste generated. To illustrate the size differences, microsized experiments only require spot-plates and eyedroppers while traditionally scaled experiments require test tubes and pipettes.

Decisionmaking Activities. In ChemCom, students are exposed to decisionmaking in three forms:

- ChemQuandary—a short exercise of 10-15 minutes conducted three to five times per unit, designed to provoke thought and discussion.

- You Decide—similar to a laboratory experiment, but involving no equipment or chemicals. This type of activity is intended as a problem-solving exercise with students working in groups. It takes from 30-50 minutes, and some of the activities involve homework. Approximately five are conducted per unit.

- Putting It All Together—a closing exercise after each unit where students sum up, review, and apply the principles learned throughout the unit. Each of these activities is intended to provide a forum to discuss/solve the societal problem introduced in the unit. They usually take two days of class time and are preceded by individual or group research.

Table 4.   Matrix of Chemistry Topics in ChemCom Units

Concept	ChemCom Unit							
	Wat.	Res.	Petro.	Food	Nuc.	Air	Health	Ind.
Metric (SI) measurement	I	U	E	U	U	U	U	U
Scale and order of magnitude	I	U	U	U	U	U	U	U
Physical & chemical properties	I	E	E	U	E	E	E	E
Solids, liquids, and gases	I	U	E		U	E	U	U
Solutions and solubility	I	E	U	U	U	U	E	U
Elements and compounds	I	E	E	E	E	U	U	U
Nomenclature	I	E	E	E	E	U	U	U
Formula and equation writing	I	E	E	E	U	U	U	U
Atomic structure	I	E	E		E			
Chemical bonding	I	U	E	E		U	E	U
Shape of molecules	I		E	U			E	U
Ionization	I	U	E		E	E	E	E
Periodicity		I/E/U						
Mole concept		I	E	U	E	E	U	U
Stoichiometry		I	E	E		U	U	U
Energy relationships		I	E	E	E	E	E	E
Acids, bases, & pH	I			E		E	E	U
Oxidation-reduction		I		U		U	U	E
Reaction rate/kinetics				I	E	U	E	U
Gas laws						I/E/U		
Equilibrium								I/U
Chemical analysis	I	E	E	E		U	U	U
Synthesis			I			U		E
Biochemistry				I	U		E	
Industrial chemistry	I	E	E	E	E	E	E	E
Organic chemistry			I	E			E	
Nuclear chemistry					I/E/U			

Legend: I = Introduced;   E = Elaborated;   U = Used.

Source: ACS (1993a, p. 571).

In addition to the decisionmaking exercises, drill and practice exercises called Your Turn are conducted approximately nine times per unit. This activity usually involves homework followed by classroom discussion (5-15 minutes). Students work individually. This type of exercise is intended to ensure that students know nomenclature and acquire the necessary skills to use the metric system, balance equations, perform computations, graph, and analyze graphs.

ChemCom Spin-offs

As shown in figure 2, ACS has applied various characteristics of ChemCom to the development of other curricula:

- ChemCom in other languages

- a college course,

- middle school units,

- a "tech prep" program, and

- a videodisc encyclopedia.

Figure 2.   American Chemical Society Projects Related to ChemCom

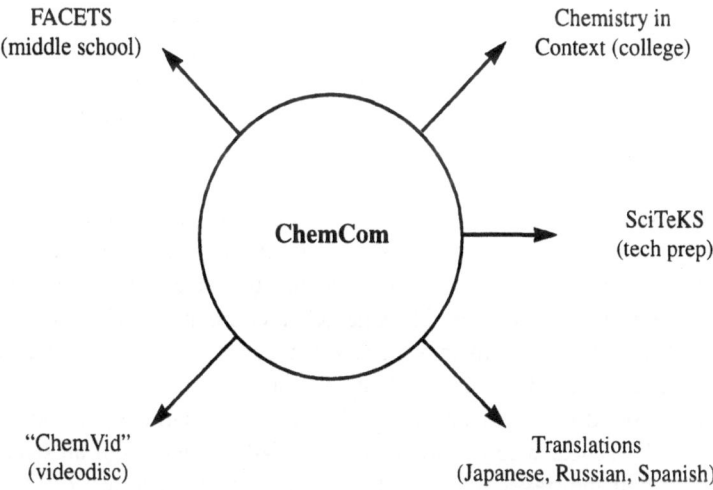

FACETS
(middle school)

Chemistry in
Context (college)

ChemCom

SciTeKS
(tech prep)

"ChemVid"
(videodisc)

Translations
(Japanese, Russian, Spanish)

Related Courses and Materials. Using the expertise of its membership, and working through its Education Division, ACS has developed a college chemistry course based on the ChemCom philosophy. *Chemistry in Context* is a one-semester course for non-science majors which meets the science requirement of many programs. Some colleges are supplementing it in order to use it as the first semester of a general chemistry course. As of early 1996, the text dominated about 30 percent of its market. The second edition will be available in 1996.

Foundations and Challenges to Encourage Technology-Based Science (FACETS) is a modular approach to a diverse set of topics meant for 6th through 8th graders. It is integrated science. As in ChemCom, this curriculum introduces science on a need-to-know basis. Since the program consists of 24 stand-alone units, an infusion model will probably predict the pattern of its dissemination.

The units focus on topics such as the marketplace, the farmlands, the cities, etc. One module based on the suburbs has students propose a new plan for energy use in their school. Another module emphasizing the ocean consists of units that explore the effects of oil spills, what happens to sunken ships and their cargo, and how the shoreline changes.

Anne Benbow, a special assistant to Ware in the ACS Education Division, is the principal investigator for the middle school curriculum project. Emmett Wright, a science educator at the University of Kansas, is the co-principal investigator. Kendall/Hunt is publishing FACETS; NSF supported it's development. There are funds for field testing, but the ACS development office is seeking funding for teacher training.

The ACS Education Division is currently developing a two-year tech prep program called SciTeKS—Science and Technology: Knowledge and Skills. Like FACETS, this is a modular course; like ChemCom, the units teach science and technology based on the need of students to know this information. SciTeKS presents the knowledge and skills required by students intending to go from school to work as technicians in a variety of chemistry—and biology—based industries. Some 20 modules are planned, each one set in a "virtual workplace." The modules consist of text, CD-ROM materials, videos, and computer programs.

The ACS Education Division is also producing a videodisc encyclopedia, "ChemVid," which will contain two- to five-minute full-motion video segments related to science in industry and in the environment. These segments are being selected to enhance the teaching of FACETS, ChemCom, and Chemistry in Context. There will be three separate scripts written around the same basic video spots.

ChemCom Overseas. The ChemCom curriculum is already available in several other countries in their respective languages. The Japanese translation is completed. The MIR Publishing Co. in Russia has produced a Russian version. Russia sponsored an ACS team to visit Moscow and train teachers. ACS sent 20 high school teachers who ran workshops for 140 teachers from all over Russia. ACS paid half the travel costs for the U.S. contingent. The rest of the funding came from school districts whose teachers were on the team. In-country expenses were picked up by the Russians. ACS has also run a workshop for teachers in Siberia. The Russian version of ChemCom was published in June 1995.

ACS has authorized the Spanish division of Addison Wesley, Iberamericano, to develop the Spanish version. To adapt the text to the Latin American situation, the publishers are looking for local examples from Mexico, Chile, Argentina, Columbia, and Venezuela. A teacher training workshop was held in Mexico during the summer of 1994; the Spanish language version should be available in 1996.

◆

Spreading ChemCom

In figure 3, we depict the saga of ChemCom as a river that moves from its head waters across a varied landscape, picks up tributaries, gains momentum, encoun-

Figure 3.   **The ChemCom Stream: Influences and Effects**

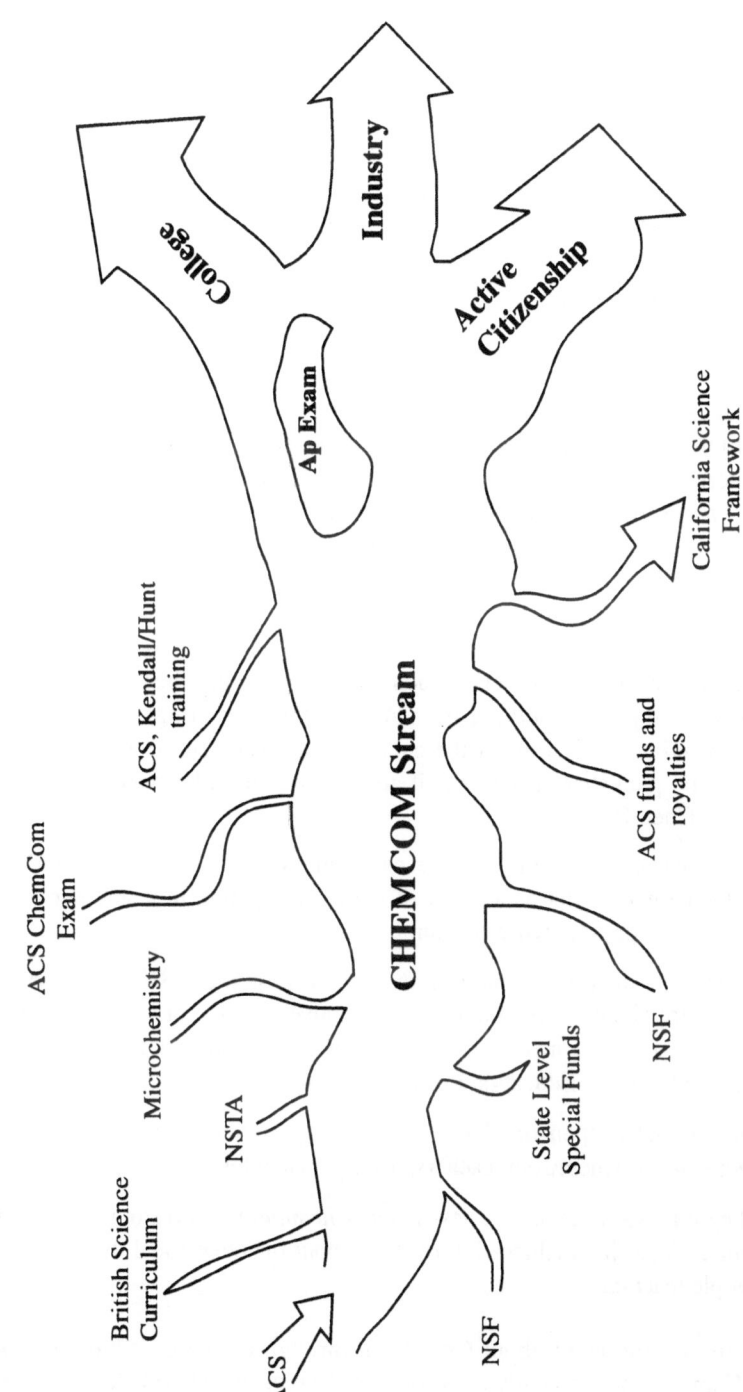

Curriculuar Influences

British Science Curriculum

NSTA

Microchemistry

ACS ChemCom Exam

ACS, Kendall/Hunt training

ACS

College

Ap Exam

Industry

Active Citizenship

California Science Framework

CHEMCOM Stream

NSF

State Level Special Funds

NSF

ACS funds and royalties

Financial Influences

ters barriers, and finally empties into three great reservoirs. In the next sections, we explore the ChemCom river. We sample its waters near the head, follow where it flows, and draw samples along its course to see how they change—i.e., how the curriculum looks the farther away one gets from the original source. We examine some of the landscapes and tributaries along the way—i.e., the growing impact of standards and frameworks on curriculum and ChemCom dissemination strategies. We see how ChemCom intersects the rising tide of tech prep programs and science, technology, and society initiatives. The river metaphor helps us see how myriad seemingly separate but ultimately relevant, topics connect to make a reasonably coherent story about how one curricular innovation was born, grew, spread, and continues. There are useful lessons to be learned on the river.

How People Learn About ChemCom

Teachers are one of the prime movers of ChemCom. The ACS oversees the conduct of three distinct kinds of teacher programs: short-term awareness activities, longer term training activities, and leadership institutes.

Awareness Activities. We asked teachers at schools we visited, as well as those we met at state, regional, and national meetings, how they first learned about ChemCom. National and/or regional meetings of NSTA and ACS were mentioned most often. At the meetings, they may have attended sessions given by a ChemCom teacher team. Some told how important it was to hear discussions among chemistry teachers at such sessions: It helped them understand why people were trying something new. Also at these meetings, they may have picked up print material about the course from presenters or at a booth sponsored by the publisher in the exhibit area. Below are other ways that teachers learned of ChemCom:

- They hear about it from other teachers, many of whom conduct workshops in their own district as a result of their participation in a summer training program of one- to two-week duration.

- A science supervisor conducts an awareness session. Science supervisors also learn about the program from the above sources. In addition, they may learn about it at meetings of their own professional group—e.g., the National Science Education Leadership Association.

- The publisher sponsors "drive in" evening or Saturday awareness workshops run by ChemCom teachers, usually in a hotel.

- They may see or receive a newsletter (*Chemunity News*) from the education arm of ACS that includes information about the course and the activities of people teaching it.

Some people learn about ChemCom through a computer bulletin board, ChemCom Listserv, recently established by two or three ChemCom teachers as a forum for discussion.

Some teachers and supervisors we interviewed commented that it was the repeated appearances of ChemCom on conference programs and in print material that finally prompted them to look into the course more closely. Teachers acted as the agents of change. They brought back information from sessions, went to supervisors and/or administrators, and urged the district or school to undertake a pilot of some sort. In this way, those who were successful in making the plea moved into the adoption stage.

The staff of the ACS Education Division produces a free newsletter, *Chemunity News*, five times a year, which contains information on ChemCom. The staff also make presentations at various teacher meetings. Within the Division's Office of High School Chemistry, there is a full-time staff member responsible for the ChemCom project. Staff of this office as well as Ware, who is now the director of the Education Division, may all run ChemCom teacher workshops.

Training Opportunities. In the transition from awareness to adoption, it is important to see what is happening in the background that facilitates the flow of the ChemCom river (see figure 3). From 1988 to 1990, ACS supported 10-day teacher workshops run, initially, by field test directors at universities around the country. These workshops were designed to train teachers to conduct ChemCom workshops. After 1990, ACS gradually replaced all the college teachers with high school teachers, reduced the workshops from 10 to 5 days in length, and dropped the "training of trainers" aspect of the workshops. As of 1996, ACS was supporting up to 10 of these workshops per year.

Teacher training sessions by ACS and orientation sessions by the publisher, Kendall/Hunt, began when ChemCom came on the market. In five-day summer workshops, teachers learn about the ChemCom "philosophy, goals, rationale, teaching/learning strategies, course content, classroom management models, and specific instructional activities. See appendix C for an illustrative workshop agenda. We quote a teacher enthusiast who put the following message on the ChemCom bulletin board:

> Kendall/Hunt Publishing Company has a week-long seminar for teachers preparing to teach ChemCom for the first time. I highly recommend it. If your school purchases 150+ textbooks, it is free (for one teacher). There you get a hands-on feel for what the course is like, and how to teach it. There were no quizzes and tests, and no one tried to find out what you didn't know about chemistry. It was a very positive experience and I doubt I would have had as successful a start if I hadn't gone.

Contributed funds from the ACS capital campaign and profits from textbook sales continue to fuel ChemCom teacher training efforts. About 160 teachers, selected from the pool of applicants, were trained in eight five-day workshops in the summer of 1995 in Ohio, New Jersey, Texas, Florida, Illinois, Oregon, California, and North Carolina. In 1995, the ACS Board of Directors, recognizing the importance of ChemCom teacher workshops, authorized a sum of up to

$500,000 from ACS reserves to match any external funds raised by the Education Division to support ChemCom teacher development. Another eight workshops are planned for summer 1996.

Leadership Institutes. ACS uses "master" ChemCom teachers to disseminate the ChemCom message to more teachers around the country. ACS initially used 10-day training sessions as a means for teachers to learn how to give in-service workshops to other teachers in their regions as well as how to deal with some of the contents.

ACS trained 237 resource teachers during 12 workshops held in the summers from 1988 to 1990. Workshop sites were spread across the United States in Washington, California, Colorado, Texas, Wisconsin, Ohio, and New York (ACS 1991). About 50 percent of these trained resource teachers taught approximately 7,300 other teachers in their localities about how they could use ChemCom to teach chemistry to their students.

ChemCom "Internetworking"

Recently, some prospective ChemCom users have learned about the program from an electronic bulletin board. ACS initiated its own electronic bulletin board at the time it began to train ChemCom resource teachers in 1988, but teachers did not use the board enough to justify its continuance. ACS attributes this to technical problems, but low use has been a repeated problem in other content areas—e.g., physics, where technological orientation is not a major teacher issue.

In July 1994, a few ChemCom teachers established a bulletin board on the Internet for discussion among ChemCom teachers and advocates. They had attended a ChemCom workshop in 1992, and felt that the bulletin board would be useful and beneficial as a way to maintain communication. It is unmoderated and open to discussions concerning questions, lesson ideas, and labs.[3] One of the list owners' first postings for the new discussion group reads:

> We thought that it may be useful if teachers had an established electronic way to communicate on the subject of ChemCom throughout the school year and beyond. I know that the ACS has trained hundreds of teachers at their sessions at various universities. This will be a good way for those people (of which I am one) to stay in touch.

In addition to ChemCom teachers, the 156 current subscribers to the list include chemistry students and teachers at the university level in the United States and abroad, ChemCom workshop presenters, engineers, chemists, and other high school science teachers. Though discussion started off slowly due to confusion about the list's purpose, teachers eventually began to put out questions about suitable use of ChemCom as a high school chemistry course.

---

[3]Subscription to the list is open to the world. To subscribe, send the following command in the body of E-mail to LISTSERV@UBVM on BITNET or LISTSERV@UBVM.CC.BUFFA-CLO.EDU on the internet: SUB ChemCom yourfirstname yourlastname.

Help!! My school is preparing to pilot ChemCom this fall. There is much discussion concerning who should take this class, what criteria to use in selecting students, and students' perceptions of this class vs. a traditional chemistry class. If anyone out there would talk to me about their experiences, I would be most appreciative. Did your traditional course become nonexistent because all students elect the "easier" ChemCom course? Can students who take ChemCom move into AP Chemistry successfully? Do you screen students going into ChemCom by their math level? What math do you use?

A few teachers in the United States and abroad appeared to find out about the text through the discussion list, as illustrated by the three teacher comments below.

I am trying to learn about ChemCom from participating in this list. Our school system is implementing Tech Prep and we are looking for new courses. I have not seen any of the course materials. I was wondering if someone could get a little more specific about how ChemCom is more practical and beneficial than traditional chemistry courses. What, specifically, are some of the topics/objectives in this curriculum?

I have a shortage of demos and labs that really work *and* make a relevant point. I know that there are lots "out there," but as time goes on, they seem less relevant . . . Where can I get a copy of the ChemCom text (teacher a/o student)? I'm trying to get more "real world" oriented.

Hello, I am from the Netherlands. I'm a teacher of chemistry, chemometrics and quality assurance. I participate in a number of projects about automation in labs. I was looking for chemistry lists and found this one. Can somebody tell me what "ChemCom" is? It looks like chemistry and household? Has the ACS written text on this?

Several subscribers state that they are thinking about adopting ChemCom in their schools and they want more information about it from experienced ChemCom teachers.

Just wondering if any of you has dealt with ChemCom as a "suitable" first year course for those that want to take AP or second-year (college-level) chemistry . . . We will probably adopt ChemCom next year as our first-year, "college prep" text. One of our chem teachers (an older fellow . . .) has expressed reservations about ChemCom being "meaty" enough for those kids that will go on to take AP Chem in a year or two. Also, does anyone know the date(s) for ChemCom summer workshops?

I teach Chem 1 in a magnet program. The program is "a break the mold" high school. I have thought about using ChemCom for some time but I have been reluctant because of what other chemistry [teachers] might say. I thought my students would not be prepared for college. Then I thought about my own education and concluded it is dumb for students to learn about moles. The concept has no relation to their life. I remember when I took my first biology course in college it was a repeat of what I learned in high school. My first chemistry course was a repeat of what I learned in high school. I believe that chemistry should be made relevant to their lives. I also feel that students should learn the process of science and not memorize the periodic table. It seems to me that my students will be prepared for college if I use the ChemCom curriculum. I also believe that they will learn more from it!! Consequently, I plan on using this wonderful curriculum!

The Local Adoption Process

Most, but not all, of the districts we visited went through a two-stage adoption process. First, they run a pilot year in which two or three teachers, usually the enthusiasts who brought the course to district attention, try out the program. At the end of the year, these teachers make a recommendation to adopt or not based on their experience, as illustrated by the following comments from a teacher:

> I attended a two-week training workshop. I became so impressed with Chem-Com that I used my influence in the district to pilot one class. The surveys from the students were so positive that they decided to open up more sections. The following year, the other high school in our district started using the ChemCom textbook.

Once a district makes a decision to conduct a pilot study, it is common for some of the teachers to attend an ACS-sanctioned workshop. A district can apply to send a team to one of the ACS summer in-service programs. Workshop leaders are ChemCom teachers who have also been part of an ACS leadership training program. The classroom experience of the ChemCom teacher workshop leaders and the workshop curriculum they carry out using the ACS guidelines is effective on two counts:

- There is a specified agenda implemented by skilled practitioners.

- The credibility of the claims for the program content and recommended pedagogical strategies is high because the teacher-leaders are experienced.

ACS provides food and lodging at the workshop site and a stipend of $250. The schools are asked to pick up any transportation costs; they often use Eisenhower funds (federal block grant monies to the states which are then dispersed on a largely formulaic basis to districts) for this purpose.

Pilot teachers often conduct awareness workshops for fellow teachers in the district. They are sometimes instrumental in persuading others to take part in a ChemCom in-service program of shorter duration, usually three to five days. In two large districts, two trainer teachers left the classroom for a full or part-time assignment to help teachers implement the program. They returned to the classroom after one year of implementation activities.

We noted that a number of districts use Eisenhower funds to support training for the ChemCom implementation. One Florida supervisor talked about planning programmatic support. Rather than disperse Eisenhower monies over a miscellaneous assortment of in-service ventures, he concentrates the funds to support targeted programs. Each summer, he sends a few additional teachers for ChemCom training. After three years, all the teachers will be trained-provided they accept the invitation to go for training. Teachers can elect to attend or not; they cannot be required to take that particular opportunity.

We found two instances where schools bought the ChemCom text in classroom lots and installed the course without any particular training or pilot stage. The catalog description of the course matched some of the school's objectives. As one teacher said, "Why do you need a workshop that lasts several days? Chemistry is chemistry." While we were not in a position to judge whether the quality of what they taught suffered from the lack of training, we did note that these teachers did fewer of the ChemQuandary and You Decide activities.

Recent changes in high school graduation requirements, occurring in most states, have motivated some ChemCom adoption decisions. States have increased the number of science credits needed, and laboratories must be part of most science courses offered to meet this requirement. Teachers need courses that appeal to more diverse student populations. According to our interviewees, more females and minorities are indeed taking chemistry than would otherwise be the case: "We wanted to get more students to take chemistry, especially females. There are a lot more females taking my classes now that I am teaching ChemCom than in previous years." Another teacher remarked, "In our school, science enrollment has gone way up since we included ChemCom just because it gives them another avenue. I think a lot of students who would really be afraid of chemistry are not so afraid of ChemCom."

After the first year or two of offering ChemCom, a school often changes what prerequisites, if any, it wants students to satisfy as a condition for taking the course. In one district, for example, students initially had to take beginning algebra. Now, however, there is no algebra requirement. The teacher, who was a ChemCom enthusiast, is rather discouraged. He claims the time to get through the modules is almost doubled because he has to stop to teach the mathematics necessary to do the most basic computations.

> I've been teaching ChemCom as a course for four years, actually. I taught bits and pieces of it, I guess, that fifth year before I decided to adopt the program at our high school. It's been a very good experience, although this last experience wasn't as good. I need to tell you that, too. We found out there was a definite math limitation with the program. We've tried to open up this course to all kinds of students and found out that certain kids still cannot get the math that we utilize in ChemCom. I found their grades went down a lot this year because we're bringing in a new kind of student and we thought they could handle it, but their algebra skills, if they had them, were very poor. We found out that they had a hard time mastering the math we had. So we modified the ChemCom course around that and even dropped the math level a little lower than they had in the book.

There is a general consensus among the teachers we interviewed that the recommendations for time needed to do each module or section as specified in the *Teacher's Guide* is very misleading. No one could cover the content as rapidly as recommended. When teachers discover this in the pilot stage, it sometimes prompts them to oppose adoption.

Adoption of ChemCom in some states depends on the course unit status given to it by the public university system for the state or by the state board of education. The issue is whether universities will accept ChemCom as a high school science credit that fulfills their admission requirements. In California, for example, teachers found that the university system did not give credit for consumer courses. The ChemCom title led university registrars to treat it as a consumer course. Many ChemCom advocates in California were able to get this sort of misconception clarified and the course now qualifies, but the course is simply identified on transcripts as chemistry. In New York, a number of teachers received permission from the State Board of Education to use ChemCom in Regent's Chemistry classrooms for the 1994-95 school year.

A special ChemCom examination has been produced by the ACS Examination Institute at Clemson University (this is discussed later in more detail). This allows increased understanding on the part of teachers and college faculty of what students are expected to have learned in a ChemCom course.

The number of ChemCom textbooks sold is one impact indicator of awareness and adoption activities. Ware told us that over 300,000 copies have been sold as of February 1995. (This number reflects purchases for the first as well as the second editions.) That would suggest an estimate of 10,000 classrooms that are using or have used the ChemCom text as a reasonably conservative projection of market penetration. Since schools keep books an average of five years, more than half a million students have probably taken the ChemCom course.

The Story of a Recent Adoption

At the start of this investigation, we had an opportunity to follow how one school district in the South became aware of ChemCom, did a pilot study, made an adoption decision, and carried out implementation. Most of these changes took place within the lifespan of the case study. We could identify adaptations individual teachers made and talk with them about their reactions to the course. A synopsis of this case-within-a-case provides an illustration of how a district became engaged with ChemCom.

This Florida district, with four high schools, serves a university town plus a large rural area with many small villages. There are six chemistry teachers in the district. The science supervisor is a chemistry teacher initially brought in half time to fill the post while he continued teaching. He is now full time in the central office but encumbered with the management of other programs besides science—e.g., drug education, a major voc-tech/sci-tech middle school initiative, and elementary science. He has little time or energy to devote to curriculum for the high schools. Having come from the chemistry group, he knows all six instructors well and characterizes them as strong teachers.

In 1991, two of the six teachers attended a ChemCom awareness program at a regional NSTA meeting. Experienced ChemCom teachers led the session. The two teachers also talked with other ChemCom teachers who made presentations during the NSTA meeting. They went home thinking that the course could attract more students to chemistry, and went to see the supervisor to persuade him to let them try it. Given that Florida had recently increased its laboratory science requirement for graduation, they thought ChemCom might make another choice available to students. In any case, the supervisor agreed to a pilot study the following year if the pair would attend a training course; the district would provide some support for this training.

Upon their return from the summer training session, the two chemistry teachers undertook to recruit students for pilot classes. They also went to each of the high schools to share information about the course as they experienced it first in the summer training program and then later in their own classrooms. Their intent was to share their enthusiasm and recruit the interest of fellow teachers.

All six teachers agreed that ChemCom might serve a constituency not ordinarily attracted to chemistry. They thought the issues orientation of the course would be motivating to students, and they felt the math requirement was less demanding than in regular chemistry. Meanwhile, the two teacher enthusiasts tested the students in the pilot classes. On the basis of their experience and the performance of the students, the two teachers argued for adoption of ChemCom. The district agreed and mandated a ChemCom option for each school. It is not clear how many of the chemistry teachers were involved in and actively supported the decision. In any case, there does not seem to be any objection. All agree that the course provides access to chemistry for students who would otherwise be doing their best to avoid the subject. In addition, the course satisfies some of the pressure from parents as well as students to make school work more practical for students. For the science teachers, the task at some schools was to inform and persuade counselors to schedule students into chemistry. To help garner student interest, they also put out circulars explaining how ChemCom was organized around important practical topics.

In the meantime, the science teachers in the most rural of the four high schools decided that all students in the school would take chemistry. ChemCom provided a satisfactory alternative to the standard chemistry course. For one year, ChemCom was the only chemistry offered at the beginning level. In the following year (1995-96), the school added an honors chemistry class for students who were better prepared in mathematics. (They reported that incoming freshman had better mathematical skills thanks to a strong focus on mathematics in the contributing middle schools.)

In interviews, the two teachers of the rural school report that ChemCom's applications thrust has crept into the honors course. So, within the first year, some interaction between courses was identifiable. It is a form of adaptation that can bode well (or not) depending upon how it is done. This fusion of contents was more apparent for teachers who did not take the training course but who are nevertheless competent chemistry instructors.

It was soon apparent that no one could get through all eight ChemCom sections in one year, so the schools began to choose which they would cover. That varied among the schools in the district based in part on their constituencies. So the biggest and most easily identified adaptation of the course took place according to the interests of teachers, the kinds of students they served, and the fit of the topics with other items in the curriculum. One school, for example, skipped the nutrition unit since it already had a strong health program that did the job well enough. Another dropped the energy unit on the grounds it was too complex. These decisions rested with the individual schools rather than with the supervisor or central administration.

What about training for the other teachers? For implementation of the course the science supervisor provided support for the remaining four chemistry teachers to attend a summer program. He urged them to go in the summers of 1993 or 1994. To date, he has four ChemCom-trained people. The two instructors who have not attended a summer program are highly competent, experienced teachers (with recognition to prove it). They are too occupied in summer with other agendas to have time to attend the ChemCom training: One is a Woodrow Wilson fellow; the other is completing a doctorate in chemical education. In addition, they feel the chemistry per se presents no problem for them. These two teachers admit, as have other competent chemistry instructors, that they might be more enthusiastic about some of the activities if they had taken training.

The district in this mini-case is a late adopter. Thanks to the persistent efforts of two teachers (the change agents) and the happy fact that the newly appointed science supervisor is himself a chemistry instructor, the ChemCom seed fell on fertile ground. The fact that no additional equipment demands for the course had to be met made it fiscally acceptable. After adoption, the student case load for chemistry almost doubled. More and different students enrolled. The teachers are proud of that. In the next section, we look at what happens to the course as it moves out into classrooms.

◆

ChemCom in the Classroom

What were key aspects of teachers' implementation of ChemCom in their classroom and students' views of the course? Teachers' beliefs about the purposes of chemistry instruction and their professional backgrounds influenced their approach. Many teachers added traditional chemistry topics that ChemCom does not include, but also tended to incorporate some of ChemCom's features into their standard courses. Teachers employed varied criteria for deciding which ChemCom units to cover. Most teachers saw room for improvement in the ChemCom laboratories. For their part, students generally praised the relevance of ChemCom to their lives.

Influence of Teachers' Backgrounds

Differences in teachers' backgrounds accounted for some of the curricular modifications and instructional decisions we observed. Among the people we interviewed, some could be characterized as primarily content-centered; i.e., they are teaching chemistry to students. Others were more student-centered; i.e., they are teaching students chemistry. Moreover, teachers vary in their science content background. Some we interviewed were chemistry majors, and chemistry was their preferred subject for teaching. Others had another science as their preferred and best-prepared subject, e.g., biology or earth science. Taken together, these two aspects—view of the content and perceived role of the teacher—provide a useful framework for interpreting teacher comments and changes in the curriculum. We interpret adaptations and many comments reported or observed as reflections of the differing pedagogical and curricular perspectives of the interviewees. The following teacher comment illustrates our premise.

> We've modified a lot of labs. We're working on a lot of things that are different. I think every teacher does that, because the course is open-ended. So, based upon your own expertise and your own background, you tend to do things a little differently. The other teacher here and I are different in the way that we approach things. We stay about the same as far as material is concerned, but we don't teach it exactly the same way. But that's fine, because he comes from more of a biology background, and I have a physical science-type background but I have a strong interest in biochemistry; that's why I like teaching nutrition the best.

While most of the people we interviewed have some enthusiasm for ChemCom, the non-chemistry—major teachers are on the whole more enthusiastic and less likely to meddle with the content. Chemistry specialists, for their part, are more prone to complain about content omissions. Their ChemCom classes are beginning to suffer from "creeping mathematization." On the whole, both groups view the applications or relevancy features of the course favorably, but they seem to differ in how much time or effort they are willing to devote to that aspect of the program. Chemistry majors discuss the material but are less prone to pursue any of the questions in depth. They feel there was not enough time to cover essential chemistry concepts. The town hall meeting simulation in the water unit, for example, was more likely to be run in full form (including, in two sites, a visit to the town council) by the non-chemistry—major teachers. However, the following chemistry specialist emphasizes student relevancy and active participation in social issues based on problem solving, but finds this emphasis problematic. "ChemCom is so slow, and they're so tied up with all these social things, to try to squeeze in the chemistry is really difficult. My students are really tuning out like it's a social studies thing, and I'm having a hard time with it."

Mixing ChemCom and Traditional Chemistry

Many of the teachers interviewed teach more than one kind of chemistry course, i.e., ChemCom and another chemistry course, typically an honors or traditional chemistry course. It appears that, over time, more and more mathematics creeps into ChemCom from the standard chemistry, while at the same time some of the issues explored in ChemCom migrate into and enliven the standard course. So we are seeing some blending of courses over time. As a rule, this form of adaptation does not take place in any major way until the second or third year a teacher does ChemCom. It does not happen for all teachers; one person, for example, said he did the course just the way it was written. He told some ChemCom trainers that he did not have time or interest in changing anything, and that it was not his job. The trainers remark: "This teacher told us that his school had bought the materials, and he shouldn't be expected to supplement them. In effect, he was saying: 'If we pay the money for the materials, it should all be there.'"

Most teachers in our sample have some favorite topics that they feel no chemistry course should be without. Teachers add these topics as supplements and are prone to spend more time on them. One of the teachers in our sample, a ChemCom trainer and chemistry major, loves energy and prefers to substitute a unit he developed for the energy unit in ChemCom. His unit is extensive, fairly technical, full of mathematics; and requires more instructional time. It was not being well-received by the students during the year of our study, although it went well the previous year. In fact, the teacher notes that many of the students this year, in contrast to those in the previous three years, simply "could not care less about any of the issues." (Our observer reported that six students had their heads on their desks, and two were doing something unrelated to the instruction.) The teacher was puzzled, frustrated, and worried about what was happening to young people.

Before our visit with a third year ChemCom teacher, she told us that Chem-Com was the lowest of the chemistry tracks in her school and that we shouldn't expect too much of the class. When we got to the class, the teacher told us that she was not going to be teaching out of the text that day. Instead, she was going to be supplementing the text with a section on orbitals, which is not taught in ChemCom. Earlier, in talks with teachers at a training session, several teachers also mentioned the lack of in-depth discussions of ions and orbitals, and the fact that they elected to supplement the curriculum.

While ChemCom trainers will occasionally discuss their own methods of supplementing (with videos, extra labs, etc.), they try to encourage new teachers to trust the curriculum's own approach. One trainer at the Los Angeles training in summer 1994 gave this advice to teachers who were about to use the text for the first time: "You will always have the impulse to supplement the text, especially if you have taught other chemistry courses. Even though it might at times be difficult, you have to trust that the text will cover the material that you want to cover."

We also found evidence that teachers incorporate aspects of ChemCom into standard chemistry courses. A teacher whose course load included three AP Chemistry classes and one ChemCom class was profoundly affected by the effect of ChemCom's different pedagogical approaches on students. He saw students, whom he never would have predicted would make an effort, become engaged and experience successes. While he did not have motivational problems with AP students, he had become convinced that some of them would elect to make greater strides if he incorporated some of ChemCom's features, such as real-world applications and contexts for chemistry topics, and collaborative group work.

Deciding What to Drop or Add

None of the teachers in our sample felt they could cover all eight units, so their courses differ according to the units they exclude. Indeed, the ChemCom textbook states that teachers should cover chapters 1-4 and as many of the other chapters as possible. One Florida school omits the nutrition and nuclear units, each for a different reason. The school has an extensive health program in which nutrition is thoroughly treated, so it could be dropped from the course. The nuclear unit requires mathematics and concepts felt to be more appropriate for the advanced or the honors chemistry, so it is taught there. A California teacher dropped the petroleum unit after the second year because, in a student survey, it proved unpopular:

> The first year I basically followed the book to learn what was in it, though I was unable to follow the time line it lays out. I did find, however, that I was teaching faster than my colleagues in other schools. I taught five and a half sections out of eight, whereas they taught about four. Perhaps this may have been because I skipped things which I did not feel necessary to teach. I did chapters 1-4 and 7, thus putting together the food and health sections.

We found examples of ChemCom modules being used by some teachers as supplements in other courses. In two sites, we found *ChemStudy*[4] labs added to or substituted for ChemCom labs. A teacher at one of these sites explained: "We use more labs because we go through five modules. I don't believe you can go through all eight, but you need more lab work then. You need to really augment the five units that you do teach with more lab work, I think. So we have gone and taught more of the kinds of labs I brought in from traditional chemistry."

Whether or how much to supplement the curriculum is a question that inspires a variety of different answers among teachers. Some, including those who are fans of the ChemCom approach, consider the thematic organization of the book as flexible and perfectly suited to supplementing. In fact, they argue that this flexibility is one of the strong points of the curriculum.

For example, several teachers mentioned that the use of newspaper articles in ChemCom frequently encourages students to bring to class questions and concerns that they have heard or read about. This rarely happens in an honors course. This teacher articulates why flexibility is valuable:

---

[4] *ChemStudy* is a popular traditional chemistry textbook.

Anything taught in isolation is less than it could be otherwise. We should always seek relevance in the discussion of any topic. Some fear that "digression" is equal to disorganization, the "loss" of a planned lesson. On the contrary, this is where we need to go. If a discussion on the chemistry of water leads to other interesting pursuits, all the better. If the students fail to attach significance to their learning, or are motivated solely by grades and achievement measures, their learning truly suffers. I think the STS [Science, technology, and society] approach is valuable because it gives us a framework to teach meaningful chemistry. We can certainly expand on the ChemCom curriculum as we each see fit.

About the Labs

There seems to be general agreement among the teachers in our study that the labs are exercises—i.e., prescriptive—leaving little or no latitude for investigation. While the majority of teachers found that disturbing and make changes or substitutions (often drawn from *ChemStudy*), not all feel a need or desire to change the labs.

In addition, teachers report that some labs do not work. For example, several teachers had to modify the milk lab. This was attributed to poor choice of reagents or quantities. Teachers suggested changes to remedy these problems, some of which are illustrated below.

In general the labs are OK. We're adding extra work to emphasize controls, for example, adding an indicator to the filtrate. They have only one control in the lab we're doing right now. We added proteins. All in all, we had three levels of control. But we really haven't done that many labs. The only other lab we did was the foul water lab.

Also, I didn't really think the support book was that clear. Mixing reagents was one example. I haven't been that impressed with the level of detail. It's really sloppy in some places. They ask you to mix cupric sulfate and hydrogen sulfate without telling you the amounts. Sometimes ChemCom labs are logistically difficult.

Also, I would like them to do more in the biochemical area and the structure of protein and its functions. You could talk about the fact that sugar is water soluble—bring it into biology at the chemical level. We've already covered polar and bipolar, but we didn't do plastics. We started with the hydrocarbon structure—oxygen-hydrogen—There is a clear relation to fats-carbohydrates.

ChemCom laboratories have been a topic on the electronic bulletin board. One teacher inquired, "I am looking for a way to do a reaction rate in a different way. Any suggestions out there?" There were many responses to this inquiry. Here is a college chemistry professor's response:

While reading your request for a different way to do a rate of reaction experiment, I had a brainstorm. I don't know if somebody has already done this, but it seems like a simple experiment. It is based on the "Volcano" demonstration we do using hydrogen peroxide, potassium iodide, and soap in a graduated cylinder. We usually just "throw" the ingredients together and watch the result. One could, however, carefully place a constant volume of 3 percent hydrogen peroxide in the cylinder. Add a constant amount of soap. Then time how long it takes for the bubbles to reach the top of the cylinder after placing a constant volume of KI solution into the cylinder. One could vary the concentration of the KI and see the effect. Or, if you can get more concentrated hydrogen peroxide, you vary its concentration.

Note: I have not tested this. It was just a thought. During our demonstrations, we use 30 percent hydrogen peroxide because it gives very rapid results. If you are able to get 30 percent hydrogen peroxide, *BE CAREFUL.* It is a strong oxidizer. Be sure to wear gloves and beware of the heat generated when picking up the graduated cylinder. Also, be sure to place the cylinder in a pan to catch the overflow. One nice aspect is that the used chemicals can be diluted with water and poured down the drain.

The laboratory exercises appear in the book embedded in the text rather than in a separate manual. Teacher response to this format is mixed. Some complain that the placement of the lab is not close enough to the text material. Others worry about the textbook getting spoiled by chemicals since it has to be available on the lab desk as the students work.

The third edition of ChemCom, which is now being edited by Conrad Stanitski of the University of Central Arkansas, is replacing most of the labs with small-scale versions.

What Students Say About ChemCom

Students' reactions to ChemCom allow us to observe their growth in understanding of both science and science's applications as a result of the course. The following samples come from responses to questionnaires administered by a ChemCom teacher to her students. Many of the students' remarks reflect a new awareness about the importance of conservation from a chemical perspective. Many of these students didn't get good grades in the course, but their responses show that they still learned a lot about chemistry. These comments reply to the question: "How has taking chemistry affected your life this year?"

By taking ChemCom, I realize how important science is. I became a scientist in my own right and thought about important issues that did not seem that important before. One example is the nuclear power plant. I never knew what that was until our discussion this year. Even when I go home, I catch myself trying to recycle and being aware of our environment. I'm glad I took this class because not only will it help me in college, but I learned some things that will stick with me forever (male student who received a C).

This year I've learned more about the environment and ways to save it. I feel as though in the future when my generation is faced with having to find a better energy source, I'll have a little headstart in knowing what's best. I've also learned that there are so many options if we use them in time. Through science we learn how to make our world a better place (female student who received a B).

Chemistry has been a real challenge for me this year. But I feel that some of it got into my thick [head] like chemical names, formulas, and writing out formulas, balancing equations, making things blow up in labs. All in all chemistry this year was great. I enjoyed it a lot (male student who received an F).

I've learned a lot this year, from balancing equations to recycling. I got more out of this class than any other science class I've been in. We've covered so much stuff that I will see in my everyday life. Like nuclear power, gases in our atmosphere. I have a better understanding for what goes on in the world around me (male student who received a C+).

Another teacher asked his ChemCom students to respond to a short survey of statements about the chemistry they learned. The results of this survey include the following:

- 100 percent of his students believed that they learned chemistry that is relevant to them;

- 46 percent said they applied chemistry concepts learned to things outside of class;

- 85 percent would take ChemCom if they had to choose a chemistry course all over again; and

- 77 percent enjoyed ChemCom labs and said that these added to their understanding of chemistry concepts.

◆

ChemCom in the Curriculum

ChemCom primarily is designed for use with college-bound students who are not planning to be science specialists. But the course frequently is offered for college-bound students who *do* emphasize science, and many schools offer ChemCom for students who are *not* college-bound. Can ChemCom prepare the science specialist for AP Chemistry in high school or general chemistry in college? Who should take ChemCom and who should take standard chemistry, or should ChemCom be the course for everyone? We found some very different answers among teachers and college educators, many of whom gave us impassioned and thoughtful responses based on their experiences and beliefs.

College-Bound Versus Non-College-Bound Students

With today's increasing laboratory science requirements for graduation, one of the common courses taken to meet the science requirement is chemistry. Students who ordinarily would not have chosen to take a chemistry course in the past are now often obligated to do so. This group includes not only non-science—oriented college-bound students, but also non-college—bound students. The standard chemistry course, frequently perceived as too abstract and difficult even by science-oriented college-bound students, is daunting to non-science-oriented college-bound students who are not motivated by the content or sufficiently facile in mathematics. For non-college-bound students, the traditional chemistry is overwhelming and to be avoided at any cost.

Students who take ChemCom come from a mixture of academic, socioeconomic and ethnic backgrounds. ChemCom's original intended audience consisted primarily of those students who were college-bound but not planning to study chemistry in college. Although ACS still maintains that this is the intended audience for ChemCom, the ChemCom constituency is much wider. According to our interviewees, the course draws students who may enroll in technical schools after graduation as well as those going directly to work. It has some immediate, real-world pertinence for these students. The role that ChemCom plays in developing good citizens has the potential to apply to students of all levels, but the goal is to attract a broader array of students than is typical of chemistry enrollments. O'Brien (1990, p.3) cautions:

> Although ChemCom makes lower mathematical demands on students than a conventional chemistry course, both the conceptual complexity of STS [science, technology, and society] issues and the process of the decisionmaking activities require mental abilities and social maturity typically expected of college-bound students. ChemCom was not designed for students with below-grade reading abilities and/or significant motivational problems. ChemCom will not be likely to achieve its potential and may even prove to be counterproductive with such students, unless teachers are willing and capable of making considerable curriculum changes. This message is not being sent forcibly enough to some school administrators, guidance counselors, and science department heads and, as a result, some ChemCom teachers find themselves in the unenviable position of curriculum-student mismatch.

The fact that more students of varying abilities and interests are taking chemistry today affects how schools are importing ChemCom into their chemistry programs. As a result of its lower mathematics demand, coupled with its higher applications context, school districts, administrators, and even science teachers often regard ChemCom as a course for less able or less prepared students. Many have adopted it to fit specific needs, directly or indirectly indicating that it should be taken by lower ability level students, specifically poor mathematics students. So in some cases, it has a status problem, as the following teacher—and others—point out:

The designers of ChemCom make it very clear that it is *not* supposed to be a "lower level" chemistry course. But that's what it is in our school. We have grade-weighting, and ChemCom, while not classified as a basic-level course, does receive a grade-weight one point lower than regular chemistry and two points lower than honors.

In some schools, ChemCom is the first course in chemistry for all students, i.e., the program is "detracked." Sylvia Ware (1991, p. 3) adds that ChemCom is used as the *only* chemistry course in some schools: "Since ChemCom fits for so many kids, the approximately 90+% who do not choose a science major in college, it may be the best course if someone can offer only one."

In other schools, tracking in some form predominates. There are levels of chemistry differentiated on approach (i.e., knowledge of immediate use versus banking it for an indefinite future use); mathematical demand (high quantitative orientation or strong conceptualization); and social cliques (i.e., are your friends taking it?)

There are many students enrolling in ChemCom who are at the lower academic end of their cohort. One teacher explains that his typical ChemCom student is in the bottom 40 percent of the graduating class and that a few of his students are at the very bottom. In detracked schools, when students were asked why they took ChemCom, it turned out that they didn't know about any other chemistry when they enrolled in it. In tracked schools, they took it simply because they needed another science course for college, and ChemCom was the lowest ranked—and presumably the easiest—that was available. These tensions related to tracking may increase if the national science standards are taken seriously, since they call for heterogeneous grouping.

But the low end, nonrigorous view is certainly not unanimous, evidenced by these remarks from ChemCom teachers:

> I think that saying the abstract concepts and mathematical rigor is missing is incorrect. You can always supplement this material on your own. What is missing from most other chem texts is the application, the scope . . . I'd just as soon toss those other texts than use them for first year kids . . . and I teach chemistry to kids at a school where 85-90 percent of the students go to four-year colleges! I know several ChemCom teachers that also teach AP Chem . . . no problem, according to them!

> When we first started ChemCom, we were much surprised that the course quickly gained a reputation for being *harder* than our intensive chemistry course.

> Our school offers both courses, honors chem for the whiz kids and ChemCom for the not-quite-sure; they are all getting a much better science education because they are in courses that challenge but don't overwhelm them. We are not a "one flavor" nation in which one course fits everybody. Diversity benefits all students!

Influence of ChemCom's Mathematics Requirements

It is quite probable that the lower tracking of ChemCom in many schools is heavily influenced by the lesser amount of mathematics required in the course. Schools often screen students' mathematics capabilities or use the grades from previous science classes in order to determine in which level chemistry course students should enroll. Counselors and teachers encourage students with more mathematics background and/or good grades in previous science courses to take standard chemistry, whereas they refer more "weakly" prepared students to ChemCom. In the second year in the Florida case, a teacher reported that "more students are taking regular chemistry because they are coming in with better math preparation now."

Some teachers believe that less emphasis on mathematics in the course is a plus, as illustrated by this comment: "There is much less math in ChemCom— hardly any stoichiometry and less equilibrium constants than one would expect in a traditional chemistry class. I think students feel that they can be successful in the ChemCom class because there isn't such an emphasis on math abilities." Another teacher states: "The standard chemistry course is extremely math oriented, doesn't relate to students' lives or interests, and is not appropriate for the majority of our student body." Finally, two teachers and a university educator participating in the ChemCom bulletin board note:

> Simply because a course isn't mathematically rigorous, or taught in a pre-scribed norm, doesn't mean it is less or lacking. The problem seems to be that educators today who have been trained in one way can't adjust to the concept of new and/or innovative ways of teaching science.

> I certainly don't equate rigor with mathematical content in any sense. Chem-Com can be taught in a thoroughly rigorous fashion, though it's not terribly mathematical. On the other hand, ChemCom can also be an easy (though painfully ineffectual) way out for a poorly educated teacher.

> Being effectual as a teacher is independent of the material being taught and more a reflection of personality as well as commitment. The previous comment indicated lack of mathematical rigor as an indication of the weakness of ChemCom. I have seen Ph.D. chemists ineffectively teach "watered down" and rigorous courses. It was more their attitude toward being a teacher, not the material being taught.

Other teachers feel that students cannot afford to learn chemistry without more exposure to mathematics, and these teachers either supplement ChemCom with problems from traditional chemistry or caution against using ChemCom altogether.

> Our school teaches the traditional honors chem class and AP Chemistry. We have a physical science course that teaches half a year of chemistry that is basically math free. I prefer the mathematical rigor of a regular chemistry course

and think ChemCom to be a step backward when used for the traditional academically tracked student. I have seen the text. We need to challenge our youth, not make things easier and watered down.

Influence of Students' Peers and Parents

ChemCom's goal of creating a "chemistry for all" course has met several barriers with respect to other hierarchical issues in the school. In addition to the label given to ChemCom by the grade-weighting system in some school districts, students tend to select courses that others in their peer group take. According to one teacher, some of the "fast track" kids signed up for the ChemCom course because they were attracted by the social and environmental focus of the class, but after looking around and seeing that most of the students were "low track," most transferred out within the first week.

Whenever schools adopt a new curriculum, administrators need to assure parents that the new course will adequately fulfill college preparatory requirements. A few parents in one school expressed concern about ChemCom's ability to meet the college entrance requirements in California.

A Terminal or Preparatory Chemistry Course?

Some teachers believe ChemCom students can go on successfully in AP Chemistry courses or college chemistry. Sample comments include the following: "ChemCom students can go on successfully to take more advanced chemistry. It depends on their math background, which is true for students taking more traditional chemistry as well." "There's enough chemistry in ChemCom that motivated students can do well in a traditional general chemistry college course." "I teach Chem I, Honors Chem II, and AP Chemistry in a private school, and while I'm not thrilled with the ChemCom program, I do think it provides an adequate prep for AP."

But the following teacher disagrees quite strongly:

> As an AP Chem teacher and a traditional chem teacher, I believe the ChemCom program has some serious deficiencies—until the chem achievement test and the AP test change, I can only see the program being used for the non-science-oriented student. I have taught the course (grudgingly) for three years—both the original and the new version. While the topics may be PC, the treatment of the material is superficial at best—let's abandon politically correct topics and teach a *real* chemistry course with measurable objectives!!

Questions about whether ChemCom provides adequate preparation for doing chemistry in an AP course or in a first year college chemistry course occur often. Ware says she has no evidence to the contrary. She also points out that the greatest predictor of success in college chemistry is not the chemistry taken in high school, but the mathematics. She adds that much of the chemistry learned in 10th or 11th grade is forgotten by the time students get to college chemistry.

She also indicates that some AP Chemistry teachers are using ChemCom to supplement the more traditional text used for the second year chemistry program.

Some teachers feel ChemCom prepares and induces students to take other science courses in high school that they previously would not have considered:

> I am a former engineer/chemist from NASA [National Aeronautics and Space Administration]. Initially I was shocked at the scarcity of info in the text; now I've learned to approach the course with emphasis on qualitative learning first, then quantitative (mathematical) learning. This approach works extremely well. I also deal with social and political repercussions from science/technology. My students leave and go on to take "regular" science courses because they feel confident instead of scared. They say I make science fun and approachable. They "can really do the math." There is a long-standing phobia out there about science and math. A lot of my colleagues—academic and professional—have taken pride in being part of the elitist scientist circle.

> Unfortunately we can't afford to continue along this path, because more and more nonscientists will be making technological decisions. They need to be science literate. We (science instructors) need to be flexible in achieving that goal. I have not compromised my standards; merely found an alternate path to achieving a necessary goal.

The following three teachers argue that, more important than preparing students for future classes, ChemCom is preparing students to relate to their world:

> Unfortunately, I teach in a system that clings to the notion that traditional high school chemistry is somehow essential to preparing all students for "college." My own experience leads me to conclude that this is only true because of the nature of introductory college courses. The ChemCom curriculum, I am convinced, is actually better preparation for living and further learning.

> ChemCom makes students better consumers as well as gives them an understanding of how the importance of chemistry relates to the community, the nation, and the world. There is a heavy emphasis on cooperative learning skills and decisionmaking which would be important for anyone entering the workforce. ChemCom is setting up students for understanding chemistry and how it relates to their world.

> I am a second year teacher of biology and chemistry. I am truly concerned about what I should be teaching. My administration tells me to wait for the new curriculum guidelines coming down the pike next fall. I have had several students I taught last year stop by and tell me that I helped them with what they are learning in their college chem classes. Which made me feel good. But I had to ask myself, "What was I really preparing them for? To pass a college chemistry class?" I know what I taught them won't greatly help them in the real world. Where is the middle ground?

A Passionate Internet Debate

A very intense debate is transpiring on the ChemCom bulletin board. It further addresses several questions we already have discussed:

- Should multiple chemistry courses be offered so all types of students study the subject?

- Should high school chemistry courses, including ChemCom, primarily prepare students for college chemistry?

We provide some of the bulletin board discussion without much explanation, editing, or commentary, because individual statements often contained several interconnected points: Dissecting them could change their overall meaning. Further, we found it intriguing to see the lively interplay among participants' comments.

Chemistry for the Elite. One teacher writes:

Most high school students do *not* attend college to become chem majors . . . nor to take more than the basic science requirements. Most students do not get exposure to chem in high school because chem is usually a college prep course. So, can we create a course that allows more students to continue taking chem without being scared off . . . or is chem still for the "elite" taught by the elitists. Physics is in the same boat . . . but we don't have a quality program like ChemCom to offer. We offer both chem and ChemCom.

Following is the response posted by a physical organic chemist:

Science is not an elitist course!!! Unfortunately, not all people are born with the same talents. This does not mean that courses have to be offered at all talent levels, though. In universities throughout the nation, courses are being taught at sub-university level to remove deficiencies from entering students not only in chemistry but mathematics, English, etc., as well. Those of you responsible for preparing these students during high school are in large part failing. If curriculum such as ChemCom, or whatever it's called, are responsible, remove them from your classrooms. These students, university bound or not, face a world that requires far greater knowledge of science and mathematics than at any time in the past. They do not need to be prepared to be scientists, but they need a solid foundation in science; at the high school level, that means rigorous courseware, so that they can intelligently analyze coming public policy issues regarding scientific issues. A case in point is the "global warming problem." The truth is that scientific evidence is very thin on this issue, but many high school students accept this as fact. So teach science and math and damn the PC.

A teacher replied to the chemist:

Agreed . . . science is *not* "elitist" . . . The way science is taught may be. Complex science questions are solved by those who have succeeded in their studies of science. However, if we leave "science education" just for those who can or plan to follow a science course of study, we risk creating a "scientifically" illiterate public. This can happen when we create a science curriculum that acts as if science was only for those who can handle college prep courses. The danger in creating a scientifically ignorant population is that they will not understand nor support the need to do pure science, and solve complex problems.

From what I read . . . we have done just that . . . filtered out the population . . . leaving science for those who can handle the math and the rigors. If we are to improve the literacy of the citizenry . . . there needs to be courses designed for both those moving on to postsecondary study in science and for those who are not . . . All of us are moving into a world dominated by science and technology problems and solutions. There is room for both ChemCom and regular chemistry. And, if at some point in time, someone without the background should find themselves in a college chemistry course . . . hopefully, they will find encouragement and not a "survival of the fittest" atmosphere . . .

A college chemistry professor responded to this teacher comment:

Has she hit upon the cause of some of our problems at the college level? We require high school chemistry as a prerequisite for our first year General Chemistry course. We are finding, however, that this requirement is increasingly meaningless. Students with good grades from good high schools can be very glib about certain topics, but totally ignorant about the basics of definitions, nomenclature, measurement systems, balancing equations, and stoichiometry. There is definitely a place for a course to make future voters (taxpayers, consumers) a little sensitive to the working of the "scientific mind" (whatever that is). However, such a course should be clearly labeled as a non-college-prep course. Personally, I love teaching the Introductory Chemistry for the students who never got it in high school, but I hate identifying those same students when they are misregistered for the General Chem course. Frankly, I'm waiting for a class action lawsuit charging false advertising.

Another teacher commented:

In this discussion, everyone is talking about the suitability of chem for college prep, but isn't that what almost 80 percent of our students are today. Everybody wants to go to college (although most don't know why). The fact is, most kids and some of us teachers don't know what we will be doing 15 years from now. They need as broad and thorough a preparation as we can give them. Math is one of the tools of science, and I personally don't believe that you really have much of an understanding of chemistry without some math. Unfortunately, it seems that many of my students have a fear of math to the point of a phobia. Perhaps if we can help them overcome that, they will begin to appreciate both subjects more and be better prepared for whatever future faces them.

Chemistry for the Majority. An inner-city school teacher noted:

Check out the world of the big cities. I have taught in a high school with 800 freshmen and 89 graduating seniors. This does not sound like 80 percent desire to go to college. There is a need for a course such as ChemCom. Try doing an honest job with it. It is not a kiddy course. I have two teachers using it now. Much of the material seems above the students' ability to do any independent study.

Another teacher responded:

Agreed . . . we don't know what students will be doing 10 years from now. But does that mean that the only meaningful chem course to take is *only* the college prep one or the "watered" down first year chem course? Is taking *no* chemistry better than taking ChemCom? Some of these questions are as old as science reform. Do we design our curriculum for one group to the exclusion of another? And what are the human value judgments we are making in this whole discussion? Is there a bigger picture we are not seeing? And doesn't this argument spill over into most of the other disciplines?

And another teacher commented:

I believe it better to have students take some chem than none. I also agree that ChemCom is (1) not a high-powered course, (2) doesn't prepare adequately for freshman chem for majors in college. Most of our students are not going to college for chemistry. Everyone will be able to vote (on reaching 18), will be able to have children and spend money at local stores. All students need to be able to read a newspaper and decide if a landfill should go where the politicians want it or any of a number of other enviro-political issues. They need to be able to decipher a can label. ChemCom does this. Shouldn't we stop arguing about which course is better and realize that all of the courses have their place? How about AP for the gifted and ambitious, chemistry for the interested, and Chem-Com for the masses! If everyone was required to take a chem course, the vast majority would probably take ChemCom, but even this would raise the technical knowledge level of the population as a whole!

And, finally, a teacher essay:

The majority of people in our society have little science background. These people are often in positions to make decisions that affect all of us. This has been a sore point in numerous threads of late. It is always easy to point fingers at others, but in the end this problem comes back to us as science educators. In our current system we tend to concentrate on the elite and ignore the needs of the majority of the student population when it comes to science. We then complain when that same majority lacks the necessary scientific literacy to make wise decisions that involve science.

The theme of "Science for All Americans" in the proposed national science standards clearly addresses this issue. We cannot afford to have a scientifically illiterate general population. Only a small percent of secondary students will pursue a four-year program in science and even fewer will specialize in chemistry. Should we be surprised when the rest of the student population later become elected officials or administrators that have difficulty using chemistry in their decisionmaking process? Perhaps we as science educators failed to teach them the connection.

"Science for All Americans." Does that mean that we should create "watered down" versions of science (including chemistry) for the masses? The proposed national standards would say *absolutely not*. Science courses for the non-science major should be just as rigorous as any other science course for majors. However, the focus should be different and have an emphasis that is more relevant to the world and the kinds of decisions these young adults will be faced with when they graduate.

For example, a detailed knowledge of suborbital hybridization will be less relevant to a politician, to a motor operator, or school board member than a knowledge of the chemical effects of alcohol on human metabolic processes, brain chemistry, and workplace productivity. For non-science majors (it wouldn't hurt the majors either), we need to focus more on connecting chemistry to the real world such as making the connection between turning on a light switch and changing the chemistry of our atmosphere, which in turn changes the chemistry of buildings and living organisms.

Personally, I find it relatively easy to develop curriculum and assessment tools for a subject taught in isolation and based on memorization of specific content or applications of concepts only within the confines of a test tube and an ideal highly controlled laboratory setting. It is much more difficult to develop curriculum that will help students understand the chemistry that goes on outside the classroom walls in a world full of synergistic interactions, multiple variables, and proprietary nomenclature. It is even harder to teach students the necessary thinking skills that will allow them to apply chemistry to the complexity of their daily decisionmaking processes. I often wonder if our traditional teaching of chemistry isn't the real "watered down" approach to chemistry when it comes to the needs of non-science majors.

Recent development of chaos and complexity theories would suggest that chemistry should not be taught as an isolated subject and should include real-world applications and connections to other areas of science. I will be the first to admit that I could use more training to help non-science majors make real connections with chemistry outside the lab.

Smaller ChemCom Niches

While ChemCom typically is offered as a chemistry course for juniors and seniors in academic or general courses of study, some schools are making other uses of it.

ChemCom for Freshmen. Some schools are introducing chemistry as a freshman subject for students who had algebra in eighth grade. ChemCom seems to be the course of choice for freshman chemistry as illustrated by the following teacher comment. In these classes, the social relevancy features fit with a growing emphasis on the science, technology, and society aspects of curriculum.

One of the reasons that our high school's science department decided to look at ChemCom had to do with their perception that the curriculum fit in well with the California State Framework. It is now used as the basis for a ninth grade integrated science honors course. The course offers students a "passing acquaintance" with basic background in physics and chemistry, and is generally viewed as the preparation for Advanced Placement Biology which is offered in the second year of the honors track.

ChemCom for Tech Prep. We also found some of the ChemCom modules used for a relatively new genre of course (or an old genre in new clothes). Teachers refer to these as "tech-prep" programs. There is as yet no widely accepted curriculum or characterization of the tech-prep domain. No solution comes easily, as this teacher describes how his school decided to import ChemCom:

> The local college and the biology teachers are pushing, and the biology party, is pushing this biotechnology. What that basically means is that we haven't had a supply or equipment budget in sciences in this school for over 15 years. So, the only way to get equipment is to tie in with some type of supposedly new thing. So the new thing is biotechnology. They're trying to get some "biotechnologists" trained, who do not need four years of college, who need a two-year "technologist" program, so that they can fill into some of these industries, who need people like crazy, who have some experience, but not necessarily college-bound.

> The difficulty is that while it sounds great on paper, any student who would be a good student in that situation could easily handle the normal high school class. And when you offer a lower level class, what they try to do then is basically pump the class full of kids who either shouldn't be there because they're too low, or what they do tell the kids that could take is regular standard chemistry with all the math, and say, "You're not good enough. Why don't you take this technology program?" So the intent was to take the middle cut of the kids who don't take chemistry, and expose them to chemistry. That's the intent, but in this situation, what it works out to be is that you tell these kids that they're not up to the standard chemistry, and put them into the other class, just to fill it up their biotech stream.

◆

Assessment and Evaluation

In this section, we briefly look at how ChemCom teachers assess student learning and what kinds of evaluation of ChemCom have been done by staff and researchers outside the project.

Assessment of Student Learning

ChemCom Tests. Many teachers with whom we talked worry about how well ChemCom students could meet the demand of colleges for chemistry knowledge or how they would survive an AP course. Most of these worries come from teachers who are chemistry majors and who have been teaching the course from one to three years. Six of them are AP or honors chemistry teachers themselves but did not yet have any AP students from a ChemCom class. They want to be reassured that "smart students" who take ChemCom will not be handicapped. One teacher notes, "You can separate the sheep from the goats by how competent they are in mathematics."

Teachers for whom chemistry is their second subject—e.g., biologists teaching chemistry—mention mathematics concerns much less frequently. They feel the emphasis on relevance of the content is more important. They want to know whether ChemCom graduates think differently about the kinds of issues they studied. These teachers are more willing to spend time on the You Decide sections of units. They are curious to know what kinds of knowledge and attitudes students gained from that aspect of the program. This group embraces the *knowledge delivered on a need-to-know basis* for organizing chemistry content. Their chemistry major peers, however, are generally more comfortable with a strategy that we describe as the *banking of content knowledge against some future time of need*, e.g., in college or in an honors course. These contrasting perspectives present some challenges for test developers. Consumers vary in what they want to know.

ACS has a test development capability in its Chemical Education Examinations Institute. With the help of college and high school instructors who developed ChemCom, the institute produced a nationally normed, machine-scorable examination that takes 80 minutes to complete—i.e., two class periods. While quantitative demands may be less in ChemCom, the use of decisionmaking and problem solving in situations that do not have clear boundaries is a substantial and demanding aspect of ChemCom (e.g., in the You Decide portion of each unit). ACS had to develop a test that mirrored the untraditional. About 14,000 copies of the test have been sold. Since test booklets can be used over again, and the test has been in existence for some five years, probably some 50,000 to 75,000 students (10 to 20 percent of ChemCom students) have taken the test. A new version of the test is being constructed in 1996.

The *Teacher's Guide* for ChemCom contains tests to be given at the end of each unit. These tests include both multiple-choice and short-answer questions, as illustrated in figure 4. Answers to all items appear in the guide at the end of each test. We found little use being made of these questions by the teachers with whom we talked, despite their obvious convenience. Most teachers, it seems, prefer to write their own examinations.

Grading ChemCom. ChemCom presents some grading issues not encountered in traditional courses. For example, how should teachers appraise performance in a You Decide session? Take the one on pros and cons of aspirin use (ACS 1993a, pp. 452-53). The topic is an example of a risk/benefit analysis related to self-medication practices. The activity makes the point that even a common drug can have undesirable side effects. Indiscriminate use of over-the-counter drugs carries risks. In this aspirin example, students explore the question of whether pain relief from intensive use of aspirin is worth the danger of its

Figure 4.   **Sample Questions From a ChemCom End-of-Unit Test**

CONSERVING CHEMICAL RESOURCES

Mutliple-Choice Questions (34 questions)

8.   Which statement about relation is not true?
     (A) Reduction is always accompanied by oxidation.
     (B) Reduction involves the gaining of electron(s).
     (C) Reduction is the process whereby metals rust.
     (D) Reduction of minerals can produce elemental metals.

18. What is the mass percentage of carbon in CO_2? (atomic mass of C=12, O=16)
     (A) 75%   (B) 50%   (C) 33%   (D) 27%

26. If a strip of magnesium metal is placed into a water solution of copper (II) ions, what
     happens?
     (A) The magnesium strip reacts.   (C) More copper(II) ions form.
     (B) Carbon dioxide gas evolves.   (D) No reaction occurs.

31. Which statement about the future availability of chemical resources is true?
     (A) Factors such as the natural abundance, our ability to develop technologies to extract
          these elements, patterns of use, and the cost related to the rest of the economy will
          help determine future availability of chemical resources.
     (B) There will be no shortages of chemical resources in the future because atoms of ele-
          ments are not destroyed during chemical changes.
     (C) All of the necessary chemical resources needed in the future will be obtained from
          water and air, so it will not be necessary to think about continued availability of tra-
          ditional chemical resources.
     (D) Any "used up" chemical resources can always be replaced in the future by plactic
          substitutes, as long as consumers are willing to pay the price.

Short-Answer Questions (7 questions)

35. Balance this chemical expression. (This reaction that takes place during the flocculation
     step of water purification.)
     ____$Al_2(SO_4)_3$+____$Ca(OH)_2$ ➡ ____$Al(OH)_3(s)$+____$CaSO_4$

41. In what way are phrases such as "using up" and "throwing away" inaccurate from a
     chemical point of view? What is the real meaning behind such phrases?

possible side effects. How should teachers grade these decisionmaking sessions
where there are not precise outcomes? The question is not trivial, since there
seems to be a clear connection in the minds of students between what they
should pay attention to and what gets tested and graded. Interestingly, on the few
examples of teacher-made tests we saw, there were one or two short essay questions per test. One of the Florida teachers showed us responses from two classes
to illustrate how diverse the writing skills of her students were. As she said,
"You have to read a lot into what they wrote."

Evaluations of ChemCom

Evaluation of Summer Training. Content and sequence of topics in ChemCom summer training programs is uniform at all sites, insofar as that is possible. Workshop leaders are experienced ChemCom teachers who attended a training program for leaders. They receive a big looseleaf notebook full of directions, guides for collecting feedback, content to include, etc. Participants fill out questionnaires at the end of the training. Their responses help ChemCom staff improve training and awareness programs. In addition, Ware sends knowledgeable chemistry educators to evaluate the workshops.

On the basis of feedback from questionnaires and observers, instructors may receive helpful suggestions. For many instructors, this workshop was the first time they had tried to teach teachers like themselves. People who do well and who are receptive to feedback will generally have more paid opportunities to do other ChemCom sessions. One experienced trainer in San Jose remarks on the difference between teaching students and teaching teachers.

> It takes a little adjustment. They are adults and they have a lot of worries. They don't want to be tested, and when they don't have that to worry about, they really get involved in learning for themselves. Some of them know more chemistry than I do, but its the decisionmaking part that throws them. Most of that kind of thing is new to them. It bugs some of them that sometimes there just isn't a single right answer. Part of that is because of grades. What do you grade? We talk about that.

Research Studies on ChemCom. At Worcester Polytechnical Institute, researchers got student feedback on two ChemCom units used as infusion material in standard high school chemistry courses. They then administered two preference inventories and found that high problem—*finding*—oriented students like ChemCom. High problem—*solving*—oriented students prefer standard chemistry. Students who like both problem finding and solving may prefer one or the other depending on their teacher. Students who are *not* problem finding—or solving—oriented usually prefer ChemCom (Hogan and Reagan 1991).

Another study based at Cornell analyzed ChemCom in a sociocultural context. Researchers conducted qualitative research using text from the unit on nuclear chemistry. Their analysis suggested that:

> ChemCom frames chemistry-related problems in ways that imply that disputes are almost exclusively political, that problems rooted in technology are an inevitable consequence of modern life, and that science and industry offer the solution to problems like radioactive contamination, air pollution, and waste disposal. The text says little about controversy within the scientific community, about reducing society's appetite for chemical resources, and about the forces that produce problems that science is called upon, somewhat ironically, to solve (Carlsen, Cunningham, and Kelly 1994).

MaryAnn Martin, a teacher at Estes Park High School in Colorado, set out to study the question of how much chemistry ChemCom students learn. She used two AC- developed standardized tests, an ACS-NSTA examination and the ChemCom nationally normed test mentioned earlier. In a pre-/post-design, students in the Denver metropolitan area took the two measures. In addition to the two standardized tests, a sample of students did two other exercises; one was an essay question about an environmental issue, and the other was a concept mapping activity. Martin presented her results at the NSTA national meeting in March 1995. While there were problems with the study design, Martin (1995) reports that ChemCom students did learn chemistry as indicated by results on the traditional content measure (ACS-NSTA test). They also were able to do reasonably well with the application essay. Martin thinks more test development is needed to ferret out all the various levels of knowledge built by or needed in the ChemCom program.

At the University of Delaware, Nancy Brickhouse (1993) examined ChemCom instruction and compared it with instruction in a standard chemistry course, both of which were being taught by the same teacher. She gathered data through interviews and observations in the two kinds of chemistry classes in order to "highlight the tensions and peculiarities in one teacher's decisionmaking."

Brickhouse observed that the teacher emphasized different goals in each class based on his assumptions about the students' futures. According to her analysis, the teacher "is much clearer about the educational aims he has for honors students than for his other classes: preparation for a rigorous college chemistry course. In the case of ChemCom he felt many conflicts about how to best educate these students" (Brickhouse 1993).

In an earlier study, Sutman and Bruce (1992) at Temple University measured students' abilities to understand chemical knowledge and to apply this knowledge to social issues. They compared ChemCom students' test results with those of traditional chemistry course students. They report that results of a testing program indicate that students completing the entire year-long ChemCom course significantly outperformed students completing more traditional college preparatory chemistry on test items designed to assess both chemistry learned and applications of chemistry.

◆

Summary of Factors in ChemCom's Success

It is remarkable that ChemCom is surviving and gaining adherents, given its differences from traditional chemistry courses. These include demands for interdisciplinary connections and for teaming to explore issues, its intent to appeal to a broader audience, and its focus on the centrality of student decisionmaking processes. It is a course that might be more fruitfully advanced through collaboration with social studies teachers, or at least with other chemistry teachers. With rare exceptions, however, course coordination or joint planning does not appear to happen.

Chemistry teachers are specialists who talk with other chemistry teachers. Until the arrival of ChemCom, most of them had no need or particular incentive to recruit more diverse students to chemistry or to broaden the content as in ChemCom. Chemistry did not have an interdisciplinary thrust, so there was no motivation to seek help from fellow faculty members in other disciplines. Many schools have only one chemistry teacher, so there is no arena for cooperative planning on science curriculum content and certainly no habit of collaboration with social studies teachers. Chemistry teachers have a history of working independently, possibly coordinating use of laboratory equipment and scheduling if there are two or more chemistry faculty members at the school. Their professional contacts are typically outside the school with other chemistry teachers. This is true of the other science specialists as well—i.e., physics teachers talk to physics teachers, geologists to geologists, etc.

Add to the situation the fact that most chemistry teachers come out of an academic tradition that paid little or no attention to the impact of chemistry on society or to the organization of chemical knowledge in the service of understanding some major issues. Chemistry majors get their training in chemistry departments which often manage to convey the perception that people doing applied chemistry are in some way on a lesser path. Interdisciplinary courses, for example, are usually offered for nonspecialists. Chemistry departments at universities are busy training specialists of one kind or another—e.g., physical, organic, biochemical. Tensions between the basic and applied sciences have their origins in different conceptions of the purposes of education. Scott Montgomery (1994, pp. 3-4) contrasts old world and new world views about the purposes of education:

> . . . in 19th-century England, Germany, and France, "to educate" meant primarily to train people to assume specific social roles. According to the European view, education existed to teach people to accept their station in life . . . In these countries the curriculum was intensely classical; moreover, it ended with an extremely difficult series of exams . . . meant to restrict entry into higher level positions of sociopolitical power, especially the professions and government service.

Education in America has always been conceived in a radically different way. Here, from the very beginning, education has been viewed as a transforming experience (pp. 4-5) . . . children stood as the carriers of the immediate future, which was always uncertain. They were the means to create this future . . . (p. 7).

Into this perpetual tug of war between old world classical conceptions of science and new world Jeffersonian beliefs about empowerment through science, comes ChemCom with its issues orientation and recruitment of a wider student audience. It is not chemistry just for the elite. It is chemistry, as the preface to the second edition says, for everyone. In what follows, we summarize the main features and conditions that we think account for the survival and spread of the ChemCom program and that are likely predictors of its continuation.

Long-Term Support

Fiscal Support. From a policy perspective, public and private agencies that give grants for curriculum development and dissemination want to be assured that a project will be able to stand on its own after some reasonable time. It is not surprising that developers and funders differ on what constitutes a reasonable time. The ChemCom case makes a nice example of a project that worked as grantors hoped.

The ChemCom fiscal story began with grants to ACS for development followed by support for dissemination and teacher training. When government funding ended, ACS supported ChemCom activities out of its own funds, which are substantial in comparison to most professional societies. When the project acquired a publisher (Kendall/Hunt), money from the company's advertising budget provided additional means for outreach activities. Increasingly, districts are paying to send teachers for training. As sales increased, ACS received royalties for the textbook, which it pumped back into the ChemCom project. This reinvestment policy helps a project keep current and adaptive. In short, a steady infusion of funds is a necessary, but not sufficient, condition for survival and growth of an innovation like ChemCom.

ACS, the sponsoring and host organization, has housed and supported ChemCom for more than 10 years and plans to continue to do so. When the purposes meant to be served by a project are congruent with purposes of the sponsors and when the sponsor's intents stay basically stable over a long enough time, then a project has a good chance of survival and growth. In this case, ACS nurtured the ChemCom project by including the project's needs within its periodic capital campaign drives as well as returning royalties to it. In short, ChemCom has continuing strong fiscal support from the sponsoring organization.

Political Support. The political support of ACS has been as important as its fiscal support for this program. As the world's largest scientific society, ACS can be presumed to care about maintaining the integrity of its own discipline. In fact, ACS operates a program that approves the quality of undergraduate degrees in chemistry. The ACS governance has provided a great deal of support for ChemCom beyond the fiscal. This support has come from both academic and industrial members of the society.

ChemCom is likely to continue its interdisciplinary, issues-oriented curricular direction because such a substantial portion of its host organization's membership are industrial chemists. The issues-orientation of the project is congenial to their interests. In general, the durability of a relationship between a project and its host depends on the degree of congruence between the project's purposes and the sponsor's goals.

Stable Leadership Addressing Changing Context. An effective organizational memory, which stores and makes uses of what is learned from experience to inform planning, is an essential element of a survivor innovation. This need usually, though not always, translates into key individuals who stay in an innovative stance with the project over a long period. Changes of the magnitude ChemCom requires take time; the landscape through which the project must travel is dotted with landmines. In the ChemCom case, we came to think about an innovation's need for a person who is aware of the landscape and helps the project navigate it successfully. In our view, Sylvia Ware functions in this role. She also is a remarkably effective communicator for the ChemCom project. Many reforms are imposed from the top down; some emerge from the bottom up. Successful reforms are often said to do both. ChemCom is successful because it has a middle, capable of communicating with both the "top" and the "bottom." The ACS Education Division staff, under Ware's leadership, is operating as that middle ground. It is capable of relaying information in both directions and of acting on that information where appropriate.

Figure 3 illustrates something of the multifaceted, time-dependent, turbulent tributaries that feed and shape the course of the main ChemCom river. It includes attention to relations with other professional societies that could provide platforms for dissemination, guidance for districts on how to find funds for training from state and federal programs, awareness of competitor programs, qualifying the course for college credit, adapting to changes in long-term demand, etc. With the addition of a middle school curriculum (FACETS) and a one-semester college course, the project extends its reach in time as well as content.

Effective Dissemination

A Teacher-to-Teacher Policy. ChemCom curriculum development began with teams of teachers each accompanied by a university chemist or chemistry educators. Teachers on those teams tried out their modules in their own classrooms. In the first summer in which training was provided to additional teachers, the sessions were led by the university people. Team members quickly saw that it would be better to use the experienced teachers as workshop leaders. That has become ChemCom policy. Teachers speak with the voice of experience. They take on the role of disseminators in awareness sessions and trainers in workshops—and get paid for it. In addition, classroom teachers make presentations about their ChemCom work at professional meetings. This is a major element in the spread of the program. Many of the teachers with whom we spoke in connection with this project were the people who brought knowledge of the project to their districts and persuaded their administration to let them pilot the program. These teachers function as change agents.

Quality Assurance. Dissemination and implementation are long-term ventures. ChemCom built and maintains an infrastructure that provides access to project information, leadership, and teacher training. To ensure reasonably good quality and program fidelity in the workshops it sponsors, the project regularly evaluates the performance of instructors via questionnaires given to participants and visits of staff or consultants. This is a process that is repeated each year. The ChemCom spread depends on a multiplier strategy—i.e., teachers teaching other teachers.

Reasonable Flexibility

Accommodating Traditional Views. What role do teachers have in the ChemCom classroom? Teachers are to be discussion leaders more than they are to be lecturers. Therein lies a problem for some chemistry teachers. ChemCom urges new trainees to do the curriculum as specified. In-service program leaders try to discourage individuals from making major modifications to the course for at least the first two or three years.

ChemCom presents different challenges for teachers with conceptions of the purpose of chemistry that differ from those of the project. Teachers, for example, whose declared major area of content and teaching expertise is chemistry may find it philosophically as well as practically troublesome to spend so much time on issues and applications or to let issues dictate the choice of chemical content. This is particularly true if their teaching is based on a tightly specified curriculum and instructional repertoire. They find themselves less comfortable in the role of discussion leader than in the role of teller. Teachers, however, whose preferred content is in some other science may be more hospitable to the issues approach and the importance of student discussion, but they may lack the in-depth knowledge of chemistry that the ChemCom curriculum requires. The in-classroom roles of teachers are more varied than they are in a standard chemistry course, but nevertheless are substantively and instructionally demanding. The most effective ChemCom teachers are those with a substantive chemistry background who are flexible in their teaching approach

How does ChemCom accommodate these tensions? ACS devised training experiences so that prospective users can have an immersion in new perspectives. It shows teachers exactly where ChemCom treats key traditional topics. The project strongly encourages users to subscribe to ChemCom as is, but not dogmatically. That is, the project accepts that some teachers will bring in supplemental topics and activities.

As teachers gain experience with the curriculum, we found numerous examples of creeping infusion of standard chemistry concepts and mathematics into ChemCom—and conversely, the mention of issues from ChemCom into the standard chemistry courses. There are other adaptations as well. These are mostly related to laboratory activities. Here the oft-repeated complaint about insufficient examples of inquiry led to substitution by some teachers of laboratory investigations borrowed from other curricula or produced by the teachers themselves. In addition, with the appearance of an electronic bulletin board focused on ChemCom, we are seeing an increasing number of chemists and teachers suggesting alternative activities.

Modular Origins. The switch from a modular infusion paradigm to a course size dissemination paradigm did not totally erase the modular origins of ChemCom. That fact makes the course flexible and attractive for more purposes than originally intended. With its concepts clustered around issues and with more issues treated than can be completed in a year, teachers can make choices once they have covered the suggested first four units. Some schools, for example, found the nutrition unit to be popular. Others, e.g., in the Florida district visited, did not select the nutrition unit because its concepts were so thoroughly treated in the school health program.

In addition, we found modules from the course serving another constituency, in science/technology courses of one kind or another. Moreover we talked to teachers who were doing program planning with a science, technology, and society emphasis. To them, some of the ChemCom modules seemed pertinent.

Closing Appraisal

In what sense does ChemCom count as an innovation? What new things can we learn from the ChemCom story? Briefly put, it was an early entry into the science-for-all parade. It relies on strong applications of change literature strategies but it put a spin or two of its own on these strategies. It has a person who, among other things, functions as a communicator and mediator operating in the middle ground between top-down and bottom-up innovation. ChemCom represents an innovation we think is powerful. It uses issues as a way of organizing content. In that sense, at its first appearnce, it was new and faced more resistance from single subject chemistry majors than from science teachers for whom chemistry was the second subject. It raises as-yet-unresolved questions about the purposes of chemistry in the general education of high school students. It has grown middle school and college spin-offs. At least some of its materials are serving science, technology, and society purposes. ChemCom is, in fact, still going strong after its first decade.

References

American Chemical Society (ACS). 1988a. *ChemCom: Chemistry in the community.* 1st ed. Dubuque, IA: Kendall/Hunt.

——. 1988b. *ChemCom: Chemistry in the community. Teacher's guide.* 1st ed. Dubuque, IA: Kendall/Hunt.

——. 1988c. What drives chemistry curricula? Westminster, MD: Western Maryland College.

——. 1991a. *FACETS: Foundations and challenges to encourage technology based science*

——. 1991b. *Chemunity News* 1(1).

——. 1993a. *ChemCom: Chemistry in the community. Teacher's guide.* 2nd ed. Dubuque, IA: Kendall/Hunt.

——. 1993b. *ChemCom: Chemistry in the community.* 2nd ed. Dubuque, IA: Kendall/Hunt.

——. 1993c. *Chemistry at the K-12 level.*

American Association for the Advancement of Science (AAAS). 1993. *Project 2061: Benchmarks for science literacy.* New York: Oxford University Press.

Berry, K. O. (1988). Learning in the Laboratory Isn't Just an Idea! *ChemComments* 1 (1): 8-9.

BioCom. 1993. *BioCom Newsletter* 1. Clemson, SC: Clemson University.

Brickhouse, N. W. 1993. What counts as successful instruction? An account of a teacher's self-assessment. *Science Education* 77(2): 115-29.

Carlsen, W. S., C. M. Cunningham, and G. J. Kelly. 1994. Teaching ChemCom: Can we use the text without being used by the text? In *Science, technology, and society education,* ed. J. Solomon and G. S. Aikenhead, ch. 9. New York: Teachers College Press.

Gross, G. R. (1989). Microsizing ChemCom Labs. ChemComments 1 (1): 3.

Hogan, R. W., Jr., and M. E. Reagan. 1991. *Response to STS curriculum: A comparative study of ChemCom.* Worcester, MA: Worcester Polytechnical Institute.

Lynch, M. and E. Britton. 1993. Chemistry in the Community. In *Science and mathematics eduction in the United States: Eight Innovations.* Washington, D.C.: Organisation for Economic Co-Operation and Development.

Martin, M. V. 1995. *ChemCom: What are students learning?* Paper presented at annual meeting of National Science Teachers Association: Philadelphia.

Montgomery, S. L. 1994. *Minds for the making: The role of science in American education, 1750-1990.* New York: The Gilford Press.

O'Brien, T. 1990. One year delayed post workshop survey results from the summer 1988 ChemCom resource teachers. Binghamton, NY: State University of New York.

Schwartz, A. T., D. M. Bunce, R. G. Silberman, C. L. Stanitski, W. J. Stratton, A. P. Zipp. 1994. *Chemistry in context: Applying chemistry to society.* Dubuque, IA: Wm. C. Brown Publishers.

Silberman, R. G., W. J. Stratton, D. M. Bunce, C. L. Stanitski, A. T. Schwartz, A. P. Zipp. 1994. *Chemistry in context: Applying chemistry to society.* Lab Manual. Dubuque, IA: Wm. C. Brown Publishers.

Stanitski, C. L., J. Bieron, D. M. Bunce, A. T. Schwartz, R. G. Silberman, W. J. Stratton, S. Penhale, K. Warren, A. P. Zipp. 1994. *Chemistry in context: Applying chemistry to society.* Instructor's Resource Guide. Dubuque, IA: Wm. C. Brown Publishers.

Sutman, F. X., and M. H. Bruce. 1992. ChemCom: A five-year evaluation. *Journal of Chemical Education* 69(7).

Ware, S. 1990. ChemCom: Chemistry on a need-to-know basis. International symposium on the environment and chemistry teaching.

Ware, S. 1991. *Secondary school science in developing countries: Status and issues.* Washington, D.C.: World Bank.

Appendix A: Case Study Team

Mary Budd Rowe, to whom this volume is dedicated, was professor of science education at Stanford University and the University of Florida. Her career-long interest in finding ways to have all students experience and understand science in part prompted her to lead the case study of ChemCom, a course seeking to bring chemistry to more students. Mary's leading-edge work brought her many tributes, including the annual Journal of Research in Science Teaching award for her seminal paper that coined the term "wait time," and the NSTA's Roger Carleton Award for national leadership in science education. She later was elected president of NSTA. In her last decade, Mary focused particularily on pressing new technologies into the service of professional development for science teachers, including videotapes, CD-ROM discs, and telecommunications networks.

Michael Midling, currently a social scientist at Social Policy Research Associates, was a doctoral student in international development education at Stanford University during the ChemCom case study. His dissertation's focus on environmental education in China, as well as his interest in ChemCom's emphasis on environmental chemistry and social action, led Michael to participate in the ChemCom case study. After teaching graduate courses at Golden Gate University in research methods, applied statistics, and sustainable development in the Third World, Michael joined SPR where his research addresses domestic issues in education, employment, and training.

Tom Keating is an Assistant Professor of Science Education at Indiana University at Bloomington. His contribution to the ChemCom Chapter was shaped by his dissertation research on teacher telecommunication networks while at Stanford University. His research interests include the integration of emerging technologies in science classrooms, developing project-based science, mathematics, and technology curricula, and the teaching of evolutionary biology.

Julie Montgomery, is a graduate student in education at Stanford University. She has worked in educational development in East Africa.

Appendix B: Preface to Second Editionof Chemistry in the Community

The United States is a world leader in science, technology, and the education of scientists and engineers. Yet, overall, U.S. citizens are barely literate in science. In responding to this situation, our government and many professional groups have assigned high priority to improving the nation's science literacy.

Chemistry in the Community (ChemCom) represents a major effort to enhance science literacy through a high school chemistry course that emphasizes chemistry's impact on society. Preliminary work on *ChemCom* began in 1980 with the formation of a steering committee, staff, and groups of writing teams consisting of high school, college, and university teachers, assisted by chemists from industry and government. Developed by the American Chemical Society (ACS) with financial support from the National Science Foundation and several ACS funding sources, the themes of *ChemCom* were developed, emphasizing the application of chemistry to societal issues utilizing student-centered activities. During the summers of 1982 and 1983 the first drafts of the eight *ChemCom* units were written and carefully evaluated by content specialists, social science consultants, and field-tested in local communities. In 1984 the overall syllabus was developed, and the national field test version was written by a team of writers under the direction of Henry Heikkinen.

In 1985 the field test began in 13 centers around the country, in which 61 teachers used the materials with 2900 students. In several locations the teachers met monthly to share ideas and iron out difficulties, proposing alterations in the text and laboratory material. ChemCom's showing was very favorable, as indicated by the increase in requests for textbooks by field test teachers, and after final revision the book was published by the Kendall/Hunt Publishing Company in 1988. Since then, *ChemCom* has been successfully implemented by chemistry teachers in thousands of classrooms. Many teachers report that the program offers a motivational, engaging approach to the study of chemistry for a remarkably wide range of students. The second edition was released in 1992.

Briefly, *ChemCom* is designed to help students:

- realize the important roles that chemistry will play in their personal and professional lives.

- use chemistry knowledge to think through and make informed decisions about issues involving science and technology.

- develop a lifelong awareness of the potential and limitations of science and technology.

Each of *ChemCom*'s eight units focuses on a chemistry-related technological issue currently confronting our society and the world. The issue serves as a basis for introducing the chemistry needed to understand and analyze it. The setting for each unit is a community. This may be the school community, the town or region in which the students live, or the world community—Spaceship Earth.

The major *ChemCom* topics are: *Supplying Our Water Needs*; *Conserving Chemical Resources*; *Petroleum: To Build? To Burn?*; *Understanding Food*; *Nuclear Chemistry in Our World*; *Chemistry, Air, and Climate*; *Health: Your Risks and Choices*, and *The Chemical Industry: Promise and Challenge*.

The eight units include the major concepts, vocabulary, thinking skills, and laboratory techniques expected in any introductory chemistry course. However, the program contains a greater number and variety of student-oriented activities than is customary. In addition to numerous laboratory exercises, including many developed especially for *ChemCom*, each unit contains three levels of decision-making activities and several types of problem-solving exercises.

The first four *ChemCom* units are designed to be studied in sequence. Each new unit builds on the previous ones. For example, considering water as a resource leads to consideration of resources in general, which leads to information about another very special resource, petroleum. The two uses of petroleum, for energy production and building petrochemicals, mimic the two main uses of food, for energy production and building components. The first four units also involve systematic development of basic chemical concepts, problem-solving skills, and decision-making abilities. By contrast, the last four units have been designed independently, so that they can be studied in any sequence, based on the interests and needs of your students.

This new edition of *ChemCom*, while maintaining the overall structure and approach of the first edition, provides up-dated information on many *ChemCom* topics as well as detailed improvements based on suggestions from classroom experience. A new feature, appropriately named *Chemistry in the Community*, describes how individuals in widely diverse careers find uses for chemistry in their daily work. All second-edition changes are inteded to make *ChemCom* even more "user-friendly" for both teachers and their students.

Dozens of professionals from all segments of the chemistry community contributed their talents and energies to create *ChemCom*. Their hope is that its impact will be substantial and lasting, and that those who study *ChemCom*, will find chemistry interesting, captivating, and useful.

Source: ACS (1993, pp. xi-xii).

Appendix C: Agenda for a ChemCom Teacher Training Workshop

Time	Saturday 8/7	Sunday 8/8	Monday 8/9	Tuesday 8/10	Wednesday 8/11	Thursday 8/12
7:30 - 8:15 AM		Breakfast	Breakfast	Breakfast	Breakfast	Bkfst/KEY RETURN
8:30 AM		CC Overview	Complex Instru	Team Modeling	Team Modeling	Team Interviews w/Stanford Group
9:00 AM		Foul Water Lab Unit Review	Group Process	Team Modeling	Team Modeling	
10:00 AM		BREAK YD B.9	BREAK Simulation Debrief	BREAK Unit Overview Lab Debrief	BREAK Networking Unit Overview	Team Modeling Auth. Assessment Evaluation Certificates/Awards
11:00 AM						
Noon–1:00 PM		Lunch	Lunch	Lunch	Lunch	Lunch-Adjourn
1:15 PM	Registration (Energy Survey) (Nutrition Quiz)	Solutions Lab	Milk Lab	Computers	Combustion Lab	
2:00 PM		Frac Cryst Lab	Vit. C Lab	Computers	Penny Lab	
3:00-4:30 PM	Room Check-in	Milk Lab Team Tasks	Team Tasks	Team Teasks	RELAX	
5:30-6:30 PM	Dinner	Dinner	Dinner	Dinner	Dinner/India J	
7:00-9:00 PM	Reception/Team	YD "E" In/Out	Team Planning	Iron in Foods & Using up a Metal	Calamari Cabaret	

Appendix C: Agenda for a ChemCom Teacher Training Workshop (continued)

Detail for Sunday, August 8

Time	Activity	Leader	Comments	Site
8:30 AM	Intro to Day Concerns & Questions	GS SB	Daily Schedule Wed. Dinner Trip	NS 267
8:45 AM	ChemCom Overview Cooperative Learning Introductory Activity	GS GW GS	History, Units, Goals First Day Activity "Riverwood", pp. 2-3	
9:15 AM	Lab: "Foul Water" Lab Group I Lab Group II Demo: Distillation Lab Feedback (debriefing)	 SB GS GW SB	 Microscale Macroscale Apparatus in lab	NS 267
	BREAK			
10:30 AM	Coop Activity: YD B.9, p36 Waste Water Treatment "Town Council" role play	GS GW GS, GW	Team processes video video	NS 267
12:15 PM	**LUNCH**			Porter DC
1:30 PM	Lab: "Types of Solutions" Lab: "Frac. Crystalization"	GW	Alternative labs	NS 267
	BREAK			
3:00 PM	Lab: "Milk Analysis"	SB, GW	Unit Transition	NS 267
4:15 PM	Team Tasks	GS, SB	Assign YD, YT, CQ	
4:30 PM	**END**			
5:30 PM	**DINNER**			Porter DC
7:00 PM	Process "Energy Survey" Video: Diet for a New America			Kresge Apt

Acronyms and Abbreviations

AAAS	American Association for the Advancement of Science
ACS	American Chemical Society
AIMS	Activities that Integrate Math and Science
AP	Advanced Placement
BOARS	Board of Admissions and Relations With Schools
CIC	Curriculum Implementation Center
CLAS	California Learning Assessment System
CSIN	California Science Implementation Network
CSP	California Science Project
CSTA	California Science Teachers Association
CSU	California State University
CTC	Commission on Teacher Credentialing
ESL	English as a second language
FACETS	Foundations and Challenges to Encourage Technology-Based Science
GSE	Golden State Exam
LAN	Local area network
MASTTE	Mathematics Science and Technology Teacher Education
MOST	Minority Opportunities in Science Teaching
NCTM	National Council of Teachers of Mathematics
NGS	National Geographic Society
NRC	National Research Council
NSF	National Science Foundation
NSTA	National Science Teachers Association
NTE	National Teacher Exam
OECD	Organisation for Economic Co-operation and Development
PACES	Processes and Concepts in Elementary School Science
PEM	Program Elements Matrix
PQR	Program Quality Review
PTO	Parent-Teacher Organization
R&D	Research and development
RSL	Resources for Science Literacy
SAISD	San Antonio Independent School District
SciTeKS	Science and Technology: Knowledge and Skills
SES	Socioeconomic status
SFAA	Science for All American
SPAN	Science Partnerships for Articulation and Networking
SS&C	Scope, Sequence, and Coordination
SSI	State Systemic Initiative
STDP	Science Teaching Development Project

STELAR	Science Teachers and Educators Leading Assessment Reform
TERC	Technology Education Resource Center
USI	Urban Systemic Initiative
WASC	Western Association of School and Colleges
XBT	Expendable bathythermograph

Note: for abbreviations see List of Acronyms and Abbreviations, pp. 585–586.

592

594